Numbered Examples: Chapters Eight to Eleven

Number and Topic

8.3–1 An underdetermined set with three equations and three unknowns
8.3–2 A statically indeterminate problem
8.3–3 Three equations in three unknowns, *continued*
8.3–4 Production planning
8.3–5 Traffic engineering
8.4–1 The least-squares method
8.4–2 An overdetermined set

Chapter Nine

9.1–1 Velocity from an accelerometer
9.1–2 Evaluation of Fresnel's cosine integral
9.1–3 Double integral over a nonrectangular region
9.3–1 Response of an RC circuit
9.3–2 Liquid height in a spherical tank
9.4–1 A nonlinear pendulum model
9.5–1 Trapezoidal profile for a dc motor

Number and Topic

Chapter Ten

10.2–1 Simulink solution of $\dot{y} = 10\sin t$
10.2–2 Exporting to the MATLAB workspace
10.2–3 Simulink model for $\dot{y} = -10y + f(t)$
10.3–1 Simulink model of a two-mass suspension system
10.4–1 Simulink model of a rocket-propelled sled
10.4–2 Model of a relay-controlled motor
10.5–1 Response with a dead zone
10.6–1 Model of a nonlinear pendulum

Chapter Eleven

11.3–1 Intersection of two circles
11.3–2 Positioning a robot arm
11.5–1 Topping the Green Monster

Introduction to MATLAB® for Engineers

William J. Palm III
University of Rhode Island

The McGraw·Hill Companies

Connect
Learn
Succeed™

INTRODUCTION TO MATLAB® FOR ENGINEERS, THIRD EDITION

Published by McGraw-Hill, a business unit of The McGraw-Hill Companies, Inc., 1221 Avenue of the Americas, New York, NY 10020. Copyright © 2011 by The McGraw-Hill Companies, Inc. All rights reserved. Previous editions © 2005 and 2001. No part of this publication may be reproduced or distributed in any form or by any means, or stored in a database or retrieval system, without the prior written consent of The McGraw-Hill Companies, Inc., including, but not limited to, in any network or other electronic storage or transmission, or broadcast for distance learning.

Some ancillaries, including electronic and print components, may not be available to customers outside the United States.

This book is printed on acid-free paper containing 10% postconsumer waste.

1 2 3 4 5 6 7 8 9 0 DOC/DOC 1 0 9 8 7 6 5 4 3 2 1 0

ISBN 978-0-07-353487-9

MHID 0-07-353487-0

Vice President & Editor-in-Chief: *Martin Lange*
Vice President, EDP: *Kimberly Meriwether David*
Global Publisher: *Raghu Srinivasan*
Sponsoring Editor: *Bill Stenquist*
Marketing Manager: *Curt Reynolds*
Development Editor: *Lora Neyens*
Senior Project Manager: *Joyce Watters*
Design Coordinator: *Margarite Reynolds*
Cover Designer: *Rick D. Noel*
Photo Research: *John Leland*
Cover Image: © *Ingram Publishing/AGE Fotostock*
Production Supervisor: *Nicole Baumgartner*
Media Project Manager: *Joyce Watters*
Compositor: *MPS Limited, A Macmillan Company*
Typeface: *10/12 Times Roman*
Printer: *RRDonnelly*

All credits appearing on page or at the end of the book are considered to be an extension of the copyright page.

Library of Congress Cataloging-in-Publication Data

Palm, William J. (William John), 1944–
  Introduction to MATLAB for engineers / William J. Palm III.—3rd ed.
    p. cm.
  Includes bibliographical references and index.
  ISBN 978-0-07-353487-9
  1. MATLAB. 2. Numerical analysis—Data processing. I. Title.
  QA297.P33 2011
  518.0285—dc22
                                                                    2009051876

www.mhhe.com

To my sisters, Linda and Chris, and to my parents, Lillian and William

ABOUT THE AUTHOR

William J. Palm III   is Professor of Mechanical Engineering at the University of Rhode Island. In 1966 he received a B.S. from Loyola College in Baltimore, and in 1971 a Ph.D. in Mechanical Engineering and Astronautical Sciences from Northwestern University in Evanston, Illinois.

During his 38 years as a faculty member, he has taught 19 courses. One of these is a freshman MATLAB course, which he helped develop. He has authored eight textbooks dealing with modeling and simulation, system dynamics, control systems, and MATLAB. These include *System Dynamics*, 2nd Edition (McGraw-Hill, 2010). He wrote a chapter on control systems in the *Mechanical Engineers' Handbook* (M. Kutz, ed., Wiley, 1999), and was a special contributor to the fifth editions of *Statics* and *Dynamics,* both by J. L. Meriam and L. G. Kraige (Wiley, 2002).

Professor Palm's research and industrial experience are in control systems, robotics, vibrations, and system modeling. He was the Director of the Robotics Research Center at the University of Rhode Island from 1985 to 1993, and is the coholder of a patent for a robot hand. He served as Acting Department Chair from 2002 to 2003. His industrial experience is in automated manufacturing; modeling and simulation of naval systems, including underwater vehicles and tracking systems; and design of control systems for underwater-vehicle engine-test facilities.

CONTENTS

Preface   ix

CHAPTER **1**
An Overview of MATLAB®   3

1.1   MATLAB Interactive Sessions   4
1.2   Menus and the Toolbar   16
1.3   Arrays, Files, and Plots   18
1.4   Script Files and the Editor/Debugger   27
1.5   The MATLAB Help System   33
1.6   Problem-Solving Methodologies   38
1.7   Summary   46
Problems   47

CHAPTER **2**
Numeric, Cell, and Structure Arrays   53

2.1   One- and Two-Dimensional Numeric Arrays   54
2.2   Multidimensional Numeric Arrays   63
2.3   Element-by-Element Operations   64
2.4   Matrix Operations   73
2.5   Polynomial Operations Using Arrays   85
2.6   Cell Arrays   90
2.7   Structure Arrays   92
2.8   Summary   96
Problems   97

CHAPTER **3**
Functions and Files   113

3.1   Elementary Mathematical Functions   113
3.2   User-Defined Functions   119
3.3   Additional Function Topics   130
3.4   Working with Data Files   138
3.5   Summary   140
Problems   140

CHAPTER **4**
Programming with MATLAB   147

4.1   Program Design and Development   148
4.2   Relational Operators and Logical Variables   155
4.3   Logical Operators and Functions   157
4.4   Conditional Statements   164
4.5   `for` Loops   171
4.6   `while` Loops   183
4.7   The `switch` Structure   188
4.8   Debugging MATLAB Programs   190
4.9   Applications to Simulation   193
4.10   Summary   199
Problems   200

CHAPTER **5**
Advanced Plotting   219

5.1   *xy* Plotting Functions   219
5.2   Additional Commands and Plot Types   226
5.3   Interactive Plotting in MATLAB   241
5.4   Three-Dimensional Plots   246
5.5   Summary   251
Problems   251

CHAPTER **6**
Model Building and Regression   263

6.1   Function Discovery   263
6.2   Regression   271
6.3   The Basic Fitting Interface   282
6.4   Summary   285
Problems   286

CHAPTER **7**
Statistics, Probability, and Interpolation   295

7.1   Statistics and Histograms   296
7.2   The Normal Distribution   301
7.3   Random Number Generation   307
7.4   Interpolation   313
7.5   Summary   322
Problems   324

CHAPTER **8**
Linear Algebraic Equations   331

8.1   Matrix Methods for Linear Equations   332
8.2   The Left Division Method   335
8.3   Underdetermined Systems   341
8.4   Overdetermined Systems   350
8.5   A General Solution Program   354
8.6   Summary   356
Problems   357

CHAPTER **9**
Numerical Methods for Calculus and Differential Equations   369

9.1   Numerical Integration   370
9.2   Numerical Differentiation   377
9.3   First-Order Differential Equations   382
9.4   Higher-Order Differential Equations   389
9.5   Special Methods for Linear Equations   395
9.6   Summary   408
Problems   410

CHAPTER **10**
Simulink   419

10.1   Simulation Diagrams   420
10.2   Introduction to Simulink   421
10.3   Linear State-Variable Models   427
10.4   Piecewise-Linear Models   430
10.5   Transfer-Function Models   437
10.6   Nonlinear State-Variable Models   441

10.7   Subsystems   443
10.8   Dead Time in Models   448
10.9   Simulation of a Nonlinear Vehicle Suspension Model   451
10.10   Summary   455
Problems   456

CHAPTER **11**
MuPAD   465

11.1   Introduction to MuPAD   466
11.2   Symbolic Expressions and Algebra   472
11.3   Algebraic and Transcendental Equations   479
11.4   Linear Algebra   489
11.5   Calculus   493
11.6   Ordinary Differential Equations   501
11.7   Laplace Transforms   506
11.8   Special Functions   512
11.9   Summary   514
Problems   515

APPENDIX **A**
Guide to Commands and Functions in This Text   527

APPENDIX **B**
Animation and Sound in MATLAB   538

APPENDIX **C**
Formatted Output in MATLAB   549

APPENDIX **D**
References   553

APPENDIX **E**
**Some Project Suggestions
www.mhhe.com/palm**

Answers to Selected Problems   554

Index   557

PREFACE

Formerly used mainly by specialists in signal processing and numerical analysis, MATLAB® in recent years has achieved widespread and enthusiastic acceptance throughout the engineering community. Many engineering schools now require a course based entirely or in part on MATLAB early in the curriculum. MATLAB is programmable and has the same logical, relational, conditional, and loop structures as other programming languages, such as Fortran, C, BASIC, and Pascal. Thus it can be used to teach programming principles. In most schools a MATLAB course has replaced the traditional Fortran course, and MATLAB is the principal computational tool used throughout the curriculum. In some technical specialties, such as signal processing and control systems, it is the standard software package for analysis and design.

The popularity of MATLAB is partly due to its long history, and thus it is well developed and well tested. People trust its answers. Its popularity is also due to its user interface, which provides an easy-to-use interactive environment that includes extensive numerical computation and visualization capabilities. Its compactness is a big advantage. For example, you can solve a set of many linear algebraic equations with just three lines of code, a feat that is impossible with traditional programming languages. MATLAB is also extensible; currently more than 20 "toolboxes" in various application areas can be used with MATLAB to add new commands and capabilities.

MATLAB is available for MS Windows and Macintosh personal computers and for other operating systems. It is compatible across all these platforms, which enables users to share their programs, insights, and ideas. This text is based on MATLAB version 7.9 (R2009b). Some of the material in Chapter 9 is based on the control system toolbox, Version 8.4. Chapter 10 is based on Version 7.4 of Simulink®. Chapter 11 is based on Version 5.3 of the Symbolic Math toolbox.

TEXT OBJECTIVES AND PREREQUISITES

This text is intended as a stand-alone introduction to MATLAB. It can be used in an introductory course, as a self-study text, or as a supplementary text. The text's material is based on the author's experience in teaching a required two-credit semester course devoted to MATLAB for engineering freshmen. In addition, the text can serve as a reference for later use. The text's many tables and its referencing system in an appendix have been designed with this purpose in mind.

A secondary objective is to introduce and reinforce the use of problem-solving methodology as practiced by the engineering profession in general and

---

®MATLAB and Simulink are a registered trademarks of The MathWorks, Inc.

as applied to the use of computers to solve problems in particular. This methodology is introduced in Chapter 1.

The reader is assumed to have some knowledge of algebra and trigonometry; knowledge of calculus is not required for the first seven chapters. Some knowledge of high school chemistry and physics, primarily simple electric circuits, and basic statics and dynamics is required to understand some of the examples.

TEXT ORGANIZATION

This text is an update to the author's previous text.* In addition to providing new material based on MATLAB 7, especially the addition of the MuPAD program, the text incorporates the many suggestions made by reviewers and other users.

The text consists of 11 chapters. The first chapter gives an overview of MATLAB features, including its windows and menu structures. It also introduces the problem-solving methodology. Chapter 2 introduces the concept of an array, which is the fundamental data element in MATLAB, and describes how to use numeric arrays, cell arrays, and structure arrays for basic mathematical operations.

Chapter 3 discusses the use of functions and files. MATLAB has an extensive number of built-in math functions, and users can define their own functions and save them as a file for reuse.

Chapter 4 treats programming with MATLAB and covers relational and logical operators, conditional statements, `for` and `while` loops, and the `switch` structure. A major application of the chapter's material is in simulation, to which a section is devoted.

Chapter 5 treats two- and three-dimensional plotting. It first establishes standards for professional-looking, useful plots. In the author's experience, beginning students are not aware of these standards, so they are emphasized. The chapter then covers MATLAB commands for producing different types of plots and for controlling their appearance.

Chapter 6 covers function discovery, which uses data plots to discover a mathematical description of the data. It is a common application of plotting, and a separate section is devoted to this topic. The chapter also treats polynomial and multiple linear regression as part of its modeling coverage.

Chapter 7 reviews basic statistics and probability and shows how to use MATLAB to generate histograms, perform calculations with the normal distribution, and create random number simulations. The chapter concludes with linear and cubic spline interpolation. The following chapters are not dependent on the material in this chapter.

Chapter 8 covers the solution of linear algebraic equations, which arise in applications in all fields of engineering. This coverage establishes the terminology and some important concepts required to use the computer methods properly. The chapter then shows how to use MATLAB to solve systems of linear equations that have a unique solution. Underdetermined and overdetermined systems are also covered. The remaining chapters are independent of this chapter.

Introduction to MATLAB 7 for Engineers, McGraw-Hill, New York, 2005.

Chapter 9 covers numerical methods for calculus and differential equations. Numerical integration and differentiation methods are treated. Ordinary differential equation solvers in the core MATLAB program are covered, as well as the linear system solvers in the Control System toolbox. This chapter provides some background for Chapter 10.

Chapter 10 introduces Simulink, which is a graphical interface for building simulations of dynamic systems. Simulink has increased in popularity and has seen increased use in industry. This chapter need not be covered to read Chapter 11.

Chapter 11 covers symbolic methods for manipulating algebraic expressions and for solving algebraic and transcendental equations, calculus, differential equations, and matrix algebra problems. The calculus applications include integration and differentiation, optimization, Taylor series, series evaluation, and limits. Laplace transform methods for solving differential equations are also introduced. This chapter requires the use of the Symbolic Math toolbox, which includes MuPAD. MuPAD is a new feature in MATLAB. It provides a notebook interface for entering commands and displaying results, including plots.

Appendix A contains a guide to the commands and functions introduced in the text. Appendix B is an introduction to producing animation and sound with MATLAB. While not essential to learning MATLAB, these features are helpful for generating student interest. Appendix C summarizes functions for creating formatted output. Appendix D is a list of references. Appendix E, which is available on the text's website, contains some suggestions for course projects and is based on the author's experience in teaching a freshman MATLAB course. Answers to selected problems and an index appear at the end of the text.

All figures, tables, equations, and exercises have been numbered according to their chapter and section. For example, Figure 3.4–2 is the second figure in Chapter 3, Section 4. This system is designed to help the reader locate these items. The end-of-chapter problems are the exception to this numbering system. They are numbered 1, 2, 3, and so on to avoid confusion with the in-chapter exercises.

The first four chapters constitute a course in the essentials of MATLAB. The remaining seven chapters are independent of one another, and may be covered in any order or may be omitted if necessary. These chapters provide additional coverage and examples of plotting and model building, linear algebraic equations, probability and statistics, calculus and differential equations, Simulink, and symbolic processing, respectively.

SPECIAL REFERENCE FEATURES

The text has the following special features, which have been designed to enhance its usefulness as a reference.

■ Throughout each of the chapters, numerous tables summarize the commands and functions as they are introduced.

- Appendix A is a complete summary of all the commands and functions described in the text, grouped by category, along with the number of the page on which they are introduced.
- At the end of each chapter is a list of the key terms introduced in the chapter, with the page number referenced.
- Key terms have been placed in the margin or in section headings where they are introduced.
- The index has four sections: a listing of symbols, an alphabetical list of MATLAB commands and functions, a list of Simulink blocks, and an alphabetical list of topics.

PEDAGOGICAL AIDS

The following pedagogical aids have been included:

- Each chapter begins with an overview.
- **Test Your Understanding** exercises appear throughout the chapters near the relevant text. These relatively straightforward exercises allow readers to assess their grasp of the material as soon as it is covered. In most cases the answer to the exercise is given with the exercise. Students should work these exercises as they are encountered.
- Each chapter ends with numerous problems, grouped according to the relevant section.
- Each chapter contains numerous practical examples. The major examples are numbered.
- Each chapter has a summary section that reviews the chapter's objectives.
- Answers to many end-of-chapter problems appear at the end of the text. These problems are denoted by an asterisk next to their number (for example, **15***).

Two features have been included to motivate the student toward MATLAB and the engineering profession:

- Most of the examples and the problems deal with engineering applications. These are drawn from a variety of engineering fields and show realistic applications of MATLAB. A guide to these examples appears on the inside front cover.
- The facing page of each chapter contains a photograph of a recent engineering achievement that illustrates the challenging and interesting opportunities that await engineers in the 21st century. A description of the achievement and its related engineering disciplines and a discussion of how MATLAB can be applied in those disciplines accompanies each photo.

ONLINE RESOURCES

An Instructor's Manual is available online for instructors who have adopted this text. This manual contains the complete solutions to all the Test Your Understanding exercises and to all the chapter problems. The text website (at http://www.mhhe.com/palm) also has downloadable files containing PowerPoint slides keyed to the text and suggestions for projects.

ELECTRONIC TEXTBOOK OPTIONS

Ebooks are an innovative way for students to save money and create a greener environment at the same time. An ebook can save students about one-half the cost of a traditional textbook and offers unique features such as a powerful search engine, highlighting, and the ability to share notes with classmates using ebooks.

McGraw-Hill offers this text as an ebook. To talk about the ebook options, contact your McGraw-Hill sales rep or visit the site www.coursesmart.com to learn more.

MATLAB INFORMATION

For MATLAB® and Simulink® product information, please contact:

The MathWorks, Inc.
3 Apple Hill Drive
Natick, MA, 01760-2098 USA
Tel: 508-647-7000
Fax: 508-647-7001
E-mail: info@mathworks.com
Web: www.mathworks.com

ACKNOWLEDGMENTS

Many individuals are due credit for this text. Working with faculty at the University of Rhode Island in developing and teaching a freshman course based on MATLAB has greatly influenced this text. Email from many users contained useful suggestions. The author greatly appreciates their contributions.

The MathWorks, Inc., has always been very supportive of educational publishing. I especially want to thank Naomi Fernandes of The MathWorks, Inc., for her help. Bill Stenquist, Joyce Watters, and Lora Neyens of McGraw-Hill efficiently handled the manuscript reviews and guided the text through production.

My sisters, Linda and Chris, and my mother, Lillian, have always been there, cheering my efforts. My father was always there for support before he passed away. Finally, I want to thank my wife, Mary Louise, and my children, Aileene, Bill, and Andy, for their understanding and support of this project.

William J. Palm, III
Kingston, Rhode Island
September 2009

Engineering in the 21st Century...

Remote Exploration

It will be many years before humans can travel to other planets. In the meantime, unmanned probes have been rapidly increasing our knowledge of the universe. Their use will increase in the future as our technology develops to make them more reliable and more versatile. Better sensors are expected for imaging and other data collection. Improved robotic devices will make these probes more autonomous, and more capable of interacting with their environment, instead of just observing it.

NASA's planetary rover *Sojourner* landed on Mars on July 4, 1997, and excited people on Earth while they watched it successfully explore the Martian surface to determine wheel-soil interactions, to analyze rocks and soil, and to return images of the lander for damage assessment. Then in early 2004, two improved rovers, *Spirit* and *Opportunity,* landed on opposite sides of the planet. In one of the major discoveries of the 21st century, they obtained strong evidence that water once existed on Mars in significant amounts.

About the size of a golf cart, the new rovers have six wheels, each with its own motor. They have a top speed of 5 centimeters per second on flat, hard ground and can travel up to about 100 meters per day. Needing 100 watts to move, they obtain power from solar arrays that generate 140 watts during a 4-hour window each day. The sophisticated temperature control system must not only protect against nighttime temperatures of $-96°C$, but also prevent the rover from overheating.

The robotic arm has three joints (shoulder, elbow, and wrist), driven by five motors, and it has a reach of 90 centimeters. The arm carries four tools and instruments for geological studies. Nine cameras provide hazard avoidance, navigation, and panoramic views. The onboard computer has 128 MB of DRAM and coordinates all the subsystems including communications.

Although originally planned to last for three months, both rovers were still exploring Mars at the end of 2009.

All engineering disciplines were involved with the rovers' design and launch. The MATLAB Neural Network, Signal Processing, Image Processing, PDE, and various control system toolboxes are well suited to assist designers of probes and autonomous vehicles like the Mars rovers. ∎

CHAPTER 1

An Overview of MATLAB®*

OUTLINE

1.1 MATLAB Interactive Sessions

1.2 Menus and the Toolbar

1.3 Arrays, Files, and Plots

1.4 Script Files and the Editor/Debugger

1.5 The MATLAB Help System

1.6 Problem-Solving Methodologies

1.7 Summary

Problems

This is the most important chapter in the book. By the time you have finished this chapter, you will be able to use MATLAB to solve many kinds of problems. Section 1.1 provides an introduction to MATLAB as an interactive calculator. Section 1.2 covers the main menus and toolbar. Section 1.3 introduces arrays, files, and plots. Section 1.4 discusses how to create, edit, and save MATLAB programs. Section 1.5 introduces the extensive MATLAB Help System and Section 1.6 introduces the methodology of engineering problem solving.

How to Use This Book

The book's chapter organization is flexible enough to accommodate a variety of users. However, it is important to cover at least the first four chapters, in that order. Chapter 2 covers *arrays,* which are the basic building blocks in MATLAB. Chapter 3 covers file usage, functions built into MATLAB, and user-defined functions.

---

*MATLAB is a registered trademark of The MathWorks, Inc.

Chapter 4 covers programming using relational and logical operators, conditional statements, and loops.

Chapters 5 through 11 are independent chapters that can be covered in any order. They contain in-depth discussions of how to use MATLAB to solve several common types of problems. Chapter 5 covers two- and three-dimensional plots in greater detail. Chapter 6 shows how to use plots to build mathematical models from data. Chapter 7 covers probability, statistics and interpolation applications. Chapter 8 treats linear algebraic equations in more depth by developing methods for the overdetermined and underdetermined cases. Chapter 9 introduces numerical methods for calculus and ordinary differential equations. Simulink®*, the topic of Chapter 10, is a graphical user interface for solving differential equation models. Chapter 11 covers symbolic processing with MuPAD®*, a new feature of the MATLAB Symbolic Math toolbox, with applications to algebra, calculus, differential equations, transforms, and special functions.

Reference and Learning Aids

The book has been designed as a reference as well as a learning tool. The special features useful for these purposes are as follows.

■ Throughout each chapter margin notes identify where new terms are introduced.

■ Throughout each chapter short Test Your Understanding exercises appear. Where appropriate, answers immediately follow the exercise so you can measure your mastery of the material.

■ Homework exercises conclude each chapter. These usually require greater effort than the Test Your Understanding exercises.

■ Each chapter contains tables summarizing the MATLAB commands introduced in that chapter.

■ At the end of each chapter is
   ■ A summary of what you should be able to do after completing that chapter
   ■ A list of key terms you should know

■ Appendix A contains tables of MATLAB commands, grouped by category, with the appropriate page references.

■ The index has four parts: MATLAB symbols, MATLAB commands, Simulink blocks, and topics.

1.1 MATLAB Interactive Sessions

We now show how to start MATLAB, how to make some basic calculations, and how to exit MATLAB.

---

*Simulink and MuPAD are registered trademarks of The MathWorks, Inc.

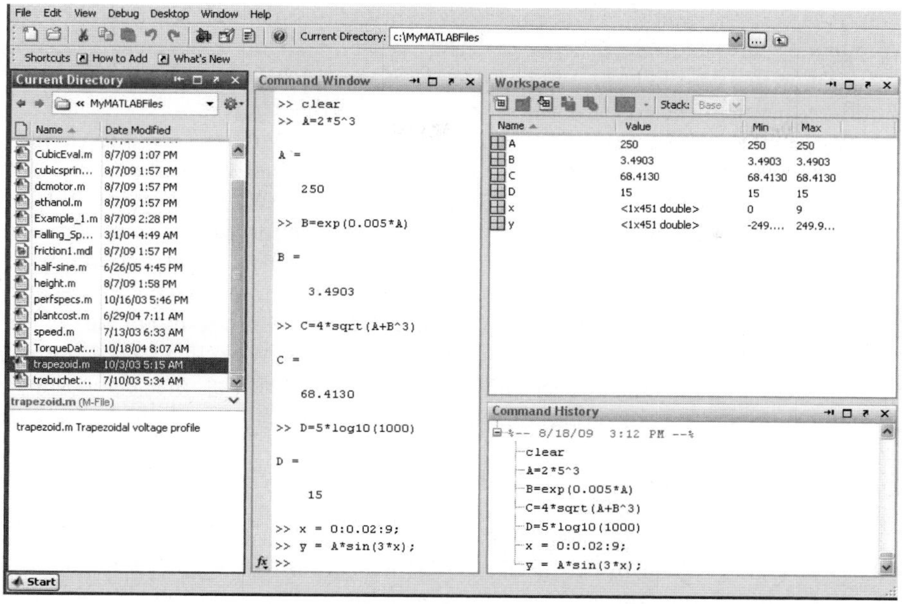

Figure 1.1–1 The default MATLAB Desktop.

Conventions

In this text we use typewriter font to represent MATLAB commands, any text that you type in the computer, and any MATLAB responses that appear on the screen, for example, y = 6*x. Variables in normal mathematics text appear in italics, for example, $y = 6x$. We use boldface type for three purposes: to represent vectors and matrices in normal mathematics text (for example, **Ax = b**), to represent a key on the keyboard (for example, **Enter**), and to represent the name of a screen menu or an item that appears in such a menu (for example, **File**). It is assumed that you press the **Enter** key after you type a command. We do not show this action with a separate symbol.

Starting MATLAB

To start MATLAB on a MS Windows system, double-click on the MATLAB icon. You will then see the MATLAB *Desktop*. The Desktop manages the Command window and a Help Browser as well as other tools. The default appearance of the Desktop is shown in Figure 1.1–1. Five windows appear. These are the Command window in the center, the Command History window in the lower right, the Workspace window in the upper right, the Details window in the lower left, and the

DESKTOP

Current Directory window in the upper left. Across the top of the Desktop are a row of menu names and a row of icons called the *toolbar*. To the right of the toolbar is a box showing the directory where MATLAB looks for and saves files. We will describe the menus, toolbar, and directories later in this chapter.

**COMMAND
WINDOW**

You use the Command window to communicate with the MATLAB program, by typing instructions of various types called *commands, functions,* and *statements*. Later we will discuss the differences between these types, but for now, to simplify the discussion, we will call the instructions by the generic name *commands*. MATLAB displays the prompt (≫) to indicate that it is ready to receive instructions. Before you give MATLAB instructions, make sure the cursor is located just after the prompt. If it is not, use the mouse to move the cursor. The prompt in the Student Edition looks like EDU ≫. We will use the normal prompt symbol ≫ to illustrate commands in this text. The Command window in Figure 1.1–1 shows some commands and the results of the calculations. We will cover these commands later in this chapter.

Four other windows appear in the default Desktop. The Current Directory window is much like a file manager window; you can use it to access files. Double-clicking on a file name with the extension .m will open that file in the MATLAB Editor. The Editor is discussed in Section 1.4. Figure 1.1–1 shows the files in the author's directory C:\MyMATLABFiles.

Underneath the Current Directory window is the . . . window. It displays any comments in the file. Note that two file types are shown in the Current Directory. These have the extensions .m and .mdl. We will cover M files in this chapter. Chapter 10 covers Simulink, which uses MDL files. You can have other file types in the directory.

The Workspace window appears in the upper right. The Workspace window displays the variables created in the Command window. Double-click on a variable name to open the Array Editor, which is discussed in Chapter 2.

The fifth window in the default Desktop is the Command History window. This window shows all the previous keystrokes you entered in the Command window. It is useful for keeping track of what you typed. You can click on a keystroke and drag it to the Command window or the Editor to avoid retyping it. Double-clicking on a keystroke executes it in the Command window.

You can alter the appearance of the Desktop if you wish. For example, to eliminate a window, just click on its Close-window button (×) in its upper right-hand corner. To undock, or separate the window from the Desktop, click on the button containing a curved arrow. An undocked window can be moved around on the screen. You can manipulate other windows in the same way. To restore the default configuration, click on the **Desktop** menu, then click on **Desktop Layout,** and select **Default.**

Entering Commands and Expressions

To see how simple it is to use MATLAB, try entering a few commands on your computer. If you make a typing mistake, just press the **Enter** key until you get the prompt, and then retype the line. Or, because MATLAB retains your previous keystrokes in a command file, you can use the up-arrow key (↑) to scroll back through the commands. Press the key once to see the previous entry, twice to see the entry before that, and so on. Use the down-arrow key (↓) to scroll forward through the commands. When you find the line you want, you can edit it using the left- and right-arrow keys (← and →), and the **Backspace** key, and the **Delete** key. Press the **Enter** key to execute the command. This technique enables you to correct typing mistakes quickly.

Note that you can see your previous keystrokes displayed in the Command History window. You can copy a line from this window to the Command window by highlighting the line with the mouse, holding down the left mouse button, and dragging the line to the Command window.

Make sure the cursor is at the prompt in the Command window. To divide 8 by 10, type 8/10 and press **Enter** (the symbol / is the MATLAB symbol for division). Your entry and the MATLAB response look like the following on the screen (we call this interaction between you and MATLAB an *interactive session,* or simply *a session*). Remember, the symbol >> automatically appears on the screen; you do not type it.

SESSION

```
>> 8/10
ans =
     0.8000
```

MATLAB indents the numerical result. MATLAB uses high precision for its computations, but by default it usually displays its results using four decimal places except when the result is an integer.

MATLAB assigns the most recent answer to a variable called ans, which is an abbreviation for *answer*. A *variable* in MATLAB is a symbol used to contain a value. You can use the variable ans for further calculations; for example, using the MATLAB symbol for multiplication (*), we obtain

VARIABLE

```
>> 5*ans
ans =
     4
```

Note that the variable ans now has the value 4.

You can use variables to write mathematical expressions. Instead of using the default variable ans, you can assign the result to a variable of your own choosing, say, r, as follows:

```
>> r=8/10
r =
     0.8000
```

Spaces in the line improve its readability; for example, you can put a space before and after the = sign if you want. MATLAB ignores these spaces when making its calculations. It also ignores spaces surrounding + and − signs.

If you now type r at the prompt and press **Enter,** you will see

```
>> r
r =
    0.8000
```

thus verifying that the variable r has the value 0.8. You can use this variable in further calculations. For example,

```
>> s=20*r
s =
    16
```

A common mistake is to forget the multiplication symbol * and type the expression as you would in algebra, as *s = 20r*. If you do this in MATLAB, you will get an error message.

MATLAB has hundreds of functions available. One of these is the *square root* function, sqrt. A pair of parentheses is used after the function's name to enclose the value—called the function's *argument*—that is operated on by the function. For example, to compute the square root of 9 and assign its value to the variable *r*, you type r = sqrt(9). Note that the previous value of r has been replaced by 3.

ARGUMENT

Order of Precedence

SCALAR

A *scalar* is a single number. A *scalar variable* is a variable that contains a single number. MATLAB uses the symbols + − * / ^ for addition, subtraction, multiplication, division, and exponentiation (power) of scalars. These are listed in Table 1.1–1. For example, typing x = 8 + 3*5 returns the answer x = 23. Typing 2^3-10 returns the answer ans = -2. The *forward slash* (/) represents *right division*, which is the normal division operator familiar to you. Typing 15/3 returns the result ans = 5.

MATLAB has another division operator, called *left division,* which is denoted by the *backslash* (\). The left division operator is useful for solving sets of linear algebraic equations, as we will see. A good way to remember the difference between the right and left division operators is to note that the slash slants toward the denominator. For example, 7/2 = 2\7 = 3.5.

Table 1.1–1 Scalar arithmetic operations

Symbol	Operation	MATLAB form
^	exponentiation: a^b	a^b
*	multiplication: ab	a*b
/	right division: $a/b = \frac{a}{b}$	a/b
\	left division: $a\backslash b = \frac{b}{a}$	a\b
+	addition: $a + b$	a+b
−	subtraction: $a - b$	a−b

Table 1.1–2 Order of precedence

Precedence	Operation
First	Parentheses, evaluated starting with the innermost pair.
Second	Exponentiation, evaluated from left to right.
Third	Multiplication and division with equal precedence, evaluated from left to right.
Fourth	Addition and subtraction with equal precedence, evaluated from left to right.

The mathematical operations represented by the symbols $+$ $-$ $*$ $/$ $\backslash$ and $\wedge$ follow a set of rules called *precedence*. Mathematical expressions are evaluated starting from the left, with the exponentiation operation having the highest order of precedence, followed by multiplication and division with equal precedence, followed by addition and subtraction with equal precedence. Parentheses can be used to alter this order. Evaluation begins with the innermost pair of parentheses and proceeds outward. Table 1.1–2 summarizes these rules. For example, note the effect of precedence on the following session.

PRECEDENCE

```
>>8 + 3*5
ans =
      23
>>(8 + 3)*5
ans =
      55
>>4^2 - 12 - 8/4*2
ans =
      0
>>4^2 - 12 - 8/(4*2)
ans =
      3
>>3*4^2 + 5
ans =
      53
>>(3*4)^2 + 5
ans =
      149
>>27^(1/3) + 32^(0.2)
ans =
      5
>>27^(1/3) + 32^0.2
ans =
      5
>>27^1/3 + 32^0.2
ans =
      11
```

To avoid mistakes, feel free to insert parentheses wherever you are unsure of the effect precedence will have on the calculation. Use of parentheses also improves the readability of your MATLAB expressions. For example, parentheses are not needed in the expression `8+(3*5)`, but they make clear our intention to multiply 3 by 5 before adding 8 to the result.

Test Your Understanding

T1.1–1 Use MATLAB to compute the following expressions.

a. $6\left(\dfrac{10}{13}\right) + \dfrac{18}{5(7)} + 5(9^2)$

b. $6(35^{1/4}) + 14^{0.35}$

(Answers: a. 410.1297 b. 17.1123.)

The Assignment Operator

The = sign in MATLAB is called the *assignment* or *replacement* operator. It works differently than the equals sign you know from mathematics. When you type $x = 3$, you tell MATLAB to assign the value 3 to the variable x. This usage is no different than in mathematics. However, in MATLAB we can also type something like this: $x = x + 2$. This tells MATLAB to add 2 to the current value of x, and to replace the current value of x with this new value. If x originally had the value 3, its new value would be 5. This use of the = operator is different from its use in mathematics. For example, the mathematics equation $x = x + 2$ is invalid because it implies that $0 = 2$.

In MATLAB the variable on the *left-hand* side of the = operator is replaced by the value generated by the *right-hand* side. Therefore, one variable, and only one variable, must be on the left-hand side of the = operator. Thus in MATLAB you cannot type $6 = x$. Another consequence of this restriction is that you cannot write in MATLAB expressions like the following:

```
>>x+2=20
```

The corresponding equation $x + 2 = 20$ is acceptable in algebra and has the solution $x = 18$, but MATLAB cannot solve such an equation without additional commands (these commands are available in the Symbolic Math toolbox, which is described in Chapter 11).

Another restriction is that the right-hand side of the = operator must have a computable value. For example, if the variable y has not been assigned a value, then the following will generate an error message in MATLAB.

```
>>x = 5 + y
```

In addition to assigning known values to variables, the assignment operator is very useful for assigning values that are not known ahead of time, or for changing the value of a variable by using a prescribed procedure. The following example shows how this is done.

Volume of a Circular Cylinder

EXAMPLE 1.1–1

The volume of a circular cylinder of height h and radius r is given by $V = \pi r^2 h$. A particular cylindrical tank is 15 m tall and has a radius of 8 m. We want to construct another cylindrical tank with a volume 20 percent greater but having the same height. How large must its radius be?

■ Solution

First solve the cylinder equation for the radius r. This gives

$$r = \sqrt{\frac{V}{\pi h}}$$

The session is shown below. First we assign values to the variables r and h representing the radius and height. Then we compute the volume of the original cylinder and increase the volume by 20 percent. Finally we solve for the required radius. For this problem we can use the MATLAB built-in constant `pi`.

```
>>r = 8;
>>h = 15;
>>V = pi*r^2*h;
>>V = V + 0.2*V;
>>r = sqrt(V/(pi*h))
r =
    8.7636
```

Thus the new cylinder must have a radius of 8.7636 m. Note that the original values of the variables `r` and `V` are replaced with the new values. This is acceptable as long as we do not wish to use the original values again. Note how precedence applies to the line `V = pi*r^2*h;`. It is equivalent to `V = pi*(r^2)*h;`.

Variable Names

The term *workspace* refers to the names and values of any variables in use in the current work session. Variable names must begin with a letter; the rest of the name can contain letters, digits, and underscore characters. MATLAB is case-sensitive. Thus the following names represent five different variables: `speed`, `Speed`, `SPEED`, `Speed_1`, and `Speed_2`. In MATLAB 7, variable names can be no longer than 63 characters.

WORKSPACE

Managing the Work Session

Table 1.1–3 summarizes some commands and special symbols for managing the work session. A semicolon at the end of a line suppresses printing the results to the screen. If a semicolon is not put at the end of a line, MATLAB displays the

Table 1.1–3 Commands for managing the work session

Command	Description
clc	Clears the Command window.
clear	Removes all variables from memory.
clear var1 var2	Removes the variables var1 and var2 from memory.
exist('name')	Determines if a file or variable exists having the name 'name'.
quit	Stops MATLAB.
who	Lists the variables currently in memory.
whos	Lists the current variables and sizes and indicate if they have imaginary parts.
:	Colon; generates an array having regularly spaced elements.
,	Comma; separates elements of an array.
;	Semicolon; suppresses screen printing; also denotes a new row in an array.
...	Ellipsis; continues a line.

results of the line on the screen. Even if you suppress the display with the semicolon, MATLAB still retains the variable's value.

You can put several commands on the same line if you separate them with a comma if you want to see the results of the previous command or semicolon if you want to suppress the display. For example,

```
>>x=2;y=6+x,x=y+7
y =
    8
x =
    15
```

Note that the first value of x was not displayed. Note also that the value of x changed from 2 to 15.

If you need to type a long line, you can use an *ellipsis,* by typing three periods, to delay execution. For example,

```
>>NumberOfApples = 10; NumberOfOranges = 25;
>>NumberOfPears = 12;
>>FruitPurchased = NumberOfApples + NumberOfOranges ...
+NumberOfPears
FruitPurchased =
     47
```

Use the arrow, **Tab,** and **Ctrl** keys to recall, edit, and reuse functions and variables you typed earlier. For example, suppose you mistakenly enter the line

```
>>volume = 1 + sqr(5)
```

MATLAB responds with an error message because you misspelled sqrt. Instead of retyping the entire line, press the up-arrow key (↑) once to display the previously typed line. Press the left-arrow key (←) several times to move the cursor and add the missing t, then press **Enter.** Repeated use of the up-arrow key recalls lines typed earlier.

Tab and Arrow Keys

You can use the *smart recall* feature to recall a previously typed function or variable whose first few characters you specify. For example, after you have entered the line starting with `volume,` typing `vol` and pressing the up-arrow key (↑) once recalls the last-typed line that starts with the function or variable whose name begins with `vol`. This feature is case-sensitive.

You can use the *tab completion* feature to reduce the amount of typing. MATLAB automatically completes the name of a function, variable, or file if you type the first few letters of the name and press the **Tab** key. If the name is unique, it is automatically completed. For example, in the session listed earlier, if you type `Fruit` and press **Tab,** MATLAB completes the name and displays `FruitPurchased`. Press **Enter** to display the value of the variable, or continue editing to create a new executable line that uses the variable `FruitPurchased`.

If there is more than one name that starts with the letters you typed, MATLAB displays these names when you press the **Tab** key. Use the mouse to select the desired name from the pop-up list by double-clicking on its name.

The left-arrow (←) and right-arrow (→) keys move left and right through a line one *character* at a time. To move through one *word* at a time, press **Ctrl** and → simultaneously to move to the *right;* press **Ctrl** and ← simultaneously to move to the *left.* Press **Home** to move to the beginning of a line; press **End** to move to the end of a line.

Deleting and Clearing

Press **Del** to delete the character at the cursor; press **Backspace** to delete the character before the cursor. Press **Esc** to clear the entire line; press **Ctrl** and **k** simultaneously to delete (*kill*) to the end of the line.

MATLAB retains the last value of a variable until you quit MATLAB or clear its value. Overlooking this fact commonly causes errors in MATLAB. For example, you might prefer to use the variable x in a number of different calculations. If you forget to enter the correct value for x, MATLAB uses the last value, and you get an incorrect result. You can use the `clear` function to remove the values of *all* variables from memory, or you can use the form `clear var1 var2` to clear the variables named `var1` and `var2`. The effect of the `clc` command is different; it clears the Command window of everything in the window display, but the values of the variables remain.

You can type the name of a variable and press **Enter** to see its current value. If the variable does not have a value (i.e., if it does not exist), you see an error message. You can also use the `exist` function. Type `exist('x')` to see if the variable x is in use. If a 1 is returned, the variable exists; a 0 indicates that it does not exist. The `who` function lists the names of all the variables in memory, but does not give their values. The form `who var1 var2` restricts the display to the variables specified. The wildcard character * can be used to display variables that match a pattern. For instance, `who A*` finds all variables in the current workspace that start with A. The `whos` function lists the variable names and their sizes and indicates whether they have nonzero imaginary parts.

The difference between a function and a command or a statement is that functions have their arguments enclosed in parentheses. Commands, such as `clear`, need not have arguments; but if they do, they are not enclosed in parentheses, for example, `clear x`. Statements cannot have arguments; for example, `clc` and `quit` are statements.

Press **Ctrl-C** to cancel a long computation without terminating the session. You can quit MATLAB by typing `quit`. You can also click on the **File** menu, and then click on **Exit MATLAB.**

Predefined Constants

MATLAB has several predefined special constants, such as the built-in constant `pi` we used in Example 1.1–1. Table 1.1–4 lists them. The symbol `Inf` stands for ∞, which in practice means a number so large that MATLAB cannot represent it. For example, typing `5/0` generates the answer `Inf`. The symbol `NaN` stands for "not a number." It indicates an undefined numerical result such as that obtained by typing `0/0`. The symbol `eps` is the smallest number which, when added to 1 by the computer, creates a number greater than 1. We use it as an indicator of the accuracy of computations.

The symbols `i` and `j` denote the imaginary unit, where $i = j = \sqrt{-1}$. We use them to create and represent complex numbers, such as `x = 5 + 8i`.

Try not to use the names of special constants as variable names. Although MATLAB allows you to assign a different value to these constants, it is not good practice to do so.

Complex Number Operations

MATLAB handles complex number algebra automatically. For example, the number $c_1 = 1 - 2i$ is entered as follows: `c1 = 1-2i`. You can also type `c1 = Complex(1, -2)`.

> **Caution:** Note that an asterisk is not needed between `i` or `j` and a number, although it is required with a variable, such as `c2 = 5 - i*c1`. This convention can cause errors if you are not careful. For example, the expressions `y = 7/2*i` and `x = 7/2i` give two different results: $y = (7/2)i = 3.5i$ and $x = 7/(2i) = -3.5i$.

Table 1.1–4 Special variables and constants

Command	Description
ans	Temporary variable containing the most recent answer.
eps	Specifies the accuracy of floating point precision.
i,j	The imaginary unit $\sqrt{-1}$.
Inf	Infinity.
NaN	Indicates an undefined numerical result.
pi	The number π.

Addition, subtraction, multiplication, and division of complex numbers are easily done. For example,

```
>>s = 3+7i;w = 5-9i;
>>w+s
ans =
    8.0000 - 2.0000i
>>w*s
ans =
    78.0000 + 8.0000i
>>w/s
ans =
    -0.8276 - 1.0690i
```

Test Your Understanding

T1.1–2 Given $x = -5 + 9i$ and $y = 6 - 2i$, use MATLAB to show that $x + y = 1 + 7i$, $xy = -12 + 64i$, and $x/y = -1.2 + 1.1i$.

Formatting Commands

The `format` command controls how numbers appear on the screen. Table 1.1–5 gives the variants of this command. MATLAB uses many significant figures in its calculations, but we rarely need to see all of them. The default MATLAB display format is the `short` format, which uses four decimal digits. You can display more by typing `format long`, which gives 16 digits. To return to the default mode, type `format short`.

You can force the output to be in scientific notation by typing `format short e`, or `format long e`, where e stands for the number 10. Thus the output `6.3792e+03` stands for the number 6.3792×10^3. The output `6.3792e-03`

Table 1.1–5 Numeric display formats

Command	Description and example
format short	Four decimal digits (the default); 13.6745.
format long	16 digits; 17.27484029463547.
format short e	Five digits (four decimals) plus exponent; 6.3792e+03.
format long e	16 digits (15 decimals) plus exponent; 6.379243784781294e−04.
format bank	Two decimal digits; 126.73.
format +	Positive, negative, or zero; +.
format rat	Rational approximation; 43/7.
format compact	Suppresses some blank lines.
format loose	Resets to less compact display mode.

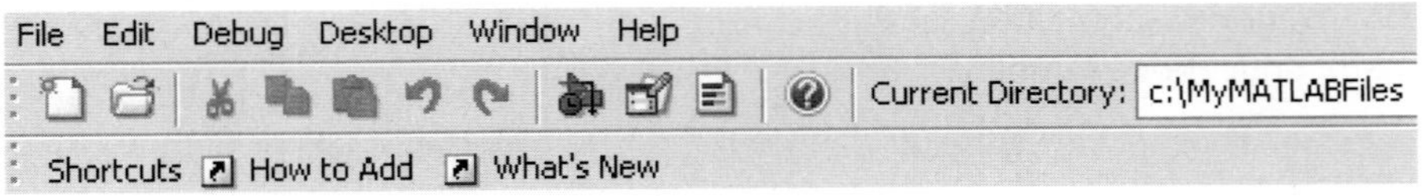

Figure 1.2–1 The top of the MATLAB Desktop.

stands for the number 6.3792×10^{-3}. Note that in this context e does *not* represent the number *e*, which is the base of the natural logarithm. Here e stands for "exponent." It is a poor choice of notation, but MATLAB follows conventional computer programming standards that were established many years ago.

Use `format bank` only for monetary calculations; it does not recognize imaginary parts.

1.2 Menus and the Toolbar

CURRENT DIRECTORY

The Desktop manages the Command window and other MATLAB tools. The default appearance of the Desktop is shown in Figure 1.1–1. Across the top of the Desktop are a row of menu names and a row of icons called the *toolbar*. To the right of the toolbar is a box showing the *current directory,* where MATLAB looks for files. See Figure 1.2–1.

Other windows appear in a MATLAB session, depending on what you do. For example, a graphics window containing a plot appears when you use the plotting functions; an editor window, called the Editor/Debugger, appears for use in creating program files. Each window type has its own menu bar, with one or more menus, at the top. *Thus the menu bar will change as you change windows.* To activate or select a menu, click on it. Each menu has several items. Click on an item to select it. *Keep in mind that menus are context-sensitive. Thus their contents change, depending on which features you are currently using.*

The Desktop Menus

Most of your interaction will be in the Command window. When the Command window is active, the default MATLAB 7 Desktop (shown in Figure 1.1–1) has six menus: **File**, **Edit**, **Debug**, **Desktop**, **Window,** and **Help.** Note that these menus change depending on what window is active. Every item on a menu can be selected with the menu open either by clicking on the item or by typing its underlined letter. Some items can be selected without the menu being open by using the shortcut key listed to the right of the item. Those items followed by three dots (**. . .**) open a submenu or another window containing a dialog box.

The three most useful menus are the **File**, **Edit,** and **Help** menus. The **Help** menu is described in Section 1.5. The **File** menu in MATLAB 7 contains the following items, which perform the indicated actions when you select them.

The File Menu in MATLAB 7

New Opens a dialog box that allows you to create a new program file, called an M-file, using a text editor called the Editor/Debugger, a new Figure, a variable in the Workspace window, Model file (a file type used by Simulink), or a new GUI (which stands for Graphical User Interface).

Open. . . Opens a dialog box that allows you to select a file for editing.

Close Command Window (or **Current Folder**) Closes the Command window or current file if one is open.

Import Data. . . Starts the Import Wizard which enables you to import data easily.

Save Workspace As. . . Opens a dialog box that enables you to save a file.

Set Path. . . Opens a dialog box that enables you to set the MATLAB search path.

Preferences. . . Opens a dialog box that enables you to set preferences for such items as fonts, colors, tab spacing, and so forth.

Page Setup Opens a dialog box that enables you to format printed output.

Print. . . Opens a dialog box that enables you to print all the Command window.

Print Selection. . . Opens a dialog box that enables you to print selected portions of the Command window.

File List Contains a list of previously used files, in order of most recently used.

Exit MATLAB Closes MATLAB.

The **New** option in the **File** menu lets you select which type of M-file to create: a blank M-file, a function M-file, or a class M-file. Select blank M-file to create an M-file of the type discussed in Section 1.4. Function M-files are discussed in Chapter 3, but class M-files are beyond the scope of this text.

The **Edit** menu contains the following items.

The Edit Menu in MATLAB 7

Undo Reverses the previous editing operation.

Redo Reverses the previous Undo operation.

Cut Removes the selected text and stores it for pasting later.

Copy Copies the selected text for pasting later, without removing it.

Paste Inserts any text on the clipboard at the current location of the cursor.

Paste to Workspace. . . Inserts the contents of the clipboard into the workspace as one or more variables.

Select All Highlights all text in the Command window.

Delete Clears the variable highlighted in the Workspace Browser.

Find. . . Finds and replaces phrases.

Find Files. . . Finds files.

Clear Command Window Removes all text from the Command window.

Clear Command History Removes all text from the Command History window.

Clear Workspace Removes the values of all variables from the workspace.

You can use the **Copy** and **Paste** selections to copy and paste commands appearing on the Command window. However, an easier way is to use the up-arrow key to scroll through the previous commands, and press **Enter** when you see the command you want to retrieve.

Use the **Debug** menu to access the Debugger, which is discussed in Chapter 4. Use the **Desktop** menu to control the configuration of the Desktop and to display toolbars. The **Window** menu has one or more items, depending on what you have done thus far in your session. Click on the name of a window that appears on the menu to open it. For example, if you have created a plot and not closed its window, the plot window will appear on this menu as **Figure 1.** However, there are other ways to move between windows (such as pressing the **Alt** and **Tab** keys simultaneously if the windows are not docked).

The **View** menu will appear to the right of the **Edit** menu if you have selected a file in the folder in the Current Folder window. This menu gives information about the selected file.

The toolbar, which is below the menu bar, provides buttons as shortcuts to some of the features on the menus. Clicking on the button is equivalent to clicking on the menu, then clicking on the menu item; thus the button eliminates one click of the mouse. The first seven buttons from the left correspond to the **New M-File**, **Open File**, **Cut**, **Copy**, **Paste**, **Undo,** and **Redo.** The eighth button activates Simulink, which is a program built on top of MATLAB. The ninth button activates the GUIDE Quick Start, which is used to create and edit graphical user interfaces (GUIs). The tenth button activates the Profiler, which can be used to optimize program performance. The eleventh button (the one with the question mark) accesses the Help System.

Below the toolbar is a button that accesses help for adding shortcuts to the toolbar and a button that accesses a list of the features added since the previous release.

1.3 Arrays, Files, and Plots

This section introduces arrays, which are the basic building blocks in MATLAB, and shows how to handle files and generate plots.

Arrays

MATLAB has hundreds of functions, which we will discuss throughout the text. For example, to compute $\sin x$, where x has a value in radians, you type $\sin(x)$. To compute $\cos x$, type $\cos(x)$. The exponential function e^x is computed from $\exp(x)$. The natural logarithm, $\ln x$, is computed by typing $\log(x)$. (Note the spelling difference between mathematics text, ln, and MATLAB syntax, $\log$.)

You compute the base-10 logarithm by typing `log10(x)`. The inverse sine, or arcsine, is obtained by typing `asin(x)`. It returns an answer in radians, not degrees. The function `asind(x)` returns degrees.

One of the strengths of MATLAB is its ability to handle collections of numbers, called *arrays,* as if they were a single variable. A numerical array is an ordered collection of numbers (a set of numbers arranged in a specific order). An example of an array variable is one that contains the numbers 0, 4, 3, and 6, in that order. We use square brackets to define the variable x to contain this collection by typing `x = [0, 4, 3, 6]`. The elements of the array may also be separated by spaces, but commas are preferred to improve readability and avoid mistakes. Note that the variable y defined as `y = [6, 3, 4, 0]` is not the same as x because the order is different. The reason for using the brackets is as follows. If you were to type `x = 0, 4, 3, 6`, MATLAB would treat this as four separate inputs and would assign the value 0 to x. The array `[0, 4, 3, 6]` can be considered to have one row and four columns, and it is a subcase of a *matrix,* which has multiple rows and columns. As we will see, matrices are also denoted by square brackets.

ARRAY

We can add the two arrays x and y to produce another array z by typing the single line `z = x + y`. To compute z, MATLAB adds all the corresponding numbers in x and y to produce z. The resulting array z contains the numbers 6, 7, 7, 6.

You need not type all the numbers in the array if they are regularly spaced. Instead, you type the first number and the last number, with the spacing in the middle, separated by colons. For example, the numbers 0, 0.1, 0.2, . . . , 10 can be assigned to the variable u by typing `u = 0:0.1:10`. In this application of the colon operator, the brackets should not be used.

To compute $w = 5 \sin u$ for $u = 0, 0.1, 0.2, \ldots, 10$, the session is

```
>>u = 0:0.1:10;
>>w = 5*sin(u);
```

The single line `w = 5*sin(u)` computed the formula $w = 5 \sin u$ 101 times, once for each value in the array u, to produce an array z that has 101 values.

You can see all the u values by typing u after the prompt; or, for example, you can see the seventh value by typing `u(7)`. The number 7 is called an *array index,* because it points to a particular element in the array.

ARRAY INDEX

```
>>u(7)
ans =
     0.6000
>>w(7)
ans =
     2.8232
```

You can use the `length` function to determine how many values are in an array. For example, continue the previous session as follows:

```
>>m = length(w)
m =
   101
```

Arrays that display on the screen as a single row of numbers with more than one column are called *row arrays*. You can create *column arrays*, which have more than one row, by using a semicolon to separate the rows.

Polynomial Roots

We can describe a polynomial in MATLAB with an array whose elements are the polynomial's coefficients, *starting with the coefficient of the highest power of x.* For example, the polynomial $4x^3 - 8x^2 + 7x - 5$ would be represented by the array [4,-8,7,-5]. The *roots* of the polynomial $f(x)$ are the values of x such that $f(x) = 0$. Polynomial roots can be found with the roots(a) function, where a is the polynomial's coefficient array. The result is a *column* array that contains the polynomial's roots. For example, to find the roots of $x^3 - 7x^2 + 40x - 34 = 0$, the session is

```
>>a = [1,-7,40,-34];
>>roots(a)
ans =
     3.0000 + 5.000i
     3.0000 - 5.000i
     1.0000
```

The roots are $x = 1$ and $x = 3 \pm 5i$. The two commands could have been combined into the single command roots([1,-7,40,-34]).

Test Your Understanding

T1.3–1 Use MATLAB to determine how many elements are in the array cos(0):0.02:log10(100). Use MATLAB to determine the 25th element. (Answer: 51 elements and 1.48.)

T1.3–2 Use MATLAB to find the roots of the polynomial $290 - 11x + 6x^2 + x^3$. (Answer: $x = -10, 2 \pm 5i.$)

Built-in Functions

We have seen several of the functions built into MATLAB, such as the sqrt and sin functions. Table 1.3–1 lists some of the commonly used built-in functions. Chapter 3 gives extensive coverage of the built-in functions. MATLAB users can create their own functions for their special needs. Creation of user-defined functions is covered in Chapter 3.

Working with Files

MATLAB uses several types of files that enable you to save programs, data, and session results. As we will see in Section 1.4, MATLAB function files and program files are saved with the extension .m, and thus are called M-files. *MAT-files*

MAT-FILES

Table 1.3–1 Some commonly used mathematical functions

Function	MATLAB syntax*
e^x	exp(x)
$\sqrt{x}$	sqrt(x)
$\ln x$	log(x)
$\log_{10} x$	log10(x)
$\cos x$	cos(x)
$\sin x$	sin(x)
$\tan x$	tan(x)
$\cos^{-1} x$	acos(x)
$\sin^{-1} x$	asin(x)
$\tan^{-1} x$	atan(x)

*The MATLAB trigonometric functions listed here use radian measure. Trigonometric functions ending in d, such as sind(x) and cosd(x), take the argument x in degrees. Inverse functions such as atand(x) return values in degrees.

have the extension .mat and are used to save the names and values of variables created during a MATLAB session.

Because they are *ASCII files,* M-files can be created using just about any word processor. MAT-files are *binary* files that are generally readable only by the software that created them. MAT-files contain a machine signature that allows them to be transferred between machine types such as MS Windows and Macintosh machines.

ASCII FILES

The third type of file we will be using is a data file, specifically an ASCII *data file,* that is, one created according to the ASCII format. You may need to use MATLAB to analyze data stored in such a file created by a spreadsheet program, a word processor, or a laboratory data acquisition system or in a file you share with someone else.

DATA FILE

Saving and Retrieving Your Workspace Variables

If you want to continue a MATLAB session at a later time, you must use the save and load commands. Typing save causes MATLAB to save the workspace variables, that is, the variable names, their sizes, and their values, in a binary file called matlab.mat, which MATLAB can read. To retrieve your workspace variables, type load. You can then continue your session as before. To save the workspace variables in another file named filename.mat, type save filename. To load the workspace variables, type load filename. If the saved MAT-file filename contains the variables A, B, and C, then loading the file filename places these variables back into the workspace and overwrites any existing variables having the same name.

To save just some of your variables, say, var1 and var2, in the file filename.mat, type save filename var1 var2. You need not type the variable names to retrieve them; just type load filename.

Directories and Path   It is important to know the location of the files you use with MATLAB. File location frequently causes problems for beginners. Suppose you use MATLAB on your home computer and save a file to a removable disk, as discussed later in this section. If you bring that disk to use with MATLAB on another computer, say, in a school's computer lab, you must make sure that MATLAB knows how to find your files. Files are stored in *directories,* called *folders* on some computer systems. Directories can have subdirectories below them. For example, suppose MATLAB was installed on drive c: in the directory c:\matlab. Then the toolbox directory is a subdirectory under the directory c:\matlab, and symbolic is a subdirectory under the toolbox directory. The *path* tells us and MATLAB how to find a particular file.

PATH

Working with Removable Disks   In Section 1.4 you will learn how to create and save M-files. Suppose you have saved the file problem1.m in the directory \homework on a disk, which you insert in drive f:. The path for this file is f:\homework. As MATLAB is normally installed, when you type problem1,

1. MATLAB first checks to see if problem1 is a variable and if so, displays its value.
2. If not, MATLAB then checks to see if problem1 is one of its own commands, and executes it if it is.
3. If not, MATLAB then looks in the current directory for a file named problem1.m and executes problem1 if it finds it.
4. If not, MATLAB then searches the directories in its *search path,* in order, for problem1.m and then executes it if found.

SEARCH PATH

You can display the MATLAB search path by typing path. If problem1 is on the disk only and if directory f: is not in the search path, MATLAB will not find the file and will generate an error message, unless you tell it where to look. You can do this by typing cd f:\homework, which stands for "change directory to f:\homework." This will change the current directory to f:\homework and force MATLAB to look in that directory to find your file. The general syntax of this command is cd dirname, where dirname is the full path to the directory.

An alternative to this procedure is to copy your file to a directory on the hard drive that is in the search path. However, there are several pitfalls with this approach: (1) if you change the file during your session, you might forget to copy the revised file back to your disk; (2) the hard drive becomes cluttered (this is a problem in public computer labs, and you might not be permitted to save your file on the hard drive); (3) the file might be deleted or overwritten if MATLAB is reinstalled; and (4) someone else can access your work!

You can determine the current directory (the one where MATLAB looks for your file) by typing pwd. To see a list of all the files in the current directory, type dir. To see the files in the directory dirname, type dir dirname.

The what command displays a list of the MATLAB-specific files in the current directory. The what dirname command does the same for the directory dirname. Type which item to display the full path name of the function

Table 1.3–2 System, directory, and file commands

Command	Description
addpath dirname	Adds the directory dirname to the search path.
cd dirname	Changes the current directory to dirname.
dir	Lists all files in the current directory.
dir dirname	Lists all the files in the directory dirname.
path	Displays the MATLAB search path.
pathtool	Starts the Set Path tool.
pwd	Displays the current directory.
rmpath dirname	Removes the directory dirname from the search path.
what	Lists the MATLAB-specific files found in the current working directory. Most data files and other non-MATLAB files are not listed. Use dir to get a list of all files.
what dirname	Lists the MATLAB-specific files in directory dirname.
which item	Displays the path name of item if item is a function or file. Identifies item as a variable if so.

item or the file item (include the file extension). If item is a variable, then MATLAB identifies it as such.

You can add a directory to the search path by using the addpath command. To remove a directory from the search path, use the rmpath command. The Set Path tool is a graphical interface for working with files and directories. Type pathtool to start the browser. To save the path settings, click on **Save** in the tool. To restore the default search path, click on **Default** in the browser.

These commands are summarized in Table 1.3–2.

Plotting with MATLAB

MATLAB contains many powerful functions for easily creating plots of several different types, such as rectilinear, logarithmic, surface, and contour plots. As a simple example, let us plot the function $y = 5 \sin x$ for $0 \leqslant x \leqslant 7$. We choose to use an increment of 0.01 to generate a large number of x values in order to produce a smooth curve. The function plot(x,y) generates a plot with the x values on the horizontal axis (the abscissa) and the y values on the vertical axis (the ordinate). The session is

```
>>x = 0:0.01:7;
>>y = 3*cos(2*x);
>>plot(x,y),xlabel('x'),ylabel('y')
```

The plot appears on the screen in a *graphics window,* named **Figure 1,** as shown in Figure 1.3–1. The xlabel function places the text in single quotes as a label on the horizontal axis. The ylabel function performs a similar function for the vertical axis. When the plot command is successfully executed, a graphics window automatically appears. If a hard copy of the plot is desired,

GRAPHICS WINDOW

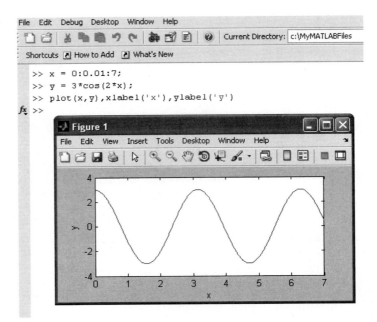

Figure 1.3–1 A graphics window showing a plot.

the plot can be printed by selecting **Print** from the **File** menu on the graphics window. The window can be closed by selecting **Close** on the **File** menu in the graphics window. You will then be returned to the prompt in the Command window.

Other useful plotting functions are `title` and `gtext`. These functions place text on the plot. Both accept text within parentheses and single quotes, as with the `xlabel` function. The `title` function places the text at the top of the plot; the `gtext` function places the text at the point on the plot where the cursor is located when you click the left mouse button.

OVERPLAY PLOT

You can create multiple plots, called *overlay plots,* by including another set or sets of values in the `plot` function. For example, to plot the functions $y = 2\sqrt{x}$ and $z = 4 \sin 3x$ for $0 \le x \le 5$ on the same plot, the session is

```
>>x = 0:0.01:5;
>>y = 2*sqrt(x);
>>z = 4*sin(3*x);
>>plot(x,y,x,z),xlabel('x'),gtext('y'),gtext('z')
```

After the plot appears on the screen, the program waits for you to position the cursor and click the mouse button, once for each `gtext` function used.

Use the `gtext` function to place the labels *y* and *z* next to the appropriate curves.

You can also distinguish curves from one another by using different line types for each curve. For example, to plot the *z* curve using a dashed line, replace the `plot(x,y,x,z)` function in the above session with `plot(x,y,x,z, '− −')`. Other line types can be used. These are discussed in Chapter 5.

Sometimes it is useful or necessary to obtain the coordinates of a point on a plotted curve. The function `ginput` can be used for this purpose. Place it at the end of all the plot and plot formatting statements, so that the plot will be in its final form. The command `[x,y] = ginput(n)` gets *n* points and returns the *x* and *y* coordinates in the vectors `x` and `y`, which have a length *n*. Position the cursor using a mouse, and press the mouse button. The returned coordinates have the same scale as the coordinates on the plot.

In cases where you are plotting *data,* as opposed to functions, you should use *data markers* to plot each data point (unless there are very many data points). To mark each point with a plus sign +, the required syntax for the `plot` function is `plot(x,y,'+')`. You can connect the data points with lines if you wish. In that case, you must plot the data twice, once with a data marker and once without a marker.

DATA MARKER

For example, suppose the data for the independent variable is x = `[15:2:23]` and the dependent variable values are y = `[20, 50, 60, 90, 70]`. To plot the data with plus signs, use the following session:

```
>>x = 15:2:23;
>>y = [20, 50, 60, 90, 70];
>>plot(x,y,'+',x,y),xlabel('x'),ylabel('y'), grid
```

The `grid` command puts grid lines on the plot. Other data markers are available. These are discussed in Chapter 5.

Table 1.3–3 summarizes these plotting commands. We will discuss other plotting functions, and the Plot Editor, in Chapter 5.

Table 1.3–3 Some MATLAB plotting commands

Command	Description
`[x,y] = ginput(n)`	Enables the mouse to get *n* points from a plot, and returns the *x* and *y* coordinates in the vectors `x` and `y`, which have a length *n*.
`grid`	Puts grid lines on the plot.
`gtext('text')`	Enables placement of text with the mouse.
`plot(x,y)`	Generates a plot of the array `y` versus the array `x` on rectilinear axes.
`title('text')`	Puts text in a title at the top of the plot.
`xlabel('text')`	Adds a text label to the horizontal axis (the abscissa).
`ylabel('text')`	Adds a text label to the vertical axis (the ordinate).

Test Your Understanding

T1.3–3 Use MATLAB to plot the function $s = 2 \sin(3t + 2) + \sqrt{5t + 1}$ over the interval $0 \le t \le 5$. Put a title on the plot, and properly label the axes. The variable s represents speed in feet per second; the variable t represents time in seconds.

T1.3–4 Use MATLAB to plot the functions $y = 4\sqrt{6x + 1}$ and $z = 5e^{0.3x} - 2x$ over the interval $0 \le x \le 1.5$. Properly label the plot and each curve. The variables y and z represent force in newtons; the variable x represents distance in meters.

Linear Algebraic Equations

You can use the left division operator (\) in MATLAB to solve sets of linear algebraic equations. For example, consider the set

$$6x + 12y + 4z = 70$$

$$7x - 2y + 3z = 5$$

$$2x + 8y - 9z = 64$$

To solve such sets in MATLAB, you must create two arrays; we will call them A and B. The array A has as many rows as there are equations and as many columns as there are variables. The rows of A must contain the coefficients of x, y, and z in that order. In this example, the first row of A must be 6, 12, 4; the second row must be 7, −2, 3; and the third row must be 2, 8, −9. The array B contains the constants on the right-hand side of the equation; it has one column and as many rows as there are equations. In this example, the first row of B is 70, the second is 5, and the third is 64. The solution is obtained by typing A\B. The session is

```
>>A = [6,12,4;7,-2,3;2,8,-9];
>>B = [70;5;64];
>>Solution = A\B
Solution =
      3
      5
     -2
```

The solution is $x = 3$, $y = 5$, and $z = -2$.

This method works fine when the equation set has a unique solution. To learn how to deal with problems having a nonunique solution (or perhaps no solution at all!), see Chapter 8.

Test Your Understanding

T1.3–5 Use MATLAB to solve the following set of equations.

$$6x - 4y + 8z = 112$$
$$-5x - 3y + 7z = 75$$
$$14x + 9y - 5z = -67$$

(Answer: $x = 2$, $y = -5$, $z = 10$.)

1.4 Script Files and the Editor/Debugger

You can perform operations in MATLAB in two ways:

1. In the interactive mode, in which all commands are entered directly in the Command window.

2. By running a MATLAB program stored in *script* file. This type of file contains MATLAB commands, so running it is equivalent to typing all the commands, one at a time, at the Command window prompt. You can run the file by typing its name at the Command window prompt.

When the problem to be solved requires many commands or a repeated set of commands, or has arrays with many elements, the interactive mode is inconvenient. Fortunately, MATLAB allows you to write your own programs to avoid this difficulty. You write and save MATLAB programs in M-files, which have the extension .m; for example, `program1.m`.

MATLAB uses two types of M-files: *script files* and *function* files. You can use the Editor/Debugger built into MATLAB to create M-files. Because they contain commands, script files are sometimes called *command* files. Function files are discussed in Chapter 3.

SCRIPT FILE

Creating and Using a Script File

The symbol % designates a *comment,* which is not executed by MATLAB. Comments are used mainly in script files for the purpose of documenting the file. The comment symbol may be put anywhere in the line. MATLAB ignores everything to the right of the % symbol. For example, consider the following session.

COMMENT

```
>>% This is a comment.
>>x = 2+3 % So is this.
x =
    5
```

Note that the portion of the line before the % sign is executed to compute x.

Here is a simple example that illustrates how to create, save, and run a script file, using the Editor/Debugger built into MATLAB. However, you may use another

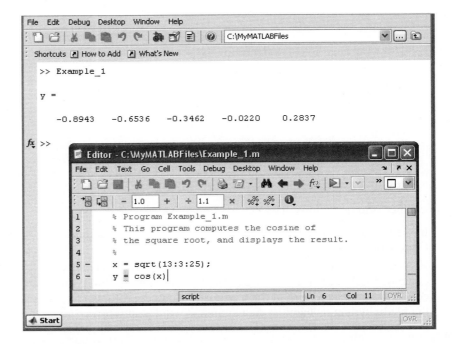

Figure 1.4–1 The MATLAB Command window with the Editor/Debugger open.

text editor to create the file. The sample file is shown below. It computes the cosine of the square root of several numbers and displays the results on the screen.

```
% Program Example_1.m
% This program computes the cosine of
% the square root and displays the result.
x = sqrt(13:3:25);
y = cos(x)
```

To create this new M-file in the Command window, select **New** from the **File** menu, then select **Blank M-file.** You will then see a new edit window. This is the Editor/Debugger window as shown in Figure 1.4–1. Type in the file as shown above. You can use the keyboard and the **Edit** menu in the Editor/Debugger as you would in most word processors to create and edit the file. When finished, select **Save** from the **File** menu in the Editor/Debugger. In the dialog box that appears, replace the default name provided (usually named Untitled) with the name Example_1, and click on **Save.** The Editor/Debugger will automatically provide the extension .m and save the file in the MATLAB current directory, which for now we will assume to be on the hard drive.

Once the file has been saved, in the MATLAB Command window type the script file's name Example_1 to execute the program. You should see the result displayed in the Command window. Figure 1.4–1 shows a screen containing the resulting Command window display and the Editor/Debugger opened to display the script file.

Effective Use of Script Files

Create script files to avoid the need to retype lengthy and commonly used procedures. Here are some other things to keep in mind when using script files:

1. The name of a script file must follow the MATLAB convention for naming variables.

2. Recall that typing a variable's name at the Command window prompt causes MATLAB to display the value of that variable. Thus, do not give a script file the same name as a variable it computes because MATLAB will not be able to execute that script file more than once, unless you clear the variable.

3. Do not give a script file the same name as a MATLAB command or function. You can check to see if a command, function, or file name already exists by using the `exist` command. For example, to see if a variable `example1` already exists, type `exist('example1')`; this will return a 0 if the variable does not exist and a 1 if it does. To see if an M-file `example1.m` already exists, type `exist('example1.m','file')` *before* creating the file; this will return a 0 if the file does not exist and a 2 if it does. Finally, to see if a built-in function `example1` already exists, type `exist ('example1', 'builtin')` *before* creating the file; this will return a 0 if the built-in function does not exist and a 5 if it does.

Note that not all functions supplied with MATLAB are built-in functions. For example, the function `mean.m` is supplied but is not a built-in function. The command `exist('mean.m', 'file')` will return a 2, but the command `exist('mean', 'builtin')` will return a 0. You may think of built-in functions as primitives that form the basis for other MATLAB functions. You cannot view the entire file of a built-in function in a text editor, only the comments.

Debugging Script Files

Debugging a program is the process of finding and removing the "bugs," or errors, in a program. Such errors usually fall into one of the following categories.

DEBUGGING

1. Syntax errors such as omitting a parenthesis or comma, or spelling a command name incorrectly. MATLAB usually detects the more obvious errors and displays a message describing the error and its location.

2. Errors due to an incorrect mathematical procedure, called *runtime errors.* They do not necessarily occur every time the program is executed; their occurrence often depends on the particular input data. A common example is division by zero.

To locate an error, try the following:

1. Always test your program with a simple version of the problem, whose answers can be checked by hand calculations.

2. Display any intermediate calculations by removing semicolons at the end of statements.

3. Use the debugging features of the Editor/Debugger, which are introduced in Chapter 4. However, one advantage of MATLAB is that it requires relatively simple programs to accomplish many types of tasks. Thus you probably will not need to use the Debugger for the problems encountered in this text.

Programming Style

Comments may be put anywhere in the script file. However, because the first comment line before any executable statement is the line searched by the `lookfor` command, discussed later in this chapter, consider putting keywords that describe the script file in this first line (called the H1 line). A suggested structure for a script file is the following.

1. *Comments section* In this section put comment statements to give
   a. The name of the program and any keywords in the first line.
   b. The date created and the creators' names in the second line.
   c. The definitions of the variable names for every input and output variable. Divide this section into at least two subsections, one for input data and one for output data. A third, optional section may include definitions of variables used in the calculations. *Be sure to include the units of measurement for all input and all output variables!*
   d. The name of every user-defined function called by the program.

2. *Input section* In this section put the input data and/or the input functions that enable data to be entered. Include comments where appropriate for documentation.

3. *Calculation section* Put the calculations in this section. Include comments where appropriate for documentation.

4. *Output section* In this section put the functions necessary to deliver the output in whatever form required. For example, this section might contain functions for displaying the output on the screen. Include comments where appropriate for documentation.

The programs in this text often omit some of these elements to save space. Here the text discussion associated with the program provides the required documentation.

Controlling Input and Output

MATLAB provides several useful commands for obtaining input from the user and for formatting the output (the results obtained by executing the MATLAB commands). Table 1.4–1 summarizes these commands.

The `disp` function (short for "display") can be used to display the value of a variable but not its name. Its syntax is `disp(A)`, where `A` represents a MATLAB variable name. The `disp` function can also display text such as a message to the user. You enclose the text within single quotes. For example, the command `disp('The predicted speed is:')` causes the message to appear on

Table 1.4–1 Input/output commands

Command	Description
disp(A)	Displays the contents, but not the name, of the array A.
disp('text')	Displays the text string enclosed within single quotes.
format	Controls the screen's output display format (see Table 1.1–5).
x = input('text')	Displays the text in quotes, waits for user input from the keyboard, and stores the value in x.
x = input('text','s')	Displays the text in quotes, waits for user input from the keyboard, and stores the input as a string in x.
k=menu('title','option1', 'option2',...	Displays a menu whose title is in the string variable 'title' and whose choices are 'option1','option2', and so on.

the screen. This command can be used with the first form of the disp function in a script file as follows (assuming the value of Speed is 63):

```
disp('The predicted speed is:')
disp(Speed)
```

When the file is run, these lines produce the following on the screen:

```
The predicted speed is:
   63
```

The input function displays text on the screen, waits for the user to enter something from the keyboard, and then stores the input in the specified variable. For example, the command x = input('Please enter the value of x:') causes the message to appear on the screen. If you type 5 and press **Enter,** the variable x will have the value 5.

A *string variable* is composed of text (alphanumeric characters). If you want to store a text input as a string variable, use the other form of the input command. For example, the command Calendar = input('Enter the day of the week:','s') prompts you to enter the day of the week. If you type Wednesday, this text will be stored in the string variable Calendar.

STRING VARIABLE

Use the menu function to generate a menu of choices for user input. Its syntax is

```
k = menu('title','option1','option2',...)
```

The function displays the menu whose title is in the string variable 'title' and whose choices are string variables 'option1', 'option2', and so on. The returned value of k is 1, 2, . . . depending on whether you click on the button for option1, option2, and so forth. For example, the following script uses a menu to select the data marker for a graph, assuming that the arrays x and y already exist.

```
k = menu('Choose a data marker','o','*','x');
type = ['o','*','x'];
plot(x,y,x,y,type(k))
```

Test Your Understanding

T1.4–1 The surface area A of a sphere depends on its radius r as follows: $A = 4\pi r^2$. Write a script file that prompts the user to enter a radius, computes the surface area, and displays the result.

Example of a Script File

The following is a simple example of a script file that shows the preferred program style. The speed v of a falling object dropped with no initial velocity is given as a function of time t by $v = gt$, where g is the acceleration due to gravity. In SI units, $g = 9.81$ m/s^2. We want to compute and plot v as a function of t for $0 \leq t \leq t_{final}$, where t_{final} is the final time entered by the user. The script file is the following.

```
% Program Falling_Speed.m: plots speed of a falling object.
% Created on March 1, 2009 by W. Palm III
%
% Input Variable:
% tfinal = final time (in seconds)
%
% Output Variables:
% t = array of times at which speed is computed (seconds)
% v = array of speeds (meters/second)

%
% Parameter Value:
g = 9.81; % Acceleration in SI units
%
% Input section:
tfinal = input('Enter the final time in seconds:');
%
% Calculation section:
dt = tfinal/500;
t = 0:dt:tfinal; % Creates an array of 501 time values.
v = g*t;
%
% Output section:
plot(t,v),xlabel('Time (seconds)'),ylabel('Speed (meters/second)')
```

After creating this file, you save it with the name `Falling_Speed.m`. To run it, you type `Falling_Speed` (without the `.m`) in the Command window at the prompt. You will then be asked to enter a value for t_{final}. After you enter a value and press **Enter,** you will see the plot on the screen.

1.5 The MATLAB Help System

To explore the more advanced features of MATLAB not covered in this book, you will need to know how to use effectively the MATLAB Help System. MATLAB has these options to get help for using MathWorks products.

1. *Function Browser* This provides quick access to the documentation for the MATLAB function.
2. *Help Browser* This graphical user interface helps you find information and view online documentation for your MathWorks products.
3. *Help Functions* The functions `help`, `lookfor`, and `doc` can be used to display syntax information for a specified function.
4. *Other Resources* For additional help, you can run demos, contact technical support, search documentation for other MathWorks products, view a list of other books, and participate in a newsgroup.

The Function Browser

To activate the Function Browser, either select Function Browser from the **Help** menu or select the **fx** icon to the left of the prompt. Figure 1.5–1 shows the resulting menu after the Graphics category has been selected. The subwindow shown opens when the **plot** function is selected. Scroll down to see the entire documentation of the **plot** function.

The Help Browser

To open the Help Browser, select **Product Help** from the **Help** menu, or click the question mark button in the toolbar. The Help Browser contains two window

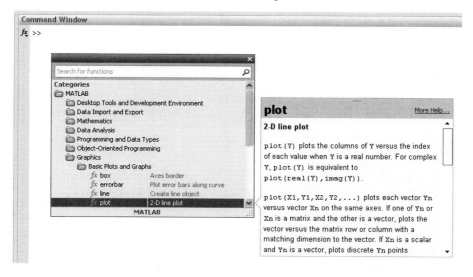

Figure 1.5–1 The Function Browser after **plot** has been selected.

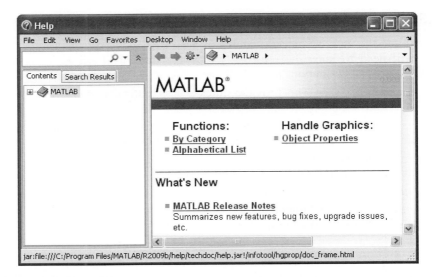

Figure 1.5–2 The MATLAB Help Browser.

"panes": the Help Navigator pane on the left and the Display pane on the right (see Figure 1.5–2). The Help Navigator contains two tabs:

■ *Contents:* a contents listing tab
■ *Search Results:* a search tab having a find function and full text search features

Use the tabs in the Help Navigator to find documentation. You view documentation in the Display pane. To open the Help Navigator pane from the display pane, click on **Help Navigator** in the **View** menu.

Finding Documentation

Figure 1.5–3 shows the result of clicking on the + sign next to MATLAB in the **Help Navigator.** A submenu appears that shows the various Help topics for MATLAB.

Viewing Documentation

After finding documentation with the Help Navigator, view the documentation in the Display pane. While viewing a page of documentation, you can

■ Scroll to see contents not currently visible in the window.
■ View the previous or next page in the document by clicking the left or right arrow at the top of the page.
■ View the previous or next item in the index by clicking the left or right arrow at the bottom of the page.

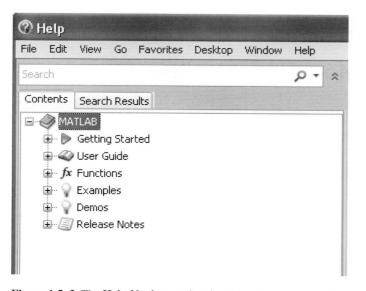

Figure 1.5–3 The **Help Navigator** showing the submenus under the MATLAB category.

- Find a phrase or specified term by typing the term in the **Search** box below the Help Browser toolbar and pressing the **Enter** key. The left-hand pane will then display all the Help pages and function documentation that contains the specified term.

Using the Contents Tab

Click the **Contents** tab in the Help Navigator to list the titles and table of contents for all product documentation. To expand the listing for an item, click the + to the left of the item. To collapse the listings for an item, click the − to the left of the item, or double-click the item. Click on an item to select it. The first page of that document appears in the Display pane. Double-clicking an item in the contents listing expands the listing for that item and shows the first page of that document in the Display pane.

The Contents pane is synchronized with the Display pane. By default, the item selected in the Contents pane always matches the documentation appearing in the Display pane. Thus, the contents tree is synchronized with the displayed document.

Using the Search Results Tab

Click the **Search Results** tab in the Help Navigator pane to find all MATLAB documents containing a specified phrase. Type the phrase in the "Search" box. Then press **Enter.** The list of documents and the heading under which the phrase is found in that document then appear in the Help Navigator pane. Select an entry from the list of results to view that document in the Display pane.

Figure 1.5–4 shows the results of typing `plot` in the "Search" box. The Display pane shows the documentation for the `plot` function (scroll down to see all of it), and the **Help Navigator** pane shows the Documentation Search results and the Demo Search results.

Help Functions

Three MATLAB functions can be used for accessing online information about MATLAB functions.

The `help` Function   The `help` function is the most basic way to determine the syntax and behavior of a particular function. For example, typing `help log10` in the Command window produces the following display:

```
LOG10 Common (base 10) logarithm.
  LOG10(X) is the base 10 logarithm of the elements of X.
  Complex results are produced if X is not positive.

  See also LOG, LOG2, EXP, LOGM.
```

Note that the display describes what the function does, warns about any unexpected results if nonstandard argument values are used, and directs the user to other related functions.

All the MATLAB functions are organized into logical groups, upon which the MATLAB directory structure is based. For instance, all elementary mathematical

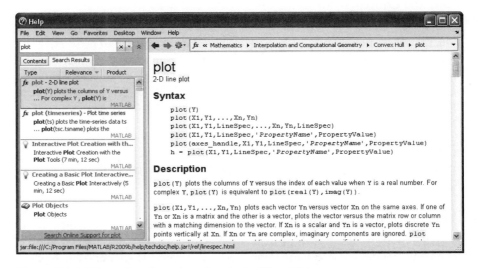

Figure 1.5–4 The results of entering `plot` in the "Search" box.

functions such as log10 reside in the elfun directory, and the polynomial functions reside in the polyfun directory. To list the names of all the functions in that directory, with a brief description of each, type help polyfun. If you are unsure of what directory to search, type help to obtain a list of all the directories, with a description of the function category each represents.

Typing helpwin topic displays the Help text for the specified topic inside the Desktop Help Browser window. Links are created to functions referenced in the "See Also" line of the Help text. You can also access the Help window by selecting the **Help** option under the **Help** menu, or by clicking the question mark button on the toolbar.

The lookfor Function The lookfor function allows you to search for functions on the basis of a keyword. It searches through the first line of Help text, known as the H1 line, for each MATLAB function, and returns all the H1 lines containing a specified keyword. For example, MATLAB does not have a function named sine. So the response from help sine is

```
sine.m not found
```

However, typing lookfor sine produces over a dozen matches, depending on which toolboxes you have installed. For example, you will see, among others,

```
ACOS        Inverse cosine, result in radians
ACOSD       Inverse cosine, result in degrees
ACOSH       Inverse hyperbolic cosine
ASIN        Inverse sine, result in radians
...
SIN         Sine of argument in radians
...
```

From this list you can find the correct name for the sine function. Note that all words containing sine are returned, such as cosine. Adding -all to the lookfor function searches the entire Help entry, not just the H1 line.

The doc Function Typing doc function displays the documentation for the MATLAB function function. Typing doc toolbox/function displays the documentation for the specified toolbox function. Typing doc toolbox displays the documentation road map page for the specified toolbox.

The MathWorks Website

If your computer is connected to the Internet, you can access The MathWorks, Inc., the home of MATLAB. You can use electronic mail to ask questions, make suggestions, and report possible bugs. You can also use a solution search engine at The MathWorks website to query an up-to-date database of technical support information. The website address is http://www.mathworks.com.

Table 1.5–1 MATLAB Help functions

Function	Use
`doc`	Displays the start page of the documentation in the Help Browser.
`doc function`	Displays the documentation for the MATLAB function `function`.
`doc toolbox/ function`	Displays the documentation for the specified toolbox function.
`doc toolbox`	Displays the documentation road map page for the specified toolbox.
`help`	Displays a list of all the function directories, with a description of the function category each represents.
`help function`	Displays in the Command window a description of the specified function `function`.
`helpwin topic`	Displays the Help text for the specified `topic` inside the Desktop Help Browser window.
`lookfor topic`	Displays in the Command window a brief description for all functions whose description includes the specified keyword `topic`.
`type filename`	Displays the M-file `filename` without opening it with a text editor.

The Help system is very powerful and detailed, so we have only described its basics. You can, and should, use the Help system to learn how to use its features in greater detail.

Table 1.5–1 summarizes the MATLAB Help functions.

1.6 Problem-Solving Methodologies

Designing new engineering devices and systems requires a variety of problem-solving skills. (This variety is what keeps engineering from becoming boring!) When you are solving a problem, it is important to plan your actions ahead of time. You can waste many hours by plunging into the problem without a plan of attack. Here we present a plan of attack, or *methodology,* for solving engineering problems in general. Because solving engineering problems often requires a computer solution and because the examples and exercises in this text require you to develop a computer solution (using MATLAB), we also discuss a methodology for solving computer problems in particular.

Steps in Engineering Problem Solving

Table 1.6–1 summarizes the methodology that has been tried and tested by the engineering profession for many years. These steps describe a general problem-solving procedure. Simplifying the problem sufficiently and applying the appropriate fundamental principles is called *modeling,* and the resulting mathematical

MODEL

Table 1.6–1 Steps in engineering problem solving

1. Understand the purpose of the problem.
2. Collect the known information. Realize that some of it might later be found unnecessary.
3. Determine what information you must find.
4. Simplify the problem only enough to obtain the required information. State any assumptions you make.
5. Draw a sketch and label any necessary variables.
6. Determine which fundamental principles are applicable.
7. Think generally about your proposed solution approach and consider other approaches before proceeding with the details.
8. Label each step in the solution process.
9. If you solve the problem with a program, hand check the results using a simple version of the problem. Checking the dimensions and units and printing the results of intermediate steps in the calculation sequence can uncover mistakes.
10. Perform a "reality check" on your answer. Does it make sense? Estimate the range of the expected result and compare it with your answer. Do not state the answer with greater precision than is justified by any of the following:
   (a) The precision of the given information.
   (b) The simplifying assumptions.
   (c) The requirements of the problem.

   Interpret the mathematics. If the mathematics produces multiple answers, do not discard some of them without considering what they mean. The mathematics might be trying to tell you something, and you might miss an opportunity to discover more about the problem.

description is called a *mathematical model,* or just a *model.* When the modeling is finished, we need to solve the mathematical model to obtain the required answer. If the model is highly detailed, we might need to solve it with a computer program. Most of the examples and exercises in this text require you to develop a computer solution (using MATLAB) to problems for which the model has already been developed. Thus we will not always need to use all the steps shown in Table 1.6–1. Greater discussion of engineering problem solving can be found in [Eide, 2008].*

Example of Problem Solving

Consider the following simple example of the steps involved in problem solving. Suppose you work for a company that produces packaging. You are told that a new packaging material can protect a package when dropped, provided that the package hits the ground at less than 25 ft/sec. The package's total weight is 20 lb, and it is rectangular with dimensions of 12 by 12 by 8 in. You must determine whether the packaging material provides enough protection when the package is carried by delivery persons.

___

* References appear in Appendix D.

The steps in the solution are as follows:

1. *Understand the purpose of the problem.* The implication here is that the packaging is intended to protect against being dropped while the delivery person is carrying it. It is not intended to protect against the package falling off a moving delivery truck. In practice, you should make sure that the person giving you this assignment is making the same assumption. Poor communication is the cause of many errors!

2. *Collect the known information.* The known information is the package's weight, dimensions, and maximum allowable impact speed.

3. *Determine what information you must find.* Although it is not explicitly stated, you need to determine the maximum height from which the package can be dropped without damage. You need to find a relationship between the speed of impact and the height at which the package is dropped.

4. *Simplify the problem only enough to obtain the required information. State any assumptions you make.* The following assumptions will simplify the problem and are consistent with the problem statement as we understand it:

   a. The package is dropped from rest with no vertical or horizontal velocity.

   b. The package does not tumble (as it might when dropped from a moving truck). The given dimensions indicate that the package is not thin and thus will not "flutter" as it falls.

   c. The effect of air drag is negligible.

   d. The greatest height from which the delivery person could drop the package is 6 ft (and thus we ignore the existence of a delivery person 8 ft tall!).

   e. The acceleration g due to gravity is constant (because the distance dropped is only 6 ft).

5. *Draw a sketch and label any necessary variables.* Figure 1.6–1 is a sketch of the situation, showing the height h of the package, its mass m, its speed v, and the acceleration due to gravity g.

6. *Determine which fundamental principles are applicable.* Because this problem involves a mass in motion, we can apply Newton's laws. From physics we know that the following relations result from Newton's laws and the basic kinematics of an object falling a short distance under the influence of gravity, with no air drag or initial velocity:

   a. Height versus time to impact t_i: $h = \frac{1}{2}gt_i^2$.

   b. Impact speed v_i versus time to impact: $v_i = gt_i$.

   c. Conservation of mechanical energy: $mgh = \frac{1}{2}mv_i^2$.

7. *Think generally about your proposed solution approach and consider other approaches before proceeding with the details.* We could solve the second equation for t_i and substitute the result into the first equation to obtain the relation between h and v_i. This approach would also allow us to find the time to drop t_i. However, this method involves more work than necessary

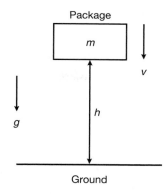

Figure 1.6–1 Sketch of the dropped-package problem.

because we need not find the value of t_i. The most efficient approach is to solve the third relation for h.

$$h = \frac{1}{2}\frac{v_i^2}{g} \tag{1.6–1}$$

Notice that the mass m cancels out of the equation. The mathematics just told us something! It told us that the mass does not affect the relation between the impact speed and the height dropped. Thus we do not need the weight of the package to solve the problem.

8. *Label each step in the solution process.* This problem is so simple that there are only a few steps to label:

   a. Basic principle: conservation of mechanical energy

$$h = \frac{1}{2}\frac{v_i^2}{g}$$

   b. Determine the value of the constant g: $g = 32.2$ ft/sec^2.

   c. Use the given information to perform the calculation and round off the result consistent with the precision of the given information:

$$h = \frac{1}{2}\frac{25^2}{32.2} = 9.7 \text{ ft}$$

Because this text is about MATLAB, we might as well use it to do this simple calculation. The session looks like this:

```
>>g=32.2;
>>vi=25;
>>h=vi^2/(2*g)
h =
    9.7050
```

9. *Check the dimensions and units.* This check proceeds as follows, using Equation (1.6–1),

$$[\text{ft}] = \left[\frac{1}{2}\right]\frac{[\text{ft/sec}]^2}{[\text{ft/sec}^2]} = \frac{[\text{ft}]^2}{[\text{sec}]^2}\frac{[\text{sec}]^2}{[\text{ft}]} = [\text{ft}]$$

which is correct.

10. *Perform a reality check and precision check on the answer.* If the computed height were negative, we would know that we did something wrong. If it were very large, we might be suspicious. However, the computed height of 9.7 ft does not seem unreasonable.

If we had used a more accurate value for g, say $g = 32.17$, then we would be justified in rounding the result to $h = 9.71$. However, given the need to be conservative here, we probably should round the answer *down* to the nearest foot. So we probably should report that the package will not be damaged if it is dropped from a height of less than 9 ft.

The mathematics told us that the package mass does not affect the answer. The mathematics did not produce multiple answers here. However, many problems involve the solution of polynomials with more than one root; in such cases we must carefully examine the significance of each.

Steps for Obtaining a Computer Solution

If you use a program such as MATLAB to solve a problem, follow the steps shown in Table 1.6–2. Greater discussion of modeling and computer solutions can be found in [Starfield, 1990] and [Jayaraman, 1991].

MATLAB is useful for doing numerous complicated calculations and then automatically generating a plot of the results. The following example illustrates the procedure for developing and testing such a program.

Table 1.6–2 Steps for developing a computer solution

1.	State the problem concisely.
2.	Specify the data to be used by the program. This is the *input.*
3.	Specify the information to be generated by the program. This is the *output.*
4.	Work through the solution steps by hand or with a calculator; use a simpler set of data if necessary.
5.	Write and run the program.
6.	Check the output of the program with your hand solution.
7.	Run the program with your input data and perform a "reality check" on the output. Does it make sense? Estimate the range of the expected result and compare it with your answer.
8.	If you will use the program as a general tool in the future, test it by running it for a range of reasonable data values; perform a reality check on the results.

Piston Motion

EXAMPLE 1.6–1

Figure 1.6–2a shows a piston, connecting rod, and crank for an internal combustion engine. When combustion occurs, it pushes the piston down. This motion causes the connecting rod to turn the crank, which causes the crankshaft to rotate. We want to develop a MATLAB program to compute and plot the distance d traveled by the piston as a function of the angle A, for given values of lengths L_1 and L_2. Such a plot would help the engineers designing the engine to select appropriate values for lengths L_1 and L_2.

We are told that typical values for these lengths are $L_1 = 1$ ft and $L_2 = 0.5$ ft. Because the mechanism's motion is symmetrical about $A = 0$, we need consider only angles in the range $0 \leq A \leq 180°$. Figure 1.6–2b shows the geometry of the motion. From this figure we can use trigonometry to write the following expression for d:

$$d = L_1 \cos B + L_2 \cos A \qquad (1.6\text{--}2)$$

Thus to compute d given the lengths L_1 and L_2 and the angle A, we must first determine the angle B. We can do so using the law of sines, as follows:

$$\frac{\sin A}{L_1} = \frac{\sin B}{L_2}$$

Solve this for B:

$$\sin B = \frac{L_2 \sin A}{L_1}$$

$$B = \sin^{-1}\left(\frac{L_2 \sin A}{L_1}\right) \qquad (1.6\text{--}3)$$

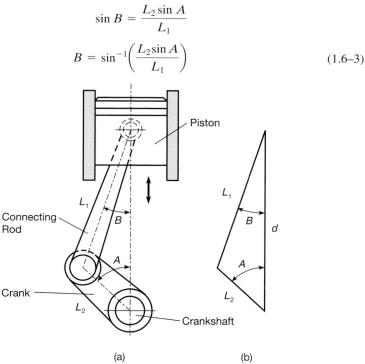

(a)   (b)

Figure 1.6–2 A piston, connecting rod, and crank for an internal combustion engine.

Equations (1.6–2) and (1.6–3) form the basis of our calculations. Develop and test a MATLAB program to plot d versus A.

■ Solution

Here are the steps in the solution, following those listed in Table 1.6–2.

1. *State the problem concisely.* Use Equations (1.6–2) and (1.6–3) to compute d; use enough values of A in the range $0 \leq A \leq 180°$ to generate an adequate (smooth) plot.
2. *Specify the input data to be used by the program.* The lengths L_1 and L_2 and the angle A are given.
3. *Specify the output to be generated by the program.* A plot of d versus A is the required output.
4. *Work through the solution steps by hand or with a calculator.* You could have made an error in deriving the trigonometric formulas, so you should check them for several cases. You can check for these errors by using a ruler and protractor to make a scale drawing of the triangle for several values of the angle A; measure the length d; and compare it to the calculated values. Then you can use these results to check the output of the program.

   Which values of A should you use for the checks? Because the triangle "collapses" when $A = 0°$ and $A = 180°$, you should check these cases. The results are $d = L_1 - L_2$ for $A = 0°$ and $d = L_1 + L_2$ for $A = 180°$. The case $A = 90°$ is also easily checked by hand, using the Pythagorean theorem; for this case $d = \sqrt{L_1^2 - L_2^2}$. You should also check one angle in the quadrant $0° < A < 90°$ and one in the quadrant $90° < A < 180°$. The following table shows the results of these calculations using the given typical values: $L_1 = 1, L_2 = 0.5$ ft.

A (degrees)	d (ft)
0	1.5
60	1.15
90	0.87
120	0.65
180	0.5

5. *Write and run the program.* The following MATLAB session uses the values $L_1 = 1, L_2 = 0.5$ ft.

```
>>L_1 = 1;
>>L_2 = 0.5;
>>R = L_2/L_1;
>>A_d = 0:0.5:180;
>>A_r = A_d*(pi/180);
>>B = asin(R*sin(A_r));
>>d = L_1*cos(B)+L_2*cos(A_r);
>>plot(A_d,d),xlabel('A (degrees)'), ...
ylabel('d (feet)'),grid
```

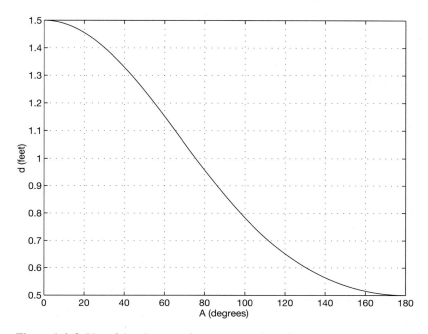

Figure 1.6–3 Plot of the piston motion versus crank angle.

Note the use of the underscore (_) in variable names to make the names more meaningful. The variable `A_d` represents the angle A in degrees. Line 4 creates an array of numbers 0, 0.5, 1, 1.5, . . . ,180. Line 5 converts these degree values to radians and assigns the values to the variable `A_r`. This conversion is necessary because MATLAB trigonometric functions use radians, not degrees. (A common oversight is to use degrees.) MATLAB provides the built-in constant `pi` to use for π. Line 6 uses the inverse sine function `asin`.

The `plot` command requires the label and grid commands to be on the same line, separated by commas. The line-continuation operator, called an *ellipsis,* consists of three periods. This operator enables you to continue typing the line after you press **Enter.** Otherwise, if you continued typing without using the ellipsis, you would not see the entire line on the screen. Note that the prompt is not visible when you press **Enter** after the ellipsis.

The `grid` command puts grid lines on the plot so that you can read values from the plot more easily. The resulting plot appears in Figure 1.6–3.

6. *Check the output of the program with your hand solution.* Read the values from the plot corresponding to the values of A given in the preceding table. You can use the `ginput` function to read values from the plot. The values should agree with one another, and they do.

7. *Run the program and perform a reality check on the output.* You might suspect an error if the plot showed abrupt changes or discontinuities. However, the plot is smooth and shows that d behaves as expected. It decreases smoothly from its maximum at $A = 0°$ to its minimum at $A = 180°$.

8. *Test the program for a range of reasonable input values.* Test the program using various values for L_1 and L_2, and examine the resulting plots to see whether they are reasonable. Something you might try on your own is to see what happens if $L_1 \leq L_2$. Should the mechanism work the same way it does when $L_1 > L_2$? What does your intuition tell you to expect from the mechanism? What does the program predict?

1.7 Summary

You should now be familiar with basic operations in MATLAB. These include

- Starting and exiting MATLAB
- Computing simple mathematical expressions
- Managing variables

You should also be familiar with the MATLAB menu and toolbar system.

The chapter gives an overview of the various types of problems MATLAB can solve. These include

- Using arrays and polynomials
- Creating plots
- Creating script files

Table 1.7–1 is a guide to the tables of this chapter. The following chapters give more details on these topics.

Table 1.7–1 Guide to commands and features introduced in this chapter

Scalar arithmetic operations	Table 1.1–1
Order of precedence	Table 1.1–2
Commands for managing the work session	Table 1.1–3
Special variables and constants	Table 1.1–4
Numeric display formats	Table 1.1–5
Some commonly used mathematical functions	Table 1.3–1
System, directory, and file commands	Table 1.3–2
Some MATLAB plotting commands	Table 1.3–3
Input/output commands	Table 1.4–1
MATLAB Help functions	Table 1.5–1

Key Terms with Page References

Argument, 8

Array, 19

Array index, 19

ASCII files, 21

Command window, 6

Comment, 27

Current directory, 16

Data file, 21

Data marker, 25

Debugging, 29

Desktop, 5

Graphics window, 23

MAT-files, 20

Model, 39

Overlay plot,   24
Path,   22
Precedence,   9
Scalar,   8
Script file,   27

Search path,   22
Session,   7
String variable,   31
Variable,   7
Workspace,   11

Problems

Answers to problems marked with an asterisk are given at the end of the text.

Section 1.1

1. Make sure you know how to start and quit a MATLAB session. Use MATLAB to make the following calculations, using the values $x = 10$, $y = 3$. Check the results by using a calculator.

    a. $u = x + y$ b. $v = xy$ c. $w = x / y$

    d. $z = \sin x$ e. $r = 8 \sin y$ f. $s = 5 \sin (2y)$

2.* Suppose that $x = 2$ and $y = 5$. Use MATLAB to compute the following.

    a. $\dfrac{yx^3}{x - y}$ b. $\dfrac{3x}{2y}$ c. $\dfrac{3}{2}xy$ d. $\dfrac{x^5}{x^5 - 1}$

3. Suppose that $x = 3$ and $y = 4$. Use MATLAB to compute the following, and check the results with a calculator.

    a. $\left(1 - \dfrac{1}{x^5}\right)^{-1}$ b. $3\pi x^2$ c. $\dfrac{3y}{4x - 8}$ d. $\dfrac{4(y - 5)}{3x - 6}$

4. Evaluate the following expressions in MATLAB for the given value of x. Check your answers by hand.

    a. $y = 6x^3 + \dfrac{4}{x}$, $x = 3$ b. $y = \dfrac{x}{4}3$, $x = 7$

    c. $y = \dfrac{(4x)^2}{25}$, $x = 9$ d. $y = 2\dfrac{\sin x}{5}$, $x = 4$

    e. $y = 7(x^{1/3}) + 4x^{0.58}$, $x = 30$

5. Assuming that the variables a, b, c, d, and f are scalars, write MATLAB statements to compute and display the following expressions. Test your statements for the values $a = 1.12$, $b = 2.34$, $c = 0.72$, $d = 0.81$, and $f = 19.83$.

    $$x = 1 + \frac{a}{b} + \frac{c}{f^2} \qquad\qquad s = \frac{b - a}{d - c}$$

    $$r = \frac{1}{\frac{1}{a} + \frac{1}{b} + \frac{1}{c} + \frac{1}{d}} \qquad\qquad y = ab\frac{1}{c}\frac{f^2}{2}$$

6.   Use MATLAB to calculate

a. $\dfrac{3}{4}\,(6)\,(7^2) + \dfrac{4^5}{7^3 - 145}$

b. $\dfrac{48.2(55) - 9^3}{53 + 14^2}$

c. $\dfrac{27^2}{4} + \dfrac{319^{4/5}}{5} + 60(14)^{-3}$

Check your answers with a calculator.

7.   The volume of a sphere is given by $V = 4\pi r^3/3$, where r is the radius. Use MATLAB to compute the radius of a sphere having a volume 40 percent greater than that of a sphere of radius 4 ft.

8.*   Suppose that $x = -7 - 5i$ and $y = 4 + 3i$. Use MATLAB to compute

a. $x + y$ b. xy c. x/y

9.   Use MATLAB to compute the following. Check your answers by hand.

a. $(3 + 6i)(-7 - 9i)$

b. $\dfrac{5 + 4i}{5 - 4i}$

c. $\dfrac{3}{2}i$

d. $\dfrac{3}{2i}$

10.   Evaluate the following expressions in MATLAB, for the values $x = 5 + 8i$, $y = -6 + 7i$. Check your answers by hand.

a. $u = x + y$ b. $v = xy$ c. $w = x/y$

d. $z = e^x$ e. $r = \sqrt{y}$ f. $s = xy^2$

11.   The *ideal gas law* provides one way to estimate the pressure exerted by a gas in a container. The law is

$$P = \dfrac{nRT}{V}$$

More accurate estimates can be made with the *van der Waals* equation

$$P = \dfrac{nRT}{V - nb} - \dfrac{an^2}{V^2}$$

where the term nb is a correction for the volume of the molecules and the term an^2/V^2 is a correction for molecular attractions. The values of a and b depend on the type of gas. The gas constant is R, the *absolute* temperature is T, the gas volume is V, and the number of gas molecules is indicated by n. If $n = 1$ mol of an ideal gas were confined to a volume of $V = 22.41$ L at 0°C (273.2 K), it would exert a pressure of 1 atm. In these units, $R = 0.08206$.

For chlorine (Cl_2), $a = 6.49$ and $b = 0.0562$. Compare the pressure estimates given by the ideal gas law and the van der Waals equation for 1 mol of Cl_2 in 22.41 L at 273.2 K. What is the main cause of the difference in the two pressure estimates, the molecular volume or the molecular attractions?

12. The *ideal gas law* relates the pressure P, volume V, absolute temperature T, and amount of gas n. The law is

$$P = \frac{nRT}{V}$$

where R is the gas constant.

     An engineer must design a large natural gas storage tank to be expandable to maintain the pressure constant at 2.2 atm. In December when the temperature is 4°F (-15°C), the volume of gas in the tank is 28 500 ft³. What will the volume of the same quantity of gas be in July when the temperature is 88°F (31°C)? (*Hint*: Use the fact that n, R, and P are constant in this problem. Note also that K = °C + 273.2.)

Section 1.3

13. Suppose x takes on the values $x = 1, 1.2, 1.4, \ldots, 5$. Use MATLAB to compute the array `y` that results from the function $y = 7 \sin(4x)$. Use MATLAB to determine how many elements are in the array `y` and the value of the third element in the array `y`.

14. Use MATLAB to determine how many elements are in the array `sin(-pi/2):0.05:cos(0)`. Use MATLAB to determine the 10th element.

15. Use MATLAB to calculate

   a. $e^{(-2.1)^3} + 3.47 \log(14) + \sqrt[4]{287}$ *b.* $(3.4)^7 \log(14) + \sqrt[4]{287}$

   c. $\cos^2\left(\dfrac{4.12\pi}{6}\right)$ *d.* $\cos\left(\dfrac{4.12\pi}{6}\right)^2$

Check your answers with a calculator.

16. Use MATLAB to calculate

   a. $6\pi \tan^{-1}(12.5) + 4$ *b.* $5 \tan[3 \sin^{-1}(13/5)]$

   c. $5 \ln(7)$ *d.* $5 \log(7)$

Check your answers with a calculator.

17. The Richter scale is a measure of the intensity of an earthquake. The energy E (in joules) released by the quake is related to the magnitude M on the Richter scale as follows.

$$E = 10^{4.4} 10^{1.5M}$$

How much more energy is released by a magnitude 7.6 quake than a 5.6 quake?

18.* Use MATLAB to find the roots of $13x^3 + 182x^2 - 184x + 2503 = 0$.

19. Use MATLAB to find the roots of the polynomial $70x^3 + 24x^2 - 10x + 20$.

20. Determine which search path MATLAB uses on your computer. If you use a lab computer as well as a home computer, compare the two search paths. Where will MATLAB look for a user-created M-file on each computer?

21. Use MATLAB to plot the function $T = 6 \ln t - 7e^{0.2t}$ over the interval $1 \leq t \leq 3$. Put a title on the plot and properly label the axes. The variable T represents temperature in degrees Celsius; the variable t represents time in minutes.

22. Use MATLAB to plot the functions $u = 2 \log_{10}(60x + 1)$ and $v = 3 \cos(6x)$ over the interval $0 \leq x \leq 2$. Properly label the plot and each curve. The variables u and v represent speed in miles per hour; the variable x represents distance in miles.

23. The Fourier series is a series representation of a periodic function in terms of sines and cosines. The Fourier series representation of the function

$$f(x) = \begin{cases} 1 & 0 < x < \pi \\ -1 & -\pi < x < 0 \end{cases}$$

is

$$\frac{4}{\pi}\left(\frac{\sin x}{1} + \frac{\sin 3x}{3} + \frac{\sin 5x}{5} + \frac{\sin 7x}{7} + \cdots\right)$$

Plot on the same graph the function $f(x)$ and its series representation, using the four terms shown.

24. A *cycloid* is the curve described by a point P on the circumference of a circular wheel of radius r rolling along the x axis. The curve is described in parametric form by the equations

$$x = r(\phi - \sin \phi)$$
$$y = r(1 - \cos \phi)$$

Use these equations to plot the cycloid for $r = 10$ in. and $0 \leq \phi \leq 4\pi$.

Section 1.4

25. A fence around a field is shaped as shown in Figure P25. It consists of a rectangle of length L and width W and a right triangle that is symmetric about the central horizontal axis of the rectangle. Suppose the width W is known (in meters) and the enclosed area A is known (in square meters). Write a MATLAB script file in terms of the given variables W and A to determine the length L required so that the enclosed area is A. Also determine the total length of fence required. Test your script for the values $W = 6$ m and $A = 80$ m^2.

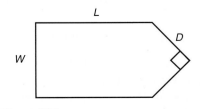

Figure P25

26. The four-sided figure shown in Figure P26 consists of two triangles having a common side a. The law of cosines for the top triangle states that

$$a^2 = b_1^2 + c_1^2 - 2b_1c_1 \cos A_1$$

and a similar equation can be written for the bottom triangle. Develop a procedure for computing the length of side c_2 if you are given the lengths of sides b_1, b_2, and c_1 and the angles A_1 and A_2 in degrees. Write a script file to implement this procedure. Test your script, using the following values: $b_1 = 180$ m, $b_2 = 165$ m, $c_1 = 115$ m, $A_1 = 120°$, and $A_2 = 100°$.

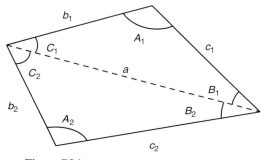

Figure P26

Section 1.5

27. Use the MATLAB Help facilities to find information about the following topics and symbols: plot, label, cos, cosine, :, and *.

28. Use the MATLAB Help facilities to determine what happens if you use the sqrt function with a negative argument.

29. Use the MATLAB Help facilities to determine what happens if you use the exp function with an imaginary argument.

Section 1.6

30. *a.* With what initial speed must you throw a ball vertically for it to reach a height of 20 ft? The ball weighs 1 lb. How does your answer change if the ball weighs 2 lb?
   b. Suppose you want to throw a steel bar vertically to a height of 20 ft. The bar weighs 2 lb. How much initial speed must the bar have to reach this height? Discuss how the length of the bar affects your answer.

31. Consider the motion of the piston discussed in Example 1.6–1. The piston *stroke* is the total distance moved by the piston as the crank angle varies from 0° to 180°.

   a. How does the piston stroke depend on L_1 and L_2?
   b. Suppose $L_2 = 0.5$ ft. Use MATLAB to plot the piston motion versus crank angle for two cases: $L_1 = 0.6$ ft and $L_1 = 1.4$ ft. Compare each plot with the plot shown in Figure 1.6–3. Discuss how the shape of the plot depends on the value of L_1.

Photo courtesy of Stratosphere Corporation.

Engineering in the 21st Century. . .

Innovative Construction

We tend to remember the great civilizations of the past in part by their public works, such as the Egyptian pyramids and the medieval cathedrals of Europe, which were technically challenging to create. Perhaps it is in our nature to "push the limits," and we admire others who do so. The challenge of innovative construction continues today. As space in our cities becomes scarce, many urban planners prefer to build vertically rather than horizontally. The newest tall buildings push the limits of our abilities, not only in structural design but also in areas that we might not think of, such as elevator design and operation, aerodynamics, and construction techniques. The photo above shows the 1149-ft-high Las Vegas Stratosphere Tower, the tallest observation tower in the United States. It required many innovative techniques in its assembly. The construction crane shown in use is 400 ft tall.

Designers of buildings, bridges, and other structures will use new technologies and new materials, some based on nature's designs. Pound for pound, spider silk is stronger than steel, and structural engineers hope to use cables of synthetic spider silk fibers to build earthquake-resistant suspension bridges. *Smart* structures, which can detect impending failure from cracks and fatigue, are now close to reality, as are *active* structures that incorporate powered devices to counteract wind and other forces. The MATLAB Financial toolbox is useful for financial evaluation of large construction projects, and the MATLAB Partial Differential Equation toolbox can be used for structural design. ■

CHAPTER **2**

Numeric, Cell, and Structure Arrays

OUTLINE

2.1   One- and Two-Dimensional Numeric Arrays

2.2   Multidimensional Numeric Arrays

2.3   Element-by-Element Operations

2.4   Matrix Operations

2.5   Polynomial Operations Using Arrays

2.6   Cell Arrays

2.7   Structure Arrays

2.8   Summary

Problems

One of the strengths of MATLAB is the capability to handle collections of items, called *arrays,* as if they were a single entity. The array-handling feature means that MATLAB programs can be very short.

The array is the basic building block in MATLAB. The following classes of arrays are available in MATLAB 7:

Array						
numeric	character	logical	cell	structure	function handle	Java

So far we have used only numeric arrays, which are arrays containing only numeric values. Within the numeric class are the subclasses *single* (single precision), *double* (double precision), *int8*, *int16*, and *int32* (signed 8-bit, 16-bit, and 32-bit integers), and *uint8*, *uint16*, and *uint32* (unsigned 8-bit, 16-bit, and 32-bit integers). A character array is an array containing strings. The elements of logical arrays are "true" or "false," which, although represented by the symbols 1 and 0, are not numeric quantities. We will study the logical arrays in Chapter 4. Cell arrays and structure arrays are covered in Sections 2.6 and 2.7. Function handles are treated in Chapter 3. The Java class is not covered in this text.

The first four sections of this chapter treat concepts that are essential to understanding MATLAB and therefore must be covered. Section 2.5 treats polynomial applications. Sections 2.6 and 2.7 introduce two types of arrays that are useful for some specialized applications.

2.1 One- and Two-Dimensional Numeric Arrays

We can represent the location of a point in three-dimensional space by three Cartesian coordinates x, y, and z. These three coordinates specify a *vector* **p**. (In mathematical text we often use boldface type to indicate vectors.) The set of *unit vectors* **i**, **j**, **k**, whose lengths are 1 and whose directions coincide with the x, y, and z axes, respectively, can be used to express the vector mathematically as follows: $\mathbf{p} = x\mathbf{i} + y\mathbf{j} + z\mathbf{k}$. The unit vectors enable us to associate the vector components x, y, z with the proper coordinate axes; therefore, when we write $\mathbf{p} = 5\mathbf{i} + 7\mathbf{j} + 2\mathbf{k}$, we know that the x, y, and z coordinates of the vector are 5, 7, and 2, respectively. We can also write the components in a specific order, separate them with a space, and identify the group with brackets, as follows: [5 7 2]. As long as we agree that the vector components will be written in the order x, y, z, we can use this notation instead of the unit-vector notation. In fact, MATLAB uses this style for vector notation. MATLAB allows us to separate the components with commas for improved readability if we desire so that the equivalent way of writing the preceding vector is [5, 7, 2]. This expression is a *row vector*, which is a horizontal arrangement of the elements.

ROW VECTOR

COLUMN VECTOR

We can also express the vector as a *column vector*, which has a vertical arrangement. A vector can have only one column, or only one row. Thus, a vector is a one-dimensional array. In general, arrays can have more than one column and more than one row.

Creating Vectors in MATLAB

The concept of a vector can be generalized to any number of components. In MATLAB a vector is simply a list of scalars, whose order of appearance in the list might be significant, as it is when specifying *xyz* coordinates. As another example, suppose we measure the temperature of an object once every hour. We can represent the measurements as a vector, and the 10th element in the list is the temperature measured at the 10th hour.

To create a row vector in MATLAB, you simply type the elements inside a pair of square brackets, separating the elements with a space or a comma. Brackets are required for arrays unless you use the colon operator to create the array. In this case you should not use brackets, but you can optionally use parentheses. The choice between a space or comma is a matter of personal preference, although the chance of an error is less if you use a comma. (You can also use a comma followed by a space for maximum readability.)

To create a column vector, you can separate the elements by semicolons; alternatively, you can create a row vector and then use the *transpose* notation (′), which converts a row vector into a column vector, or vice versa. For example:

```
>>g = [3;7;9]
g =
    3
    7
    9
>>g = [3,7,9]'
g =
    3
    7
    9
```

The third way to create a column vector is to type a left bracket ([ ) and the first element, press **Enter**, type the second element, press **Enter**, and so on until you type the last element followed by a right bracket ( ]) and **Enter**. On the screen this sequence looks like

```
>>g = [3
7
9]
g =
    3
    7
    9
```

Note that MATLAB displays row vectors horizontally and column vectors vertically.

You can create vectors by "appending" one vector to another. For example, to create the row vector u whose first three columns contain the values of r = [2,4,20] and whose fourth, fifth, and sixth columns contain the values of w = [9,-6,3], you type u = [r,w]. The result is the vector u = [2,4,20,9,-6,3].

The colon operator (:) easily generates a large vector of regularly spaced elements. Typing

```
>>x = m:q:n
```

creates a vector x of values with a spacing q. The first value is m. The last value is n if m − n is an integer multiple of q. If not, the last value is less than n. For example, typing x = 0:2:8 creates the vector x = [0,2,4,6,8], whereas typing x = 0:2:7 creates the vector x = [0,2,4,6]. To create a row vector z consisting of the values from 5 to 8 in steps of 0.1, you type z = 5:0.1:8. If the increment q is omitted, it is presumed to be 1. Thus y = -3:2 produces the vector y = [-3,-2,-1,0,1,2].

The increment q can be negative. In this case m should be greater than n. For example, $u = 10:-2:4$ produces the vector $[10,8,6,4]$.

The linspace command also creates a linearly spaced row vector, but instead you specify the number of values rather than the increment. The syntax is linspace(x1,x2,n), where x1 and x2 are the lower and upper limits and n is the number of points. For example, linspace(5,8,31) is equivalent to 5:0.1:8. If n is omitted, the spacing is 1.

The logspace command creates an array of logarithmically spaced elements. Its syntax is logspace(a,b,n), where n is the number of points between 10^a and 10^b. For example, x = logspace(-1,1,4) produces the vector x = [0.1000, 0.4642, 2.1544, 10.000]. If n is omitted, the number of points defaults to 50.

Two-Dimensional Arrays

MATRIX

An array having rows and columns is a two-dimensional array that is sometimes called a *matrix*. In mathematical text, if possible, vectors are usually denoted by boldface lowercase letters and matrices by boldface uppercase letters. An example of a matrix having three rows and two columns is

$$\mathbf{M} = \begin{bmatrix} 2 & 5 \\ -3 & 4 \\ -7 & 1 \end{bmatrix}$$

ARRAY SIZE

We refer to the *size* of an array by the number of rows and the number of columns. For example, an array with 3 rows and 2 columns is said to be a 3×2 array. *The number of rows is always stated first!* We sometimes represent a matrix $\mathbf{A}$ as $[a_{ij}]$ to indicate its elements a_{ij}. The subscripts i and j, called *indices*, indicate the row and column location of the element a_{ij}. *The row number must always come first!* For example, the element a_{32} is in row 3, column 2. Two matrices $\mathbf{A}$ and $\mathbf{B}$ are equal if they have the same size and if all their corresponding elements are equal, that is, $a_{ij} = b_{ij}$ for every value of i and j.

Creating Matrices

The most direct way to create a matrix is to type the matrix row by row, separating the elements in a given row with spaces or commas and separating the rows with semicolons. Brackets are required. For example, typing

```
>>A = [2,4,10;16,3,7];
```

creates the following matrix:

$$\mathbf{A} = \begin{bmatrix} 2 & 4 & 10 \\ 16 & 3 & 7 \end{bmatrix}$$

If the matrix has many elements, you can press **Enter** and continue typing on the next line. MATLAB knows you are finished entering the matrix when you type the closing bracket ( ]).

You can append a row vector to another row vector to create either a third row vector or a matrix (if both vectors have the same number of columns). Note the difference between the results given by [a,b] and [a;b] in the following session:

```
>>a = [1,3,5];
>>b = [7,9,11];
>>c = [a,b]
c =
    1 3 5 7 9 11
>> D = [a;b]
D =
    1 3 5
    7 9 11
```

Matrices and the Transpose Operation

The transpose operation interchanges the rows and columns. In mathematics text we denote this operation by the superscript T. For an $m \times n$ matrix $\mathbf{A}$ with m rows and n columns, $\mathbf{A}^T$ (read "A transpose") is an $n \times m$ matrix.

$$\mathbf{A} = \begin{bmatrix} -2 & 6 \\ -3 & 5 \end{bmatrix} \qquad \mathbf{A}^T = \begin{bmatrix} -2 & -3 \\ 6 & 5 \end{bmatrix}$$

If $\mathbf{A}^T = \mathbf{A}$, the matrix $\mathbf{A}$ is *symmetric*. Note that the transpose operation converts a row vector into a column vector, and vice versa.

If the array contains complex elements, the transpose operator (') produces the *complex conjugate transpose;* that is, the resulting elements are the complex conjugates of the original array's transposed elements. Alternatively, you can use the *dot transpose* operator (. ') to transpose the array without producing complex conjugate elements, for example, A . '. If all the elements are real, the operators ' and . ' give the same result.

Array Addressing

Array indices are the row and column numbers of an element in an array and are used to keep track of the array's elements. For example, the notation v(5) refers to the fifth element in the vector v, and A(2,3) refers to the element in row 2, column 3 in the matrix A. *The row number is always listed first!* This notation enables you to correct entries in an array without retyping the entire array. For example, to change the element in row 1, column 3 of a matrix **D** to 6, you can type D(1,3) = 6.

The colon operator selects individual elements, rows, columns, or "subarrays" of arrays. Here are some examples:

- v(:) represents all the row or column elements of the vector v.
- v(2:5) represents the second through fifth elements; that is v(2), v(3), v(4), v(5).

- A(:,3) denotes all the elements in the third *column* of the matrix A.
- A(3,:) denotes all the elements in the third *row* of A.
- A(:,2:5) denotes all the elements in the second through fifth columns of A.
- A(2:3,1:3) denotes all the elements in the second and third rows that are also in the first through third columns.
- v = A(:) creates a vector v consisting of all the columns of A stacked from first to last.
- A(end,:) denotes the last row in A, and A(:,end) denotes the last column.

You can use array indices to extract a smaller array from another array. For example, if you create the array **B**

$$\mathbf{B} = \begin{bmatrix} 2 & 4 & 10 & 13 \\ 16 & 3 & 7 & 18 \\ 8 & 4 & 9 & 25 \\ 3 & 12 & 15 & 17 \end{bmatrix} \qquad (2.1\text{--}1)$$

by typing

```
>>B = [2,4,10,13;16,3,7,18;8,4,9,25;3,12,15,17];
```

and then type

```
>>C = B(2:3,1:3);
```

you can produce the following array:

$$\mathbf{C} = \begin{bmatrix} 16 & 3 & 7 \\ 8 & 4 & 9 \end{bmatrix}$$

EMPTY ARRAY

The *empty array* contains no elements and is expressed as [ ]. Rows and columns can be deleted by setting the selected row or column equal to the empty array. This step causes the original matrix to collapse to a smaller one. For example, A(3,:) = [ ] deletes the third row in A, while A(:,2:4) = [ ] deletes the second through fourth columns in A. Finally, A([1 4],:) = [ ] deletes the first and fourth rows of A.

Suppose we type A = [6,9,4;1,5,7] to define the following matrix:

$$\mathbf{A} = \begin{bmatrix} 6 & 9 & 4 \\ 1 & 5 & 7 \end{bmatrix}$$

Typing A(1,5) = 3 changes the matrix to

$$\mathbf{A} = \begin{bmatrix} 6 & 9 & 4 & 0 & 3 \\ 1 & 5 & 7 & 0 & 0 \end{bmatrix}$$

Because **A** did not have five columns, its size is automatically expanded to accept the new element in column 5. MATLAB adds zeros to fill out the remaining elements.

MATLAB does not accept negative or zero indices, but you can use negative increments with the colon operator. For example, typing B = A(:,5:-1:1) reverses the order of the columns in **A** and produces

$$\mathbf{B} = \begin{bmatrix} 3 & 0 & 4 & 9 & 6 \\ 0 & 0 & 7 & 5 & 1 \end{bmatrix}$$

Suppose that C = [-4,12,3,5,8]. Then typing B(2,:) = C replaces row 2 of B with C. Thus **B** becomes

$$\mathbf{B} = \begin{bmatrix} 3 & 0 & 4 & 9 & 6 \\ -4 & 12 & 3 & 5 & 8 \end{bmatrix}$$

Suppose that D = [3,8,5;4,-6,9]. Then typing E = D([2,2,2],:) repeats row 2 of D three times to obtain

$$\mathbf{E} = \begin{bmatrix} 4 & -6 & 9 \\ 4 & -6 & 9 \\ 4 & -6 & 9 \end{bmatrix}$$

Using `clear` to Avoid Errors

You can use the `clear` command to protect yourself from accidentally reusing an array that has the wrong dimension. Even if you set new values for an array, some previous values might still remain. For example, suppose you had previously created the 2 × 2 array A = [2, 5; 6, 9], and then you create the 5 × 1 arrays x = (1:5)' and y = (2:6)'. Note that parentheses are needed here to use the transpose operator. Suppose you now redefine A so that its columns will be x and y. If you then type A(:,1) = x to create the first column, MATLAB displays an error message telling you that the number of rows in A and x must be the same. MATLAB thinks A should be a 2 × 2 matrix because A was previously defined to have only two rows and its values remain in memory. The `clear` command wipes A and all other variables from memory and avoids this error. To clear A only, type `clear A` before typing A(:,1) = x.

Some Useful Array Functions

MATLAB has many functions for working with arrays (see Table 2.1–1). Here is a summary of some of the more commonly used functions.

The `max(A)` function returns the algebraically greatest element in **A** if **A** is a vector having all real elements. It returns a row vector containing the greatest elements in each *column* if **A** is a matrix containing all real elements. If *any* of the elements are complex, `max(A)` returns the element that has the largest magnitude. The syntax [x,k] = max(A) is similar to max(A), but it stores the maximum values in the row vector **x** and their indices in the row vector **k**.

If A and B have the same size, C = max(A,B) creates an array the same size, having the maximum value from each corresponding location in A and B. For example, the following **A** and **B** matrices give the **C** matrix shown.

Table 2.1–1 Basic syntax of array functions*

Command	Description
find(x)	Computes an array containing the indices of the nonzero elements of the array **x**.
[u,v,w] = find(A)	Computes the arrays **u** and **v**, containing the row and column indices of the nonzero elements of the matrix **A**, and the array **w**, containing the values of the nonzero elements. The array **w** may be omitted.
length(A)	Computes either the number of elements of A if A is a vector or the largest value of *m* or *n* if **A** is an $m \times n$ matrix.
linspace(a,b,n)	Creates a row vector of *n* regularly spaced values between *a* and *b*.
logspace(a,b,n)	Creates a row vector of *n* logarithmically spaced values between *a* and *b*.
max(A)	Returns the algebraically largest element in **A** if **A** is a vector. Returns a row vector containing the largest elements in each column if **A** is a matrix. If any of the elements are complex, max(A) returns the elements that have the largest magnitudes.
[x,k] = max(A)	Similar to max(A) but stores the maximum values in the row vector **x** and their indices in the row vector **k**.
min(A)	Same as max(A) but returns minimum values.
[x,k] = min(A)	Same as [x,k] = max(A) but returns minimum values.
norm(x)	Computes a vector's geometric length $\sqrt{x_1^2 + x_2^2 + \cdots + x_n^2}$.
size(A)	Returns a row vector [m n] containing the sizes of the $m \times n$ array **A**.
sort(A)	Sorts each column of the array **A** in ascending order and returns an array the same size as **A**.
sum(A)	Sums the elements in each column of the array **A** and returns a row vector containing the sums.

*Many of these functions have extended syntax. See the text and MATLAB help for more discussion.

$$\mathbf{A} = \begin{bmatrix} 1 & 6 & 4 \\ 3 & 7 & 2 \end{bmatrix} \quad \mathbf{B} = \begin{bmatrix} 3 & 4 & 7 \\ 1 & 5 & 8 \end{bmatrix} \quad \mathbf{C} = \begin{bmatrix} 3 & 6 & 7 \\ 3 & 7 & 8 \end{bmatrix}$$

The functions min(A) and [x,k] = min(A) are the same as max(A) and [x,k] = max(A) except that they return minimum values.

The function size(A) returns a row vector [m n] containing the sizes of the $m \times n$ array **A**. The length(A) function computes either the number of elements of A if A is a vector or the largest value of *m* or *n* if **A** is an $m \times n$ matrix. For example, if

$$\mathbf{A} = \begin{bmatrix} 6 & 2 \\ -10 & -5 \\ 3 & 0 \end{bmatrix}$$

then max(A) returns the vector [6,2]; min(A) returns the vector [-10,-5]; size(A) returns [3,2]; and length(A) returns 3.

The sum(A) function sums the elements in each *column* of the array **A** and returns a row vector containing the sums. The sort(A) function sorts each *column* of the array **A** in ascending order and returns an array the same size as **A**.

If A has one or more complex elements, the max, min, and sort functions act on the absolute values of the elements and return the element that has the largest magnitude.

For example, if

$$\mathbf{A} = \begin{bmatrix} 6 & 2 \\ -10 & -5 \\ 3+4i & 0 \end{bmatrix}$$

then `max(A)` returns the vector `[-10,-5]` and `min(A)` returns the vector `[3+4i,0]`. (The magnitude of $3 + 4i$ is 5.)

The sort will be done in descending order if the form `sort(A, 'descend')` is used. The `min`, `max`, and `sort` functions can be made to act on the rows instead of the columns by transposing the array.

The complete syntax of the `sort` function is `sort(A, dim, mode)`, where `dim` selects a dimension along which to sort and `mode` selects the direction of the sort, `'ascend'` for ascending order and `'descend'` for descending order. So, for example, `sort(A,2, 'descend')` would sort the elements in each row of **A** in descending order.

The `find(x)` command computes an array containing the indices of the *nonzero* elements of the vector **x**. The syntax `[u,v,w] = find(A)` computes the arrays **u** and **v**, containing the row and column indices of the nonzero elements of the matrix **A**, and the array **w**, containing the values of the nonzero elements. The array **w** may be omitted.

For example, if

$$\mathbf{A} = \begin{bmatrix} 6 & 0 & 3 \\ 0 & 4 & 0 \\ 2 & 7 & 0 \end{bmatrix}$$

then the session

```
>>A = [6, 0, 3; 0, 4, 0; 2, 7, 0];
>>[u, v, w] = find(A)
```

returns the vectors

$$\mathbf{u} = \begin{bmatrix} 1 \\ 3 \\ 2 \\ 3 \\ 1 \end{bmatrix} \qquad \mathbf{v} = \begin{bmatrix} 1 \\ 1 \\ 2 \\ 2 \\ 3 \end{bmatrix} \qquad \mathbf{w} = \begin{bmatrix} 6 \\ 2 \\ 4 \\ 7 \\ 3 \end{bmatrix}$$

The vectors **u** and **v** give the (row, column) indices of the nonzero values, which are listed in **w**. For example, the second entries in **u** and **v** give the indices (3, 1), which specifies the element in row 3, column 1 of **A**, whose value is 2.

These functions are summarized in Table 2.1–1.

Magnitude, Length, and Absolute Value of a Vector

The terms *magnitude, length,* and *absolute value* are often loosely used in everyday language, but you must keep their precise meaning in mind when using MATLAB.

The MATLAB `length` command gives the number of elements in the vector. The *magnitude* of a vector **x** having real elements $x_1, x_2, \ldots, x_n$ is a scalar, given by $\sqrt{x_1^2 + x_2^2 + \cdots + x_n^2}$, and is the same as the vector's geometric length. The *absolute value* of a vector **x** is a vector whose elements are the absolute values of the elements of **x**. For example, if x = [2,-4,5], its length is 3; its magnitude is $\sqrt{2^2 + (-4)^2 + 5^2} = 6.7082$; and its absolute value is [2,4,5]. The length, magnitude, and absolute value of x are computed by `length(x)`, `norm(x)`, and `abs(x)`, respectively.

Test Your Understanding

T2.1–1   For the matrix **B**, find the array that results from the operation [B;B']. Use MATLAB to determine what number is in row 5, column 3 of the result.

$$\mathbf{B} = \begin{bmatrix} 2 & 4 & 10 & 13 \\ 16 & 3 & 7 & 18 \\ 8 & 4 & 9 & 25 \\ 3 & 12 & 15 & 17 \end{bmatrix}$$

T2.1–2   For the same matrix **B**, use MATLAB to (a) find the largest and smallest elements in **B** and their indices and (b) sort each column in **B** to create a new matrix **C**.

The Variable Editor

The MATLAB Workspace Browser provides a graphical interface for managing the workspace. You can use it to view, save, and clear workspace variables. It includes the *Variable Editor,* a graphical interface for working with variables, including arrays. To open the Workspace Browser, type `workspace` at the Command window prompt. The browser appears as shown in Figure 2.1–1.

Keep in mind that the Desktop menus are context-sensitive. Thus their contents will change depending on which features of the browser and Variable Editor you are currently using. The Workspace Browser shows the name of each variable, its value, array size, and class. The icon for each variable illustrates its class.

Figure 2.1–1 The Workspace Browser.

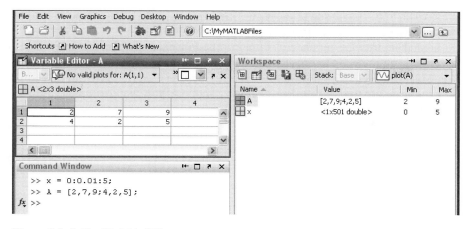

Figure 2.1–2 The Variable Editor.

From the Workspace Browser you can open the Variable Editor to view and edit a visual representation of two-dimensional numeric arrays, with the rows and columns numbered. To open the Variable Editor from the Workspace Browser, double-click on the variable you want to open. The Variable Editor opens, displaying the values for the selected variable. The Variable Editor appears as shown in Figure 2.1–2.

To open a variable, you can also right-click it and use the **Context** menu. Repeat the steps to open additional variables into the Variable Editor. In the Variable Editor, access each variable via its tab at the bottom of the window, or use the **Window** menu. You can also open the Variable Editor directly from the Command window by typing `open('var')`, where `var` is the name of the variable to be edited. Once an array is displayed in the Variable Editor, you can change a value in the array by clicking on its location, typing in the new value, and pressing **Enter**.

Right-clicking on a variable brings up the **Context** menu, which can be used to edit, save, or clear the selected variable, or to plot the rows of the variable versus its columns (this type of plot is discussed in Chapter 5).

You can also clear a variable from theWorkspace Browser by first highlighting it in the Browser, then clicking on **Delete** in the **Edit** menu.

2.2 Multidimensional Numeric Arrays

MATLAB supports multidimensional arrays. For more information, type `help datatypes`.

A three-dimensional array has the dimension $m \times n \times q$. A four-dimensional array has the dimension $m \times n \times q \times r$, and so forth. The first two dimensions are the row and column, as with a matrix. The higher dimensions are called *pages*. You can think of a three-dimensional array as layers of matrices. The first layer is page 1; the second layer is page 2, and so on. If A is a $3 \times 3 \times 2$ array, you can access the element in row 3, column 2 of page 2 by typing A(3,2,2). To access all of page 1, type A(:,:,1). To access all of page 2, type A(:,:,2). The ndims command returns the number of dimensions. For example, for the array A just described, ndims(A) returns the value 3.

You can create a multidimensional array by first creating a two-dimensional array and then extending it. For example, suppose you want to create a three-dimensional array whose first two pages are

$$\begin{bmatrix} 4 & 6 & 1 \\ 5 & 8 & 0 \\ 3 & 9 & 2 \end{bmatrix} \quad \begin{bmatrix} 6 & 2 & 9 \\ 0 & 3 & 1 \\ 4 & 7 & 5 \end{bmatrix}$$

To do so, first create page 1 as a 3 × 3 matrix and then add page 2, as follows:

```
>>A = [4,6,1;5,8,0;3,9,2];
>>A(:,:,2) = [6,2,9;0,3,1;4,7,5];
```

Another way to produce such an array is with the cat command. Typing cat(n,A,B,C,...) creates a new array by concatenating the arrays A, B, C, and so on along the dimension n. Note that cat(1,A,B) is the same as [A;B] and that cat(2,A,B) is the same as [A,B]. For example, suppose we have the 2 × 2 arrays **A** and **B**:

$$\mathbf{A} = \begin{bmatrix} 8 & 2 \\ 9 & 5 \end{bmatrix} \quad \mathbf{B} = \begin{bmatrix} 4 & 6 \\ 7 & 3 \end{bmatrix}$$

Then C = cat(3,A,B) produces a three-dimensional array composed of two layers; the first layer is the matrix A, and the second layer is the matrix B. The element C(m,n,p) is located in row m, column n, and layer p. Thus the element C(2,1,1) is 9, and the element C(2,2,2) is 3.

Multidimensional arrays are useful for problems that involve several parameters. For example, if we have data on the temperature distribution in a rectangular object, we could represent the temperatures as an array T with three dimensions.

2.3 Element-by-Element Operations

To increase the magnitude of a vector, multiply it by a scalar. For example, to double the magnitude of the vector r = [3,5,2], multiply each component by 2 to obtain [6,10,4]. In MATLAB you type v = 2*r.

Multiplying a matrix **A** by a scalar w produces a matrix whose elements are the elements of **A** multiplied by w. For example:

$$3 \begin{bmatrix} 2 & 9 \\ 5 & -7 \end{bmatrix} = \begin{bmatrix} 6 & 27 \\ 15 & -21 \end{bmatrix}$$

This multiplication is performed in MATLAB as follows:

```
>>A = [2,9;5,-7];
>>3*A
```

Thus multiplication of an array by a scalar is easily defined and easily carried out. However, multiplication of two *arrays* is not so straightforward. In fact, MATLAB uses two definitions of multiplication: (1) array multiplication and (2) matrix multiplication. Division and exponentiation must also be carefully defined when you are dealing with operations between two arrays. MATLAB has two forms of arithmetic operations on arrays. In this section we introduce one form, called *array operations,* which are also called *element-by-element* operations. In the next section we introduce *matrix* operations. Each form has its own applications, which we illustrate by examples.

ARRAY OPERATIONS

Array Addition and Subtraction

Array addition can be done by adding the corresponding components. To add the arrays $r = [3,5,2]$ and $v = [2,-3,1]$ to create w in MATLAB, you type $w = r + v$. The result is $w = [5,2,3]$.

When two arrays have identical size, their sum or difference has the same size and is obtained by adding or subtracting their corresponding elements. Thus $\mathbf{C} = \mathbf{A} + \mathbf{B}$ implies that $c_{ij} = a_{ij} + b_{ij}$ if the arrays are matrices. The array $\mathbf{C}$ has the same size as $\mathbf{A}$ and $\mathbf{B}$. For example,

$$\begin{bmatrix} 6 & -2 \\ 10 & 3 \end{bmatrix} + \begin{bmatrix} 9 & 8 \\ -12 & 14 \end{bmatrix} = \begin{bmatrix} 15 & 6 \\ -2 & 17 \end{bmatrix} \qquad (2.3\text{–}1)$$

Array subtraction is performed in a similar way.

The addition shown in Equation (2.3–1) is performed in MATLAB as follows:

```
>>A = [6,-2;10,3];
>>B = [9,8;-12,14]
>>A+B
ans =
     15    6
     -2   17
```

Array addition and subtraction are associative and commutative. For addition these properties mean that

$$(\mathbf{A} + \mathbf{B}) + \mathbf{C} = \mathbf{A} + (\mathbf{B} + \mathbf{C}) \qquad (2.3\text{–}2)$$

$$\mathbf{A} + \mathbf{B} + \mathbf{C} = \mathbf{B} + \mathbf{C} + \mathbf{A} = \mathbf{A} + \mathbf{C} + \mathbf{B} \qquad (2.3\text{–}3)$$

Array addition and subtraction require that both arrays be the same size. The only exception to this rule in MATLAB occurs when we add or subtract a *scalar* to or from an array. In this case the scalar is added or subtracted from each element in the array. Table 2.3–1 gives examples.

Element-by-Element Multiplication

MATLAB defines element-by-element multiplication only for arrays that are the same size. The definition of the product $x.*y$, where x and y each have n elements, is

$$x.*y = [x(1)y(1), \quad x(2)y(2) \; . \; . \; . \; , \; x(n)y(n)]$$

Table 2.3–1 Element-by-element operations

Symbol	Operation	Form	Example
+	Scalar-array addition	A + b	[6,3]+2=[8,5]
−	Scalar-array subtraction	A − b	[8,3]−5=[3,−2]
+	Array addition	A + B	[6,5]+[4,8]=[10,13]
−	Array subtraction	A − B	[6,5]−[4,8]=[2,−3]
.*	Array multiplication	A.*B	[3,5].*[4,8]=[12,40]
./	Array right division	A./B	[2,5]./[4,8]=[2/4,5/8]
.\	Array left division	A.\B	[2,5].\[4,8]=[2\4,5\8]
.^	Array exponentiation	A.^B	[3,5].^2=[3^2,5^2]
			2.^[3,5]=[2^3,2^5]
			[3,5].^[2,4]=[3^2,5^4]

if x and y are row vectors. For example, if

$$\mathbf{x} = [2, 4, -5] \qquad \mathbf{y} = [-7, 3, -8] \tag{2.3–4}$$

then $z = x.*y$ gives

$$\mathbf{z} = [2(-7), 4(3), -5(-8)] = [-14, 12, 40]$$

This type of multiplication is sometimes called *array* multiplication.

If u and v are column vectors, the result of $u.*v$ is a column vector.

Note that x' is a column vector with size 3×1 and thus does not have the same size as y, whose size is 1×3. Thus for the vectors x and y the operations $x'.*y$ and $y.*x'$ are not defined in MATLAB and will generate an error message. With element-by-element multiplication, it is important to remember that the dot (.) and the asterisk (*) form *one* symbol (.*). It might have been better to have defined a single symbol for this operation, but the developers of MATLAB were limited by the selection of symbols on the keyboard.

The generalization of array multiplication to arrays with more than one row or column is straightforward. Both arrays must be the same size. The array operations are performed between the elements in corresponding locations in the arrays. For example, the array multiplication operation A.*B results in a matrix C that has the same size as A and B and has the elements $c_{ij} = a_{ij}b_{ij}$. For example, if

$$\mathbf{A} = \begin{bmatrix} 11 & 5 \\ -9 & 4 \end{bmatrix} \qquad \mathbf{B} = \begin{bmatrix} -7 & 8 \\ 6 & 2 \end{bmatrix}$$

then $C = A.*B$ gives this result:

$$\mathbf{C} = \begin{bmatrix} 11(-7) & 5(8) \\ -9(6) & 4(2) \end{bmatrix} = \begin{bmatrix} -77 & 40 \\ -54 & 8 \end{bmatrix}$$

Vectors and Displacement

EXAMPLE 2.3–1

Suppose two divers start at the surface and establish the following coordinate system: x is to the west, y is to the north, and z is down. Diver 1 swims 55 ft west, 36 ft north, and then dives 25 ft. Diver 2 dives 15 ft, then swims east 20 ft and then north 59 ft. (a) Find the distance between diver 1 and the starting point. (b) How far in each direction must diver 1 swim to reach diver 2? How far in a straight line must diver 1 swim to reach diver 2?

■ Solution

(a) Using the xyz coordinates selected, the position of diver 1 is $\mathbf{r} = 55\mathbf{i} + 36\mathbf{j} + 25\mathbf{k}$, and the position of diver 2 is $\mathbf{r} = -20\mathbf{i} + 59\mathbf{j} + 15\mathbf{k}$. (Note that diver 2 swam east, which is in the negative x direction.) The distance from the origin of a point xyz is given by $\sqrt{x^2 + y^2 + z^2}$, that is, by the magnitude of the vector pointing from the origin to the point xyz. This distance is computed in the following session.

```
>>r = [55,36,25];w = [-20,59,15];
>>dist1 = sqrt(sum(r.*r))
dist1 =
   70.3278
```

The distance is approximately 70 ft. The distance could also have been computed from `norm(r)`.

(b) The location of diver 2 relative to diver 1 is given by the vector $\mathbf{v}$ pointing from diver 1 to diver 2. We can find this vector using vector subtraction: $\mathbf{v} = \mathbf{w} - \mathbf{r}$. Continue the above MATLAB session as follows:

```
>>v = w-r
v =
   -75      23     -10
>>dist2 = sqrt(sum(v.*v))
dist2 =
   79.0822
```

Thus to reach diver 2 by swimming along the coordinate directions, diver 1 must swim 75 ft east, 23 ft north, and 10 ft up. The straight-line distance between them is approximately 79 ft.

Vectorized Functions

The built-in MATLAB functions such as `sqrt(x)` and `exp(x)` automatically operate on array arguments to produce an array result the same size as the array argument `x`. Thus these functions are said to be *vectorized* functions.

Thus, when multiplying or dividing these functions, or when raising them to a power, you must use element-by-element operations if the arguments are arrays. For example, to compute $z = (e^y \sin x)\cos^2 x$, you must type `z = exp(y).* sin(x).*(cos(x)).^2`. Obviously, you will get an error message if the size of `x` is not the same as the size of `y`. The result `z` will have the same size as `x` and `y`.

EXAMPLE 2.3–2   Aortic Pressure Model

The following equation is a specific case of one model used to describe the blood pressure in the aorta during systole (the period following the closure of the heart's aortic valve). The variable t represents time in seconds, and the dimensionless variable y represents the pressure difference across the aortic valve, normalized by a constant reference pressure.

$$y(t) = e^{-8t} \sin\left(9.7t + \frac{\pi}{2}\right)$$

Plot this function for $t \geq 0$.

■ **Solution**

Note that if t is a vector, the MATLAB functions `exp(-8*t)` and `sin(9.7*t+pi/2)` will also be vectors the same size as t. Thus we must use element-by-element multiplication to compute $y(t)$.

We must decide on the proper spacing to use for the vector t and its upper limit. The sine function $\sin(9.7t + \pi/2)$ oscillates with a frequency of 9.7 rad/sec, which is $9.7/(2\pi) = 1.5$ Hz. Thus its period is $1/1.5 = 2/3$ sec. The spacing of t should be a small fraction of the period in order to generate enough points to plot the curve. Thus we select a spacing of 0.003 to give approximately 200 points per period.

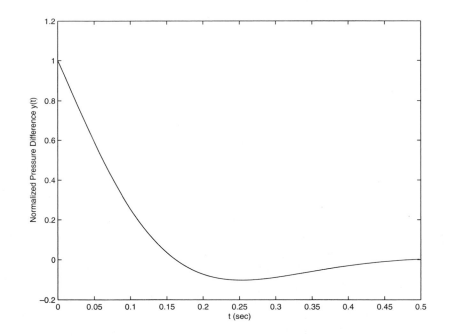

Figure 2.3–1 Aortic pressure response for Example 2.3–2.

The amplitude of the sine wave decays with time because the sine is multiplied by the decaying exponential e^{-8t}. The exponential's initial value is $e^0 = 1$, and it will be 2 percent of its initial value at $t = 0.5$ (because $e^{-8\,(0.5)} = 0.02$). Thus we select the upper limit of t to be 0.5. The session is

```
>>t = 0:0.003:0.5;
>>y = exp(-8*t).*sin(9.7*t+pi/2);
>>plot(t,y),xlabel('t (sec)'), . . .
    ylabel('Normalized Pressure Difference y(t)')
```

The plot is shown in Figure 2.3–1. Note that we do not see much of an oscillation despite the presence of a sine wave. This is so because the period of the sine wave is greater than the time it takes for the exponential e^{-8t} to become essentially zero.

Element-by-Element Division

The definition of element-by-element division, also called array division, is similar to the definition of array multiplication except, of course, that the elements of one array are divided by the elements of the other array. Both arrays must be the same size. The symbol for array right division is ./. For example, if

$$\mathbf{x} = [8, 12, 15] \qquad \mathbf{y} = [-2, 6, 5]$$

then z = x./y gives

$$\mathbf{z} = [8/(-2), 12/6, 15/5] = [-4, 2, 3]$$

Also, if

$$\mathbf{A} = \begin{bmatrix} 24 & 20 \\ -9 & 4 \end{bmatrix} \qquad \mathbf{B} = \begin{bmatrix} -4 & 5 \\ 3 & 2 \end{bmatrix}$$

then C = A./B gives

$$\mathbf{C} = \begin{bmatrix} 24/(-4) & 20/5 \\ -9/3 & 4/2 \end{bmatrix} = \begin{bmatrix} -6 & 4 \\ -3 & 2 \end{bmatrix}$$

The array left division operator (.\ ) is defined to perform element-by-element division using left division. Refer to Table 2.3–1 for examples. Note that A.\B is not equivalent to A./B.

EXAMPLE 2.3–3   Transportation Route Analysis

The following table gives data for the distance traveled along five truck routes and the corresponding time required to traverse each route. Use the data to compute the average speed required to drive each route. Find the route that has the highest average speed.

	1	2	3	4	5
Distance (mi)	560	440	490	530	370
Time (hr)	10.3	8.2	9.1	10.1	7.5

■ Solution

For example, the average speed on the first route is $560/10.3 = 54.4$ mi/hr. First we define the row vectors d and t from the distance and time data. Then, to find the average speed on each route using MATLAB, we use array division. The session is

```
>>d = [560, 440, 490, 530, 370]
>>t = [10.3, 8.2, 9.1, 10.1, 7.5]
>>speed = d./t
speed =
   54.3689    53.6585    53.8462    52.4752    49.3333
```

The results are in miles per hour. Note that MATLAB displays more significant figures than is justified by the three-significant-figure accuracy of the given data, so we should round the results to three significant figures before using them.

To find the highest average speed and the corresponding route, continue the session as follows:

```
>>[highest_speed, route] = max(speed)
highest_speed =
   54.3689
route =
   1
```

The first route has the highest speed.

If we did not need the speeds for every route, we could have solved this problem by combining two lines as follows: [highest_speed, route] = max(d./t).

Element-by-Element Exponentiation

MATLAB enables us not only to raise arrays to powers but also to raise scalars and arrays to *array* powers. To perform exponentiation on an element-by-element basis, we must use the .^ symbol. For example, if $x = [3, 5, 8]$, then typing $x.^3$ produces the array $[3^3, 5^3, 8^3] = [27, 125, 512]$. If $x = 0:2:6$, then typing $x.^2$ returns the array $[0^2, 2^2, 4^2, 6^2] = [0, 4, 16, 36]$. If

$$\mathbf{A} = \begin{bmatrix} 4 & -5 \\ 2 & 3 \end{bmatrix}$$

then B = A.^3 gives this result:

$$\mathbf{B} = \begin{bmatrix} 4^3 & (-5)^3 \\ 2^3 & 3^3 \end{bmatrix} = \begin{bmatrix} 64 & -125 \\ 8 & 27 \end{bmatrix}$$

We can raise a scalar to an array power. For example, if p = [2, 4, 5], then typing 3.^p produces the array $[3^2, 3^4, 3^5]$ = [9, 81, 243]. This example illustrates a common situation in which it helps to remember that .^ is a *single* symbol; the dot in 3.^p is not a decimal point associated with the number 3. The following operations, with the value of p given here, are equivalent and give the correct answer:

```
3.^p
3.0.^p
3..^p
(3).^p
3.^[2,4,5]
```

With array exponentiation, the power may be an array if the base is a scalar or if the power's dimensions are the same as the base dimensions. For example if, x = [1,2,3] and y = [2,3,4], then y.^x gives the answer 2964. If A = [1,2; 3,4], then 2.^A gives the array [2,4;8,16].

Test Your Understanding

T2.3–1 Given the matrices

$$\mathbf{A} = \begin{bmatrix} 21 & 27 \\ -18 & 8 \end{bmatrix} \qquad \mathbf{B} = \begin{bmatrix} -7 & -3 \\ 9 & 4 \end{bmatrix}$$

find (a) their array product, (b) their array right division (**A** divided by **B**), and (c) **B** raised to the third power element by element.
(Answers: (a) [-147, -81; -162, 32], (b) [-3, -9; -2, 2], and (c) [-343, -27; 729, 64].)

Current and Power Dissipation in Resistors

EXAMPLE 2.3–4

The current i passing through an electrical resistor having a voltage v across it is given by Ohm's law, $i = v/R$, where R is the resistance. The power dissipated in the resistor is given by v^2/R. The following table gives data for the resistance and voltage for five resistors. Use the data to compute (a) the current in each resistor and (b) the power dissipated in each resistor.

	1	2	3	4	5
$R\ (\Omega)$	10^4	2×10^4	3.5×10^4	10^5	2×10^5
$v\ (V)$	120	80	110	200	350

■ **Solution**

(a) First we define two row vectors, one containing the resistance values and one containing the voltage values. To find the current $i = v/R$ using MATLAB, we use array division. The session is

```
>>R = [10000, 20000, 35000, 100000, 200000];
>>v = [120, 80, 110, 200, 350];
>>current = v./R
current =
    0.0120   0.0040   0.0031   0.0020   0.0018
```

The results are in amperes and should be rounded to three significant figures because the voltage data contains only three significant figures.

(b) To find the power $P = v^2/R$, use array exponentiation and array division. The session continues as follows:

```
>>power = v.^2./R
power =
    1.4400   0.3200   0.3457   0.4000   0.6125
```

These numbers are the power dissipation in each resistor in watts. Note that the statement `v.^2./R` is equivalent to `(v.^2)./R`. Although the rules of precedence are unambiguous here, we can always put parentheses around quantities if we are unsure how MATLAB will interpret our commands.

EXAMPLE 2.3–5   A Batch Distillation Process

Consider a system for heating a liquid benzene/toluene solution to distill a pure benzene vapor. A particular batch distillation unit is charged initially with 100 mol of a 60 percent mol benzene/40 percent mol toluene mixture. Let L (mol) be the amount of liquid remaining in the still, and let x (mol B/mol) be the benzene mole fraction in the remaining liquid. Conservation of mass for benzene and toluene can be applied to derive the following relation [Felder, 1986].

$$L = 100 \left(\frac{x}{0.6}\right)^{0.625} \left(\frac{1-x}{0.4}\right)^{-1.625}$$

Determine what mole fraction of benzene remains when $L = 70$. Note that it is difficult to solve this equation directly for x. Use a plot of x versus L to solve the problem.

■ **Solution**

This equation involves both array multiplication and array exponentiation. Note that MATLAB enables us to use decimal exponents to evaluate L. It is clear that L must be in the range $0 \leq L \leq 100$; however, we do not know the range of x, except that

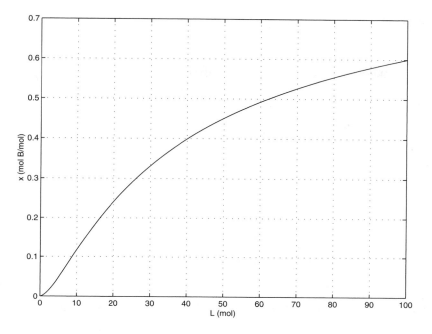

Figure 2.3–2 Plot for Example 2.3–5.

$x \geq 0$. Therefore, we must make a few guesses for the range of x, using a session like the following. We find that $L > 100$ if $x > 0.6$, so we choose x = 0:0.001:0.6. We use the `ginput` function to find the value of x corresponding to $L = 70$.

```
>>x = 0:0.001:0.6;
>>L = 100*(x/0.6).^(0.625).*((1-x)/0.4).^(-1.625);
>>plot(L,x),grid,xlabel('L(mol)'),ylabel('x (mol B/mol)'), ...
  [L,x] = ginput(1)
```

The plot is shown in Figure 2.3–2. The answer is $x = 0.52$ if $L = 70$. The plot shows that the remaining liquid becomes leaner in benzene as the liquid amount becomes smaller. Just before the still is empty ($L = 0$), the liquid is pure toluene.

2.4 Matrix Operations

Matrix addition and subtraction are identical to element-by-element addition and subtraction. The corresponding matrix elements are summed or subtracted. However, matrix multiplication and division are not the same as element-by-element multiplication and division.

Multiplication of Vectors

Recall that vectors are simply matrices with one row or one column. Thus matrix multiplication and division procedures apply to vectors as well, and we will introduce matrix multiplication by considering the vector case first.

The *vector dot product* $\mathbf{u} \cdot \mathbf{w}$ of the vectors $\mathbf{u}$ and $\mathbf{w}$ is a scalar and can be thought of as the perpendicular projection of $\mathbf{u}$ onto $\mathbf{w}$. It can be computed from $|\mathbf{u}||\mathbf{w}| \cos \theta$, where θ is the angle between the two vectors and $|\mathbf{u}|$, $|\mathbf{w}|$ are the magnitudes of the vectors. Thus if the vectors are parallel and in the same direction, $\theta = 0$ and $\mathbf{u} \cdot \mathbf{w} = |\mathbf{u}||\mathbf{w}|$. If the vectors are perpendicular, $\theta = 90°$ and thus $\mathbf{u} \cdot \mathbf{w} = 0$. Because the unit vectors $\mathbf{i}$, $\mathbf{j}$, and $\mathbf{k}$ have unit length,

$$\mathbf{i} \cdot \mathbf{i} = \mathbf{j} \cdot \mathbf{j} = \mathbf{k} \cdot \mathbf{k} = 1 \qquad (2.4\text{--}1)$$

Because the unit vectors are perpendicular,

$$\mathbf{i} \cdot \mathbf{j} = \mathbf{i} \cdot \mathbf{k} = \mathbf{j} \cdot \mathbf{k} = 0 \qquad (2.4\text{--}2)$$

Thus the vector dot product can be expressed in terms of unit vectors as

$$\mathbf{u} \cdot \mathbf{w} = (u_1\mathbf{i} + u_2\mathbf{j} + u_3\mathbf{k}) \cdot (w_1\mathbf{i} + w_2\mathbf{j} + w_3\mathbf{k})$$

Carrying out the multiplication algebraically and using the properties given by (2.4–1) and (2.4–2), we obtain

$$\mathbf{u} \cdot \mathbf{w} = u_1w_1 + u_2w_2 + u_3w_3$$

The *matrix* product of a *row* vector $\mathbf{u}$ with a *column* vector $\mathbf{w}$ is defined in the same way as the vector dot product; the result is a scalar that is the sum of the products of the corresponding vector elements; that is,

$$[u_1 \quad u_2 \quad u_3] \begin{bmatrix} w_1 \\ w_2 \\ w_3 \end{bmatrix} = u_1w_1 + u_2w_2 + u_3w_3$$

if each vector has three elements. Thus the result of multiplying a 1×3 vector by a 3×1 vector is a 1×1 array, that is, a scalar. This definition applies to vectors having any number of elements, as long as both vectors have the same number of elements.

Thus the result of multiplying a $1 \times n$ vector by an $n \times 1$ vector is a 1×1 array, that is, a scalar.

<table>
<tr><td>EXAMPLE 2.4–1</td><td style="text-align:right">Miles Traveled</td></tr>
</table>

Table 2.4–1 gives the speed of an aircraft on each leg of a certain trip and the time spent on each leg. Compute the miles traveled on each leg and the total miles traveled.

■ **Solution**

We can define a row vector s containing the speeds and a row vector t containing the times for each leg. Thus s = [200, 250, 400, 300] and t = [2, 5, 3, 4].

Table 2.4–1 Aircraft speeds and times per leg

	Leg			
	1	2	3	4
Speed (mi/hr)	200	250	400	300
Time (hr)	2	5	3	4

To find the miles traveled on each leg, we multiply the speed by the time. To do so, we use the MATLAB symbol `.*`, which *specifies* the multiplication `s.*t` to produce the row vector whose elements are the products of the corresponding elements in `s` and `t`:

$$s.*t = [200(2), 250(5), 400(3), 300(4)] = [400, 1250, 1200, 1200]$$

This vector contains the miles traveled by the aircraft on each leg of the trip.

To find the total miles traveled, we use the matrix product, denoted by `s*t'`. In this definition the product is the *sum* of the individual element products; that is,

$$s*t' = [200(2) + 250(5) + 400(3) + 300(4)] = 4050$$

These two examples illustrate the difference between *array* multiplication `s.*t` and *matrix* multiplication `s*t'`.

Vector-Matrix Multiplication

Not all matrix products are scalars. To generalize the preceding multiplication to a column vector multiplied by a matrix, think of the matrix as being composed of row vectors. The scalar result of each row-column multiplication forms an element in the result, which is a column vector. For example:

$$\begin{bmatrix} 2 & 7 \\ 6 & -5 \end{bmatrix} \begin{bmatrix} 3 \\ 9 \end{bmatrix} = \begin{bmatrix} 2(3) + 7(9) \\ 6(3) - 5(9) \end{bmatrix} = \begin{bmatrix} 69 \\ -27 \end{bmatrix} \tag{2.4–3}$$

Thus the result of multiplying a 2×2 matrix by a 2×1 vector is a 2×1 array, that is, a column vector. Note that the definition of multiplication requires that the number of columns in the matrix be equal to the number of rows in the vector. In general, the product $\mathbf{Ax}$, where $\mathbf{A}$ has p columns, is defined only if $\mathbf{x}$ has p rows. If $\mathbf{A}$ has m rows and $\mathbf{x}$ is a column vector, the result of $\mathbf{Ax}$ is a column vector with m rows.

Matrix-Matrix Multiplication

We can expand this definition of multiplication to include the product of two matrices $\mathbf{AB}$. The number of columns in $\mathbf{A}$ must equal the number of rows in $\mathbf{B}$. The row-column multiplications form column vectors, and these column vectors form the matrix result. The product **AB** has the same number of rows as **A** and the same number of columns as **B**. For example,

$$\begin{bmatrix} 6 & -2 \\ 10 & 3 \\ 4 & 7 \end{bmatrix} \begin{bmatrix} 9 & 8 \\ -5 & 12 \end{bmatrix} = \begin{bmatrix} (6)(9) + (-2)(-5) & (6)(8) + (-2)(12) \\ (10)(9) + (3)(-5) & (10)(8) + (3)(12) \\ (4)(9) + (7)(-5) & (4)(8) + (7)(12) \end{bmatrix}$$

$$= \begin{bmatrix} 64 & 24 \\ 75 & 116 \\ 1 & 116 \end{bmatrix} \tag{2.4-4}$$

Use the operator * to perform matrix multiplication in MATLAB. The following MATLAB session shows how to perform the matrix multiplication shown in (2.4–4).

```
>>A = [6,-2;10,3;4,7];
>>B = [9,8;-5,12];
>>A*B
```

Element-by-element multiplication is defined for the following product:

$$[3 \quad 1 \quad 7][4 \quad 6 \quad 5] = [12 \quad 6 \quad 35]$$

However, this product is *not* defined for *matrix* multiplication, because the first matrix has three columns, but the second matrix does not have three rows. Thus if we were to type [3, 1, 7]*[4, 6, 5] in MATLAB, we would receive an error message.

The following product is defined in matrix multiplication and gives the result shown:

$$\begin{bmatrix} x_1 \\ x_2 \\ x_3 \end{bmatrix} [y_1 \quad y_2 \quad y_3] = \begin{bmatrix} x_1 y_1 & x_1 y_2 & x_1 y_3 \\ x_2 y_1 & x_2 y_2 & x_2 y_3 \\ x_3 y_1 & x_3 y_2 & x_3 y_3 \end{bmatrix}$$

The following product is also defined:

$$[10 \quad 6] \begin{bmatrix} 7 & 4 \\ 5 & 2 \end{bmatrix} = [10(7) + 6(5) \quad 10(4) + 6(2)] = [100 \quad 52]$$

Evaluating Multivariable Functions

To evaluate a function of two variables, say, $z = f(x, y)$, for the values $x = x_1, x_2, \ldots, x_m$ and $y = y_1, y_2, \ldots, y_n$, define the $m \times n$ matrices:

$$\mathbf{x} = \begin{bmatrix} x_1 & x_1 & \cdots & x_1 \\ x_2 & x_2 & \cdots & x_2 \\ \vdots & \vdots & \vdots & \vdots \\ x_m & x_m & \cdots & x_m \end{bmatrix} \qquad \mathbf{y} = \begin{bmatrix} y_1 & y_2 & \cdots & y_n \\ y_1 & y_2 & \cdots & y_n \\ \vdots & \vdots & \vdots & \vdots \\ y_1 & y_2 & \cdots & y_n \end{bmatrix}$$

When the function $z = f(x, y)$ is evaluated in MATLAB using array operations, the resulting $m \times n$ matrix $\mathbf{z}$ has the elements $z_{ij} = f(x_i, y_j)$. We can extend this technique to functions of more than two variables by using multidimensional arrays.

Height versus Velocity

EXAMPLE 2.4–2

The maximum height h achieved by an object thrown with a speed v at an angle θ to the horizontal, neglecting drag, is

$$h = \frac{v^2 \sin^2 \theta}{2g}$$

Create a table showing the maximum height for the following values of v and θ:

$$v = 10, 12, 14, 16, 18, 20 \text{ m/s} \qquad \theta = 50°, 60°, 70°, 80°$$

The rows in the table should correspond to the speed values, and the columns should correspond to the angles.

■ Solution

The program is shown below.

```
g = 9.8; v = 10:2:20;
theta = 50:10:80;
h = (v'.^2)*(sind(theta).^2)/(2*g);
table = [0, theta; v', h]
```

The arrays v and theta contain the given velocities and angles. The array v is 1×6 and the array theta is 1×4. Thus the term v'.^2 is a 6×1 array, and the term sind(theta).^2 is a 1×4 array. The product of these two arrays, h, is a matrix product and is a $(6 \times 1)(1 \times 4) = (6 \times 4)$ matrix.

The array [0, theta] is 1×5 and the array [v', h] is 6×5, so the matrix table is 7×5. The following table shows the matrix table rounded to one decimal place. From this table we can see that the maximum height is 8.8 m if $v = 14$ m/s and $\theta = 70°$.

0	50	60	70	80
10	3.0	3.8	4.5	4.9
12	4.3	5.5	6.5	7.1
14	5.9	7.5	8.8	9.7
16	7.7	9.8	11.5	12.7
18	9.7	12.4	14.6	16.0
20	12.0	15.3	18.0	19.8

Test Your Understanding

T2.4–1 Use MATLAB to compute the dot product of the following vectors:

$$\mathbf{u} = 6\mathbf{i} - 8\mathbf{j} + 3\mathbf{k}$$
$$\mathbf{w} = 5\mathbf{i} + 3\mathbf{j} - 4\mathbf{k}$$

Check your answer by hand. (Answer: -6.)

T2.4–2 Use MATLAB to show that

$$
\begin{bmatrix} 7 & 4 \\ -3 & 2 \\ 5 & 9 \end{bmatrix} \begin{bmatrix} 1 & 8 \\ 7 & 6 \end{bmatrix} = \begin{bmatrix} 35 & 80 \\ 11 & -12 \\ 68 & 94 \end{bmatrix}
$$

EXAMPLE 2.4–3	Manufacturing Cost Analysis

Table 2.4–2 shows the hourly cost of four types of manufacturing processes. It also shows the number of hours required of each process to produce three different products. Use matrices and MATLAB to solve the following. (a) Determine the cost of each process to produce 1 unit of product 1. (b) Determine the cost to make 1 unit of each product. (c) Suppose we produce 10 units of product 1, 5 units of product 2, and 7 units of product 3. Compute the total cost.

■ **Solution**

(a) The basic principle we can use here is that cost equals the hourly cost times the number of hours required. For example, the cost of using the lathe for product 1 is ($10/hr)(6 hr) = $60, and so forth for the other three processes. If we define the row vector of hourly costs to be `hourly_costs` and define the row vector of hours required for product 1 to be `hours_1`, then we can compute the costs of each process for product 1 using *element-by-element* multiplication. In MATLAB the session is

```
>>hourly_cost = [10, 12, 14, 9];
>>hours_1 = [6, 2, 3, 4];
>>process_cost_1 = hourly_cost.*hours_1
process_cost_1 =
    60    24    42    36
```

These are the costs of each of the four processes to produce 1 unit of product 1.

(b) To compute the total cost of 1 unit of product 1, we can use the vectors `hourly_costs` and `hours_1` but apply *matrix* multiplication instead of element-by-element multiplication, because matrix multiplication sums the individual products. The matrix multiplication gives

Table 2.4–2 Cost and time data for manufacturing processes

Process	Hourly cost ($)	Hours required to produce one unit		
		Product 1	**Product 2**	**Product 3**
Lathe	10	6	5	4
Grinding	12	2	3	1
Milling	14	3	2	5
Welding	9	4	0	3

$$[10 \ 12 \ 14 \ 9] \begin{bmatrix} 6 \\ 2 \\ 3 \\ 4 \end{bmatrix} = 10(6) + 12(2) + 14(3) + 9(4) = 162$$

We can perform similar multiplication for products 2 and 3, using the data in the table. For product 2:

$$[10 \ 12 \ 14 \ 9] \begin{bmatrix} 5 \\ 3 \\ 2 \\ 0 \end{bmatrix} = 10(5) + 12(2) + 14(3) + 9(0) = 114$$

For product 3:

$$[10 \ 12 \ 14 \ 9] \begin{bmatrix} 4 \\ 1 \\ 5 \\ 3 \end{bmatrix} = 10(4) + 12(1) + 14(5) + 9(3) = 149$$

These three operations could have been accomplished in one operation by defining a matrix whose columns are formed by the data in the last three columns of the table:

$$[10 \ 12 \ 14 \ 9] \begin{bmatrix} 6 & 5 & 4 \\ 2 & 3 & 1 \\ 3 & 2 & 5 \\ 4 & 0 & 3 \end{bmatrix} = \begin{bmatrix} 60 + 24 + 42 + 36 \\ 50 + 36 + 28 + 0 \\ 40 + 12 + 70 + 27 \end{bmatrix} = [162 \ 114 \ 149]$$

In MATLAB the session continues as follows. Remember that we must use the transpose operation to convert the row vectors into column vectors.

```
>>hours_2 = [5, 3, 2, 0];
>>hours_3 = [4, 1, 5, 3];
>>unit_cost = hourly_cost*[hours_1', hours_2', hours_3']
unit_cost =
    162    114    149
```

Thus the costs to produce 1 unit each of products 1, 2, and 3 are $162, $114, and $149, respectively.

(c) To find the total cost to produce 10, 5, and 7 units, respectively, we can use matrix multiplication:

$$[10 \ 5 \ 7] \begin{bmatrix} 162 \\ 114 \\ 149 \end{bmatrix} = 1620 + 570 + 1043 = 3233$$

In MATLAB the session continues as follows. Note the use of the transpose operator on the vector `unit_cost`.

```
>>units = [10, 5, 7];
>>total_cost = units*unit_cost'
total_cost =
   3233
```

The total cost is $3233.

The General Matrix Multiplication Case

We can state the general result for matrix multiplication as follows: Suppose **A** has dimension $m \times p$ and **B** has dimension $p \times q$. If **C** is the product **AB**, then **C** has dimension $m \times q$ and its elements are given by

$$c_{ij} = \sum_{k=1}^{p} a_{ik} b_{kj} \qquad (2.4\text{--}5)$$

for all $i = 1, 2, \ldots, m$ and $j = 1, 2, \ldots, q$. For the product to be defined, the matrices **A** and **B** must be *conformable;* that is, the number of *rows* in **B** must equal the number of *columns* in **A**. The product has the same number of rows as **A** and the same number of columns as **B**.

Matrix multiplication does not have the commutative property; that is, in general, $\mathbf{AB} \neq \mathbf{BA}$. Reversing the order of matrix multiplication is a common and easily made mistake.

The associative and distributive properties hold for matrix multiplication. The associative property states that

$$\mathbf{A(B + C)} = \mathbf{AB} + \mathbf{AC} \qquad (2.4\text{--}6)$$

The distributive property states that

$$\mathbf{(AB)C} = \mathbf{A(BC)} \qquad (2.4\text{--}7)$$

Applications to Cost Analysis

Project cost data stored in tables must often be analyzed in several ways. The elements in MATLAB matrices are similar to the cells in a spreadsheet, and MATLAB can perform many spreadsheet-type calculations for analyzing such tables.

EXAMPLE 2.4–4	Product Cost Analysis

Table 2.4–3 shows the costs associated with a certain product, and Table 2.4–4 shows the production volume for the four quarters of the business year. Use MATLAB to find the quarterly costs for materials, labor, and transportation; the total material, labor, and transportation costs for the year; and the total quarterly costs.

Table 2.4–3 Product costs

Product	Unit costs ($ \times 10^3$)		
	Materials	Labor	Transportation
1	6	2	1
2	2	5	4
3	4	3	2
4	9	7	3

Table 2.4–4 Quarterly production volume

Product	Quarter 1	Quarter 2	Quarter 3	Quarter 4
1	10	12	13	15
2	8	7	6	4
3	12	10	13	9
4	6	4	11	5

■ **Solution**

The costs are the product of the unit cost and the production volume. Thus we define two matrices: U contains the unit costs in Table 2.4–3 in thousands of dollars, and P contains the quarterly production data in Table 2.4–4.

```
>>U = [6, 2, 1;2, 5, 4;4, 3, 2;9, 7, 3];
>>P = [10, 12, 13, 15;8, 7, 6, 4;12, 10, 13, 9;6, 4, 11, 5];
```

Note that if we multiply the first column in U by the first column in P, we obtain the total materials cost for the first quarter. Similarly, multiplying the first column in U by the *second* column in P gives the total materials cost for the *second* quarter. Also, multiplying the second column in U by the first column in P gives the total *labor* cost for the first quarter, and so on. Extending this pattern, we can see that we must multiply the *transpose* of U by P. This multiplication gives the cost matrix C.

```
>>C = U'*P
```

The result is

$$C = \begin{bmatrix} 178 & 162 & 241 & 179 \\ 138 & 117 & 172 & 112 \\ 84 & 72 & 96 & 64 \end{bmatrix}$$

Each column in **C** represents one quarter. The total first-quarter cost is the sum of the elements in the first column, the second-quarter cost is the sum of the second column, and so on. Thus because the sum command sums the columns of a matrix, the quarterly costs are obtained by typing

```
>>Quarterly_Costs = sum(C)
```

The resulting vector, containing the quarterly costs in thousands of dollars, is [400 351 509 355]. Thus the total costs in each quarter are $400,000; $351,000; $509,000; and $355,000.

The elements in the first row of **C** are the material costs for each quarter; the elements in the second row are the labor costs, and those in the third row are the transportation costs. Thus to find the total material costs, we must sum across the first row of **C**. Similarly, the total labor and total transportation costs are the sums across the second and third rows of **C**. Because the sum command sums *columns,* we must use the transpose of **C**. Thus we type the following:

```
>>Category_Costs = sum(C')
```

The resulting vector, containing the category costs in thousands of dollars, is [760 539 316]. Thus the total material costs for the year are $760 000; the labor costs are $539 000; and the transportation costs are $316 000.

We displayed the matrix **C** only to interpret its structure. If we need not display **C**, the entire analysis would consist of only four command lines.

```
>>U = [6, 2, 1;2, 5, 4;4, 3, 2;9, 7, 3];
>>P = [10, 12, 13, 15;8, 7, 6, 4;12, 10, 13, 9;6, 4, 11, 5];
>>Quarterly_Costs = sum(U'*P)
Quarterly_Costs =
    400    351    509    355
>>Category_Costs = sum((U'*P)')
Category_Costs =
    760    539    316
```

This example illustrates the compactness of MATLAB commands.

Special Matrices

NULL MATRIX

IDENTITY MATRIX

Two exceptions to the noncommutative property are the *null matrix,* denoted by **0**, and the *identity,* or *unity, matrix,* denoted by **I**. The null matrix contains all zeros and is not the same as the empty matrix [ ], which has no elements. The identity matrix is a square matrix whose diagonal elements are all equal to 1, with the remaining elements equal to 0. For example, the 2 × 2 identity matrix is

$$\mathbf{I} = \begin{bmatrix} 1 & 0 \\ 0 & 1 \end{bmatrix}$$

These matrices have the following properties:

$$\mathbf{0A} = \mathbf{A0} = \mathbf{0}$$
$$\mathbf{IA} = \mathbf{AI} = \mathbf{A}$$

MATLAB has specific commands to create several special matrices. Type help specmat to see the list of special matrix commands; also check Table 2.4–5. The identity matrix **I** can be created with the eye(n) command, where n is the desired dimension of the matrix. To create the 2 × 2 identity matrix, you type eye(2). Typing eye(size(A)) creates an identity matrix having the same dimension as the matrix **A**.

I = eye(4)

x = I*A

Table 2.4–5 Special matrices

Command	Description
eye(n)	Creates an $n \times n$ identity matrix.
eye(size(A))	Creates an identity matrix the same size as the matrix **A**.
ones(n)	Creates an $n \times n$ matrix of 1s.
ones(m,n)	Creates an $m \times n$ array of 1s.
ones(size(A))	Creates an array of 1s the same size as the array **A**.
zeros(n)	Creates an $n \times n$ matrix of 0s.
zeros(m,n)	Creates an $m \times n$ array of 0s.
zeros(size(A))	Creates an array of 0s the same size as the array **A**.

Sometimes we want to initialize a matrix to have all zero elements. The zeros command creates a matrix of all zeros. Typing zeros(n) creates an $n \times n$ matrix of zeros, whereas typing zeros(m,n) creates an $m \times n$ matrix of zeros, as will typing A(m,n) = 0. Typing zeros(size(A)) creates a matrix of all zeros having the same dimension as the matrix **A**. This type of matrix can be useful for applications in which we do not know the required dimension ahead of time. The syntax of the ones command is the same, except that it creates arrays filled with 1s.

For example, to create and plot the function

$$f(x) = \begin{cases} 10 & 0 \le x \le 2 \\ 0 & 2 < x < 5 \\ -3 & 5 \le x \le 7 \end{cases}$$

the script file is

```
x1 = 0:0.01:2;
f1 = 10*ones(size(x1));
x2 = 2.01:0.01:4.99;
f2 = zeros(size(x2));
x3 = 5:0.01:7;
f3 = -3*ones(size(x3));
f = [f1, f2, f3];
x = [x1, x2, x3];
plot(x,f),xlabel('x'),ylabel('y')
```

(Consider what the plot would look like if the command plot(x,f) were replaced with the command plot(x1,f1,x2,f2,x3,f3).)

Matrix Division and Linear Algebraic Equations

Matrix division uses both the right and left division operators, / and \, for various applications, a principal one being the solution of sets of linear algebraic equations. Chapter 8 covers a related topic, the matrix inverse.

You can use the left division operator (\\) in MATLAB to solve sets of linear algebraic equations. For example, consider the set

$$6x + 12y + 4z = 70$$
$$7x - 2y + 3z = 5$$
$$2x + 8y - 9z = 64$$

To solve such sets in MATLAB you must create two arrays; we will call them A and B. The array A has as many rows as there are equations and as many columns as there are variables. The rows of A must contain the coefficients of x, y, and z in that order. In this example, the first row of A must be 6, 12, 4; the second row must be 7, -2, 3; and the third row must be 2, 8, -9. The array B contains the constants on the right-hand side of the equation; it has one column and as many rows as there are equations. In this example, the first row of B is 70, the second is 5, and the third is 64. The solution is obtained by typing A\\B. The session is

```
>>A = [6,12,4;7,-2,3;2,8,-9];
>>B = [70;5;64];
>>Solution = A\B
Solution =
     3
     5
    -2
```

The solution is $x = 3$, $y = 5$, and $z = -2$.

LEFT DIVISION METHOD

The *left division method* works fine when the equation set has a unique solution. To learn how to deal with problems having a nonunique solution (or perhaps no solution at all!), see Chapter 8.

Test Your Understanding

T2.4–3 Use MATLAB to solve the following set of equations.

$$6x - 4y + 8z = 112$$
$$-5x - 3y + 7z = 75$$
$$14x + 9y - 5z = -67$$

(Answer: $x = 2$, $y = -5$, $z = 10$.)

Matrix Exponentiation

Raising a matrix to a power is equivalent to repeatedly multiplying the matrix by itself, for example, $\mathbf{A}^2 = \mathbf{AA}$. This process requires the matrix to have the same number of rows as columns; that is, it must be a *square* matrix. MATLAB uses the symbol ^ for matrix exponentiation. To find $\mathbf{A}^2$, type A^2.

We can raise a scalar n to a matrix power $\mathbf{A}$, if $\mathbf{A}$ is square, by typing `n^A`, but the applications for such a procedure are in advanced courses. However, raising a matrix to a matrix power—that is, $\mathbf{A^B}$—is not defined, even if $\mathbf{A}$ and $\mathbf{B}$ are square.

Special Products

Many applications in physics and engineering use the cross product and dot product; for example, calculations to compute moments and force components use these special products. If $\mathbf{A}$ and $\mathbf{B}$ are vectors with three elements, the cross product command `cross(A,B)` computes the three-element vector that is the cross product $\mathbf{A} \times \mathbf{B}$. If $\mathbf{A}$ and $\mathbf{B}$ are $3 \times n$ matrices, `cross(A,B)` returns a $3 \times n$ array whose columns are the cross products of the corresponding columns in the $3 \times n$ arrays $\mathbf{A}$ and $\mathbf{B}$. For example, the moment $\mathbf{M}$ with respect to a reference point O due to the force $\mathbf{F}$ is given by $\mathbf{M} = \mathbf{r} \times \mathbf{F}$, where $\mathbf{r}$ is the position vector from the point O to the point where the force $\mathbf{F}$ is applied. To find the moment in MATLAB, you type `M = cross(r,F)`.

The dot product command `dot(A,B)` computes a row vector of length n whose elements are the dot products of the corresponding columns of the $m \times n$ arrays $\mathbf{A}$ and $\mathbf{B}$. To compute the component of the force $\mathbf{F}$ along the direction given by the vector $\mathbf{r}$, you type `dot(F,r)`.

2.5 Polynomial Operations Using Arrays

MATLAB has some convenient tools for working with polynomials. Type `help polyfun` for more information on this category of commands. We will use the following notation to describe a polynomial:

$$f(x) = a_1 x^n + a_2 x^{n-1} + a_3 x^{n-2} + \cdots + a_{n-1} x^2 + a_n x + a_{n+1}$$

We can describe a polynomial in MATLAB with a row vector whose elements are the polynomial's coefficients, *starting with the coefficient of the highest power of x*. This vector is $[a_1, a_2, a_3, \ldots, a_{n-1}, a_n, a_{n+1}]$. For example, the vector `[4,-8,7,-5]` represents the polynomial $4x^3 - 8x^2 + 7x - 5$.

Polynomial roots can be found with the `roots(a)` function, where a is the array containing the polynomial coefficients. For example, to obtain the roots of $x^3 + 12x^2 + 45x + 50 = 0$, you type `y = roots([1,12,45,50])`. The answer (`y`) is a *column* array containing the values $-2, -5, -5$.

The `poly(r)` function computes the coefficients of the polynomial whose roots are specified by the array `r`. The result is a *row* array that contains the polynomial's coefficients. For example, to find the polynomial whose roots are 1 and $3 \pm 5i$, the session is

```
>>p = poly([1,3+5i, 3-5i])
p =
     1     -7     40     -34
```

Thus the polynomial is $x^3 - 7x^2 + 40x - 34$.

Polynomial Addition and Subtraction

To add two polynomials, add the arrays that describe their coefficients. If the polynomials are of different degrees, add zeros to the coefficient array of the lower-degree polynomial. For example, consider

$$f(x) = 9x^3 - 5x^2 + 3x + 7$$

whose coefficient array is f = [9,−5,3,7] and

$$g(x) = 6x^2 - x + 2$$

whose coefficient array is g = [6,-1,2]. The degree of $g(x)$ is 1 less that of $f(x)$. Therefore, to add $f(x)$ and $g(x)$, we append one zero to g to "fool" MATLAB into thinking $g(x)$ is a third-degree polynomial. That is, we type g = [0 g] to obtain [0,6,-1,2] for g. This vector represents $g(x) = 0x^3 + 6x^2 - x + 2$. To add the polynomials, type h = f+g. The result is h = [9,1,2,9], which corresponds to $h(x) = 9x^3 + x^2 + 2x + 9$. Subtraction is done in a similar way.

Polynomial Multiplication and Division

To multiply a polynomial by a scalar, simply multiply the coefficient array by that scalar. For example, $5h(x)$ is represented by [45,5,10,45].

Multiplication and division of polynomials are easily done with MATLAB. Use the conv function (it stands for "convolve") to multiply polynomials and use the deconv function (deconv stands for "deconvolve") to perform synthetic division. Table 2.5–1 summarizes these functions, as well as the poly, polyval, and roots functions.

Table 2.5–1 Polynomial functions

Command	Description
conv(a,b)	Computes the product of the two polynomials described by the coefficient arrays a and b. The two polynomials need not be of the same degree. The result is the coefficient array of the product polynomial.
[q,r] = deconv(num,den)	Computes the result of dividing a numerator polynomial, whose coefficient array is num, by a denominator polynomial represented by the coefficient array den. The quotient polynomial is given by the coefficient array q, and the remainder polynomial is given by the coefficient array r.
poly(r)	Computes the coefficients of the polynomial whose roots are specified by the vector r. The result is a *row* vector that contains the polynomial's coefficients arranged in descending order of power.
polyval(a,x)	Evaluates a polynomial at specified values of its independent variable x, which can be a matrix or a vector. The polynomial's coefficients of descending powers are stored in the array a. The result is the same size as x.
roots(a)	Computes the roots of a polynomial specified by the coefficient array a. The result is a *column* vector that contains the polynomial's roots.

The product of the polynomials $f(x)$ and $g(x)$ is

$$f(x)g(x) = (9x^3 - 5x^2 + 3x + 7)(6x^2 - x + 2)$$
$$= 54x^5 - 39x^4 + 41x^3 + 29x^2 - x + 14$$

Dividing $f(x)$ by $g(x)$ using synthetic division gives a quotient of

$$\frac{f(x)}{g(x)} = \frac{9x^3 - 5x^2 + 3x + 7}{6x^2 - x + 2} = 1.5x - 0.5833$$

with a remainder of $-0.5833x + 8.1667$. Here is the MATLAB session to perform these operations.

```
>>f = [9,-5,3,7];
>>g = [6,-1,2];
>>product = conv(f,g)
product =
     54    -39     41     29     -1     14
>>[quotient, remainder] = deconv(f,g)
quotient =
    1.5    -0.5833
remainder =
     0     0    -0.5833        8.1667
```

The `conv` and `deconv` functions do not require that the polynomials be of the same degree, so we did not have to fool MATLAB as we did when adding the polynomials.

Plotting Polynomials

The `polyval(a,x)` function evaluates a polynomial at specified values of its independent variable x, which can be a matrix or a vector. The polynomial's coefficient array is a. The result is the same size as x. For example, to evaluate the polynomial $f(x) = 9x^3 - 5x^2 + 3x + 7$ at the points $x = 0, 2, 4, \ldots, 10$, type

```
>>f = polyval([9,-5,3,7],[0:2:10]);
```

The resulting vector f contains six values that correspond to $f(0), f(2), f(4), \ldots, f(10)$.

The `polyval` function is very useful for plotting polynomials. To do this, you should define an array that contains many values of the independent variable x in order to obtain a smooth plot. For example, to plot the polynomial $f(x) = 9x^3 - 5x^2 + 3x + 7$ for $-2 \le x \le 5$, you type

```
>>x = -2:0.01:5;
>>polyval([9,-5,3,7], x);
>>plot(x,f),xlabel('x'),ylabel('f(x)'),grid
```

Polynomial derivatives and integrals are covered in Chapter 9.

Test Your Understanding

T2.5–1 Use MATLAB to obtain the roots of
$$x^3 + 13x^2 + 52x + 6 = 0$$
Use the `poly` function to confirm your answer.

T2.5–2 Use MATLAB to confirm that
$$(20x^3 - 7x^2 + 5x + 10)(4x^2 + 12x - 3)$$
$$= 80x^5 + 212x^4 - 124x^3 + 121x^2 + 105x - 30$$

T2.5–3 Use MATLAB to confirm that
$$\frac{12x^3 + 5x^2 - 2x + 3}{3x^2 - 7x + 4} = 4x + 11$$
with a remainder of $59x - 41$.

T2.5–4 Use MATLAB to confirm that
$$\frac{6x^3 + 4x^2 - 5}{12x^3 - 7x^2 + 3x + 9} = 0.7108$$
when $x = 2$.

T2.5–5 Plot the polynomial
$$y = x^3 + 13x^2 + 52x + 6$$
over the range $-7 \leq x \leq 1$.

EXAMPLE 2.5–1	Earthquake-Resistant Building Design

Buildings designed to withstand earthquakes must have natural frequencies of vibration that are not close to the oscillation frequency of the ground motion. A building's natural frequencies are determined primarily by the masses of its floors and by the lateral stiffness of its supporting columns (which act as horizontal springs). We can find these frequencies by solving for the roots of a polynomial called the structure's *characteristic* polynomial (characteristic polynomials are discussed further in Chapter 9). Figure 2.5–1 shows the exaggerated motion of the floors of a three-story building. For such a building, if each floor has a mass m and the columns have stiffness k, the polynomial is

$$(\alpha - f^2)[(2\alpha - f^2)^2 - \alpha^2] + \alpha^2 f^2 - 2\alpha^3$$

where $\alpha = k/4m\pi^2$ (models such as these are discussed in greater detail in [Palm, 2010]). The building's natural frequencies in cycles per second are the positive roots of this equation. Find the building's natural frequencies in cycles per second for the case where $m = 1000$ kg and $k = 5 \times 10^6$ N/m.

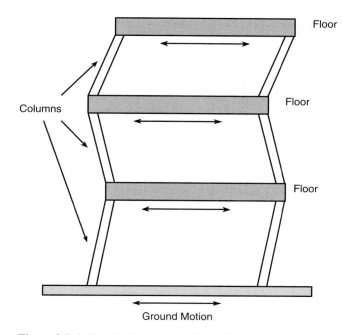

Floor

Columns

Floor

Floor

Ground Motion

Figure 2.5–1 Simple vibration model of a building subjected to ground motion.

■ Solution

The characteristic polynomial consists of sums and products of lower-degree polynomials. We can use this fact to have MATLAB do the algebra for us. The characteristic polynomial has the form

$$p_1\left(p_2^2 - \alpha^2\right) + p_3 = 0$$

where

$$p_1 = \alpha - f^2 \qquad p_2 = 2\alpha - f^2 \qquad p_3 = \alpha^2 f^2 - 2\alpha^3$$

The MATLAB script file is

```
k = 5e+6;m = 1000;
alpha = k/(4*m*pi^2);
p1 = [-1,0,alpha];
p2 = [-1,0,2*alpha];
p3 = [alpha^2,0,-2*alpha^3];
p4 = conv(p2,p2)-(0,0,0,0,alpha^2];
p5 = conv(p1,p4);
p6 = p5+[0,0,0,0,p3];
r = roots(p6)
```

The resulting positive roots and thus the frequencies, rounded to the nearest integer, are 20, 14, and 5 Hz.

2.6 Cell Arrays

The *cell array* is an array in which each element is a *bin,* or *cell,* which can contain an array. You can store different classes of arrays in a cell array, and you can group data sets that are related but have different dimensions. You access cell arrays using the same indexing operations used with ordinary arrays.

This is the only section in the text that uses cell arrays. Coverage of this section is therefore optional. Some more advanced MATLAB applications, such as those found in some of the toolboxes, do use cell arrays.

Creating Cell Arrays

CELL INDEXING

CONTENT INDEXING

You can create a cell array by using assignment statements or by using the `cell` function. You can assign data to the cells by using either *cell indexing* or *content indexing.* To use cell indexing, enclose in parentheses the cell subscripts on the left side of the assignment statement and use the standard array notation. Enclose the cell contents on the right side of the assignment statement in braces {}.

EXAMPLE 2.6–1   An Environment Database

Suppose you want to create a 2 × 2 cell array A, whose cells contain the location, the date, the air temperature (measured at 8 A.M., 12 noon, and 5 P.M.), and the water temperatures measured at the same time in three different points in a pond. The cell array looks like the following.

Walden Pond	June 13, 1997
[60 72 65]	$\begin{bmatrix} 55 & 57 & 56 \\ 54 & 56 & 55 \\ 52 & 55 & 53 \end{bmatrix}$

■ Solution
You can create this array by typing the following either in interactive mode or in a script file and running it.

```
A(1,1) = {'Walden Pond'};
A(1,2) = {'June 13, 1997'};
A(2,1) = {[60,72,65]};
A(2,2) = {[55,57,56;54,56,55;52,55,53]};
```

If you do not yet have contents for a particular cell, you can type a pair of empty braces { } to denote an empty cell, just as a pair of empty brackets [ ] denotes an empty numeric array. This notation creates the cell but does not store any contents in it.

To use content indexing, enclose in braces the cell subscripts on the left side, using the standard array notation. Then specify the cell contents on the right side of the assignment operator. For example:

```
A{1,1} = 'Walden Pond';
A{1,2} = 'June 13, 1997';
A{2,1} = [60,72,65];
A{2,2} = [55,57,56;54,56,55;52,55,53];
```

Type A at the command line. You will see

```
A =
    'Walden Pond'    'June 13, 1997'
    [1x3 double]     [3x3 double]
```

You can use the celldisp function to display the full contents. For example, typing celldisp(A) displays

```
A{1,1} =
    Walden Pond
A{2,1} =
    60   72   65
    .
    .
    .
    etc.
```

The cellplot function produces a graphical display of the cell array's contents in the form of a grid. Type cellplot(A) to see this display for the cell array A. Use commas or spaces with braces to indicate columns of cells and use semicolons to indicate rows of cells (just as with numeric arrays). For example, typing

```
B = {[2,4], [6,-9;3,5]; [7;2], 10};
```

creates the following 2 × 2 cell array:

$$\begin{array}{|c|c|} \hline [2 \quad 4] & \begin{bmatrix} 6 & -9 \\ 3 & 5 \end{bmatrix} \\ \hline [7 \quad 2] & 10 \\ \hline \end{array}$$

You can preallocate empty cell arrays of a specified size by using the cell function. For example, type C = cell(3,5) to create the 3 × 5 cell array C and fill it with empty matrices. Once the array has been defined in this way, you can use assignment statements to enter the contents of the cells. For example, type C(2,4) = {[6,-3,7]} to put the 1 × 3 array in cell (2,4) and type C(1,5) = {1:10} to put the numbers from 1 to 10 in cell (1,5). Type C(3,4) = {'30 mph'} to put the string in cell (3,4).

Accessing Cell Arrays

You can access the contents of a cell array by using either cell indexing or content indexing. To use cell indexing to place the contents of cell (3,4) of the array C in the new variable `Speed`, type `Speed = C(3,4)`. To place the contents of the cells in rows 1 to 3, columns 2 to 5 in the new cell array D, type `D = C(1:3,2:5)`. The new cell array D will have three rows, four columns, and 12 arrays. To use content indexing to access some of or all the contents in a *single cell,* enclose the cell index expression in braces to indicate that you are assigning the contents, not the cells themselves, to a new variable. For example, typing `Speed = C{3,4}` assigns the contents `'30 mph'` in cell (3,4) to the variable `Speed`. You cannot use content indexing to retrieve the contents of more than one cell at a time. For example, the statements `G = C{1,:}` and `C{1,:} = var`, where `var` is some variable, are both invalid.

You can access subsets of a cell's contents. For example, to obtain the second element in the 1 × 3-row vector in the (2,4) cell of array C and assign it to the variable r, you type `r = C{2,4}(1,2)`. The result is `r = -3`.

2.7 Structure Arrays

FIELD

Structure arrays are composed of *structures*. This class of arrays enables you to store dissimilar arrays together. The elements in structures are accessed using *named fields*. This feature distinguishes them from cell arrays, which are accessed using the standard array indexing operations.

Structure arrays are used in this text only in this section. Some MATLAB toolboxes do use structure arrays.

A specific example is the best way to introduce the terminology of structures. Suppose you want to create a database of students in a course, and you want to include each student's name, Social Security number, email address, and test scores. Figure 2.7–1 shows a diagram of this data structure. Each type of data (name, Social Security number, and so on) is a *field,* and its name is the *field name.* Thus our database has four fields. The first three fields each contain a text string, while

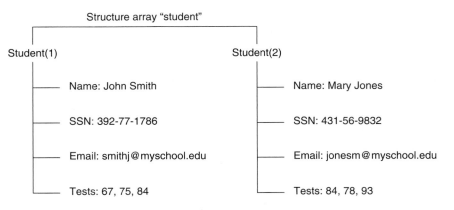

Figure 2.7–1 Arrangement of data in the structure array `student`.

the last field (the test scores) contains a vector having numerical elements. A *structure* consists of all this information for a single student. A *structure array* is an array of such structures for different students. The array shown in Figure 2.7–1 has two structures arranged in one row and two columns.

Creating Structures

You can create a structure array by using assignment statements or by using the `struct` function. The following example uses assignment statements to build a structure. Structure arrays use the dot notation (.) to specify and to access the fields. You can type the commands either in the interactive mode or in a script file.

A Student Database

EXAMPLE 2.7–1

Create a structure array to contain the following types of student data:

- Student name.
- Social Security number.
- Email address.
- Test scores.

Enter the data shown in Figure 2.7–1 into the database.

■ Solution

You can create the structure array by typing the following either in the interactive mode or in a script file. Start with the data for the first student.

```
student.name = 'John Smith';
student.SSN = '392-77-1786';
student.email = 'smithj@myschool.edu';
student.tests = [67,75,84];
```

If you then type

```
>>student
```

at the command line, you will see the following response:

```
name: 'John Smith'
SSN: = '392-77-1786'
email: = 'smithj@myschool.edu'
tests: = [67 75 84]
```

To determine the size of the array, type `size(student)`. The result is `ans = 1 1`, which indicates that it is a 1×1 structure array.

To add a second student to the database, use a subscript 2 enclosed in parentheses after the structure array's name and enter the new information. For example, type

```
student(2).name = 'Mary Jones';
student(2).SSN = '431-56-9832';
student(2).email = 'jonesm@myschool.edu';
student(2).tests = [84,78,93];
```

This process "expands" the array. Before we entered the data for the second student, the *dimension* of the structure array was 1×1 (it was a single structure). Now it is a 1×2 array consisting of two structures, arranged in one row and two columns. You can confirm this information by typing size(student), which returns ans = 1 2. If you now type length(student), you will get the result ans = 2, which indicates that the array has two elements (two structures). When a structure array has more than one structure, MATLAB does not display the individual field contents when you type the structure array's name. For example, if you now type student, MATLAB displays

```
>>student =
1x2 struct array with fields:
    name
    SSN
    email
    tests
```

You can also obtain information about the fields by using the fieldnames function (see Table 2.7–1). For example:

```
>>fieldnames(student)
ans =
    'name'
    'SSN'
    'email'
    'tests'
```

As you fill in more student information, MATLAB assigns the same number of fields and the same field names to each element. If you do not enter some information—for example, suppose you do not know someone's email address—MATLAB assigns an empty matrix to that field for that student.

Table 2.7–1 Structure functions

Function	Description
names = fieldnames(S)	Returns the field names associated with the structure array S as names, a cell array of strings.
isfield(S,'field')	Returns 1 if 'field' is the name of a field in the structure array S and 0 otherwise.
isstruct(S)	Returns 1 if the array S is a structure array and 0 otherwise.
S = rmfield(S,'field')	Removes the field 'field' from the structure array S.
S = struct('f1','v1','f2', 'v2', ...)	Creates a structure array with the fields 'f1', 'f2', ... having the values 'v1', 'v2', ...

The fields can he different sizes. For example, each name field can contain a different number of characters, and the arrays containing the test scores can be different sizes, as would be the case if a certain student did not take the second test.

In addition to the assignment statement, you can build structures using the `struct` function, which lets you "preallocate" a structure array. To build a structure array named `sa_1`, the syntax is

```
sa_1 = struct('field1','values1','field2',values2', ...)
```

where the arguments are the field names and their values. The values arrays `values1, values2,...` must all be arrays of the same size, scalar cells, or single values. The elements of the values arrays are inserted into the corresponding elements of the structure array. The resulting structure array has the same size as the values arrays, or is 1×1 if none of the values arrays is a cell. For example, to preallocate a 1×1 structure array for the student database, you type

```
student = struct('name','John Smith', 'SSN', ...
'392-77-1786','email','smithj@myschool.edu', ...
'tests',[67,75,84])
```

Accessing Structure Arrays

To access the contents of a particular field, type a period after the structure array name, followed by the field name. For example, typing `student(2).name` displays the value `'Mary Jones'`. Of course, we can assign the result to a variable in the usual way. For example, typing `name2 = student(2).name` assigns the value `'Mary Jones'` to the variable `name2`. To access elements within a field, for example, John Smith's second test score, type `student(1).tests(2)`. This entry returns the value `75`. In general, if a field contains an array, you use the array's subscripts to access its elements. In this example the statement `student(1).tests(2)` is equivalent to `student(1,1).tests(2)` because `student` has one row.

To store all the information for a particular structure—say, all the information about Mary Jones—in another structure array named `M`, you type `M = student(2)`. You can also assign or change values of field elements. For example, typing `student(2).tests(2) = 81` changes Mary Jones's second test score from 78 to 81.

Modifying Structures

Suppose you want to add phone numbers to the database. You can do this by typing the first student's phone number as follows:

```
student(1).phone = '555-1653'
```

All the other structures in the array will now have a `phone` field, but these fields will contain the empty array until you give them values.

To delete a field from every structure in the array, use the `rmfield` function. Its basic syntax is

```
new_struc = rmfield(array,'field');
```

where `array` is the structure array to be modified, `'field'` is the field to be removed, and `new_struc` is the name of the new structure array so created by the removal of the field. For example, to remove the Social Security field and call the new structure array `new_student`, type

```
new_student = rmfield(student,'SSN');
```

Using Operators and Functions with Structures

You can apply the MATLAB operators to structures in the usual way. For example, to find the maximum test score of the second student, you type `max(student(2).tests)`. The answer is 93.

The `isfield` function determines whether a structure array contains a particular field. Its syntax is `isfield(S, 'field')`. It returns a value of 1 (which means "true") if `'field'` is the name of a field in the structure array S. For example, typing `isfield(student, 'name')` returns the result `ans = 1`.

The `isstruct` function determines whether an array is a structure array. Its syntax is `isstruct(S)`. It returns a value of 1 if S is a structure array and 0 otherwise. For example, typing `isstruct(student)` returns the result `ans = 1`, which is equivalent to "true."

Test Your Understanding

T2.7–1 Create the structure array `student` shown in Figure 2.7–1 and add the following information about a third student: name: Alfred E. Newman; SSN: 555-12-3456; e-mail: newmana@myschool.edu; tests: 55, 45, 58.

T2.7–2 Edit your structure array to change Newman's second test score from 45 to 53.

T2.7–3 Edit your structure array to remove the SSN field.

2.8 Summary

You should now be able to perform basic operations and use arrays in MATLAB. For example, you should be able to

■   Create, address, and edit arrays.
■   Perform array operations including addition, subtraction, multiplication, division, and exponentiation.
■   Perform matrix operations including addition, subtraction, multiplication, division, and exponentiation.

Table 2.8–1 Guide to commands introduced in Chapter 2

Special characters	Use
'	Transposes a matrix, creating complex conjugate elements.
. '	Transposes a matrix without creating complex conjugate elements.
;	Suppresses screen printing; also denotes a new row in an array.
:	Represents an entire row or column of an array.

Tables	
Array functions	Table 2.1–1
Element-by-element operations	Table 2.3–1
Special matrices	Table 2.4–5
Polynomial functions	Table 2.5–1
Structure functions	Table 2.7–1

■ Perform polynomial algebra.
■ Create databases using cell and structure arrays.

Table 2.8–1 is a reference guide to all the MATLAB commands introduced in this chapter.

Key Terms with Page References

Absolute value,   61
Array addressing,   57
Array operations,   65
Array size,   56
Cell array,   90
Cell indexing,   90
Column vector,   54
Content indexing,   90
Empty array,   58
Field,   92

Identity matrix,   82
Left division method,   84
Length,   61
Magnitude,   61
Matrix,   56
Matrix operations,   73
Null matrix,   82
Row vector,   54
Structure arrays,   92
Transpose,   55

Problems

You can find the answers to problems marked with an asterisk at the end of the text.

Section 2.1

1. *a.* Use two methods to create the vector **x** having 100 regularly spaced values starting at 5 and ending at 28.

    b. Use two methods to create the vector **x** having a regular spacing of 0.2 starting at 2 and ending at 14.

    c. Use two methods to create the vector **x** having 50 regularly spaced values starting at -2 and ending at 5.

2.   *a.* Create the vector **x** having 50 logarithmically spaced values starting at 10 and ending at 1000.

    b. Create the vector **x** having 20 logarithmically spaced values starting at 10 and ending at 1000.

3.*   Use MATLAB to create a vector **x** having six values between 0 and 10 (including the endpoints 0 and 10). Create an array **A** whose first row contains the values $3x$ and whose second row contains the values $5x - 20$.

4.   Repeat Problem 3 but make the first column of **A** contain the values $3x$ and the second column contain the values $5x - 20$.

5.   Type this matrix in MATLAB and use MATLAB to carry out the following instructions.

$$\mathbf{A} = \begin{bmatrix} 3 & 7 & -4 & 12 \\ -5 & 9 & 10 & 2 \\ 6 & 13 & 8 & 11 \\ 15 & 5 & 4 & 1 \end{bmatrix}$$

    a. Create a vector **v** consisting of the elements in the second column of **A**.

    b. Create a vector **w** consisting of the elements in the second row of **A**.

6.   Type this matrix in MATLAB and use MATLAB to carry out the following instructions.

$$\mathbf{A} = \begin{bmatrix} 3 & 7 & -4 & 12 \\ -5 & 9 & 10 & 2 \\ 6 & 13 & 8 & 11 \\ 15 & 5 & 4 & 1 \end{bmatrix}$$

    a. Create a 4 × 3 array **B** consisting of all elements in the second through fourth columns of **A**.

    b. Create a 3 × 4 array **C** consisting of all elements in the second through fourth rows of **A**.

    c. Create a 2 × 3 array **D** consisting of all elements in the first two rows and the last three columns of **A**.

7.*   Compute the length and absolute value of the following vectors:

    a. **x** = [2, 4, 7]

    b. **y** = [2, −4, 7]

    c. **z** = [5 + 3*i*, −3 + 4*i*, 2 − 7*i*]

8.   Given the matrix

$$A = \begin{bmatrix} 3 & 7 & -4 & 12 \\ -5 & 9 & 10 & 2 \\ 6 & 13 & 8 & 11 \\ 15 & 5 & 4 & 1 \end{bmatrix}$$

a. Find the maximum and minimum values in each column.
b. Find the maximum and minimum values in each row.

9.   Given the matrix

$$A = \begin{bmatrix} 3 & 7 & -4 & 12 \\ -5 & 9 & 10 & 2 \\ 6 & 13 & 8 & 11 \\ 15 & 5 & 4 & 1 \end{bmatrix}$$

a. Sort each column and store the result in an array **B**.
b. Sort each row and store the result in an array **C**.
c. Add each column and store the result in an array **D**.
d. Add each row and store the result in an array **E**.

10.   Consider the following arrays.

$$A = \begin{bmatrix} 1 & 4 & 2 \\ 2 & 4 & 100 \\ 7 & 9 & 7 \\ 3 & \pi & 42 \end{bmatrix} \qquad B = \ln(A)$$

Write MATLAB expressions to do the following.
a. Select just the second row of **B**.
b. Evaluate the sum of the second row of **B**.
c. Multiply the second column of **B** and the first column of **A** element by element.
d. Evaluate the maximum value in the vector resulting from element-by-element multiplication of the second column of **B** with the first column of **A**.
e. Use element-by-element division to divide the first row of **A** by the first three elements of the third column of **B**. Evaluate the sum of the elements of the resulting vector.

Section 2.2

11.* *a.* Create a three-dimensional array **D** whose three "layers" are these matrices:

$$A = \begin{bmatrix} 3 & -2 & 1 \\ 6 & 8 & -5 \\ 7 & 9 & 10 \end{bmatrix} \quad B = \begin{bmatrix} 6 & 9 & -4 \\ 7 & 5 & 3 \\ -8 & 2 & 1 \end{bmatrix} \quad C = \begin{bmatrix} -7 & -5 & 2 \\ 10 & 6 & 1 \\ 3 & -9 & 8 \end{bmatrix}$$

b. Use MATLAB to find the largest element in each layer of **D** and the largest element in **D**.

Section 2.3

12.* Given the matrices

$$A = \begin{bmatrix} -7 & 11 \\ 4 & 9 \end{bmatrix} \quad B = \begin{bmatrix} 4 & -5 \\ 12 & -2 \end{bmatrix} \quad C = \begin{bmatrix} -3 & -9 \\ 7 & 8 \end{bmatrix}$$

Use MATLAB to
a. Find **A** + **B** + **C**.
b. Find **A** − **B** + **C**.
c. Verify the associative law

$$(A + B) + C = A + (B + C)$$

d. Verify the commutative law

$$A + B + C = B + C + A = A + C + B$$

13.* Given the matrices

$$A = \begin{bmatrix} 56 & 32 \\ 24 & -16 \end{bmatrix} \quad B = \begin{bmatrix} 14 & -4 \\ 6 & -2 \end{bmatrix}$$

Use MATLAB to
a. Find the result of **A** times **B** using the array product.
b. Find the result of **A** divided by **B** using array right division.
c. Find **B** raised to the third power element by element.

14.* The mechanical work W done in using a force F to push a block through a distance D is $W = FD$. The following table gives data on the amount of force used to push a block through the given distance over five segments of a certain path. The force varies because of the differing friction properties of the surface.

	Path segment				
	1	2	3	4	5
Force (N)	400	550	700	500	600
Distance (m)	3	0.5	0.75	1.5	5

Use MATLAB to find (*a*) the work done on each segment of the path and (*b*) the total work done over the entire path.

15. Plane A is heading southwest at 300 mi/hr, while plane B is heading west at 150 mi/hr. What are the velocity and the speed of plane A relative to plane B?

16. The following table shows the hourly wages, hours worked, and output (number of widgets produced) in one week for five widget makers.

	Worker				
	1	2	3	4	5
Hourly wage ($)	5	5.50	6.50	6	6.25
Hours worked	40	43	37	50	45
Output (widgets)	1000	1100	1000	1200	1100

Use MATLAB to answer these questions:
a. How much did each worker earn in the week?
b. What is the total salary amount paid out?
c. How many widgets were made?
d. What is the average cost to produce one widget?
e. How many hours does it take to produce one widget on average?
f. Assuming that the output of each worker has the same quality, which worker is the most efficient? Which is the least efficient?

17. Two divers start at the surface and establish the following coordinate system: *x* is to the west, *y* is to the north, and *z* is down. Diver 1 swims 60 ft east, then 25 ft south, and then dives 30 ft. At the same time, diver 2 dives 20 ft, swims east 30 ft and then south 55 ft.
a. Compute the distance between diver 1 and the starting point.
b. How far in each direction must diver 1 swim to reach diver 2?
c. How far in a straight line must diver 1 swim to reach diver 2?

18. The potential energy stored in a spring is $kx^2/2$, where k is the spring constant and x is the compression in the spring. The force required to compress the spring is kx. The following table gives the data for five springs:

	Spring				
	1	2	3	4	5
Force (N)	11	7	8	10	9
Spring constant k (N/m)	1000	600	900	1300	700

Use MATLAB to find (*a*) the compression *x* in each spring and (*b*) the potential energy stored in each spring.

19. A company must purchase five kinds of material. The following table gives the price the company pays per ton for each material, along with the number of tons purchased in the months of May, June, and July:

Material	Price ($/ton)	Quantity purchased (tons)		
		May	**June**	**July**
1	300	5	4	6
2	550	3	2	4
3	400	6	5	3
4	250	3	5	4
5	500	2	4	3

Use MATLAB to answer these questions:

a. Create a 5×3 matrix containing the amounts spent on each item for each month.
b. What is the total spent in May? in June? in July?
c. What is the total spent on each material in the three-month period?
d. What is the total spent on all materials in the three-month period?

20. A fenced enclosure consists of a rectangle of length L and width $2R$, and a semicircle of radius R, as shown in Figure P20. The enclosure is to be built to have an area A of 1600 ft^2. The cost of the fence is $40/ft for the curved portion and $30/ft for the straight sides. Use the min function to determine with a resolution of 0.01 ft the values of R and L required to minimize the total cost of the fence. Also compute the minimum cost.

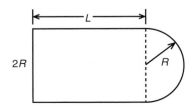

Figure P20

21. A water tank consists of a cylindrical part of radius r and height h, and a hemispherical top. The tank is to be constructed to hold 500 m^3 of fluid when filled. The surface area of the cylindrical part is $2\pi rh$, and its volume is $\pi r^2 h$. The surface area of the hemispherical top is given by $2\pi r^2$, and its volume is given by $2\pi r^3/3$. The cost to construct the cylindrical part of the tank is $300/m^2 of surface area; the hemispherical part costs $400/m^2. Plot the cost versus r for $2 \le r \le 10$ m, and determine the radius that results in the least cost. Compute the corresponding height h.

22. Write a MATLAB assignment statement for each of the following functions, assuming that w, x, y, and z are row vectors of equal length and that c and d are scalars.

$$f = \frac{1}{\sqrt{2\pi c/x}} \qquad E = \frac{x + w/(y + z)}{x + w/(y - z)}$$

$$A = \frac{e^{-c/(2x)}}{(\ln y)\sqrt{dz}} \qquad S = \frac{x(2.15 + 0.35y)^{1.8}}{z(1 - x)^y}$$

23. *a.* After a dose, the concentration of medication in the blood declines due to metabolic processes. The *half-life* of a medication is the time required after an initial dosage for the concentration to be reduced by one-half. A common model for this process is

$$C(t) = C(0)e^{-kt}$$

where $C(0)$ is the initial concentration, t is time (in hours), and k is called the *elimination rate constant*, which varies among individuals. For a particular bronchodilator, k has been estimated to be in the range $0.047 \leq k \leq 0.107$ per hour. Find an expression for the half-life in terms of k, and obtain a plot of the half-life versus k for the indicated range.

b. If the concentration is initially zero and a constant delivery rate is started and maintained, the concentration as a function of time is described by

$$C(t) = \frac{a}{k}(1 - e^{-kt})$$

where a is a constant that depends on the delivery rate. Plot the concentration after 1 hr, $C(1)$, versus k for the case where $a = 1$ and k is in the range $0.047 \leq k \leq 0.107$ per hour.

24. A cable of length L_c supports a beam of length L_b, so that it is horizontal when the weight W is attached at the beam end. The principles of statics can be used to show that the tension force T in the cable is given by

$$T = \frac{L_b L_c W}{D\sqrt{L_b^2 - D^2}}$$

where D is the distance of the cable attachment point to the beam pivot. See Figure P24.

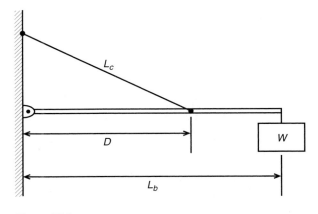

Figure P24

a. For the case where $W = 400$ N, $L_b = 3$ m, and $L_c = 5$ m, use element-by-element operations and the `min` function to compute the value of D that minimizes the tension T. Compute the minimum tension value.

b. Check the sensitivity of the solution by plotting T versus D. How much can D vary from its optimal value before the tension T increases 10 percent above its minimum value?

Section 2.4

25.* Use MATLAB to find the products **AB** and **BA** for the following matrices:

$$\mathbf{A} = \begin{bmatrix} 11 & 5 \\ -9 & -4 \end{bmatrix} \quad \mathbf{B} = \begin{bmatrix} -7 & -8 \\ 6 & 2 \end{bmatrix}$$

26. Given the matrices

$$\mathbf{A} = \begin{bmatrix} 4 & -2 & 1 \\ 6 & 8 & -5 \\ 7 & 9 & 10 \end{bmatrix} \quad \mathbf{B} = \begin{bmatrix} 6 & 9 & -4 \\ 7 & 5 & 3 \\ -8 & 2 & 1 \end{bmatrix} \quad \mathbf{C} = \begin{bmatrix} -4 & -5 & 2 \\ 10 & 6 & 1 \\ 3 & -9 & 8 \end{bmatrix}$$

Use MATLAB to

a. Verify the associative property

$$\mathbf{A(B + C) = AB + AC}$$

b. Verify the distributive property

$$\mathbf{(AB)C = A(BC)}$$

27. The following tables show the costs associated with a certain product and the production volume for the four quarters of the business year. Use MATLAB to find (*a*) the quarterly costs for materials, labor, and transportation;

(*b*) the total material, labor, and transportation costs for the year; and (*c*) the total quarterly costs.

Product	Unit product costs ($ \times 10^3$)		
	Materials	**Labor**	**Transportation**
1	7	3	2
2	3	1	3
3	9	4	5
4	2	5	4
5	6	2	1

Product	Quarterly production volume			
	Quarter 1	**Quarter 2**	**Quarter 3**	**Quarter 4**
1	16	14	10	12
2	12	15	11	13
3	8	9	7	11
4	14	13	15	17
5	13	16	12	18

28.* Aluminum alloys are made by adding other elements to aluminum to improve its properties, such as hardness or tensile strength. The following table shows the composition of five commonly used alloys, which are known by their alloy numbers (2024, 6061, and so on) [Kutz, 1999]. Obtain a matrix algorithm to compute the amounts of raw materials needed to produce a given amount of each alloy. Use MATLAB to determine how much raw material of each type is needed to produce 1000 tons of each alloy.

	Composition of aluminum alloys				
Alloy	**%Cu**	**%Mg**	**%Mn**	**%Si**	**%Zn**
2024	4.4	1.5	0.6	0	0
6061	0	1	0	0.6	0
7005	0	1.4	0	0	4.5
7075	1.6	2.5	0	0	5.6
356.0	0	0.3	0	7	0

29. Redo Example 2.4–4 as a script file to allow the user to examine the effects of labor costs. Allow the user to input the four labor costs in the following table. When you run the file, it should display the quarterly costs and the category costs. Run the file for the case where the unit labor costs are $3000, $7000, $4000, and $8000, respectively.

Product costs

Product	Unit costs ($\times 10^3$)		
	Materials	**Labor**	**Transportation**
1	6	2	1
2	2	5	4
3	4	3	2
4	9	7	3

Quarterly production volume

Product	Quarter 1	Quarter 2	Quarter 3	Quarter 4
1	10	12	13	15
2	8	7	6	4
3	12	10	13	9
4	6	4	11	5

30. Vectors with three elements can represent position, velocity, and acceleration. A mass of 5 kg, which is 3 m away from the x axis, starts at $x = 2$ m and moves with a speed of 10 m/s parallel to the y axis. Its velocity is thus described by $\mathbf{v} = [0, 10, 0]$, and its position is described by $\mathbf{r} = [2, 10t + 3, 0]$. Its angular momentum vector $\mathbf{L}$ is found from $\mathbf{L} = m(\mathbf{r} \times \mathbf{v})$, where m is the mass. Use MATLAB to

  a. Compute a matrix $\mathbf{P}$ whose 11 rows are the values of the position vector $\mathbf{r}$ evaluated at the times $t = 0, 0.5, 1, 1.5, \ldots, 5$ s.
  b. What is the location of the mass when $t = 5$ s?
  c. Compute the angular momentum vector $\mathbf{L}$. What is its direction?

31.* The *scalar triple product* computes the magnitude M of the moment of a force vector $\mathbf{F}$ about a specified line. It is $M = (\mathbf{r} \times \mathbf{F}) \cdot \mathbf{n}$, where $\mathbf{r}$ is the position vector from the line to the point of application of the force and $\mathbf{n}$ is a unit vector in the direction of the line.

  Use MATLAB to compute the magnitude M for the case where $\mathbf{F} = [12, -5, 4]$ N, $\mathbf{r} = [-3, 5, 2]$ m, and $\mathbf{n} = [6, 5, -7]$.

32. Verify the identity

$$\mathbf{A} \times (\mathbf{B} \times \mathbf{C}) = \mathbf{B}\,(\mathbf{A} \cdot \mathbf{C}) - \mathbf{C}(\mathbf{A} \cdot \mathbf{B})$$

for the vectors $\mathbf{A} = 7\mathbf{i} - 3\mathbf{j} + 7\mathbf{k}$, $\mathbf{B} = -6\mathbf{i} + 2\mathbf{j} + 3\mathbf{k}$, and $\mathbf{C} = 2\mathbf{i} + 8\mathbf{j} - 8\mathbf{k}$.

33. The area of a parallelogram can be computed from $|\mathbf{A} \times \mathbf{B}|$, where $\mathbf{A}$ and $\mathbf{B}$ define two sides of the parallelogram (see Figure P33). Compute the area of a parallelogram defined by $\mathbf{A} = 5\mathbf{i}$ and $\mathbf{B} = \mathbf{i} + 3\mathbf{j}$.

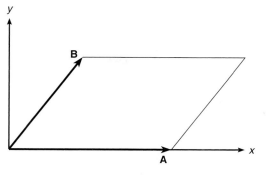

Figure P33

34. The volume of a parallelepiped can be computed from |**A** · (**B** ×**C**)|, where **A**, **B**, and **C** define three sides of the parallelepiped (see Figure P34). Compute the volume of a parallelepiped defined by **A** = 5**i**, **B** = 2**i** + 4**j**, and **C** = 3**i** − 2**k**.

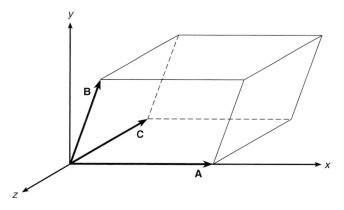

Figure P34

Section 2.5

35. Use MATLAB to plot the polynomials $y = 3x^4 - 6x^3 + 8x^2 + 4x + 90$ and $z = 3x^3 + 5x^2 - 8x + 70$ over the interval $-3 \le x \le 3$. Properly label the plot and each curve. The variables y and z represent current in milliamperes; the variable x represents voltage in volts.

36. Use MATLAB to plot the polynomial $y = 3x^4 - 5x^3 - 28x^2 - 5x + 200$ on the interval $-1 \le x \le 1$. Put a grid on the plot and use the `ginput` function to determine the coordinates of the peak of the curve.

37. Use MATLAB to find the following product:

$$(10x^3 - 9x^2 - 6x + 12)(5x^3 - 4x^2 - 12x + 8)$$

38.* Use MATLAB to find the quotient and remainder of

$$\frac{14x^3 - 6x^2 + 3x + 9}{5x^2 + 7x - 4}$$

39.* Use MATLAB to evaluate

$$\frac{24x^3 - 9x^2 - 7}{10x^3 + 5x^2 - 3x - 7}$$

at $x = 5$.

40. The *ideal gas law* provides one way to estimate the pressures and volumes of a gas in a container. The law is

$$P = \frac{RT}{\hat{V}}$$

More accurate estimates can be made with the *van der Waals* equation

$$P = \frac{RT}{\hat{V} - b} - \frac{a}{\hat{V}^2}$$

where the term b is a correction for the volume of the molecules and the term $a/\hat{V}^2$ is a correction for molecular attractions. The values of a and b depend on the type of gas. The gas constant is R, the *absolute* temperature is T, and the gas specific volume is $\hat{V}$. If 1 mol of an ideal gas were confined to a volume of 22.41 L at 0°C (273.2 K), it would exert a pressure of 1 atm. In these units, $R = 0.08206$.

For chlorine (Cl_2), $a = 6.49$ and $b = 0.0562$. Compare the specific volume estimates $\hat{V}$ given by the ideal gas law and the van der Waals equation for 1 mol of Cl_2 at 300 K and a pressure of 0.95 atm.

41. Aircraft A is flying east at 320 mi/hr, while aircraft B is flying south at 160 mi/hr. At 1:00 P.M. the aircraft are located as shown in Figure P41.

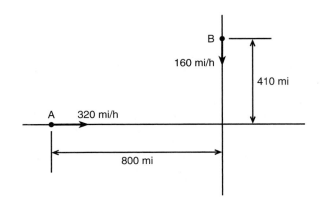

Figure P41

a. Obtain the expression for the distance D between the aircraft as a function of time. Plot D versus time until D reaches its minimum value.

b. Use the `roots` function to compute the time when the aircraft are first within 30 mi of each other.

42. The function

$$y = \frac{3x^2 - 12x + 20}{x^2 - 7x + 10}$$

approaches ∞ as $x \to 2$ and as $x \to 5$. Plot this function over the range $0 \le x \le 7$. Choose an appropriate range for the y axis.

43. The following formulas are commonly used by engineers to predict the lift and drag of an airfoil:

$$L = \frac{1}{2}\rho C_L S V^2$$

$$D = \frac{1}{2}\rho C_D S V^2$$

where L and D are the lift and drag forces, V is the airspeed, S is the wing span, ρ is the air density, and C_L and C_D are the *lift* and *drag* coefficients. Both C_L and C_D depend on α, the angle of attack, the angle between the relative air velocity and the airfoil's chord line.

Wind tunnel experiments for a particular airfoil have resulted in the following formulas.

$$C_L = 4.47 \times 10^{-5}\alpha^3 + 1.15 \times 10^{-3}\alpha^2 + 6.66 \times 10^{-2}\alpha + 1.02 \times 10^{-1}$$

$$C_D = 5.75 \times 10^{-6}\alpha^3 + 5.09 \times 10^{-4}\alpha^2 + 1.8 \times 10^{-4}\alpha + 1.25 \times 10^{-2}$$

where α is in degrees.

Plot the lift and drag of this airfoil versus V for $0 \le V \le 150$ mi/hr (you must convert V to ft /sec; there is 5280 ft/mi). Use the values $\rho = 0.002378$ slug/ft^3 (air density at sea level), $\alpha = 10°$, and $S = 36$ ft. The resulting values of L and D will be in pounds.

44. The lift-to-drag ratio is an indication of the effectiveness of an airfoil. Referring to Problem 43, the equations for lift and drag are

$$L = \frac{1}{2}\,\rho C_L S V^2$$

$$D = \frac{1}{2}\,\rho C_D S V^2$$

where, for a particular airfoil, the lift and drag coefficients versus angle of attack α are given by

$$C_L = 4.47 \times 10^{-5}\alpha^3 + 1.15 \times 10^{-3}\alpha^2 + 6.66 \times 10^{-2}\alpha + 1.02 \times 10^{-1}$$

$$C_D = 5.75 \times 10^{-6}\alpha^3 + 5.09 \times 10^{-4}\alpha^2 + 1.81 \times 10^{-4}\alpha + 1.25 \times 10^{-2}$$

Using the first two equations, we see that the lift-to-drag ratio is given simply by the ratio C_L/C_D.

$$\frac{L}{D} = \frac{\frac{1}{2}\rho C_L S V^2}{\frac{1}{2}\rho C_D S V^2} = \frac{C_L}{C_D}$$

Plot L/D versus α for $-2° \le \alpha \le 22°$. Determine the angle of attack that maximizes L/D.

Section 2.6

45. *a.* Use both cell indexing and content indexing to create the following 2×2 cell array.

Motor 28C	Test ID 6
$\begin{bmatrix} 3 & 9 \\ 7 & 2 \end{bmatrix}$	[6 5 1]

b. What are the contents of the (1,1) element in the (2,1) cell in this array?

46. The capacitance of two parallel conductors of length L and radius r, separated by a distance d in air, is given by

$$C = \frac{\pi \epsilon L}{\ln[(d - r)/r]}$$

where ϵ is the permittivity of air ($\epsilon = 8.854 \times 10^{-12}$ F/m). Create a cell array of capacitance values versus d, L, and r for $d = 0.003, 0.004, 0.005,$ and 0.01 m; $L = 1, 2, 3$ m; and $r = 0.001, 0.002, 0.003$ m. Use MATLAB to determine the capacitance value for $d = 0.005$, $L = 2$, and $r = 0.001$.

Section 2.7

47. *a.* Create a structure array that contains the conversion factors for converting units of mass, force, and distance between the metric SI system and the British Engineering System.

b. Use your array to compute the following:

- The number of meters in 48 ft.
- The number of feet in 130 m.
- The number of pounds equivalent to 36 N.
- The number of newtons equivalent to 10 lb.
- The number of kilograms in 12 slugs.
- The number of slugs in 30 kg.

48. Create a structure array that contains the following information fields concerning the road bridges in a town: bridge location, maximum load (tons), year built, year due for maintenance. Then enter the following data into the array:

Location	Max. load	Year built	Due for maintenance
Smith St.	80	1928	2011
Hope Ave.	90	1950	2013
Clark St.	85	1933	2012
North Rd.	100	1960	2012

49. Edit the structure array created in Problem 48 to change the maintenance data for the Clark St. bridge from 2012 to 2018.

50. Add the following bridge to the structure array created in Problem 48.

Location	Max. load	Year built	Due for maintenance
Shore Rd.	85	1997	2014

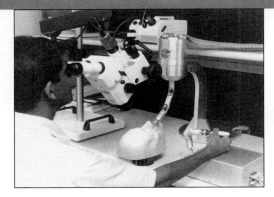

Courtesy Jet Propulsion Lab/NASA.

Engineering in the 21st Century...

Robot-Assisted Microsurgery

Y ou need not be a medical doctor to participate in the exciting developments now taking place in the health field. Many advances in medicine and surgery are really engineering achievements, and many engineers are contributing their talents in this area. Recent achievements include

- Laparoscopic surgery in which a fiber-optic scope guides a small surgical device. This technology eliminates the need for large incisions and the resulting long recuperation.
- Computerized axial tomography (CAT) scans and magnetic resonance imaging (MRI), which provide noninvasive tools for diagnosing medical problems.
- Medical instrumentation, such as a fingertip sensor for continuously measuring oxygen in the blood and automatic blood pressure sensors.

As we move into the 21st century, an exciting challenge will be the development of robot-assisted surgery in which a robot, supervised by a human, performs operations requiring precise, steady motions. Robots have already assisted in hip surgery on animals, but much greater development is needed. Another developing technology is *telesurgery* in which a surgeon uses a television interface to remotely guide a surgical robot. This technology would allow delivery of medical services to remote areas.

Robot-assisted microsurgery, which uses a robot capable of very small, precise motions, shows great promise. In some applications the robotic device is used to filter out tremors normally present in the human hand. Designing such devices requires geometric analysis, control system design, and image processing. The MATLAB Image Processing toolbox and the several MATLAB toolboxes dealing with control system design are useful for such applications. ■

CHAPTER 3

Functions and Files

OUTLINE

3.1   Elementary Mathematical Functions

3.2   User-Defined Functions

3.3   Additional Function Topics

3.4   Working with Data Files

3.5   Summary

Problems

MATLAB has many built-in functions, including trigonometric, logarithmic, and hyperbolic functions, as well as functions for processing arrays. These functions are summarized in Section 3.1. In addition, you can define your own functions with a *function* file, and you can use them just as conveniently as the built-in functions. We explain this technique in Section 3.2. Section 3.3 covers additional topics in function programming, including function handles, anonymous functions, subfunctions, and nested functions. Another type of file that is useful in MATLAB is the data file. Importing and exporting such files is covered in Section 3.4.

Sections 3.1 and 3.2 contain essential topics and must be covered. The material in Section 3.3 is useful for creating large programs. The material in Section 3.4 is useful for readers who must work with large data sets.

3.1 Elementary Mathematical Functions

You can use the `lookfor` command to find functions that are relevant to your application. For example, type `lookfor imaginary` to get a list of the functions that deal with imaginary numbers. You will see listed

```
imag  Complex imaginary part
i     Imaginary unit
j     Imaginary unit
```

Note that `imaginary` is not a MATLAB function, but the word is found in the Help descriptions of the MATLAB function `imag` and the special symbols `i` and `j`. Their names and brief descriptions are displayed when you type `lookfor imaginary`. If you know the correct spelling of a MATLAB function, for example, `disp`, you can type `help disp` to obtain a description of the function.

Some of the functions, such as `sqrt` and `sin`, are built in. These are stored as image files and are not M-files. They are part of the MATLAB core so they are very efficient, but the computational details are not readily accessible. Some functions are implemented in M-files. You can see the code and even modify it, although this is not recommended.

Exponential and Logarithmic Functions

Table 3.1–1 summarizes some of the common elementary functions. An example is the square root function `sqrt`. To compute $\sqrt{9}$, you type `sqrt(9)` at the command line. When you press **Enter,** you see the result `ans = 3`. You can use functions with variables. For example, consider the session

```
>>x = -9; y = sqrt(x)
y =
   0 + 3.0000i
```

Note that the `sqrt` function returns the positive root only.

Table 3.1–1   Some common mathematical functions

Exponential	
`exp(x)`	Exponential; e^x.
`sqrt(x)`	Square root; $\sqrt{x}$.
Logarithmic	
`log(x)`	Natural logarithm; $\ln x$.
`log10(x)`	Common (base-10) logarithm; $\log x = \log_{10} x$.
Complex	
`abs(x)`	Absolute value; x.
`angle(x)`	Angle of a complex number x.
`conj(x)`	Complex conjugate.
`imag(x)`	Imaginary part of a complex number x.
`real(x)`	Real part of a complex number x.
Numeric	
`ceil(x)`	Round to the nearest integer toward ∞.
`fix(x)`	Round to the nearest integer toward zero.
`floor(x)`	Round to the nearest integer toward $-\infty$.
`round(x)`	Round toward the nearest integer.
`sign(x)`	Signum function: $+1$ if $x > 0$; 0 if $x = 0$; -1 if $x < 0$.

One of the strengths of MATLAB is that it will treat a variable as an array automatically. For example, to compute the square roots of 5, 7, and 15, type

```
>>x = [5,7,15]; y = sqrt(x)
y =
    2.2361    2.6358    3.8730
```

The square root function operates on every element in the array x.

Similarly, we can type $\texttt{exp(2)}$ to obtain $e^2 = 7.3891$, where e is the base of the natural logarithms. Typing $\texttt{exp(1)}$ gives 2.7183, which is e. Note that in mathematics text, $\ln x$ denotes the *natural* logarithm, where $x = e^y$ implies that

$$\ln x = \ln(e^y) = y \ln e = y$$

because $\ln e = 1$. However, this notation has not been carried over into MATLAB, which uses $\texttt{log(x)}$ to represent $\ln x$.

The *common* (base-10) logarithm is denoted in text by $\log x$ or $\log_{10} x$. It is defined by the relation $x = 10^y$; that is,

$$\log_{10} x = \log_{10} 10^y = y \log_{10} 10 = y$$

because $\log_{10} 10 = 1$. The MATLAB common logarithm function is $\texttt{log10(x)}$. A common mistake is to type $\texttt{log(x)}$, instead of $\texttt{log10(x)}$.

Another common error is to forget to use the array multiplication operator $\texttt{.*}$. Note that in the MATLAB expression $\texttt{y = exp(x).*log(x)}$, we need to use the operator $\texttt{.*}$ if x is an array because both $\texttt{exp(x)}$ and $\texttt{log(x)}$ will be arrays.

Complex Number Functions

Chapter 1 explained how MATLAB easily handles complex number arithmetic. In the *rectangular* representation the number $a + ib$ represents a point in the xy plane. The number's real part a is the x coordinate of the point, and the imaginary part b is the y coordinate. The *polar* representation uses the distance M of the point from the origin, which is the length of the hypotenuse, and the angle θ the hypotenuse makes with the *positive real* axis. The pair (M, θ) is simply the polar coordinates of the point. From the Pythagorean theorem, the length of the hypotenuse is given by $M = \sqrt{a^2 + b^2}$ which is called the *magnitude* of the number. The angle θ can be found from the trigonometry of the right triangle. It is $\theta = \arctan(b/a)$.

Adding and subtracting complex numbers by hand is easy when they are in the rectangular representation. However, the polar representation facilitates multiplication and division of complex numbers by hand. We must enter complex numbers in MATLAB using the rectangular form, and its answers will be given in that form. We can obtain the rectangular representation from the polar representation as follows:

$$a = M \cos \theta \qquad b = M \sin \theta$$

The MATLAB `abs(x)` and `angle(x)` functions calculate the magnitude M and angle θ of the complex number x. The functions `real(x)` and `imag(x)` return the real and imaginary parts of x. The function `conj(x)` computes the complex conjugate of x.

The magnitude of the product z of two complex numbers x and y is equal to the product of their magnitudes: $|z| = |x||y|$. The angle of the product is equal to the sum of the angles: $\angle z = \angle x + \angle y$. These facts are demonstrated below.

```
>>x = -3 + 4i; y = 6 - 8i;
>>mag_x = abs(x)
mag_x =
    5.0000
>>mag_y = abs(y)
mag_y =
    10.0000
>>mag_product = abs(x*y)
    50.0000
>>angle_x = angle(x)
angle_x =
    2.2143
>>angle_y = angle(y)
angle_y =
    -0.9273
>>sum_angles = angle_x + angle_y
sum_angles =
    1.2870
>>angle_product = angle(x*y)
angle_product =
    1.2870
```

Similarly, for division, if $z = x/y$, then $|z| = |x|/|y|$ and $\angle z = \angle x - \angle y$.

Note that when `x` is a vector of *real* values, `abs(x)` does not give the geometric length of the vector. This length is given by `norm(x)`. If `x` is a complex number representing a geometric vector, then `abs(x)` gives its geometric length.

Numeric Functions

The `round` function rounds to the nearest integer. If `y=[2.3,2.6,3.9]`, typing `round(y)` gives the results 2, 3, 4. The `fix` function truncates to the nearest integer toward zero. Typing `fix(y)` gives the results 2, 2, 3. The `ceil` function (which stands for "ceiling") rounds to the nearest integer toward ∞. Typing `ceil(y)` produces the answers 3, 3, 4.

Suppose `z = [-2.6,-2.3,5.7]`. The `floor` function rounds to the nearest integer toward $-\infty$. Typing `floor(z)` produces the result $-3, -3, 5$. Typing `fix(z)` produces the answer $-2, -2, 5$. The `abs` function computes the absolute value. Thus `abs(z)` produces 2.6, 2.3, 5.7.

Test Your Understanding

T3.1–1 For several values of x and y, confirm that $\ln(xy) = \ln x + \ln y$.

T3.1–2 Find the magnitude, angle, real part, and imaginary part of the number $\sqrt{2 + 6i}$. (Answers: magnitude = 2.5149, angle = 0.6245 rad, real part = 2.0402, imaginary part = 1.4705)

Function Arguments

When writing mathematics in text, we use parentheses ( ), brackets [ ], and braces { } to improve the readability of expressions, and we have much latitude over their use. For example, we can write sin 2 in text, but MATLAB requires parentheses surrounding the 2 (which is called the *function argument* or *parameter*). Thus to evaluate sin 2 in MATLAB, we type `sin(2)`. The MATLAB function name must be followed by a pair of parentheses that surround the argument. To express in text the sine of the second element of the array x, we would type sin[$x(2)$]. However, in MATLAB you cannot use brackets or braces in this way, and you must type `sin(x(2))`.

You can include expressions and other functions as arguments. For example, if `x` is an array, to evaluate $\sin(x^2 + 5)$, you type `sin(x.^2 + 5)`. To evaluate $\sin(\sqrt{x} + 1)$, you type `sin(sqrt(x)+1)`. Be sure to check the order of precedence and the number and placement of parentheses when typing such expressions. Every left-facing parenthesis requires a right-facing mate. However, this condition does not guarantee that the expression is correct!

Another common mistake involves expressions such as $\sin^2 x$, which means $(\sin x)^2$. In MATLAB, if `x` is a scalar we write this expression as `(sin(x))^2`, *not* as `sin^2(x)`, `sin^2x`, or `sin(x^2)`!

Trigonometric Functions

Other commonly used functions are `cos(x)`, `tan(x)`, `sec(x)`, and `csc(x)`, which return $\cos x$, $\tan x$, $\sec x$, and $\csc x$, respectively. Table 3.1–2 lists the MATLAB trigonometric functions that operate in radian mode. Thus `sin(5)` computes the sine of 5 rad, not the sine of 5°. Similarly, the inverse trigonometric functions return an answer in radians. The functions that operate in degree mode have the letter d appended to their names. For example, `sind(x)` accepts the value of `x` in degrees. To compute the inverse sine in radians, type `asin(x)`. For example, `asin(0.5)` returns the answer 0.5236 rad. *Note*: In MATLAB, `sin(x)^(-1)` does not give $\sin^{-1}(x)$; it gives $1/\sin(x)$!

MATLAB has two inverse tangent functions. The function `atan(x)` computes arctan x—the arctangent or inverse tangent—and returns an angle between $-\pi/2$ and $\pi/2$. Another correct answer is the angle that lies in the opposite quadrant. The user must be able to choose the correct answer. For example, `atan(1)` returns the answer 0.7854 rad, which corresponds to 45°. Thus $\tan 45° = 1$. However, $\tan(45° + 180°) = \tan 225° = 1$ also. Thus arctan(1) = 225° is also correct.

Table 3.1–2 Trigonometric functions

Trigonometric*	
`cos(x)`	Cosine; cos x.
`cot(x)`	Cotangent; cot x.
`csc(x)`	Cosecant; csc x.
`sec(x)`	Secant; sec x.
`sin(x)`	Sine; sin x.
`tan(x)`	Tangent; tan x.
Inverse trigonometric†	
`acos(x)`	Inverse cosine; arccos $x = \cos^{-1} x$.
`acot(x)`	Inverse cotangent; arccot $x = \cot^{-1} x$.
`acsc(x)`	Inverse cosecant; arccsc $x = \csc^{-1} x$.
`asec(x)`	Inverse secant; arcsec $x = \sec^{-1} x$.
`asin(x)`	Inverse sine; arcsin $x = \sin^{-1} x$.
`atan(x)`	Inverse tangent; arctan $x = \tan^{-1} x$.
`atan2(y,x)`	Four-quadrant inverse tangent.

*These functions accept x in radians.
†These functions return a value in radians.

MATLAB provides the `atan2(y,x)` function to determine the arctangent unambiguously, where `x` and `y` are the coordinates of a point. The angle computed by `atan2(y,x)` is the angle between the positive real axis and line from the origin (0, 0) to the point (x, y). For example, the point $x = 1$, $y = -1$ corresponds to $-45°$ or -0.7854 rad, and the point $x = -1$, $y = 1$ corresponds to $135°$ or 2.3562 rad. Typing `atan2(-1,1)` returns -0.7854, while typing `atan2(1,-1)` returns 2.3562. The `atan2(y,x)` function is an example of a function that has two arguments. The order of the arguments is important for such functions. At present there is no `atan2d` function.

Test Your Understanding

T3.1–3 For several values of x, confirm that $e^{ix} = \cos x + i \sin x$.

T3.1–4 For several values of x in the range $0 \leq x \leq 2\pi$, confirm that $\sin^{-1} x + \cos^{-1} x = \pi/2$.

T3.1–5 For several values of x in the range $0 \leq x \leq 2\pi$, confirm that $\tan(2x) = 2\tan x/(1 - \tan^2 x)$.

Hyperbolic Functions

The *hyperbolic functions* are the solutions of some common problems in engineering analysis. For example, the *catenary* curve, which describes the shape of a hanging cable supported at both ends, can be expressed in terms of the hyperbolic cosine, cosh x, which is defined as

$$\cosh x = \frac{e^x + e^{-x}}{2}$$

Table 3.1–3 Hyperbolic functions

Hyperbolic	
`cosh(x)`	Hyperbolic cosine; $\cosh x = (e^x + e^{-x})/2$.
`coth(x)`	Hyperbolic cotangent; $\cosh x / \sinh x$.
`csch(x)`	Hyperbolic cosecant; $1/\sinh x$.
`sech(x)`	Hyperbolic secant; $1/\cosh x$.
`sinh(x)`	Hyperbolic sine; $\sinh x = (e^x - e^{-x})/2$.
`tanh(x)`	Hyperbolic tangent; $\sinh x / \cosh x$.
Inverse hyperbolic	
`acosh(x)`	Inverse hyperbolic cosine
`acoth(x)`	Inverse hyperbolic cotangent
`acsch(x)`	Inverse hyperbolic cosecant
`asech(x)`	Inverse hyperbolic secant
`asinh(x)`	Inverse hyperbolic sine
`atanh(x)`	Inverse hyperbolic tangent

The hyperbolic sine, $\sinh x$, is defined as

$$\sinh x = \frac{e^x - e^{-x}}{2}$$

The inverse hyperbolic sine, $\sinh^{-1} x$, is the value y that satisfies $\sinh y = x$.

Several other hyperbolic functions have been defined. Table 3.1–3 lists these hyperbolic functions and the MATLAB commands to obtain them.

Test Your Understanding

T3.1–6   For several values of x in the range $0 \le x \le 5$, confirm that $\sin(ix) = i \sinh x$.

T3.1–7   For several values of x in the range $-10 \le x \le 10$, confirm that $\sinh^{-1} x = \ln(x + \sqrt{x^2 + 1})$.

3.2 User-Defined Functions

Another type of M-file is a *function file*. Unlike a script file, all the variables in a function file are *local variables*, which means their values are available only within the function. Function files are useful when you need to repeat a set of commands several times. They are the building blocks of larger programs.

To create a function file, open the Editor/Debugger as described in Chapter 1. The first line in a function file must begin with a *function definition line* that has a list of inputs and outputs. This line distinguishes a function M-file from a script M-file. Its syntax is as follows:

```
function [output variables] = function_name(input variables)
```

The output variables are those variables whose values are computed by the function, using the given values of the input variables. Note that the output

FUNCTION FILE

LOCAL VARIABLE

FUNCTION DEFINITION LINE

variables are enclosed in *square brackets* (which are optional if there is only one output), while the input variables must be enclosed with *parentheses.* The function_name should be the same as the file name in which it is saved (with the .m extension). That is, if we name a function drop, it should be saved in the file drop.m. The function is "called" by typing its name (for example, drop) at the command line. The word function in the function definition line must be *lowercase.* Before naming a function, you can use the exist function to see if another function has the same name.

Some Simple Function Examples

Functions operate on variables within their own workspace (called *local variables*), which is separate from the workspace you access at the MATLAB command prompt. Consider the following user-defined function fun.

```
function z = fun(x,y)
u = 3*x;
z = u + 6*y.^2;
```

Note the use of the array exponentiation operator (.^). This enables the function to accept y as an array. Now consider what happens when you call this function in various ways in the Command window. Call the function with its output argument:

```
>>x = 3; y = 7;
>>z = fun(x,y)
z =
    303
```

or

```
>>z = fun(3,7)
z =
    303
```

The function uses $x = 3$ and $y = 7$ to compute z.

Call the function without its output argument and try to access its value. You see an error message.

```
>>clear z, fun(3,7)
ans =
    303
>>z
???   Undefined function or variable 'z'.
```

Assign the output argument to another variable:

```
>>q = fun(3,7)
q =
    303
```

You can suppress the output by putting a semicolon after the function call. For example, if you type q = fun(3,7); the value of q will be computed but not displayed.

The variables x and y are *local* to the function fun, so unless you pass their values by naming them x and y, their values will not be available in the workspace outside the function. The variable u is also local to the function. For example,

```
>>x = 3; y = 7; q = fun(x,y);
>>u
??? Undefined function or variable 'u'.
```

Compare this to

```
>>q = fun(3,7);
>>x
??? Undefined function or variable 'x'.
>>y
??? Undefined function or variable 'y'.
```

Only the order of the arguments is important, not the names of the arguments:

```
>>a = 7;b = 3;
>>z = fun(b,a)     % This is equivalent to z = fun(3,7)
z =
   303
```

You can use arrays as input arguments:

```
>>r = fun([2:4],[7:9])
r =
   300     393     498
```

A function may have more than one output. These are enclosed in square brackets. For example, the function circle computes the area A and circumference C of a circle, given its radius as an input argument.

```
function [A, C] = circle(r)
A = pi*r.^2;
C = 2*pi*r;
```

The function is called as follows, if $r = 4$.

```
>>[A, C] = circle(4)
A =
   50.2655
C =
   25.1327
```

A function may have no input arguments and no output list. For example, the following user-defined function show_date computes and stores the date in the variable today, and displays the value of today.

```
function show_date
today = date
```

Variations in the Function Line

The following examples show permissible variations in the format of the function line. The differences depend on whether there is no output, a single output, or multiple outputs.

Function definition line	File name
1. `function [area_square] = square(side);`	square.m
2. `function area_square = square(side);`	square.m
3. `function volume_box = box(height,width,length);`	box.m
4. `function [area_circle,circumf] = circle(radius);`	circle.m
5. `function sqplot(side);`	sqplot.m

Example 1 is a function with one input and one output. The square brackets are optional when there is only one output (see example 2). Example 3 has one output and three inputs. Example 4 has two outputs and one input. Example 5 has no output variable (for example, a function that generates a plot). In such cases the equal sign may be omitted.

Comment lines starting with the % sign can be placed anywhere in the function file. However, if you use `help` to obtain information about the function, MATLAB displays all comment lines immediately following the function definition line up to the first blank line or first executable line. The first comment line can be accessed by the `lookfor` command.

We can call both built-in and user-defined functions either with the output variables explicitly specified, as in examples 1 through 4, or without any output variables specified. For example, we can call the function `square` as `square (side)` if we are not interested in its output variable `area_square`. (The function might perform some other operation that we want to occur, such as producing a plot.) Note that if we omit the semicolon at the end of the function call statement, the first variable in the output variable list will be displayed using the default variable name `ans`.

Variations in Function Calls

The following function, called `drop`, computes a falling object's velocity and distance dropped. The input variables are the acceleration g, the initial velocity v_0, and the elapsed time t. Note that we must use the element-by-element operations for any operations involving function inputs that are arrays. Here we anticipate that t will be an array, so we use the element-by-element operator $(.\char94)$.

```
function [dist,vel] = drop(g,v0,t);
% Computes the distance traveled and the
% velocity of a dropped object, as functions
% of g, the initial velocity v0, and the time t.
```

```
vel = g*t + v0;
dist = 0.5*g*t.^2 + v0*t;
```

The following examples show various ways to call the function `drop`:

1. The variable names used in the function definition may, but need not, be used when the function is called:

```
a = 32.2;
initial_speed = 10;
time = 5;
[feet_dropped,speed] = drop(a,initial_speed,time)
```

2. The input variables need not be assigned values outside the function prior to the function call:

```
[feet_dropped,speed] = drop(32.2,10,5)
```

3. The inputs and outputs may be arrays:

```
[feet_dropped,speed]=drop(32.2,10,0:1:5)
```

This function call produces the arrays `feet_dropped` and `speed`, each with six values corresponding to the six values of time in the array `0:1:5`.

Local Variables

The names of the input variables given in the function definition line are local to that function. This means that other variable names can be used when you call the function. All variables inside a function are erased after the function finishes executing, except when the same variable names appear in the output variable list used in the function call.

For example, when using the `drop` function in a program, we can assign a value to the variable `dist` before the function call, and its value will be unchanged after the call because its name was not used in the output list of the call statement (the variable `feet_dropped` was used in the place of `dist`). This is what is meant by the function's variables being "local" to the function. This feature allows us to write generally useful functions using variables of our choice, without being concerned that the calling program uses the same variable names for other calculations. This means that our function files are "portable" and need not be rewritten every time they are used in a different program.

You might find the M-file Debugger to be useful for locating errors in function files. Runtime errors in functions are more difficult to locate because the function's local workspace is lost when the error forces a return to the MATLAB base workspace. The Debugger provides access to the function workspace and allows you to change values. It also enables you to execute lines one at a time and to set *breakpoints,* which are specific locations in the file where execution is temporarily halted. The applications in this text will probably not require use of the Debugger, which is useful mainly for very large programs. For more information, see Chapter 4 of this text and also Chapter 4 of [Palm, 2005].

Global Variables

The `global` command declares certain variables global, and therefore their values are available to the basic workspace and to other functions that declare these variables global. The syntax to declare the variables A, X, and Q is `global A X Q`. Use a space, not a comma, to separate the variables. Any assignment to those variables, in any function or in the base workspace, is available to all the other functions declaring them global. If the global variable doesn't exist the first time you issue the `global` statement, it will be initialized to the empty matrix. If a variable with the same name as the global variable already exists in the current workspace, MATLAB issues a warning and changes the value of that variable to match the global. In a user-defined function, make the global command the first executable line. Place the same command in the calling program. It is customary, but not required, to capitalize the names of global variables and to use long names, to make them easily recognizable.

The decision to declare a variable global is not always clear-cut. It is recommended to avoid using global variables. This can often be done by using anonymous and nested functions, as discussed in Section 3.3.

Function Handles

A *function handle* is a way to reference a given function. First introduced in MATLAB 6.0, function handles have become widely used and frequently appear in examples throughout the MATLAB documentation. You can create a function handle to any function by using the @ sign before the function name. You can then give the handle a name, if you wish, and you can use the handle to reference the function.

For example, consider the following user-defined function, which computes $y = x + 2e^{-x} - 3$.

```
function y = f1(x)
y = x + 2*exp(-x) - 3;
```

To create a handle to this function and name the handle `fh1`, you type `fh1 = @f1`.

Function Functions

Some MATLAB functions act on functions. These commands are called *function functions*. If the function acted upon is not a simple function, it is more convenient to define the function in an M-file. You can pass the function to the calling function by using a function handle.

Finding the Zeros of a Function   You can use the `fzero` function to find the zero of a function of a single variable, which is denoted by x. Its basic syntax is

```
fzero(@function, x0)
```

where `@function` is a function handle and x0 is a user-supplied guess for the zero. The `fzero` function returns a value of x that is near x0. It identifies only points where the function crosses the x axis, not points where the function just touches the axis. For example, `fzero(@cos,2)` returns the value $x = 1.5708$. As another example, $y = x^2$ is a parabola that touches the x axis at $x = 0$. Because the function never crosses the x axis, however, no zero will be found.

The function `fzero(@function,x0)` tries to find a zero of `function` near `x0`, if `x0` is a scalar. The value returned by `fzero` is near a point where `function` changes sign, or `NaN` if the search fails. In this case, the search terminates when the search interval is expanded until an `Inf`, `NaN`, or a complex value is found (`fzero` cannot find complex zeros). If `x0` is a *vector* of length 2, `fzero` assumes that `x0` is an interval where the sign of `function(x0(1))` differs from the sign of `function(x0(2))`. An error occurs if this is not true. Calling `fzero` with such an interval guarantees that `fzero` will return a value near a point where `function` changes sign. Plotting the function first is a good way to get a value for the vector `x0`. If the function is not continuous, `fzero` might return values that are discontinuous points instead of zeros. For example, `x = fzero(@tan,1)` returns `x = 1.5708`, a discontinuous point in $\tan(x)$.

Functions can have more than one zero, so it helps to plot the function first and then use `fzero` to obtain an answer that is more accurate than the answer read off the plot. Figure 3.2–1 shows the plot of the function $y = x + 2e^{-x} - 3$, which has two zeros, one near $x = -0.5$ and one near $x = 3$. Using the function file `f1` created earlier to find the zero near $x = -0.5$, type `x = fzero(@f1,-0.5)`. The answer is $x = -0.5831$. To find the zero near $x = 3$, type `x = fzero(@f1,3)`. The answer is $x = 2.8887$.

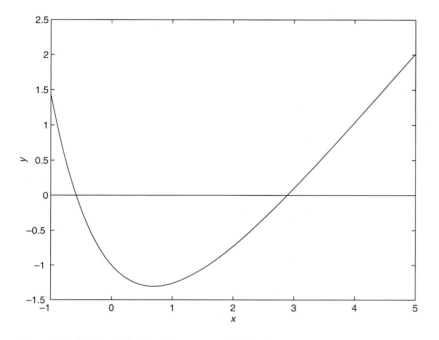

Figure 3.2–1 Plot of the function $y = x + 2e^{-x} - 3$.

The syntax `fzero(@f1,-0.5)` is preferred to the older syntax `fzero ('f1', -0.5)`.

Minimizing a Function of One Variable   The `fminbnd` function finds the minimum of a function of a single variable, which is denoted by x. Its basic syntax is

```
fminbnd(@function, x1, x2)
```

where `@function` is a function handle. The `fminbnd` function returns a value of x that minimizes the function in the interval $x1 \leq x \leq x2$. For example, `fminbnd(@cos,0,4)` returns the value $x = 3.1416$.

However, to use this function to find the minimum of more-complicated functions, it is more convenient to define the function in a function file. For example, if $y = 1 - xe^{-x}$, define the following function file:

```
function y = f2(x)
y = 1-x.*exp(-x);
```

To find the value of x that gives a minimum of y for $0 \leq x \leq 5$, type `x = fminbnd(@f2,0,5)`. The answer is $x = 1$. To find the minimum value of y, type `y = f2(x)`. The result is `y = 0.6321`.

Whenever we use a minimization technique, we should check that the solution is a true minimum. For example, consider the polynomial $y = 0.025x^5 - 0.0625x^4 - 0.333x^3 + x^2$. Its plot is shown in Figure 3.2–2. The function has two minimum points in the interval $-1 < x < 4$. The minimum near $x = 3$ is called a *relative* or *local* minimum because it forms a valley whose lowest point is higher than the minimum at $x = 0$. The minimum at $x = 0$ is the true minimum and is also called the *global* minimum. First create the function file

```
function y = f3(x)
y = polyval([0.025, -0.0625, -0.333, 1, 0, 0], x);
```

To specify the interval $-1 \leq x \leq 4$, type `x = fminbnd(@f3, -1, 4)`. MATLAB gives the answer `x = 2.0438e-006`, which is essentially 0, the true minimum point. If we specify the interval $0.1 \leq x \leq 2.5$, MATLAB gives the answer `x = 0.1001`, which corresponds to the minimum value of y on the interval $0.1 \leq x \leq 2.5$. Thus we will miss the true minimum point if our specified interval does not include it.

Also `fminbnd` can give misleading answers. If we specify the interval $1 \leq x \leq 4$, MATLAB (R2009 b) gives the answer `x = 2.8236`, which corresponds to the "valley" shown in the plot, but which is not the minimum point on the interval $1 \leq x \leq 4$. On this interval the minimum point is at the boundary $x = 1$. The `fminbnd` procedure looks for a minimum point corresponding to a zero slope. In practice, the best use of the `fminbnd` function is to determine precisely the location of a minimum point whose approximate location was found by other means, such as by plotting the function.

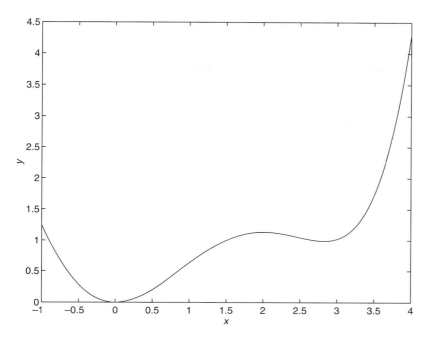

Figure 3.2–2 Plot of the function $y = 0.025x^5 - 0.0625x^4 - 0.333x^3 + x^2$.

Minimizing a Function of Several Variables   To find the minimum of a function of more than one variable, use the fminsearch function. Its basic syntax is

```
fminsearch(@function, x0)
```

where @function is a function handle. The vector x0 is a guess that must be supplied by the user. For example, to use the function $f = xe^{-x^2-y^2}$, first define it in an M-file, using the vector x whose elements are x(1) = x and x(2) = y.

```
function f = f4(x)
f = x(1).*exp(-x(1).^2-x(2).^2);
```

Suppose we guess that the minimum is near $x = y = 0$. The session is

```
>>fminsearch(@f4, [0, 0])
ans =
    -0.7071    0.000
```

Thus the minimum occurs at $x = -0.7071$, $y = 0$.

The fminsearch function can often handle discontinuities, particularly if they do not occur near the solution. The fminsearch function might give local solutions only, and it minimizes over the real numbers only; that is, x must consist of real variables only, and the function must return real

Table 3.2–1 Minimization and root-finding functions

Function	Description
`fminbnd(@function,x1,x2)`	Returns a value of x in the interval $x1 \leq x \leq x2$ that corresponds to a minimum of the single-variable function described by the handle `@function`.
`fminsearch(@function,x0)`	Uses the starting vector $x0$ to find a minimum of the multivariable function described by the handle `@function`.
`fzero(@function,x0)`	Uses the starting value $x0$ to find a zero of the single-variable function described by the handle `@function`.

numbers only. When x has complex values, they must be split into real and imaginary parts.

Table 3.2–1 summarizes the basic syntax of the `fminbnd`, `fminsearch`, and `fzero` commands.

These functions have extended syntax not described here. With these forms you can specify the accuracy required for the solution as well as the number of steps to use before stopping. Use the `help` facility to find out more about these functions.

EXAMPLE 3.2–1

Optimization of an Irrigation Channel

Figure 3.2–3 shows the cross section of an irrigation channel. A preliminary analysis has shown that the cross-sectional area of the channel should be 100 ft² to carry the desired water flow rate. To minimize the cost of concrete used to line the channel, we want to minimize the length of the channel's perimeter. Find the values of *d*, *b*, and θ that minimize this length.

■ **Solution**

The perimeter length *L* can be written in terms of the base *b*, depth *d*, and angle θ as follows:

$$L = b + \frac{2d}{\sin \theta}$$

The area of the trapezoidal cross section is

$$100 = db + \frac{d^2}{\tan \theta}$$

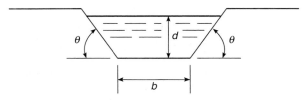

Figure 3.2–3 Cross section of an irrigation channel.

The variables to be selected are b, d, and θ. We can reduce the number of variables by solving the latter equation for b to obtain

$$b = \frac{1}{d}\left(100 - \frac{d^2}{\tan \theta}\right)$$

Substitute this expression into the equation for L. The result is

$$L = \frac{100}{d} - \frac{d}{\tan \theta} + \frac{2d}{\sin \theta}$$

We must now find the values of d and θ to minimize L.

First define the function file for the perimeter length. Let the vector $\mathbf{x}$ be $[d\ \theta]$.

```
function L = channel(x)
L = 100./x(1) - x(1)./tan(x(2)) + 2*x(1)./sin(x(2));
```

Then use the `fminsearch` function. Using a guess of $d = 20$ and $\theta = 1$ rad, the session is

```
>>x = fminsearch (@channel,[20,1])
x =
   7.5984    1.0472
```

Thus the minimum perimeter length is obtained with $d = 7.5984$ ft and $\theta = 1.0472$ rad, or $\theta = 60°$. Using a different guess, $d = 1$, $\theta = 0.1$, produces the same answer. The value of the base b corresponding to these values is $b = 8.7738$.

However, using the guess $d = 20$, $\theta = 0.1$ produces the physically meaningless result $d = -781$, $\theta = 3.1416$. The guess $d = 1$, $\theta = 1.5$ produces the physically meaningless result $d = 3.6058$, $\theta = -3.1416$.

The equation for L is a function of the two variables d and θ, and it forms a surface when L is plotted versus d and θ on a three-dimensional coordinate system. This surface might have multiple peaks, multiple valleys, and "mountain passes" called saddle points that can fool a minimization technique. Different initial guesses for the solution vector can cause the minimization technique to find different valleys and thus report different results. We can use the surface-plotting functions covered in Chapter 5 to look for multiple valleys, or we can use a large number of initial values for d and θ, say, over the physically realistic ranges $0 < d < 30$ and $0 < \theta < \pi/2$. If all the physically meaningful answers are identical, then we can be reasonably sure that we have found the minimum.

Test Your Understanding

T3.2–1 The equation $e^{-0.2x} \sin(x + 2) = 0.1$ has three solutions in the interval $0 < x < 10$. Find these three solutions. (Answers: $x = 1.0187, 4.5334, 7.0066$)

T3.2–2 The function $y = 1 + e^{-0.2x} \sin(x + 2)$ has two minimum points in the interval $0 < x < 10$. Find the values of x and y at each minimum. (Answers: $(x, y) = (2.5150, 0.4070), (9.0001, 0.8347)$)

T3.2–3 Find the depth d and angle θ to minimize the perimeter length of the channel shown in Figure 3.2–3 to provide an area of 200 ft^2. (Answer: $d = 10.7457$ ft, $\theta = 60°$.)

3.3 Additional Function Topics

In addition to function handles, *anonymous functions, subfunctions,* and *nested functions* are some of the newer features of MATLAB. This section covers the basic features of these new types of functions.

Methods for Calling Functions

There are four ways to invoke, or "call," a function into action:

1. As a character string identifying the appropriate function M-file
2. As a function handle
3. As an "inline" function object
4. As a string expression

Examples of these ways follow for the fzero function used with the user-defined function fun1, which computes $y = x^2 - 4$.

1. As a character string identifying the appropriate function M-file, which is

```
function y = fun1(x)
y = x.^2-4;
```

The function may be called as follows, to compute the zero over the range $0 \le x \le 3$:

```
>>x = fzero('fun1',[0, 3])
```

2. As a function handle to an existing function M-file:

```
>>x = fzero(@fun1,[0, 3])
```

3. As an "inline" function object:

```
>>fun1 = 'x.^2-4';
>>fun_inline = inline(fun1);
>>x = fzero(fun_inline,[0, 3])
```

4. As a string expression:

```
>>fun1 = 'x.^2-4';
>>x = fzero(fun1,[0, 3])
```

or as

```
>>x = fzero('x.^2-4',[0, 3])
```

Method 2 was not available prior to MATLAB 6.0, and it is now preferred over method 1. The third method is not discussed in this text because it is a slower method than the first two. The third and fourth methods are equivalent because they both utilize the inline function; the only difference is that with the fourth method MATLAB determines that the first argument of fzero is a string variable and calls `inline` to convert the string variable to an inline function object. The function handle method (method 2) is the fastest method, followed by method 1.

In addition to speed improvement, another advantage of using a function handle is that it provides access to subfunctions, which are normally not visible outside of their defining M-file. This is discussed later in this section.

Types of Functions

At this point it is helpful to review the types of functions provided for in MAT-LAB. MATLAB provides built-in functions, such as `clear`, `sin`, and `plot`, which are not M-files, and some functions that are M-files, such as the function `mean`. In addition, the following types of *user-defined* functions can be created in MATLAB.

- The *primary function* is the first function in an M-file and typically contains the main program. Following the primary function in the same file can be any number of subfunctions, which can serve as subroutines to the primary function. Usually the primary function is the only function in an M-file that you can call from the MATLAB command line or from another M-file function. You invoke this function by using the name of the M-file in which it is defined. We normally use the same name for the function and its file, but if the function name differs from the file name, you must use the file name to invoke the function.

  PRIMARY FUNCTION

- *Anonymous functions* enable you to create a simple function without needing to create an M-file for it. You can construct an anonymous function either at the MATLAB command line or from within another function or script. Thus, anonymous functions provide a quick way of making a function from any MATLAB expression without the need to create, name, and save a file.

  ANONYMOUS FUNCTIONS

- *Subfunctions* are placed in the primary function and are called by the primary function. You can use multiple functions within a single primary function M-file.

  SUBFUNCTIONS

- *Nested functions* are functions defined within another function. They can help to improve the readability of your program and also give you more flexible access to variables in the M-file. The difference between nested functions and subfunctions is that subfunctions normally cannot be accessed outside of their primary function file.

  NESTED FUNCTIONS

- *Overloaded* functions are functions that respond differently to different types of input arguments. They are similar to overloaded functions in any object-oriented language. For example, an overloaded function can be created to treat integer inputs differently than inputs of class double.

- *Private functions* enable you to restrict access to a function. They can be called only from an M-file function in the parent directory.

  PRIVATE FUNCTION

Anonymous Functions

Anonymous functions enable you to create a simple function without needing to create an M-file for it. You can construct an anonymous function either at the MATLAB command line or from within another function or script. The syntax for creating an anonymous function from an expression is

```
fhandle = @(arglist) expr
```

where `arglist` is a comma-separated list of input arguments to be passed to the function and `expr` is any single, valid MATLAB expression. This syntax creates the function handle `fhandle`, which enables you to invoke the function. Note that this syntax is different from that used to create other function handles, `fhandle = @functionname`. The handle is also useful for passing the anonymous function in a call to some other function in the same way as any other function handle.

For example, to create a simple function called `sq` to calculate the square of a number, type

```
sq = @(x) x.^2;
```

To improve readability, you may enclose the expression in parentheses, as `sq = @(x) (x.^2);`. To execute the function, type the name of the function handle, followed by any input arguments enclosed in parentheses. For example,

```
>>sq(5)
ans =
    25
>>sq([5,7])
ans =
    25    49
```

You might think that this particular anonymous function will not save you any work because typing `sq([5,7])` requires nine keystrokes, one more than is required to type `[5,7].^2`. Here, however, the anonymous function protects you from forgetting to type the period (.) required for array exponentiation. Anonymous functions are useful, however, for more complicated functions involving numerous keystrokes.

You can pass the handle of an anonymous function to other functions. For example, to find the minimum of the polynomial $4x^2 - 50x + 5$ over the interval $[-10, 10]$, you type

```
>>poly1 = @(x) 4*x.^2 - 50*x + 5;
>>fminbnd(poly1, -10, 10)
ans =
    6.2500
```

If you are not going to use that polynomial again, you can omit the handle definition line and type instead

```
>>fminbnd(@(x) 4*x.^2 - 50*x + 5, -10, 10)
```

Multiple-Input Arguments You can create anonymous functions having more than one input. For example, to define the function $\sqrt{x^2 + y^2}$, type

```
>>sqrtsum = @(x,y) sqrt(x.^2 + y.^2);
```

Then

```
>>sqrtsum(3, 4)
ans =
     5
```

As another example, consider the function $z = Ax + By$ defining a plane. The scalar variables A and B must be assigned values before you create the function handle. For example,

```
>>A = 6; B = 4:
>>plane = @(x,y) A*x + B*y;
>>z = plane(2,8)
z =
    44
```

No-Input Arguments To construct a handle for an anonymous function that has no input arguments, use empty parentheses for the input argument list, as shown by the following: `d = @() date;`.

Use empty parentheses when invoking the function, as follows:

```
>>d()
ans =
     01-Mar-2007
```

You must include the parentheses. If you do not, MATLAB just identifies the handle; it does not execute the function.

Calling One Function within Another One anonymous function can call another to implement function composition. Consider the function $5 \sin(x^3)$. It is composed of the functions $g(y) = 5 \sin(y)$ and $f(x) = x^3$. In the following session the function whose handle is `h` calls the functions whose handles are `f` and `g` to compute $5 \sin(2^3)$.

```
>>f = @(x) x.^3;
>>g = @(x) 5*sin(x);
>>h = @(x) g(f(x));
>>h(2)
ans =
     4.9468
```

To preserve an anonymous function from one MATLAB session to the next, save the function handle to a MAT-file. For example, to save the function associated with the handle `h`, type `save anon.mat h`. To recover it in a later session, type `load anon.mat h`.

Variables and Anonymous Functions   Variables can appear in anonymous functions in two ways:

- As variables specified in the argument list, such as $f = @(x) x.^3;$.
- As variables specified in the body of the expression, such as with the variables A and B in $plane = @(x,y) A*x + B*y$. In this case, when the function is created, MATLAB captures the values of these variables and retains those values for the lifetime of the function handle. In this example, if the values of A or B are changed after the handle is created, their values associated with the handle do not change. This feature has both advantages and disadvantages, so you must keep it in mind.

Subfunctions

A function M-file may contain more than one user-defined function. The first defined function in the file is called the *primary function,* whose name is the same as the M-file name. All other functions in the file are called *subfunctions.* Subfunctions are normally "visible" only to the primary function and to other subfunctions in the same file; that is, they normally cannot be called by programs or functions outside the file. However, this limitation can be removed with the use of function handles, as we will see later in this section.

Create the primary function first with a function definition line and its defining code, and name the file with this function name as usual. Then create each subfunction with its own function definition line and defining code. The order of the subfunctions does not matter, but function names must be unique within the M-file.

The order in which MATLAB checks for functions is very important. When a function is called from within an M-file, MATLAB first checks to see if the function is a built-in function such as sin. If not, it checks to see if it is a *subfunction* in the file, then checks to see if it is a *private* function (which is a function M-file residing in the private subdirectory of the calling function). Then MATLAB checks for a standard M-file on your search path. Thus, because MATLAB checks for a subfunction before checking for private and standard M-file functions, you may use subfunctions with the same name as another existing M-file. This feature allows you to name subfunctions without being concerned about whether another function exists with the *same name,* so you need not choose long function names to avoid conflict. This feature also protects you from using another function unintentionally.

Note that you may even supercede a MATLAB M-function in this way. The following example shows how the MATLAB M-function mean can be superceded by our own definition of the mean, one which gives the root-mean-square value. The function mean is a subfunction. The function subfun_demo is the primary function.

```
function y = subfun_demo(a)
y = a - mean(a);
%
function w = mean(x)
w = sqrt(sum(x.^2))/length(x);
```

A sample session follows.

```
>>y = subfn_demo([4, -4])
y =
   1.1716    -6.8284
```

If we had used the MATLAB M-function `mean`, we would have obtained a different answer, that is,

```
>>a=[4,-4];
>>b = a - mean(a)
b =
   4    -4
```

Thus the use of subfunctions enables you to reduce the number of files that define your functions. For example, if it were not for the subfunction `mean` in the previous example, we would have had to define a separate M-file for our `mean` function and give it a different name so as not to confuse it with the MATLAB function of the same name.

Subfunctions are normally visible only to the primary function and other subfunctions in the same file. However, we can use a function handle to allow access to the subfunction from outside the M-file, as the following example shows. Create the following M-file with the primary function `fn_demo1(range)` and the subfunction `testfun(x)` to compute the zeros of the function $(x^2 - 4) \cos x$ over the range specified in the input variable range. Note the use of a function handle in the second line.

```
function yzero = fn_demo1(range)
fun = @testfun;
[yzero,value] = fzero(fun,range);
%
function y = testfun(x)
y = (x.^2-4).*cos(x);
```

A test session gives the following results.

```
>>yzero = fn_demo1([3, 6])
yzero =
   4.7124
```

So the zero of $(x^2 - 4) \cos x$ over $3 \le x \le 6$ occurs at $x = 4.7124$.

Nested Functions

With MATLAB 7 you can now place the definitions of one or more functions within another function. Functions so defined are said to be *nested* within the main function. You can also nest functions within other nested functions. Like any M-file function, a nested function contains the usual components of an M-file function. You must, however, always terminate a nested function with an `end` statement. In fact, if an M-file contains at least one nested function, you must terminate *all* functions, including subfunctions, in the file with an `end` statement, whether or not they contain nested functions.

The following example constructs a function handle for a nested function `p(x)` and then passes the handle to the MATLAB function `fminbnd` to find the minimum point on the parabola. The `parabola` function constructs and returns a function handle `f` for the nested function `p` that evaluates the parabola $ax^2 + bx + c$. This handle gets passed to `fminbnd`.

```
function f = parabola(a, b, c)
f = @p;
   % Nested function
   function y = p(x)
     y = polyval ([a,b,c],x);
   end
end
```

In the Command window type

```
>>f = parabola(4, -50, 5);
>>fminbnd(f, -10, 10)
ans =
     6.2500
```

Note than the function `p(x)` can see the variables `a`, `b`, and `c` in the calling function's workspace.

Contrast this approach to that required using global variables. First create the function `p(x)`.

```
function y = p(x)
global a b c
y = polyval ([a, b, c], x);
```

Then, in the Command window, type

```
>>global a b c
>>a = 4; b = -50; c = 5;
>> fminbnd (@p, -10, 10)
```

Nested functions might seem to be the same as subfunctions, but they are not. Nested functions have two unique properties:

1. A nested function can access the workspaces of all functions inside of which it is nested. So, for example, a variable that has a value assigned to it by the primary function can be read or overwritten by a function nested at any level within the main function. In addition, a variable assigned in a nested function can be read or overwritten by any of the functions containing that function.

2. If you construct a function handle for a nested function, the handle not only stores the information needed to access the nested function, but also stores

the values of all variables shared between the nested function and those functions that contain it. This means that these variables persist in memory between calls made by means of the function handle.

Consider the following representation of some functions named A, B, . . ., E.

```
function A(x, y)      % The primary function
B(x, y);
D(y);

      function B(x, y)      % Nested in A
      C(x);
      D(y);

            function C(x)      % Nested in B
            D(x);
            end    % This terminates C
      end      % This terminates B

      function D(x)      % Nested in A
      E(x);

            function E      % Nested in D
            . . .
            end    % This terminates E
      end        % This terminates D
end        % This terminates A
```

You call a nested function in several ways.

1. You can call it from the level immediately above it. (In the previous code, function A can call B or D, but not C or E.)
2. You can call it from a function nested at the same level within the same parent function. (Function B can call D, and D can call B.)
3. You can call it from a function at any lower level. (Function C can call B or D, but not E.)
4. If you construct a function handle for a nested function, you can call the nested function from any MATLAB function that has access to the handle.

You can call a subfunction from any nested function in the same M-file.

Private Functions

Private functions reside in subdirectories with the special name private, and they are visible only to functions in the parent directory. Assume the directory rsmith is on the MATLAB search path. A subdirectory of rsmith called private may contain functions that only the functions in rsmith can call. Because private functions are invisible outside the parent directory `rsmith`, they can use the same names as functions in other directories. This is useful if the main directory is used by several individuals including R. Smith, but R. Smith wants to create a personal version of a particular function while retaining the original in the main directory. Because MATLAB looks for private functions before standard M-file functions, it will find a private function named, say, `cylinder.m` before a nonprivate M-file named `cylinder.m`.

Primary functions and subfunctions can be implemented as private functions. Create a private directory by creating a subdirectory called `private` using the standard procedure for creating a directory or a folder on your computer, but do not place the private directory on your path.

3.4 Working with Data Files

An ASCII data file may have one or more lines of text, called the header, at the beginning. These might be comments that describe what the data represent, the date they were created, and who created the data, for example. One or more lines of data, arranged in rows and columns, follow the header. The numbers in each row might be separated by spaces or by commas.

If it is inconvenient to edit the data file, the MATLAB environment provides many ways to bring data created by other applications into the MATLAB workspace, a process called *importing data,* and to package workspace variables so that they can be exported to other applications.

If the file has a header or the data are separated by commas, MATLAB will produce an error message. To correct this situation, first load the data file into a text editor, remove the header, and replace the commas with spaces. To retrieve these data into MATLAB, type `load filename`. If the file has m lines with n values in each line, the data will be assigned to an $m \times n$ matrix having the same name as the file with the extension stripped off. Your data file can have any extension except `.mat,` so that MATLAB will not try to load the file as a workspace file.

Importing Spreadsheet Files

Some spreadsheet programs store data in the `.wk1` format. You can use the command `M = wk1read('filename')` to import these data into MATLAB and store them in the matrix M. The command `A = xlsread('filename')` imports the Microsoft Excel workbook file `filename.xls` into the array A. The command `[A, B] = xlsread('filename')` imports all numeric data into the array A and all text data into the cell array B.

The Import Wizard

You can use the Import Wizard to import many types of ASCII data formats, including data on the clipboard. The Import Wizard presents a series of dialog boxes in which you specify the name of the file, the delimiter used in the file, and the variables that you want to import.

Do the following to import this sample tab-delimited, ASCII data file, `testdata.txt`:

```
1     2     3     4     5;
17    12    8     15    25;
```

1. Activate the Import Wizard either by typing `uiimport` or by selecting the **Import Data** option on the MATLAB Desktop **File** menu. The Import Wizard displays a dialog box that asks you to specify the name of the file you want to import.

2. The Import Wizard processes the contents of the file and displays tabs identifying the variables it recognizes in the file, and displays a portion of the data in a grid, similar to a spreadsheet. The Import Wizard uses the space character as the default delimiter. After you click **Next,** the Import Wizard attempts to identify the delimiter (see Figure 3.4–1).

3. In the next dialog box, the Import Wizard displays a list of the variables it found in the file. It also displays the contents of the first variable in the list. In this example there is only one variable, named `testdata`.

4. Choose the variables you want to import by clicking the check boxes next to their names. By default, all variables are checked for import. After you select the variables you want to import, click the **Finish** button to import the data into the MATLAB workspace.

To import data from the clipboard, select **Paste Special** from the **Edit** menu. Then proceed with step 2. The default variable name is `A_pastespecial`.

Exporting ASCII Data Files

You might want to export a MATLAB matrix as an ASCII data file where the rows and columns are represented as space-delimited, numeric values. To export a MATLAB matrix as a delimited ASCII data file, you can use either the `save`

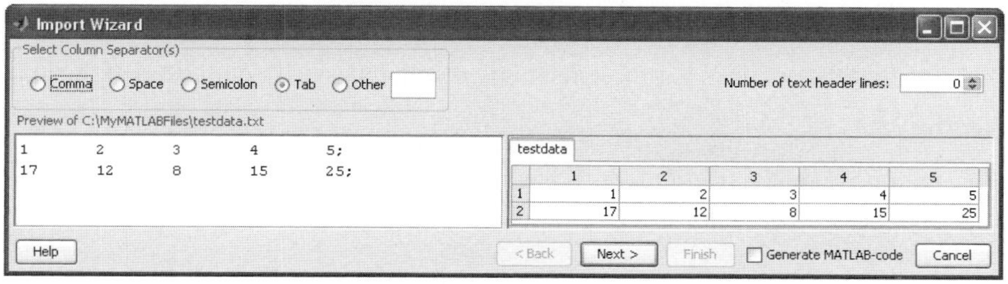

Figure 3.4–1 The first screen in the Import Wizard.

command, specifying the -ASCII qualifier, or the dlmwrite function. The save command is easy to use; however, the dlmwrite function provides greater flexibility, allowing you to specify any character as a delimiter and to export subsets of an array by specifying a range of values.

Suppose you have created the array A = [1 2 3 4; 5 6 7 8] in MATLAB. To export the array using the save command, type the following in the Command window.

```
>>save my_data.out A -ASCII
```

By default, save uses spaces as delimiters, but you can use tabs instead of spaces by specifying the -tab qualifier.

3.5 Summary

In Section 3.1 we introduced just some of the most commonly used mathematical functions. You should now be able to use the MATLAB Help to find other functions you need. If necessary, you can create your own functions, using the methods of Section 3.2. This section also covered function handles and their use with function functions.

Anonymous functions, subfunctions, and nested functions extend the capabilities of MATLAB. These topics were treated in Section 3.3. In addition to function files, data files are useful for many applications. Section 3.4 shows how to import and export such files in MATLAB.

Key Terms with Page References

Anonymous functions, 131	Local variable, 119
Function argument, 117	Nested functions, 131
Function definition line, 119	Primary function, 131
Function file, 119	Private function, 131
Function handle, 124	Subfunctions, 131
Global variables, 124	

Problems

You can find the answers to problems marked with an asterisk at the end of the text.

Section 3.1

1.* Suppose that $y = -3 + ix$. For $x = 0$, 1, and 2, use MATLAB to compute the following expressions. Hand-check the answers.

    a. $|y|$ *b.* $\sqrt{y}$

    c. $(-5 - 7i)y$ *d.* $\dfrac{y}{6 - 3i}$

2.* Let $x = -5 - 8i$ and $y = 10 - 5i$. Use MATLAB to compute the following expressions. Hand-check the answers.

a. The magnitude and angle of xy.

b. The magnitude and angle of $\frac{x}{y}$.

3.* Use MATLAB to find the angles corresponding to the following coordinates. Hand-check the answers.

a. $(x, y) = (5, 8)$ b. $(x, y) = (-5, 8)$

c. $(x, y) = (5, -8)$ d. $(x, y) = (-5, -8)$

4. For several values of x, use MATLAB to confirm that $\sinh x = (e^x - e^{-x})/2$.

5. For several values of x, use MATLAB to confirm that $\cosh^{-1} x = \ln (x + \sqrt{x^2 - 1}), x \geq 1$

6. The capacitance of two parallel conductors of length L and radius r, separated by a distance d in air, is given by

$$C = \frac{\pi \epsilon L}{\ln [(d - r)/r]}$$

where ϵ is the permittivity of air ($\epsilon = 8.854 \times 10^{-12}$ F/m).

Write a script file that accepts user input for d, L, and r and computes and displays C. Test the file with the values $L = 1$ m, $r = 0.001$ m, and $d = 0.004$ m.

7.* When a belt is wrapped around a cylinder, the relation between the belt forces on each side of the cylinder is

$$F_1 = F_2 e^{\mu \beta}$$

where β is the angle of wrap of the belt and μ is the friction coefficient. Write a script file that first prompts a user to specify β, μ, and F_2 and then computes the force F_1. Test your program with the values $\beta = 130°$, $\mu = 0.3$, and $F_2 = 100$ N. (*Hint*: Be careful with β!)

Section 3.2

8. The output of the MATLAB `atan2` function is in radians. Write a function called `atan2d` that produces an output in degrees.

9. Write a function that accepts temperature in degrees Fahrenheit (°F) and computes the corresponding value in degrees Celsius (°C). The relation between the two is

$$T°C = \frac{5}{9}(T°F - 32)$$

Be sure to test your function.

10.* An object thrown vertically with a speed v_0 reaches a height h at time t, where

$$h = v_0 t - \frac{1}{2} g t^2$$

Write and test a function that computes the time t required to reach a specified height h, for a given value of v_0. The function's inputs should be h, v_0, and g. Test your function for the case where $h = 100$ m, $v_0 = 50$ m/s, and $g = 9.81$ m/s^2. Interpret both answers.

11. A water tank consists of a cylindrical part of radius r and height h and a hemispherical top. The tank is to be constructed to hold 600 m^3 when filled. The surface area of the cylindrical part is $2\pi rh$, and its volume is $\pi r^2 h$. The surface area of the hemispherical top is given by $2\pi r^2$, and its volume is given by $2\pi r^3/3$. The cost to construct the cylindrical part of the tank is $400 per square meter of surface area; the hemispherical part costs $600 per square meter. Use the fminbnd function to compute the radius that results in the least cost. Compute the corresponding height h.

12. A fence around a field is shaped as shown in Figure P12. It consists of a rectangle of length L and width W, and a right triangle that is symmetrical about the central horizontal axis of the rectangle. Suppose the width W is known (in meters) and the enclosed area A is known (in square meters). Write a user-defined function file with W and A as inputs. The outputs are the length L required so that the enclosed area is A and the total length of fence required. Test your function for the values $W = 6$ m and $A = 80$ m^2.

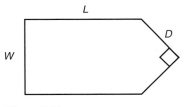

Figure P12

13. A fenced enclosure consists of a rectangle of length L and width $2R$ and a semicircle of radius R, as shown in Figure P13. The enclosure is to be built to have an area A of 2000 ft^2. The cost of the fence is $50 per foot for the curved portion and $40 per foot for the straight sides. Use the fminbnd function to determine with a resolution of 0.01 ft the values of

R and L required to minimize the total cost of the fence. Also compute the minimum cost.

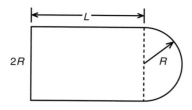

Figure P13

14. Using estimates of rainfall, evaporation, and water consumption, the town engineer developed the following model of the water volume in the reservoir as a function of time

$$V(t) = 10^9 + 10^8(1 - e^{-t/100}) - rt$$

where V is the water volume in liters, t is time in days, and r is the town's consumption rate in liters per day. Write two user-defined functions. The first function should define the function $V(t)$ for use with the `fzero` function. The second function should use `fzero` to compute how long it will take for the water volume to decrease to x percent of its initial value of 10^9 L. The inputs to the second function should be x and r. Test your functions for the case where $x = 50$ percent and $r = 10^7$ L/day.

15. The volume V and paper surface area A of a conical paper cup are given by

$$V = \frac{1}{3}\pi r^2 h \qquad A = \pi r\sqrt{r^2 + h^2}$$

where r is the radius of the base of the cone and h is the height of the cone.

a. By eliminating h, obtain the expression for A as a function of r and V.
b. Create a user-defined function that accepts R as the only argument and computes A for a given value of V. Declare V to be global within the function.
c. For $V = 10$ in.^{3}, use the function with the `fminbnd` function to compute the value of r that minimizes the area A. What is the corresponding value of the height h? Investigate the sensitivity of the solution by plotting V versus r. How much can R vary about its optimal value before the area increases 10 percent above its minimum value?

16.   A torus is shaped like a doughnut. If its inner radius is a and its outer radius is b, its volume and surface area are given by

$$V = \frac{1}{4}\pi^2(a+b)(b-a)^2 \qquad A = \pi^2(b^2-a^2)$$

   a. Create a user-defined function that computes V and A from the arguments a and b.
   b. Suppose that the outer radius is constrained to be 2 in. greater than the inner radius. Write a script file that uses your function to plot A and V versus a for $0.25 \le a \le 4$ in.

17.   Suppose it is known that the graph of the function $y = ax^3 + bx^2 + cx + d$ passes through four given points (x_i, y_i), $i = 1, 2, 3, 4$. Write a user-defined function that accepts these four points as input and computes the coefficients a, b, c, and d. The function should solve four linear equations in terms of the four unknowns a, b, c, and d. Test your function for the case where $(x_i, y_i) = (-2, -20)$, $(0, 4)$, $(2, 68)$, and $(4, 508)$, whose answer is $a = 7$, $b = 5$, $c = -6$, and $d = 4$.

Section 3.3

18.   Create an anonymous function for $10e^{-2x}$ and use it to plot the function over the range $0 \le x \le 2$.

19.   Create an anonymous function for $20x^2 - 200x + 3$ and use it
   a. To plot the function to determine the approximate location of its minimum
   b. With the `fminbnd` function to precisely determine the location of the minimum

20.   Create four anonymous functions to represent the function $6e^{3\cos x^2}$, which is composed of the functions $h(z) = 6e^z$, $g(y) = 3\cos y$, and $f(x) = x^2$. Use the anonymous functions to plot $6e^{3\cos x^2}$ over the range $0 \le x \le 4$.

21.   Use a primary function with a subfunction to compute the zeros of the function $3x^3 - 12x^2 - 33x + 80$ over the range $-10 \le x \le 10$.

22.   Create a primary function that uses a function handle with a nested function to compute the minimum of the function $20x^2 - 200x + 12$ over the range $0 \le x \le 10$.

Section 3.4

23. Use a text editor to create a file containing the following data. Then use the `load` function to load the file into MATLAB, and use the `mean` function to compute the mean value of each column.

55	42	98
49	39	95
63	51	92
58	45	90

24. Enter and save the data given in Problem 23 in a spreadsheet. Then import the spreadsheet file into the MATLAB variable A. Use MATLAB to compute the sum of each column.

25. Use a text editor to create a file from the data given in Problem 23, but separate each number with a semicolon. Then use the Import Wizard to load and save the data in the MATLAB variable A.

Courtesy Henry Guckel (Late)/Dept. of Electrical Engineering, University of Wisconsin.

Engineering in the 21st Century. . .

Nanotechnology

While large-scale technology is attracting much public attention, many of the engineering challenges and opportunities in the 21st century will involve the development of extremely small devices and even the manipulation of individual atoms. This technology is called *nanotechnology* because it involves processing materials whose size is about 1 nanometer (nm), which is 10^{-9} m, or $1/1\,000\,000$ mm. The distance between atoms in single-crystal silicon is 0.5 nm.

Nanotechnology is in its infancy, although some working devices have been created. The micromotor with gear train shown above has a dimension of approximately 10^{-4} m. This device converts electrical input power into mechanical motion. It was constructed using the magnetic properties of electroplated metal films.

While we are learning how to make such devices, another challenge is to develop innovative applications for them. Many of the applications proposed thus far are medical; small pumps for drug delivery and surgical tools are two examples. Researchers at the Lawrence Livermore Laboratory have developed a microgripper tool to treat brain aneurysms. It is about the size of a grain of sand and was constructed from silicon cantilever beams powered by a shape-memory alloy actuator. To design and apply these devices, engineers must first model the appropriate mechanical and electrical properties. The features of MATLAB provide excellent support for such analyses. ■

Programming with MATLAB

OUTLINE

4.1   Program Design and Development

4.2   Relational Operators and Logical Variables

4.3   Logical Operators and Functions

4.4   Conditional Statements

4.5   `for` Loops

4.6   `while` Loops

4.7   The `switch` Structure

4.8   Debugging MATLAB Programs

4.9   Applications to Simulation

4.10   Summary

Problems

The MATLAB interactive mode is very useful for simple problems, but more-complex problems require a script file. Such a file can be called a *computer program,* and writing such a file is called *programming.* Section 4.1 presents a general and efficient approach to the design and development of programs.

The usefulness of MATLAB is greatly increased by the use of decision-making functions in its programs. These functions enable you to write programs whose operations depend on the results of calculations made by the program. Sections 4.2, 4.3, and 4.4 deal with these decision-making functions.

MATLAB can also repeat calculations a specified number of times or until some condition is satisfied. This feature enables engineers to solve problems of great complexity or requiring numerous calculations. These "loop" structures are covered in Sections 4.5 and 4.6.

The `switch` structure enhances the MATLAB decision-making capabilities. This topic is covered in Section 4.7. Use of the MATLAB Editor/Debugger for debugging programs is covered in Section 4.8.

Section 4.9 discusses "simulation," a major application of MATLAB programs that enables us to study the operation of complicated systems, processes, and organizations. Tables summarizing the MATLAB commands introduced in this chapter appear throughout the chapter, and Table 4.10–1 will help you locate the information you need.

4.1 Program Design and Development

In this chapter we introduce *relational operators,* such as > and ==, and the two types of loops used in MATLAB, the `for` loop and the `while` loop. These features, plus MATLAB functions and the *logical operators* to be introduced in Section 4.3, form the basis for constructing MATLAB programs to solve complex problems. Design of computer programs to solve complex problems needs to be done in a systematic manner from the start to avoid time-consuming and frustrating difficulties later in the process. In this section we show how to structure and manage the program design process.

Algorithms and Control Structures

An *algorithm* is an ordered sequence of precisely defined instructions that performs some task in a finite amount of time. An ordered sequence means that the instructions can be numbered, but an algorithm often must have the ability to alter the order of its instructions using what is called a *control structure.* There are three categories of algorithmic operations:

> *Sequential operations.*   These instructions are executed in order.
>
> *Conditional operations.*   These control structures first ask a question to be answered with a true/false answer and then select the next instruction based on the answer.
>
> *Iterative operations (loops).*   These control structures repeat the execution of a block of instructions.

Not every problem can be solved with an algorithm, and some potential algorithmic solutions can fail because they take too long to find a solution.

Structured Programming

Structured programming is a technique for designing programs in which a hierarchy of *modules* is used, each having a single entry and a single exit point, and in which control is passed downward through the structure without unconditional branches to higher levels of the structure. In MATLAB these modules can be built-in or user-defined functions.

Control of the program flow uses the same three types of control structures used in algorithms: sequential, conditional, and iterative. In general, any computer program can be written with these three structures. This realization led to the development of structured programming. Languages suitable for structured programming, such as MATLAB, thus do not have an equivalent to the `goto` statement that you might have seen in the BASIC and FORTRAN languages. An unfortunate result of the `goto` statement was confusing code, called *spaghetti code,* composed of a complex tangle of branches.

Structured programming, if used properly, results in programs that are easy to write, understand, and modify. The advantages of structured programming are as follows.

1. Structured programs are easier to write because the programmer can study the overall problem first and deal with the details later.
2. Modules (functions) written for one application can be used for other applications (this is called *reusable code*).
3. Structured programs are easier to debug because each module is designed to perform just one task, and thus it can be tested separately from the other modules.
4. Structured programming is effective in a teamwork environment because several people can work on a common program, each person developing one or more modules.
5. Structured programs are easier to understand and modify, especially if meaningful names are chosen for the modules and if the documentation clearly identifies the module's task.

Top-down Design and Program Documentation

A method for creating structured programs is *top-down design,* which aims to describe a program's intended purpose at a very high level initially and then partition the problem repeatedly into more detailed levels, one level at a time, until enough is understood about the program structure to enable it to be coded. Table 4.1–1, which is repeated from Chapter 1, summarizes the process of

Table 4.1–1 Steps for developing a computer solution

1. State the problem concisely.
2. Specify the data to be used by the program. This is the *input.*
3. Specify the information to be generated by the program. This is the *output.*
4. Work through the solution steps by hand or with a calculator; use a simpler set of data if necessary.
5. Write and run the program.
6. Check the output of the program with your hand solution.
7. Run the program with your input data and perform a "reality check" on the output. Does it make sense? Estimate the range of the expected result and compare it with you answer.
8. If you will use the program as a general tool in the future, test it by running it for a range of reasonable data values; perform a reality check on the results.

top-down design. In step 4 you create the algorithms used to obtain the solution. Note that step 5, Write and run the program, is only part of the top-down design process. In this step you create the necessary modules and test them separately.

Two types of charts aid in developing structured programs and in documenting them. These are *structure charts* and *flowcharts*. A structure chart is a graphical description showing how the different parts of the program are connected. This type of diagram is particularly useful in the initial stages of top-down design.

A structure chart displays the organization of a program without showing the details of the calculations and decision processes. For example, we can create program modules using function files that do specific, readily identifiable tasks. Larger programs are usually composed of a main program that calls on the modules to do their specialized tasks as needed. A structure chart shows the connection between the main program and the modules.

For example, suppose you want to write a program that plays a game, say, Tic-Tac-Toe. You need a module to allow the human player to input a move, a module to update and display the game grid, and a module that contains the computer's strategy for selecting its moves. Figure 4.1–1 shows the structure chart of such a program.

Flowcharts are useful for developing and documenting programs that contain conditional statements, because they can display the various paths (called *branches*) that a program can take, depending on how the conditional statements are executed. The flowchart representation of the verbal description of the `if` statement (covered in Section 4.3) is shown in Figure 4.1–2. Flowcharts use the diamond symbol to indicate decision points.

The usefulness of structure charts and flowcharts is limited by their size. For large, more complicated programs, it might be impractical to draw such charts. Nevertheless, for smaller projects, sketching a flowchart and/or a structure chart might help you organize your thoughts before you begin to write the specific MATLAB code. Because of the space required for such charts we do not use them in this text. You are encouraged, however, to use them when solving problems.

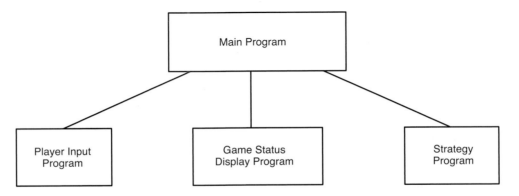

Figure 4.1–1 Structure chart of a game program.

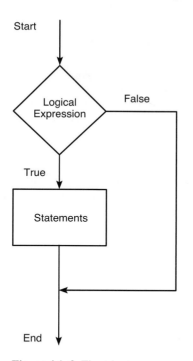

Figure 4.1–2 Flowchart
representation of the verbal
description of the `if` statement.

Documenting programs properly is very important, even if you never give your programs to other people. If you need to modify one of your programs, you will find that it is often very difficult to recall how it operates if you have not used it for some time. Effective documentation can be accomplished with the use of

1. Proper selection of variable names to reflect the quantities they represent.
2. Comments within the program.
3. Structure charts.
4. Flowcharts.
5. A verbal description of the program, often in *pseudocode*.

The advantage of using suitable variable names and comments is that they reside with the program; anyone who gets a copy of the program will see such documentation. However, they often do not provide enough of an overview of the program. The latter three elements can provide such an overview.

Pseudocode

Use of natural language, such as English, to describe algorithms often results in a description that is too verbose and is subject to misinterpretation. To avoid dealing immediately with the possibly complicated syntax of the programming language, we can instead use *pseudocode,* in which natural language and mathematical expressions are used to construct statements that look like computer statements but without detailed syntax. Pseudocode may also use some simple MATLAB syntax to explain the operation of the program.

As its name implies, pseudocode is an imitation of the actual computer code. The pseudocode can provide the basis for comments within the program. In addition to providing documentation, pseudocode is useful for outlining a program before writing the detailed code, which takes longer to write because it must conform to the strict rules of MATLAB.

Each pseudocode instruction may be numbered, but should be unambiguous and computable. Note that MATLAB does not use line numbers except in the Debugger. Each of the following examples illustrates how pseudocode can document each of the control structures used in algorithms: sequential, conditional, and iterative operations.

Example 1. Sequential Operations   Compute the perimeter p and the area A of a triangle whose sides are a, b, and c. The formulas are

$$p = a + b + c \qquad s = \frac{p}{2} \qquad A = \sqrt{s(s - a)(s - b)(s - c)}$$

1. Enter the side lengths a, b, and c.
2. Compute the perimeter p.

$$p = a + b + c$$

3. Compute the semiperimeter s.

$$s = \frac{p}{2}$$

4. Compute the area A.

$$A = \sqrt{s(s - a)(s - b)(s - c)}$$

5. Display the results p and A.
6. Stop.

The program is

```
a = input('Enter the value of side a: ');
b = input('Enter the value of side b: ');
c = input('Enter the value of side c: ');
p = a + b + c;
s = p/2;
A = sqrt(s*(s-a)*(s-b)*(s-c));
disp('The perimeter is:')
p
disp('The area is:')
A
```

Example 2. Conditional Operations   Given the (x, y) coordinates of a point, compute its polar coordinates (r, θ), where

$$r = \sqrt{x^2 + y^2} \quad \theta = \tan^{-1}\left(\frac{y}{x}\right)$$

1. Enter the coordinates x and y.
2. Compute the hypoteneuse r.

   ```
   r = sqrt(x^2+y^2)
   ```

3. Compute the angle θ.

   3.1 If $x \geq 0$

   ```
   theta = atan(y/x)
   ```

   3.2 Else

   ```
   theta = atan(y/x) + pi
   ```

4. Convert the angle to degrees.

   ```
   theta = theta*(180/pi)
   ```

5. Display the results `r` and `theta`.
6. Stop.

Note the use of the numbering scheme 3.1 and 3.2 to indicate subordinate clauses. Note also that MATLAB syntax may be used for clarity where needed. The following program implements the pseudocode using some of the MATLAB features to be introduced in this chapter. It uses the relational operator >=, which means "greater than or equal to" (in Section 4.2). The program also uses the "if-else-end" construct, which is covered in Section 4.3.

```
x = input('Enter the value of x: ');
y = input('Enter the value of y: ');
r = sqrt(x^2+y^2);
if x >= 0
   theta = atan(y/x);
else
   theta = atan(y/x) + pi;
end
disp('The hypoteneuse is:')
disp(r)
theta = theta*(180/pi);
disp('The angle is degrees is:')
disp(theta)
```

Example 3. Iterative Operations   Determine how many terms are required for the sum of the series $10k^2 - 4k + 2$, $k = 1, 2, 3, \ldots$ to exceed 20 000. What is the sum for this many terms?

   Because we do not know how many times we must evaluate the expression $10k^2 - 4k + 2$, we use a `while` loop, which is covered in Section 4.6.

1. Initialize the total to zero.
2. Initialize the counter to zero.
3. While the total is less than 20 000 compute the total.
   3.1 Increment the counter by 1.
   ```
   k = k + 1
   ```
   3.2 Update the total.
   ```
   total = 10*k^2 - 4*k + 2 + total
   ```
4. Display the current value of the counter.
5. Display the value of the total.
6. Stop.

The following program implements the pseudocode. The statements in the while loop are executed until the variable total equals or exceeds 2×10^4.

```
total = 0;
k = 0;
while total < 2e+4
    k = k+1;
    total = 10*k^2 - 4*k + 2 + total;
end
disp('The number of terms is:')
disp(k)
disp('The sum is:')
disp(total)
```

Finding Bugs

Debugging a program is the process of finding and removing the "bugs," or errors, in a program. Such errors usually fall into one of the following categories.

1. Syntax errors such as omitting a parenthesis or comma, or spelling a command name incorrectly. MATLAB usually detects the more obvious errors and displays a message describing the error and its location.
2. Errors due to an incorrect mathematical procedure. These are called *runtime errors*. They do not necessarily occur every time the program is executed; their occurrence often depends on the particular input data. A common example is division by zero.

The MATLAB error messages usually enable you to find syntax errors. However, runtime errors are more difficult to locate. To locate such an error, try the following:

1. Always test your program with a simple version of the problem, whose answers can be checked by hand calculations.
2. Display any intermediate calculations by removing semicolons at the end of statements.

Table 4.2–1 Relational operators

Relational operator	Meaning
$<$	Less than.
$<=$	Less than or equal to.
$>$	Greater than.
$>=$	Greater than or equal to.
$==$	Equal to.
$\sim=$	Not equal to.

3. To test user-defined functions, try commenting out the `function` line and running the file as a script.

4. Use the debugging features of the Editor/Debugger, which is discussed in Section 4.8.

4.2 Relational Operators and Logical Variables

MATLAB has six *relational* operators to make comparisons between arrays. These operators are shown in Table 4.2–1. Note that the *equal to* operator consists of two $=$ signs, not a single $=$ sign as you might expect. The single $=$ sign is the *assignment,* or *replacement,* operator in MATLAB.

The result of a comparison using the relational operators is either 0 (if the comparison is *false*) or 1 (if the comparison is *true*), and the result can be used as a variable. For example, if $x = 2$ and $y = 5$, typing $z = x < y$ returns the value $z = 1$ and typing $u = x == y$ returns the value $u = 0$. To make the statements more readable, we can group the logical operations using parentheses. For example, $z = (x < y)$ and $u = (x == y)$.

When used to compare arrays, the relational operators compare the arrays on an element-by-element basis. The arrays being compared must have the same dimension. The only exception occurs when we compare an array to a scalar. In that case all the elements of the array are compared to the scalar. For example, suppose that $x = [6,3,9]$ and $y = [14,2,9]$. The following MATLAB session shows some examples.

```
>>z = (x < y)
z =
    1    0    0
>>z = (x ~= y)
z =
    1    1    0
>>z = (x > 8)
z =
    0    0    1
```

The relational operators can be used for array addressing. For example, with $x = [6,3,9]$ and $y = [14,2,9]$, typing $z = x(x < y)$ finds all the elements in x that are less than the corresponding elements in y. The result is $z = 6$.

The arithmetic operators $+$, $-$, $*$, $/$, and $\backslash$ have precedence over the relational operators. Thus the statement $z = 5 > 2 + 7$ is equivalent to $z = 5 > (2+7)$ and returns the result $z = 0$. We can use parentheses to change the order of precedence; for example, $z = (5 > 2) + 7$ evaluates to $z = 8$.

The relational operators have equal precedence among themselves, and MATLAB evaluates them in order from left to right. Thus the statement

```
z = 5 > 3 ~= 1
```

is equivalent to

```
z = (5 > 3) ~= 1
```

Both statements return the result $z = 0$.

With relational operators that consist of more than one character, such as $==$ or $>=$, be careful not to put a space between the characters.

The `logical` Class

When the relational operators are used, such as $x = (5 > 2)$, they create a logical variable, in this case, x. Prior to MATLAB 6.5, `logical` was an attribute of any numeric data type. Now `logical` is a first-class data type and a MATLAB class, and so `logical` is now equivalent to other first-class types such as character and cell arrays. Logical variables may have only the values 1 (true) and 0 (false).

Just because an array contains only 0s and 1s, however, it is not necessarily a logical array. For example, in the following session k and w appear the same, but k is a logical array and w is a numeric array, and thus an error message is issued.

```
>>x = -2:2
x =
    -2   -1   0   1   2
>>k = (abs(x)>1)
k =
     1   0   0   0   1
>>z = x(k)
z =
    -2   2
>>w = [1,0,0,0,1];
>>v = x(w)
??? Subscript indices must either be real positive...
    integers or logicals.
```

The `logical` Function

Logical arrays can be created with the relational and logical operators and with the `logical` function. The `logical` function returns an array that can be used for logical indexing and logical tests. Typing $B = logical(A)$, where A is a numeric array, returns the logical array B. So to correct the error in the previous session, you may type instead w = logical([1,0,0,0,1]) before typing v = x(w).

When a finite, real value other than 1 or 0 is assigned to a logical variable, the value is converted to logical 1 and a warning message is issued. For example, when you type y = logical(9), y will be assigned the value logical 1 and a warning will be issued. You may use the double function to convert a logical array to an array of class double. For example, x = (5 > 3); y = double(x);. Some arithmetic operations convert a logical array to a double array. For example, if we add zero to each element of B by typing B = B + 0, then B will be converted to a numeric (double) array. However, not all mathematical operations are defined for logical variables. For example, typing

```
>>x = ([2, 3] > [1, 6]);
>>y = sin(x)
```

will generate an error message. This is not an important issue because it hardly makes sense to compute the sine of logical data or logical variables.

Accessing Arrays Using Logical Arrays

When a logical array is used to address another array, it extracts from that array the elements in the locations where the logical array has 1s. So typing A(B), where B is a logical array of the same size as A, returns the values of A at the indices where B is 1.

Given A = [5,6,7;8,9,10;11,12,13] and B = logical (eye(3)), we can extract the diagonal elements of A by typing C = A(B) to obtain C = [5;9;13]. Specifying array subscripts with logical arrays extracts the elements that correspond to the true (1) elements in the logical array.

Note, however, that using the *numeric* array eye(3), as C = A(eye(3)), results in an error message because the elements of eye(3) do not correspond to locations in A. If the numeric array values correspond to valid locations, you may use a numeric array to extract the elements. For example, to extract the diagonal elements of A with a numeric array, type C = A([1,5,9]).

MATLAB data types are preserved when indexed assignment is used. So now that logical is a MATLAB data type, if A is a logical array, for example, A = logical(eye(4)), then typing A(3,4) = 1 does not change A to a double array. However, typing A(3,4) = 5 will set A(3,4) to logical 1 and cause a warning to be issued.

4.3 Logical Operators and Functions

MATLAB has five *logical operators,* which are sometimes called *Boolean* operators (see Table 4.3–1). These operators perform element-by-element operations. With the exception of the NOT operator (~), they have a lower precedence than the arithmetic and relational operators (see Table 4.3–2). The NOT symbol is called the *tilde.*

Table 4.3–1 Logical operators

Operator	Name	Definition
~	NOT	~A returns an array of the same dimension as A; the new array has 1s where A is 0 and 0s where A is nonzero.
&	AND	A & B returns an array of the same dimension as A and B; the new array has 1s where both A and B have nonzero elements and 0s where either A or B is 0.
|	OR	A | B returns an array of the same dimension as A and B; the new array has 1s where at least one element in A or B is nonzero and 0s where A and B are both 0.
&&	Short-Circuit AND	Operator for scalar logical expressions. A && B returns true if both A and B evaluate to true, and false if they do not.
||	Short-Circuit OR	Operator for scalar logical expressions. A || B returns true if either A or B or both evaluate to true, and false if they do not.

Table 4.3–2 Order of precedence for operator types

Precedence	Operator type
First	Parentheses; evaluated starting with the innermost pair.
Second	Arithmetic operators and logical NOT (~); evaluated from left to right.
Third	Relational operators; evaluated from left to right.
Fourth	Logical AND.
Fifth	Logical OR.

The NOT operation ~A returns an array of the same dimension as A; the new array has 1s where A is 0 and 0s where A is nonzero. If A is logical, then ~A replaces 1s with 0s and 0s with 1s. For example, if x = [0,3,9] and y = [14,-2,9], then z = ~x returns the array z = [1,0,0] and the statement u = ~x > y returns the result u = [0,1,0]. This expression is equivalent to u = (~x) > y, whereas v = ~(x > y) gives the result v = [1,0,1]. This expression is equivalent to v = (x <= y).

The & and | operators compare two arrays of the same dimension. The only exception, as with the relational operators, is that an array can be compared to a scalar. The AND operation A & B returns 1s where both A and B have nonzero elements and 0s where any element of A or B is 0. The expression z = 0 & 3 returns z = 0; z = 2 & 3 returns z = 1; z = 0 & 0 returns z = 0, and z = [5,-3,0,0] & [2,4,0,5] returns z = [1,1,0,0]. Because of operator precedence, z = 1 & 2 + 3 is equivalent to z = 1 & (2 + 3), which returns z = 1. Similarly, z = 5 < 6 & 1 is equivalent to z = (5 < 6) & 1, which returns z = 1.

Let x = [6,3,9] and y = [14,2,9] and let a = [4,3,12]. The expression

z = (x > y) & a

gives z = [0,1,0], and

z = (x > y) & (x > a)

returns the result z = [0,0,0]. This is equivalent to

```
z = x > y & x > a
```

which is much less readable.

Be careful when using the logical operators with inequalities. For example, note that ~(x > y) is equivalent to x <= y. It is *not* equivalent to x < y. As another example, the relation $5 < x < 10$ must be written as

```
(5 < x) & (x < 10)
```

in MATLAB.

The OR operation A|B returns 1s where at least one of A and B has nonzero elements and 0s where both A and B are 0. The expression z = 0|3 returns z = 1; the expression z = 0|0 returns z = 0; and

```
z = [5,-3,0,0]|[2,4,0,5]
```

returns z = [1,1,0,1]. Because of operator precedence,

```
z = 3 < 5|4 == 7
```

is equivalent to

```
z = (3 < 5)|(4 == 7)
```

which returns z = 1. Similarly, z = 1|0 & 1 is equivalent to z = (1|0) & 1, which returns z = 1, while z = 1|0 & 0 returns z = 0, and z = 0 & 0|1 returns z = 1.

Because of the precedence of the NOT operator, the statement

```
z = ~3 == 7|4 == 6
```

returns the result z = 0, which is equivalent to

```
z = ((~3) == 7)|(4 == 6)
```

The exclusive OR function xor(A,B) returns 0s where A and B are either both nonzero or both 0, and 1s where either A or B is nonzero, *but not both*. The function is defined in terms of the AND, OR, and NOT operators as follows.

```
function z = xor(A,B)
z = (A|B) & ~(A & B);
```

The expression

```
z = xor([3,0,6],[5,0,0])
```

returns z = [0,0,1], whereas

```
z = [3,0,6]|[5,0,0]
```

returns z = [1,0,1].

Table 4.3–3 is a *truth table* that defines the operations of the logical operators and the function xor. Until you acquire more experience with the logical

TRUTH TABLE

Table 4.3–3 Truth table

x	y	~x	x|y	x & y	xor(x,y)
true	true	false	true	true	false
true	false	false	true	false	true
false	true	true	true	false	true
false	false	true	false	false	false

operators, you should use this table to check your statements. Remember that *true* is equivalent to logical 1, and *false* is equivalent to logical 0. We can test the truth table by building its numerical equivalent as follows. Let x and y represent the first two columns of the truth table in terms of 1s and 0s.

The following MATLAB session generates the truth table in terms of 1s and 0s.

```
>>x = [1,1,0,0]';
>>y = [1,0,1,0]';
>>Truth_Table = [x,y,~x,x|y,x & y,xor(x,y)]
Truth_Table =
   1  1  0  1  1  0
   1  0  0  1  0  1
   0  1  1  1  0  1
   0  0  1  0  0  0
```

Starting with MATLAB 6, the AND operator (&) was given a higher precedence than the OR operator (|). This was not true in earlier versions of MATLAB, so if you are using code created in an earlier version, you should make the necessary changes before using it in MATLAB 6 or higher. For example, now the statement y = 1|5 & 0 is evaluated as y = 1|(5 & 0), yielding the result y = 1, whereas in MATLAB 5.3 and earlier, the statement would have been evaluated as y = (1|5) & 0, yielding the result y = 0. To avoid potential problems due to precedence, it is important to use parentheses in statements containing arithmetic, relational, or logical operators, even where parentheses are optional. MATLAB now provides a feature to enable the system to produce either an error message or a warning for any expression containing & and | that would be evaluated differently than in earlier versions. If you do not use this feature, MATLAB will issue a warning as the default. To activate the error feature, type feature('OrAndError',1). To reinstate the default, type feature('OrAndError',0).

Short-Circuit Operators

The following operators perform AND and OR operations on logical expressions containing *scalar* values only. They are called short-circuit operators because they evaluate their second operand only when the result is not fully determined by the first operand. They are defined as follows in terms of the two logical variables A and B.

A && B   Returns true (logical 1) if both A and B evaluate to true, and false (logical 0) if they do not.

A | | B   Returns true (logical 1) if either A or B or both evaluate to true, and false (logical 0) if they do not.

Thus in the statement A && B, if A equals logical zero, then the entire expression will evaluate to false, regardless of the value of B, and therefore there is no need to evaluate B.

For A | | B, if A is true, regardless of the value of B, the statement will evaluate to true.

Table 4.3–4 lists several useful logical functions.

Table 4.3–4 Logical functions

Logical function	Definition
all(x)	Returns a scalar, which is 1 if all the elements in the vector x are nonzero and 0 otherwise.
all(A)	Returns a row vector having the same number of columns as the matrix A and containing 1s and 0s, depending on whether the corresponding column of A has all nonzero elements.
any(x)	Returns a scalar, which is 1 if any of the elements in the vector x is nonzero and 0 otherwise.
any(A)	Returns a row vector having the same number of columns as A and containing 1s and 0s, depending on whether the corresponding column of the matrix A contains any nonzero elements.
find(A)	Computes an array containing the indices of the nonzero elements of the array A.
[u,v,w] = find(A)	Computes the arrays u and v containing the row and column indices of the nonzero elements of the array A and computes the array w containing the values of the nonzero elements. The array w may be omitted.
finite(A)	Returns an array of the same dimension as A with 1s where the elements of are finite and 0s elsewhere.
ischar(A)	Returns a 1 if A is a character array and 0 otherwise.
isempty(A)	Returns a 1 if A is an empty matrix and 0 otherwise.
isinf(A)	Returns an array of the same dimension as A, with 1s where A has 'inf' and 0s elsewhere.
isnan(A)	Returns an array of the same dimension as A with 1s where A has 'NaN' and 0s elsewhere. ('NaN' stands for "not a number," which means an undefined result.)
isnumeric(A)	Returns a 1 if A is a numeric array and 0 otherwise.
isreal(A)	Returns a 1 if A has no elements with imaginary parts and 0 otherwise.
logical(A)	Converts the elements of the array A into logical values.
xor(A,B)	Returns an array the same dimension as A and B; the new array has 1s where either A or B is nonzero, but not both, and 0s where A and B are either both nonzero or both zero.

Logical Operators and the find Function

The find function is very useful for creating decision-making programs, especially when combined with the relational or logical operators. The function find(x) computes an array containing the indices of the nonzero elements of the array x. For example, consider the session

```
>>x = [-2, 0, 4];
>>y = find(x)
y =
    1    3
```

The resulting array y = [1, 3] indicates that the first and third elements of x are nonzero. Note that the find function returns the *indices,* not the *values.* In the following session, note the difference between the result obtained by x(x < y) and the result obtained by find(x < y).

```
>>x = [6, 3, 9, 11];y = [14, 2, 9, 13];
>>values = x(x < y)
values =
      6     11
>>how_many = length (values)
how_many =
      2
>>indices = find(x < y)
indices =
      1   4
```

Thus two values in the array x are less than the *corresponding* values in the array y. They are the first and fourth values, 6 and 11. To find out how many, we could also have typed length(indices).

The find function is also useful when combined with the logical operators. For example, consider the session

```
>>x = [5, -3, 0, 0, 8]; y = [2, 4, 0, 5, 7];
>>z = find(x & y)
z =
    1    2    5
```

The resulting array z = [1, 2, 5] indicates that the first, second, and fifth elements of both x and y are nonzero. Note that the find function returns the *indices,* and not the *values.* In the following session, note the difference between the result obtained by y(x & y) and the result obtained by find(x & y) above.

```
>>x = [5, -3, 0, 0, 8];y = [2, 4, 0, 5, 7];
>>values = y(x & y)
values =
    2    4    7
>>how_many = length(values)
how_many =
    3
```

Thus there are three nonzero values in the array y that correspond to nonzero values in the array x. They are the first, second, and fifth values, which are 2, 4, and 7.

In the above examples, there were only a few numbers in the arrays x and y, and thus we could have obtained the answers by visual inspection. However, these MATLAB methods are very useful either where there are so many data that visual inspection would be very time-consuming, or where the values are generated internally in a program.

Test Your Understanding

T4.3–1 If x = [5,-3,18,4] and y = [-9,13,7,4], what will be the result of the following operations? Use MATLAB to check your answer.

 a. z = ~y > x
 b. z = x & y
 c. z = x | y
 d. z = xor(x,y)

T4.3–2 Suppose that x = [-9, -6, 0, 2, 5] and y = [-10, -6 2, 4, 6]. What is the result of the following operations? Determine the answers by hand, and then use MATLAB to check your answers.

 a. z = (x < y)
 b. z = (x > y)
 c. z = (x ~= y)
 d. z = (x == y)
 e. z = (x > 2)

T4.3–3 Suppose that x = [-4, -1, 0, 2, 10] and y = [-5, -2, 2, 5, 9]. Use MATLAB to find the values and the indices of the elements in x that are greater than the corresponding elements in y.

Height and Speed of a Projectile

EXAMPLE 4.3–1

The height and speed of a projectile (such as a thrown ball) launched with a speed of v_0 at an angle A to the horizontal are given by

$$h(t) = v_0 t \sin A - 0.5gt^2$$
$$v(t) = \sqrt{v_0^2 - 2v_0 gt \sin A + g^2 t^2}$$

where g is the acceleration due to gravity. The projectile will strike the ground when $h(t) = 0$, which gives the time to hit $t_{hit} = 2(v_0/g)\sin A$. Suppose that $A = 40°$, $v_0 = 20$ m/s, and $g = 9.81$ m/s². Use the MATLAB relational and logical operators to find the times when the height is no less than 6 m and the speed is simultaneously no greater than 16 m/s. In addition, discuss another approach to obtaining a solution.

■ **Solution**

The key to solving this problem with relational and logical operators is to use the find command to determine the times at which the logical expression (h >= 6) & (v <= 16) is true. First we must generate the vectors h and v corresponding to times t_1 and t_2 between $0 \leq t \leq t_{hit}$, using a spacing for time t that is small enough to achieve sufficient accuracy for our purposes. We will choose a spacing of $t_{hit}/100$, which provides 101 values of time. The program follows. When computing the times t_1 and t_2, we must subtract 1 from u(1) and from length(u) because the first element in the array t corresponds to $t = 0$ (that is, t(1) is 0).

```
% Set the values for initial speed, gravity, and angle.
v0 = 20; g = 9.81; A = 40*pi/180;
% Compute the time to hit.
t_hit = 2*v0*sin(A)/g;
% Compute the arrays containing time, height, and speed.
t = 0:t_hit/100:t_hit;
h = v0*t*sin(A) - 0.5*g*t.^2;
v = sqrt(v0^2 - 2*v0*g*sin(A)*t + g^2*t.^2);
% Determine when the height is no less than 6
% and the speed is no greater than 16.
u = find(h >= 6 & v <= 16);
% Compute the corresponding times.
t_1 = (u(1)- 1)*(t_hit/100)
t_2 = u(length(u)- 1)*(t_hit/100)
```

The results are $t_1 = 0.8649$ and $t_2 = 1.7560$. Between these two times $h \geq 6$ m and $v \leq 16$ m/s.

We could have solved this problem by plotting $h(t)$ and $v(t)$, but the accuracy of the results would be limited by our ability to pick points off the graph; in addition, if we had to solve many such problems, the graphical method would be more time-consuming.

Test Your Understanding

T4.3–4 Consider the problem given in Example 4.3–1. Use relational and logical operators to find the times for which either the projectile's height is less than 4 m or the speed is greater than 17 m/s. Plot $h(t)$ and $v(t)$ to confirm your answer.

4.4 Conditional Statements

In everyday language we describe our decision making by using conditional phrases such as "If I get a raise, I will buy a new car." If the statement "I get a raise" is true, the action indicated (buy a new car) will be executed. Here is another example: If I get at least a $100 per week raise, I will buy a new car; else, I will put the raise into savings. A slightly more involved example is: If I get at least a $100 per week raise, I will buy a new car; else, if the raise is greater than $50, I will buy a new stereo; otherwise, I will put the raise into savings.

We can illustrate the logic of the first example as follows:

```
If I get a raise,
    I will buy a new car
. (period)
```

Note how the period marks the end of the statement.

The second example can be illustrated as follows:

```
If I get at least a $100 per week raise,
    I will buy a new car;
else,
    I will put the raise into savings
. (period)
```

The third example follows.

```
If I get at least a $100 per week raise,
    I will buy a new car;
else, if the raise is greater than $50,
    I will buy a new stereo;
otherwise,
    I will put the raise into savings
. (period)
```

The MATLAB *conditional statements* enable us to write programs that make decisions. Conditional statements contain one or more of the if, else, and elseif statements. The end statement denotes the end of a conditional statement, just as the period was used in the preceding examples. These conditional statements have a form similar to the examples, and they read somewhat like their English-language equivalents.

The if Statement

The if statement's basic form is

```
if logical expression
    statements
end
```

Every if statement must have an accompanying end statement. The end statement marks the end of the *statements* that are to be executed if the *logical expression* is true. A space is required between the if and the *logical expression,* which may be a scalar, a vector, or a matrix.

For example, suppose that x is a scalar and that we want to compute $y = \sqrt{x}$ only if $x \geq 0$. In English, we could specify this procedure as follows: If x is greater than or equal to zero, compute y from $y = \sqrt{x}$. The following if statement implements this procedure in MATLAB, assuming x already has a scalar value.

```
if x >= 0
    y = sqrt(x)
end
```

If x is negative, the program takes no action. The *logical expression* here is $x \geq 0$, and the *statement* is the single line $y = \text{sqrt}(x)$.

The `if` structure may be written on a single line; for example,

```
if x >= 0, y = sqrt(x), end
```

However, this form is less readable than the previous form. The usual practice is to indent the *statements* to clarify which statements belong to the `if` and its corresponding `end` and thereby improve readability.

The *logical expression* may be a compound expression; the *statements* may be a single command or a series of commands separated by commas or semicolons or on separate lines. For example, if x and y have scalar values,

```
z = 0;w = 0;
if (x >= 0)&(y >= 0)
    z = sqrt(x) + sqrt(y)
    w = sqrt(x*y)
end
```

The values of z and w are computed only if both x and y are nonnegative. Otherwise, z and w retain their values of zero. The flowchart is shown in Figure 4.4–1.

We may "nest" `if` statements, as shown by the following example.

```
if logical expression 1
    statement group 1
    if logical expression 2
        statement group 2
    end
end
```

Note that each `if` statement has an accompanying `end` statement.

The `else` Statement

When more than one action can occur as a result of a decision, we can use the `else` and `elseif` statements along with the `if` statement. The basic structure for the use of the `else` statement is

```
if logical expression
    statement group 1
else
    statement group 2
end
```

Figure 4.4–2 shows the flowchart of this structure.

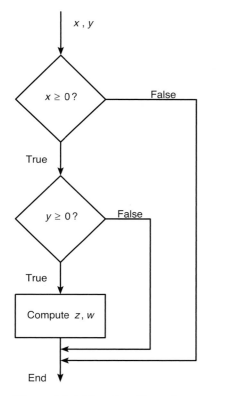

Figure 4.4–1 Flowchart illustrating two logical tests.

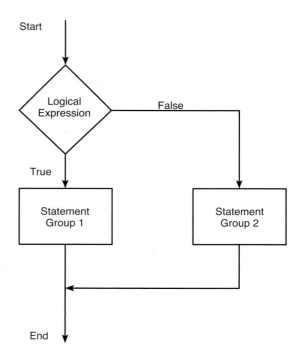

Figure 4.4–2 Flowchart of the `else` structure.

For example, suppose that $y = \sqrt{x}$ for $x \geq 0$ and that $y = e^x - 1$ for $x < 0$. The following statements will calculate y, assuming that x already has a scalar value.

```
if x >= 0
    y = sqrt(x)
else
    y = exp(x) - 1
end
```

When the test, if *logical expression,* is performed, where the logical expression may be an *array,* the test returns a value of true only if *all* the elements of the logical expression are true! For example, if we fail to recognize how the test works, the following statements do not perform the way we might expect.

```
x = [4,-9,25];
if x < 0
    disp('Some of the elements of x are negative.')
else
    y = sqrt(x)
end
```

When this program is run, it gives the result

```
y =
   2      0 + 3.000i      5
```

The program does not test each element in x in sequence. Instead it tests the truth of the vector relation $x < 0$. The test `if x < 0` returns a false value because it generates the vector $[0,1,0]$. Compare the preceding program with the following program.

```
x = [4,-9,25];
if x >= 0
    y = sqrt(x)
else
    disp('Some of the elements of x are negative.')
end
```

When executed, it produces the following result: `Some of the elements of x are negative`. The test `if x < 0` is false, and the test `if x >= 0` also returns a false value because $x >= 0$ returns the vector $[1,0,1]$.

We sometimes must choose between a program that is concise, but perhaps more difficult to understand, and one that uses more statements than is necessary. For example, the statements

`if` *logical expression 1*
    `if` *logical expression 2*
        statements
    `end`
`end`

can be replaced with the more concise program

`if` *logical expression 1* & *logical expression 2*
    statements
`end`

The `elseif` Statement

The general form of the `if` statement is

`if` *logical expression 1*
    statement group 1
`elseif` *logical expression 2*
    statement group 2
`else`
    statement group 3
`end`

The `else` and `elseif` statements may be omitted if not required. However, if both are used, the `else` statement must come after the `elseif` statement to take care of all conditions that might be unaccounted for. Figure 4.4–3 is the flowchart for the general `if` structure.

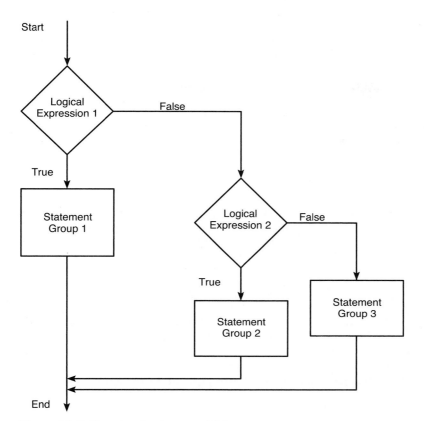

Figure 4.4–3 Flowchart for the general `if` structure.

For example, suppose that $y = \ln x$ if $x \geq 5$ and that $y = \sqrt{x}$ if $0 \leq x < 5$. The following statements will compute y if x has a scalar value.

```
if x >= 5
   y = log(x)
else
   if x >= 0
      y = sqrt(x)
   end
end
```

If $x = -2$, for example, no action will be taken. If we use an `elseif`, we need fewer statements. For example,

```
if x >= 5
   y = log(x)
elseif x >= 0
   y = sqrt(x)
end
```

Note that the `elseif` statement does not require a separate `end` statement.

The `else` statement can be used with `elseif` to create detailed decision-making programs. For example, suppose that $y = \ln x$ for $x > 10$, $y = \sqrt{x}$ for $0 \leq x \leq 10$, and $y = e^x - 1$ for $x < 0$. The following statements will compute y if x already has a scalar value.

```
if x > 10
   y = log(x)
elseif x >= 0
   y = sqrt(x)
else
   y = exp(x) - 1
end
```

Decision structures may be *nested;* that is, one structure can contain another structure, which in turn can contain another, and so on. Indentations are used to emphasize the statement groups associated with each `end` statement.

Test Your Understanding

T4.4-1 Given a number x and the quadrant q ($q = 1, 2, 3, 4$), write a program to compute $\sin^{-1}(x)$ in degrees, taking into account the quadrant. The program should display an error message if $|x| > 1$.

Checking the Number of Input and Output Arguments

Sometimes you will want to have a function act differently depending on how many inputs it has. You can use the function `nargin`, which stands for "number of input arguments." Within the function you can use conditional statements to direct the flow of the computation depending on how many input arguments there are. For example, suppose you want to compute the square root of the input if there is only one, but compute the square root of the average if there are two inputs. The following function does this.

```
function z = sqrtfun(x, y)
if (nargin == 1)
  z = sqrt(x);
elseif (nargin == 2)
  z = sqrt((x + y)/2);
end
```

The `nargout` function can be used to determine the number of output arguments.

Strings and Conditional Statements

A string is a variable that contains characters. Strings are useful for creating input prompts and messages and for storing and operating on data such as names and addresses. To create a string variable, enclose the characters in single quotes. For example, the string variable `name` is created as follows:

```
>>name = 'Leslie Student'
name =
     Leslie Student
```

The following string, called `number`,

```
>>number = '123'
number =
     123
```

is *not* the same as the variable `number` created by typing `number = 123`.

Strings are stored as row vectors in which each column represents a character. For example, the variable `name` has 1 row and 14 columns (each blank space occupies one column). We can access any column the way we access any other vector. For example, the letter S in the name Leslie Student occupies the eighth column in the vector `name`. It can be accessed by typing `name(8)`.

One of the most important applications for strings is to create input prompts and output messages. The following prompt program uses the `isempty(x)` function, which returns a 1 if the array `x` is empty and 0 otherwise. It also uses the `input` function, whose syntax is

```
x = input('prompt', 'string')
```

This function displays the string *prompt* on the screen, waits for input from the keyboard, and returns the entered value in the string variable `x`. The function returns an empty matrix if you press the **Enter** key without typing anything.

The following prompt program is a script file that allows the user to answer Yes by typing either Y or y or by pressing the **Enter** key. Any other response is treated as a No answer.

```
response = input('Do you want to continue? Y/N [Y]: ','s');
if (isempty(response))|(response == 'Y')|(response == 'y')
   response = 'Y'
else
   response = 'N'
end
```

Many more string functions are available in MATLAB. Type `help strfun` to obtain information on these.

4.5 for Loops

A *loop* is a structure for repeating a calculation a number of times. Each repetition of the loop is a *pass*. MATLAB uses two types of explicit loops: the `for` loop, when the number of passes is known ahead of time, and the `while` loop, when the looping process must terminate when a specified condition is satisfied and thus the number of passes is not known in advance.

A simple example of a `for` loop is

```
for k = 5:10:35
    x = k^2
end
```

The *loop variable* k is initially assigned the value 5, and x is calculated from
x = k^2. Each successive pass through the loop increments k by 10 and calculates x until k exceeds 35. Thus k takes on the values 5, 15, 25, and 35; and x
takes on the values 25, 225, 625, and 1225. The program then continues to execute any statements following the `end` statement.

The typical structure of a `for` loop is

```
for loop variable = m:s:n
    statements
end
```

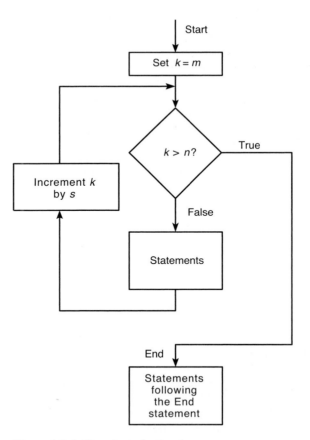

Figure 4.5–1 Flowchart of a `for` loop.

The expression m:s:n assigns an initial value of m to the loop variable, which is incremented by the value s, called the *step value* or *incremental value*. The *statements* are executed once during each pass, using the current value of the loop variable. The looping continues until the loop variable exceeds the *terminating value* n. For example, in the expression for k = 5:10:36, the final value of k is 35. Note that we need not place a semicolon after the for m:s:n statement to suppress printing k. Figure 4.5–1 shows the flowchart of a for loop.

Note that a for statement needs an accompanying end statement. The end statement marks the end of the *statements* that are to be executed. A space is required between the for and the *loop variable,* which may be a scalar, a vector, or a matrix, although the scalar case is by far the most common.

The for loop may be written on a single line; for example,

```
for x = 0:2:10, y = sqrt(x), end
```

However, this form is less readable than the previous form. The usual practice is to indent the *statements* to clarify which statements belong to the for and its corresponding end and thereby improve readability.

Series Calculation with a for Loop

EXAMPLE 4.5–1

Write a script file to compute the sum of the first 15 terms in the series $5k^2 - 2k$, $k = 1, 2, 3, \ldots, 15$.

■ Solution

Because we know how many times we must evaluate the expression $5k^2 - 2k$, we can use a for loop. The script file is the following:

```
total = 0;
for k = 1:15
    total = 5*k^2 - 2*k + total;
end
disp ('The sum for 15 terms is:')
disp (total)
```

The answer is 5960.

Plotting with a for Loop

EXAMPLE 4.5–2

Write a script file to plot the function

$$y = \begin{cases} 15\sqrt{4x} + 10 & x \geq 9 \\ 10x + 10 & 0 \leq x < 9 \\ 10 & x < 0 \end{cases}$$

for $-5 \leq x \leq 30$.

■ Solution

We choose a spacing $dx = 35/300$ to obtain 301 points, which is sufficient to obtain a smooth plot. The script file is the following:

```
dx = 35/300;
x = -5:dx:30;
for k = 1:length(x)
   if x(k) >= 9
     y(k) = 15*sqrt(4*x(k)) + 10;
   elseif x(k) >= 0
     y(k) = 10*x(k) + 10;
   else
     y(k) = 10;
   end
end
plot (x,y), xlabel('x'), ylabel('y')
```

Note that we must use the index k to refer to x within the loop, as x(k).

NESTED LOOPS

We may nest loops and conditional statements, as shown by the following example. (Note that each for and if statement needs an accompanying end statement.)

Suppose we want to create a special square matrix that has 1s in the first row and first column, and whose remaining elements are the sum of two elements, the element above and the element to the left, if the sum is less than 20. Otherwise, the element is the maximum of those two element values. The following function creates this matrix. The row index is r; the column index is c. Note how indenting improves the readability.

```
function A = specmat(n)
A = ones(n);
for r = 1:n
   for c = 1:n
      if (r > 1) & (c > 1)
         s = A(r-1,c) + A(r,c-1);
         if s < 20
            A(r,c) = s;
         else
            A(r,c) = max(A(r-1,c),A(r,c-1));
         end
      end
   end
end
```

Typing `specmat(5)` produces the following matrix:

$$\begin{bmatrix} 1 & 1 & 1 & 1 & 1 \\ 1 & 2 & 3 & 4 & 5 \\ 1 & 3 & 6 & 10 & 15 \\ 1 & 4 & 10 & 10 & 15 \\ 1 & 5 & 15 & 15 & 15 \end{bmatrix}$$

Test Your Understanding

T4.5–1 Write a script file using conditional statements to evaluate the following function, assuming that the scalar variable x has a value. The function is $y = \sqrt{x^2 + 1}$ for $x < 0$, $y = 3x + 1$ for $0 \leq x < 10$, and $y = 9 \sin(5x - 50) + 31$ for $x \geq 10$. Use your file to evaluate y for $x = -5$, $x = 5$, and $x = 15$, and check the results by hand.

T4.5–2 Use a `for` loop to determine the sum of the first 20 terms in the series $3k^2$, $k = 1, 2, 3, \ldots, 20$. (Answer: 8610.)

T4.5–3 Write a program to produce the following matrix:

$$\mathbf{A} = \begin{bmatrix} 4 & 8 & 12 \\ 10 & 14 & 18 \\ 16 & 20 & 24 \\ 22 & 26 & 30 \end{bmatrix}$$

Note the following rules when using `for` loops with the loop variable expression `k = m:s:n`:

■ The step value `s` may be negative. For example, `k = 10:-2:4` produces $k = 10, 8, 6, 4$.

■ If `s` is omitted, the step value defaults to 1.

■ If `s` is positive, the loop will not be executed if `m` is greater than `n`.

■ If `s` is negative, the loop will not be executed if `m` is less than `n`.

■ If `m` equals `n`, the loop will be executed only once.

■ If the step value `s` is not an integer, round-off errors can cause the loop to execute a different number of passes than intended.

When the loop is completed, `k` retains its last value. You should not alter the value of the loop variable `k` within the *statements*. Doing so can cause unpredictable results.

A common practice in traditional programming languages such as BASIC and FORTRAN is to use the symbols `i` and `j` as loop variables. However, this convention is not good practice in MATLAB, which uses these symbols for the imaginary unit $\sqrt{-1}$.

The break and continue Statements

It is permissible to use an if statement to "jump" out of the loop before the loop variable reaches its terminating value. The break command, which terminates the loop but does not stop the entire program, can be used for this purpose. For example,

```
for k = 1:10
    x = 50 - k^2;
    if x < 0
        break
    end
    y = sqrt(x)
end
% The program execution jumps to here
% if the break command is executed.
```

However, it is usually possible to write the code to avoid using the break command. This can often be done with a while loop as explained in the next section.

The break statement stops the execution of the loop. There can be applications where we want to not execute the case producing an error but continue executing the loop for the remaining passes. We can use the continue statement to do this. The continue statement passes control to the next iteration of the for or while loop in which it appears, skipping any remaining statements in the body of the loop. In nested loops, continue passes control to the next iteration of the for or while loop enclosing it.

For example, the following code uses a continue statement to avoid computing the logarithm of a negative number.

```
x = [10,1000,-10,100];
y = NaN*x;
for k = 1:length(x)
    if x(k) < 0
        continue
    end
    kvalue(k) = k;
    y(k) = log10(x(k));
end
kvalue
y
```

The results are k = 1, 2, 0, 4 and y = 1, 3, NaN, 2.

Using an Array as a Loop Index

It is permissible to use a matrix expression to specify the number of passes. In this case the loop variable is a vector that is set equal to the successive columns of the matrix expression during each pass. For example,

```
A = [1,2,3;4,5,6];
for v = A
    disp(v)
end
```

is equivalent to

```
A = [1,2,3;4,5,6];
n = 3;
for k = 1:n
    v = A(:,k)
end
```

The common expression `k = m:s:n` is a special case of a matrix expression in which the columns of the expression are scalars, not vectors.

For example, suppose we want to compute the distance from the origin to a set of three points specified by their xy coordinates $(3,7)$, $(6,6)$, and $(2,8)$. We can arrange the coordinates in the array `coord` as follows.

$$\begin{bmatrix} 3 & 6 & 2 \\ 7 & 6 & 8 \end{bmatrix}$$

Then `coord = [3,6,2;7,6,8]`. The following program computes the distance and determines which point is farthest from the origin. The first time through the loop the index `coord` is `[3, 7]'`. The second time the index is `[6, 6]'`, and during the final pass it is `[2, 8]'`.

```
k = 0;
for coord = [3,6,2;7,6,8]
    k = k + 1;
    distance(k) = sqrt(coord'*coord)
end
[max_distance,farthest] = max(distance)
```

The previous program illustrates the use of an array index, but the problem can be solved more concisely with the following program, which uses the `diag` function to extract the diagonal elements of an array.

```
coord = [3,6,2;7,6,8];
distance = sqrt(diag(coord'*coord))
[max_distance,farthest] = max(distance)
```

Implied Loops

Many MATLAB commands contain *implied loops*. For example, consider these statements.

```
x = [0:5:100];
y = cos(x);
```

To achieve the same result using a `for` loop, we must type

```
for k = 1:21
    x = (k - 1)*5;
    y(k) = cos(x);
end
```

The find command is another example of an implied loop. The statement y = find(x>0) is equivalent to

```
m = 0;
for k = 1:length(x)
    if x(k) > 0
        m = m + 1;
        y(m) = k;
    end
end
```

If you are familiar with a traditional programming language such as FORTRAN or BASIC, you might be inclined to solve problems in MATLAB using loops, instead of using the powerful MATLAB commands such as find. To use these commands and to maximize the power of MATLAB, you might need to adopt a new approach to problem solving. As the preceding example shows, you often can save many lines of code by using MATLAB commands, instead of using loops. Your programs will also run faster because MATLAB was designed for high-speed vector computations.

Test Your Understanding

T4.5–4 Write a for loop that is equivalent to the command sum(A), where A is a matrix.

EXAMPLE 4.5–3   Data Sorting

A vector x has been obtained from measurements. Suppose we want to consider any data value in the range $-0.1 < x < 0.1$ as being erroneous. We want to remove all such elements and replace them with zeros at the end of the array. Develop two ways of doing this. An example is given in the following table.

	Before	After
x(1)	1.92	1.92
x(2)	0.05	−2.43
x(3)	−2.43	0.85
x(4)	−0.02	0
x(5)	0.09	0
x(6)	0.85	0
x(7)	−0.06	0

■ **Solution**

The following script file uses a for loop with conditional statements. Note how the null array [ ] is used.

```
x = [1.92,0.05,-2.43,-0.02,0.09,0.85,-0.06];
y = [];z = [];
for k = 1:length(x)
    if abs(x(k)) >= 0.1
        y = [y,x(k)];
    else
        z = [z,x(k)];
    end
end
xnew = [y,zeros(size(z))]
```

The next script file uses the find function.

```
x = [1.92,0.05,-2.43,-0.02,0.09,0.85,-0.06];
y = x(find(abs(x) >= 0.1));
z = zeros(size(find(abs(x) < 0.1)));
xnew = [y,z]
```

Use of Logical Arrays as Masks

Consider the array **A**.

$$\mathbf{A} = \begin{bmatrix} 0 & -1 & 4 \\ 9 & -14 & 25 \\ -34 & 49 & 64 \end{bmatrix}$$

The following program computes the array **B** by computing the square roots of all the elements of **A** whose value is no less than 0 and adding 50 to each element that is negative.

```
A = [0, -1, 4; 9, -14, 25; -34, 49, 64];
for m = 1:size(A,1)
    for n = 1:size(A,2)
        if A(m,n) >= 0
            B(m,n) = sqrt(A(m,n));
        else
            B(m,n) = A(m,n) + 50;
        end
    end
end
B
```

The result is

$$\mathbf{B} = \begin{bmatrix} 0 & 49 & 2 \\ 3 & 36 & 5 \\ 16 & 7 & 8 \end{bmatrix}$$

When a logical array is used to address another array, it extracts from that array the elements in the locations where the logical array has 1s. We can often avoid the use of loops and branching and thus create simpler and faster programs

by using a logical array as a *mask* that selects elements of another array. Any elements not selected will remain unchanged.

The following session creates the logical array C from the numeric array A given previously.

```
>>A = [0, -1, 4; 9, -14, 25; -34, 49, 64];
>>C = (A >= 0);
```

The result is

$$C = \begin{bmatrix} 1 & 0 & 1 \\ 1 & 0 & 1 \\ 0 & 1 & 1 \end{bmatrix}$$

We can use this technique to compute the square root of only those elements of A given in the previous program that are no less than 0 and add 50 to those elements that are negative. The program is

```
A = [0, -1, 4; 9, -14, 25; -34, 49, 64];
C = (A >= 0);
A(C)  = sqrt(A(C))
A(~C) = A(~C) + 50
```

The result after the third line is executed is

$$A = \begin{bmatrix} 0 & -1 & 2 \\ 3 & -14 & 25 \\ -34 & 49 & 64 \end{bmatrix}$$

The result after the last line is executed is

$$A = \begin{bmatrix} 0 & 49 & 2 \\ 3 & 36 & 5 \\ 16 & 7 & 8 \end{bmatrix}$$

EXAMPLE 4.5-4   Flight of an Instrumented Rocket

All rockets lose weight as they burn fuel; thus the mass of the system is variable. The following equations describe the speed v and height h of a rocket launched vertically, neglecting air resistance. They can be derived from Newton's law.

$$v(t) = u \ln \frac{m_0}{m_0 - qt} - gt \qquad (4.5-1)$$

$$h(t) = \frac{u}{q}(m_0 - qt) \ln(m_0 - qt)$$

$$+ u(\ln m_0 + 1)t - \frac{gt^2}{2} - \frac{m_0 u}{q} \ln m_0 \qquad (4.5-2)$$

where m_0 is the rocket's initial mass, q is the rate at which the rocket burns fuel mass, u is the exhaust velocity of the burned fuel relative to the rocket, and g is the acceleration due to gravity. Let b be the *burn time,* after which all the fuel is consumed. Thus the rocket's mass without fuel is $m_e = m_0 - qb$.

For $t > b$ the rocket engine no longer produces thrust, and the speed and height are given by

$$v(t) = v(b) - g(t - b) \qquad (4.5-3)$$

$$h(t) = h(b) + v(b)(t - b) - \frac{g(t - b)^2}{2} \qquad (4.5-4)$$

The time t_p to reach the peak height is found by setting $v(t) = 0$. The result is $t_p = b + v(b)/g$. Substituting this expression into the expression (4.5–4) for $h(t)$ gives the following expression for the peak height: $h_p = h(b) + v^2(b)/(2g)$. The time at which the rocket hits the ground is $t_{\text{hit}} = t_p + \sqrt{2h_p/g}$.

Suppose the rocket is carrying instruments to study the upper atmosphere, and we need to determine the amount of time spent above 50 000 ft as a function of the burn time b (and thus as a function of the fuel mass qb). Assume that we are given the following values: $m_e = 100$ slugs, $q = 1$ slug/sec, $u = 8000$ ft/sec, and $g = 32.2$ ft/sec^2. If the rocket's maximum fuel load is 100 slugs, the maximum value of b is $100/q = 100$. Write a MATLAB program to solve this problem.

Table 4.5–1 Pseudocode for Example 4.5–4

Enter data.
Increment burn time from 0 to 100. For each burn time value:
    Compute m_0, v_b, h_b, h_p.
    If $h_p \geq h_{\text{desired}}$,
        Compute t_p, t_{hit}.
        Increment time from 0 to t_{hit}.
           Compute height as a function of time, using
           the appropriate equation, depending on whether
           burnout has occurred.
        Compute the duration above desired height.
        End of the time loop.
    If $h_p < h_{\text{desired}}$, set duration equal to zero.
End of the burn time loop.
Plot the results.

■ Solution

Pseudocode for developing the program appears in Table 4.5–1. A `for` loop is a logical choice to solve this problem because we know the burn time b and t_{hit}, the time it takes to hit the ground. A MATLAB program to solve this problem appears in Table 4.5–2. It has two nested `for` loops. The inner loop is over time and evaluates the equations of motion at times spaced $1/10$ sec apart. This loop calculates the duration above 50 000 ft for a specific value of the burn time b. We can obtain greater accuracy by using a smaller value of the time increment dt. The outer loop varies the burn time in integer values from $b = 1$ to $b = 100$. The final result is the vector of durations for the various burn times. Figure 4.5–2 gives the resulting plot.

Table 4.5–2 MATLAB program for Example 4.5–4

```
% Script file rocket1.m
% Computes flight duration as a function of burn time.
% Basic data values.
m_e = 100; q = 1; u = 8000; g = 32.2;
dt = 0.1; h_desired = 50 000;
for b = 1:100 % Loop over burn time.
    burn_time(b) = b;
    % The following lines implement the formulas in the text.
    m_0 = m_e + q*b; v_b = u*log(m_0/m_e) - g*b;
    h_b = ((u*m_e)/q)*log(m_e/(m_e+q*b))+u*b - 0.5*g*b^2;
    h_p = h_b + v_b^2/(2*g);
    if h_p >= h_desired
    % Calculate only if peak height > desired height.
        t_p = b + v_b/g; % Compute peak time.
        t_hit = t_p + sqrt(2*h_p/g); % Compute time to hit.
        for p = 0:t_hit/dt
            % Use a loop to compute the height vector.
            k = p + 1; t = p*dt; time(k) = t;
            if t <= b
                % Burnout has not yet occurred.
                h(k) = (u/q)*(m_0 - q*t)*log(m_0 - q*t)...
                    + u*(log(m_0) + 1)*t - 0.5*g*t^2 ...
                    - (m_0*u/q)*log(m_0);
            else
                % Burnout has occurred.
                h(k) = h_b - 0.5*g*(t - b)^2 + v_b*(t - b);
            end
        end
        % Compute the duration.
        duration(b) = length(find(h>=h_desired))*dt;
    else
        % Rocket did not reach the desired height.
        duration(b) = 0;
    end
end % Plot the results.
plot(burn_time,duration),xlabel('Burn Time (sec)'),...
ylabel('Duration (sec)'),title('Duration Above 50 000 Feet')
```

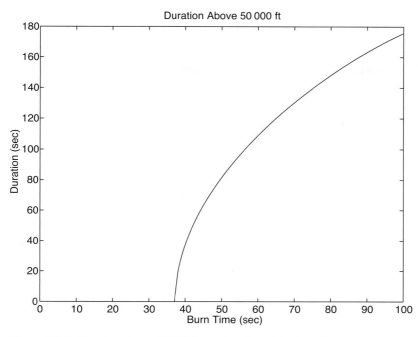

Figure 4.5–2 Duration above 50 000 ft as a function of the burn time.

4.6 `while` Loops

The `while` loop is used when the looping process terminates because a specified condition is satisfied, and thus the number of passes is not known in advance. A simple example of a `while` loop is

```
x = 5;
while x < 25
    disp(x)
    x = 2*x - 1;
end
```

The results displayed by the `disp` statement are 5, 9, and 17. The *loop variable* x is initially assigned the value 5, and it has this value until the statement x = 2*x - 1 is encountered the first time. The value then changes to 9. Before each pass through the loop, x is checked to see whether its value is less than 25. If so, the pass is made. If not, the loop is skipped and the program continues to execute any statements following the `end` statement.

A principal application of `while` loops is when we want the loop to continue as long as a certain statement is true. Such a task is often more difficult to do with a `for` loop. The typical structure of a `while` loop follows.

```
while logical expression
   statements
end
```

MATLAB first tests the truth of the *logical expression*. A loop variable must be included in the *logical expression*. For example, x is the loop variable in the statement while x < 25. If the *logical expression* is true, the *statements* are executed. For the while loop to function properly, the following two conditions must occur:

1. The loop variable must have a value before the while statement is executed.

2. The loop variable must be changed somehow by the *statements.*

The *statements* are executed once during each pass, using the current value of the loop variable. The looping continues until the *logical expression* is false. Figure 4.6–1 shows the flowchart of the while loop.

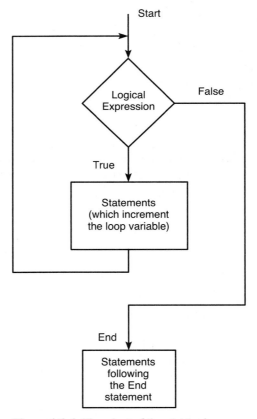

Figure 4.6–1 Flowchart of the while loop.

Each while statement must be matched by an accompanying end. As with for loops, the *statements* should be indented to improve readability. You may nest while loops, and you may nest them with for loops and if statements.

Always make sure that the loop variable has a value assigned to it before the start of the loop. For example, the following loop can give unintended results if x has an overlooked previous value.

```
while x < 10
    x = x + 1;
    y = 2*x;
end
```

If x has not been assigned a value prior to the loop, an error message will occur. If we intend x to start at zero, then we should place the statement x = 0; before the while statement.

It is possible to create an *infinite loop,* which is a loop that never ends. For example,

```
x = 8;
while x ~= 0
    x = x - 3;
end
```

Within the loop the variable x takes on the values 5, 2, −1, −4, . . . , and the condition x ~= 0 is always satisfied, so the loop never stops. If such a loop occurs, press **Ctrl-C** to stop it.

Series Calculation with a while Loop

EXAMPLE 4.6–1

Write a script file to determine the number of terms required for the sum of the series $5k^2 - 2k$, $k = 1, 2, 3, . . .$, to exceed 10 000. What is the sum for this many terms?

■ Solution

Because we do not know how many times we must evaluate the expression $5k^2 - 2k$, we use a while loop. The script file is the following:

```
total = 0;
k = 0;
while total < 1e+4
    k = k + 1;
    total = 5*k^2 - 2*k + total;
end
disp('The number of terms is:')
disp(k)
disp('The sum is:')
disp(total)
```

The sum is 10 203 after 18 terms.

| EXAMPLE 4.6–2 | Growth of a Bank Account |

Determine how long it will take to accumulate at least $10 000 in a bank account if you deposit $500 initially and $500 at the end of each year, if the account pays 5 percent annual interest.

■ Solution

Because we do not know how many years it will take, a `while` loop should be used. The script file is the following.

```
amount = 500;
k=0;
while amount < 10000
    k = k+1;
    amount = amount*1.05 + 500;
end
amount
k
```

The final results are `amount   =   1.0789e+004`, or $10 789, and `k   =   14`, or 14 years.

| EXAMPLE 4.6–3 | Time to Reach a Specified Height |

Consider the variable-mass rocket treated in Example 4.5–4. Write a program to determine how long it takes for the rocket to reach 40 000 ft if the burn time is 50 sec.

■ Solution

The pseudocode appears in Table 4.6–1. Because we do not know the time required, a `while` loop is convenient to use. The program in Table 4.6–2 performs the task and is a modification of the program in Table 4.5–2. Note that the new program allows for the possibility that the rocket might not reach 40 000 ft. It is important to write your programs to handle all such foreseeable circumstances. The answer given by the program is 53 sec.

Table 4.6–1 Pseudocode for Example 4.6–3

Enter data.
Compute m_0, v_b, h_b, h_p.
If $h_p \geq h_{\text{desired}}$,
    Use a `while` loop to increment time and compute height until desired height is reached.
        Compute height as a function of time, using the appropriate equation, depending on whether burnout has occurred.
    End of the time loop.
    Display the results.
If $h_p < h_{\text{desired}}$, rocket cannot reach desired height.

Table 4.6–2 MATLAB program for Example 4.6–3

```
% Script file rocket2.m
% Computes time to reach desired height.
% Set the data values.
h_desired = 40000; m_e = 100; q = 1;
u = 8000; g = 32.2; dt = 0.1; b = 50;
% Compute values at burnout, peak time, and height.
m_0 = m_e + q*b; v_b = u*log(m_0/m_e) - g*b;
h_b = ((u*m_e)/q)*log(m_e/(m_e+q*b))+u*b - 0.5*g*b^2;
t_p = b + v_b/g;
h_p = h_b + v_b^2/(2*g);
% If h_p > h_desired, compute time to reached h_desired.
if h_p > h_desired
   h = 0; k = 0;
   while h < h_desired % Compute h until h = h_desired.
     t = k*dt; k = k + 1;
     if t <= b
        % Burnout has not yet occurred.
        h = (u/q)*(m_0 - q*t)*log(m_0 - q*t)...
           + u*(log(m_0) + 1)*t - 0.5*g*t^2 ...
           - (m_0*u/q)*log(m_0);
     else
        % Burnout has occurred.
        h = h_b - 0.5*g*(t - b)^2 + v_b*(t - b);
     end
   end
   % Display the results.
   disp('The time to reach the desired height is:')
   disp(t)
else
   disp('Rocket cannot achieve the desired height.')
end
```

Test Your Understanding

T4.6–1 Use a while loop to determine how many terms in the series $3k^2$, $k = 1, 2, 3, \ldots$, are required for the sum of the terms to exceed 2000. What is the sum for this number of terms? (Answer: 13 terms, with a sum of 2457.)

T4.6–2 Rewrite the following code, using a while loop to avoid using the break command.

```
for k = 1:10
   x = 50 - k^2;
   if x < 0
      break
   end
   y = sqrt(x)
end
```

T4.6–3 Find to two decimal places the largest value of x before the error in the series approximation $e^x \approx 1 + x + x^2/2 + x^3/6$ exceeds 1 percent. (Answer: $x = 0.83$.)

4.7 The `switch` Structure

The `switch` structure provides an alternative to using the `if`, `elseif`, and `else` commands. Anything programmed using `switch` can also be programmed using `if` structures. However, for some applications the `switch` structure is more readable than code using the `if` structure. The syntax is

```
switch input expression (scalar or string)
    case value1
        statement group 1
    case value2
        statement group 2
        .
        .
        .
    otherwise
        statement group n
end
```

The *input expression* is compared to each `case` value. If they are the same, then the statements following that `case` statement are executed and processing continues with any statements after the `end` statement. If the *input expression* is a string, then it is equal to the `case` *value* if `strcmp` returns a value of 1 (true). Only the *first* matching `case` is executed. If no match occurs, the statements following the `otherwise` statement are executed. However, the `otherwise` statement is optional. If it is absent, execution continues with the statements following the `end` statement if no match exists. Each `case` *value* statement must be on a single line.

For example, suppose the variable `angle` has an integer value that represents an angle measured in degrees from North. The following `switch` block displays the point on the compass that corresponds to that angle.

```
switch angle
    case 45
        disp('Northeast')
    case 135
        disp('Southeast')
    case 225
        disp('Southwest')
    case 315
        disp('Northwest')
    otherwise
        disp('Direction Unknown')
end
```

The use of a string variable for the *input expression* can result in very readable programs. For example, in the following code the numeric vector x has values, and the user enters the value of the string variable `response`; its intended values are `min`, `max`, or `sum`. The code then either finds the minimum or maximum value of x or sums the elements of x, as directed by the user.

```
t = [0:100]; x = exp(-t).*sin(t);
response = input('Type min, max, or sum.','s')
response = lower('response');
switch response
    case min
        minimum = min(x)
    case max
        maximum = max(x)
    case sum
        total = sum(x)
    otherwise
        disp('You have not entered a proper choice.')
end
```

The `switch` statement can handle multiple conditions in a single `case` statement by enclosing the `case` *value* in a cell array. For example, the following `switch` block displays the corresponding point on the compass, given the integer angle measured from North.

```
switch angle
    case {0,360}
        disp('North')
    case {-180,180}
        disp('South')
    case {-270,90}
        disp('East')
    case {-90,270}
        disp('West')
    otherwise
        disp('Direction Unknown')
end
```

Test Your Understanding

T4.7–1 Write a program using the `switch` structure to input one angle, whose value may be 45, −45, 135, or −135°, and display the quadrant (1, 2, 3, or 4) containing the angle.

EXAMPLE 4.7–1   Using the `switch` Structure for Calendar Calculations

Use the `switch` structure to compute the total elapsed days in a year, given the number (1–12) of the month, the day, and an indication of whether the year is a leap year.

■ Solution

Note that February has an extra day if the year is a leap year. The following function computes the total elapsed number of days in a year, given the month, the day of the month, and the value of `extra_day`, which is 1 for a leap year and 0 otherwise.

```
function total_days = total(month,day,extra_day)
total_days = day;
for k = 1:month - 1
   switch k
      case {1,3,5,7,8,10,12}
         total_days = total_days + 31;
      case {4,6,9,11}
         total_days = total_days + 30;
      case 2
         total_days = total_days + 28 + extra_day;
   end
end
```

The function can be used as shown in the following program.

```
month = input('Enter month (1 - 12): ');
day = input('Enter day (1 - 31): ');
extra_day = input('Enter 1 for leap year; 0 otherwise: ');
total_days = total(month,day,extra_day)
```

One of the chapter problems for Section 4.4 (Problem 19) asks you to write a program to determine whether a given year is a leap year.

4.8 Debugging MATLAB Programs

Use of the MATLAB Editor/Debugger as an M-file *editor* was discussed in Section 1.4. Figure 1.4–1 shows the Editor/Debugger screen. Figure 4.8–1 shows the Debugger containing two programs to be analyzed. Here we discuss its use as a *debugger*. Before you use the Debugger, try to debug your program using the commonsense guidelines presented under **Debugging Script Files** in Section 1.4. MATLAB programs are often short because of the power of the commands, and you may not need to use the Debugger unless you are writing large programs. However, the cell mode discussed in this section is useful even for short programs. The Editor/Debugger menu bar contains the following items: **File, Edit, Text, Go, Cell, Tools, Debug, Desktop, Window,** and **Help.** The **File, Edit, Desktop,**

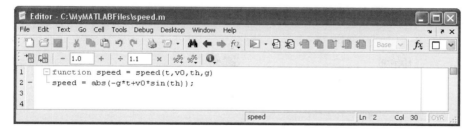

Figure 4.8–1 The Editor/Debugger containing two programs to be analyzed.

Window, and **Help** menus are similar to those in the Desktop. The **Go** menu lets you go backward or forward in the file. The **Cell** menu will be discussed shortly. The **Tools** menu involves advanced topics that will not be treated in this text. The **Desktop** menu is similar to that in the Command window. It enables you to dock and undock windows, arrange the Editor window, and turn the Editor toolbar on and off.

Below the menu bar is the Editor/Debugger toolbar. It enables you to access several of the items in the menus with one click of the mouse. Hold the mouse cursor over a button on the toolbar to see its function. For example, clicking the button with the binoculars icon is equivalent to selecting **Find and Replace** from the **Edit** menu. One item on the toolbar that is not in the menus is the function button with the script f icon (f). Use this button to go to a particular function in the M-file. The list of functions that you will see includes only those functions whose function statements are in the program. The list does not include functions that are called from the M-file.

The **Text** menu supplements the **Edit** menu for creating M-files. With the **Text** menu you can insert or remove comments, increase or decrease the amount of indenting, turn on smart indenting, and evaluate and display the values of selected variables in the Command window. Click *anywhere* in a previously typed line, and then click **Comment** in the **Text** menu. This makes the entire line a comment. To turn a commented line into an executable line, click *anywhere* in the line, and then click **Uncomment** in the **Text** menu.

Cell Mode

The cell mode can be used to debug programs. It can also be used to generate a report. See the end of Section 5.2 for a discussion of the latter usage. A cell is a group of commands. (Such a cell should not be confused with the cell array data type covered in Section 2.6.) Type the double percent character (%%) to mark the beginning of a new cell; it is called a *cell divider*. The cell mode is enabled when you first enter your program into the Editor. To disable cell mode, click the **Cell** button and select **Disable Cell Mode.** The cell toolbar is shown in Figure 4.8–2.

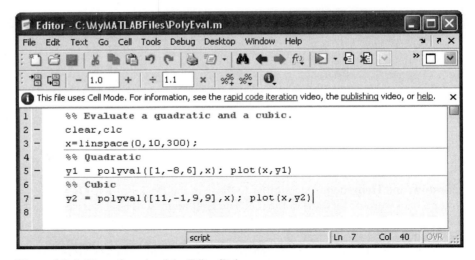

Figure 4.8–2 The cell mode of the Editor/Debugger.

Consider the following simple program that plots either a quadratic or a cubic function.

```
%% Evaluate a quadratic and a cubic.
clear, clc
x = linspace(0, 10, 300);
%% Quadratic
y1 = polyval([1, -8, 6], x); plot(x,y1)
%% Cubic
y2 = polyval([1, -11, 9, 9], x); plot(x,y2)
```

After entering and saving the program, you can click on one of the evaluation icons shown on the left-hand side of the cell toolbar (see Figure 4.8–2). These enable you to evaluate the current single cell (where the cursor is currently), to evaluate the current cell and advance to the next cell, or to evaluate the entire program.

A useful feature of cell mode is that it enables you to evaluate the results of changing a parameter. For example, in Figure 4.8–2, suppose the cursor is next to the number -8. If you click the plus $(+)$ or minus $(-)$ sign in the cell toolbar, the parameter (-8) will be decremented or incremented by the increment shown in the window (1.0 is the default, which you can change). If you have already run the program and the quadratic plot is on the screen, click the minus sign once to change the parameter from -8 to -9 and watch the plot change.

You can also change the parameter by a divisive or multiplicative factor (1.1 is the default). Click the divide or multiply symbol on the cell toolbar.

The Debug Menu

BREAKPOINT

Breakpoints are points in the file where execution stops temporarily so that you can examine the values of the variables up to that point. You set breakpoints with the **Set/Clear Breakpoint** item on the **Debug** menu. Use the **Step,**

Step In, and **Step Out** items on the **Debug** menu to step through your file after you have set breakpoints and run the file. Click **Step** to watch the script execute one step at a time. Click **Step In** to step into the first executable line in a function being called. Click **Step Out** in a called function to run the rest of the function and then return to the calling program.

The solid green arrow to the left of the line text indicates the next line to be executed. When this arrow changes to a hollow green arrow, MATLAB control is now in a function being called. Execution returns to the line with the solid green arrow after the function completes its operation. The arrow turns yellow at a line where execution pauses or where a function completes its operation. When the program pauses, you can assign new values to a variable, using either the Command window or the Array Editor.

Click on the **Go Until Cursor** item to run the file until it reaches the line where the cursor is; this process sets a temporary breakpoint at the cursor. You can save and execute your program directly from the **Debug** menu if you want, by clicking on **Run** (or **Save and Run** if you have made changes). You can also click on the Run icon. You need not set any breakpoints beforehand. Click **Exit Debug Mode** to return to normal editing. To save any changes you have made to the program, first exit the debug mode, and then save the file.

Using Breakpoints

Most debugging sessions start by setting a breakpoint. A breakpoint stops M-file execution at a specified line and allows you to view or change values in the function's workspace before resuming execution. To set a breakpoint, position the cursor in the line of text and click on the breakpoint icon in the toolbar or select **Set/Clear Breakpoints** from the **Debug** menu. You can also set a breakpoint by right-clicking on the line of text to bring up the **Context** menu and choosing **Set/Clear Breakpoint.** A red circle next to a line indicates that a breakpoint is set at that line. If the line selected for a breakpoint is not an executable line, then the breakpoint is set at the next executable line. The **Debug** menu enables you to clear all the breakpoints (select **Clear Breakpoints in All Files**). The **Debug** menu also lets you halt M-file execution if your code generates a warning, an error, or an `NaN` or `Inf` value (select **Stop if Errors/ Warnings**).

4.9 Applications to Simulation

Simulation is the process of building and analyzing the output of computer programs that describe the operations of an organization, process, or physical system. Such a program is called a *computer model*. Simulation is often used in *operations research,* which is the quantitative study of an organization in action, to find ways to improve the functioning of the organization. Simulation enables engineers to study the past, present, and future actions of the organization for this purpose. Operations research techniques are useful in all engineering fields. Common examples include airline scheduling, traffic flow studies, and production lines. The MATLAB logical operators and loops are excellent tools for building simulation programs.

OPERATIONS RESEARCH

EXAMPLE 4.9–1

A College Enrollment Model: Part I

As an example of how simulation can be used for operations research, consider the following college enrollment model. A certain college wants to analyze the effect of admissions and freshman retention rate on the college's enrollment so that it can predict the future need for instructors and other resources. Assume that the college has estimates of the percentages of students repeating a grade or leaving school before graduating. Develop a matrix equation on which to base a simulation model that can help in this analysis.

■ **Solution**

Suppose that the current freshman enrollment is 500 students and the college decides to admit 1000 freshmen per year from now on. The college estimates that 10 percent of the freshman class will repeat the year. The number of freshmen in the following year will be $0.1(500) + 1000 = 1050$, then it will be $0.1(1050) + 1000 = 1105$, and so on. Let $x_1(k)$ be the number of freshmen in year k, where $k = 1, 2, 3, 4, 5, 6, \ldots$. Then in year $k + 1$, the number of freshmen is given by

$$
\begin{aligned}
x_1(k + 1) = \ & 10 \text{ percent of previous freshman class} \\
& \text{repeating freshman year} \\
& + 1000 \text{ new freshmen} \\
= \ & 0.1x_1(k) + 1000
\end{aligned}
\tag{4.9–1}
$$

Because we know the number of freshmen in the first year of our analysis (which is 500), we can solve this equation step by step to predict the number of freshmen in the future.

Let $x_2(k)$ be the number of sophomores in year k. Suppose that 15 percent of the freshmen do not return and that 10 percent repeat freshman year. Thus 75 percent of the freshman class returns as sophomores. Suppose also 5 percent of the sophomores repeat the sophomore year and that 200 sophomores each year transfer from other schools. Then in year $k + 1$, the number of sophomores is given by

$$
x_2(k + 1) = 0.75x_1(k) + 0.05x_2(k) + 200
$$

To solve this equation, we need to solve the "freshman" equation (4.9–1) at the same time, which is easy to do with MATLAB. Before we solve these equations, let us develop the rest of the model.

Let $x_3(k)$ and $x_4(k)$ be the number of juniors and seniors, respectively, in year k. Suppose that 5 percent of the sophomores and juniors leave school and that 5 percent of the sophomores, juniors, and seniors repeat the grade. Thus 90 percent of the sophomores and juniors return and advance in grade. The models for the juniors and seniors are

$$
x_3(k + 1) = 0.9x_2(k) + 0.05x_3(k)
$$
$$
x_4(k + 1) = 0.9x_3(k) + 0.05x_4(k)
$$

These four equations can be written in the following matrix form:

$$\begin{bmatrix} x_1(k+1) \\ x_2(k+1) \\ x_3(k+1) \\ x_4(k+1) \end{bmatrix} = \begin{bmatrix} 0.1 & 0 & 0 & 0 \\ 0.75 & 0.05 & 0 & 0 \\ 0 & 0.9 & 0.05 & 0 \\ 0 & 0 & 0.9 & 0.05 \end{bmatrix} \begin{bmatrix} x_1(k) \\ x_2(k) \\ x_3(k) \\ x_4(k) \end{bmatrix} + \begin{bmatrix} 1000 \\ 200 \\ 0 \\ 0 \end{bmatrix}$$

In Example 4.9–2 we will see how to use MATLAB to solve such equations.

Test Your Understanding

T4.9–1 Suppose that 70 percent of the freshmen, instead of 75 percent, return for the sophomore year. How does the previous equation change?

A College Enrollment Model: Part II

EXAMPLE 4.9–2

To study the effects of admissions and transfer policies, generalize the enrollment model in Example 4.9–1 to allow for varying admissions and transfers.

■ Solution
Let $a(k)$ be the number of new freshmen admitted in the spring of year k for the following year $k+1$, and let $d(k)$ be the number of transfers into the following year's sophomore class. Then the model becomes

$$x_1(k+1) = c_{11}x_1(k) + a(k)$$
$$x_2(k+1) = c_{21}x_1(k) + c_{22}x_2(k) + d(k)$$
$$x_3(k+1) = c_{32}x_2(k) + c_{33}x_3(k)$$
$$x_4(k+1) = c_{43}x_3(k) + c_{44}x_4(k)$$

where we have written the coefficients c_{21}, c_{22}, and so on in symbolic, rather than numerical, form so that we can change their values if desired.

This model can be represented graphically by a *state transition diagram,* like the one shown in Figure 4.9–1. Such diagrams are widely used to represent time-dependent and probabilistic processes. The arrows indicate how the model's calculations are updated for

STATE TRANSITION DIAGRAM

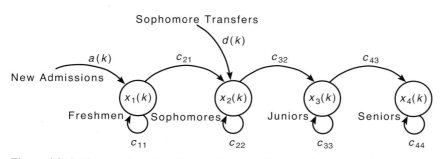

Figure 4.9–1 The state transition diagram for the college enrollment model.

each new year. The enrollment at year k is described completely by the values of $x_1(k)$, $x_2(k)$, $x_3(k)$, and $x_4(k)$, that is, by the vector $\mathbf{x}(k)$, which is called the *state vector*. The elements of the state vector are the *state variables*. The state transition diagram shows how the new values of the state variables depend on both the previous values and the inputs $a(k)$ and $d(k)$.

The four equations can be written in the following matrix form:

$$
\begin{bmatrix} x_1(k+1) \\ x_2(k+1) \\ x_3(k+1) \\ x_4(k+1) \end{bmatrix} = \begin{bmatrix} c_{11} & 0 & 0 & 0 \\ c_{21} & c_{22} & 0 & 0 \\ 0 & c_{32} & c_{33} & 0 \\ 0 & 0 & c_{43} & c_{44} \end{bmatrix} \begin{bmatrix} x_1(k) \\ x_2(k) \\ x_3(k) \\ x_4(k) \end{bmatrix} + \begin{bmatrix} a(k) \\ d(k) \\ 0 \\ 0 \end{bmatrix}
$$

or more compactly as

$$
\mathbf{x}(k+1) = \mathbf{C}\mathbf{x}(k) + \mathbf{b}(k)
$$

where

$$
\mathbf{x}(k) = \begin{bmatrix} x_1(k) \\ x_2(k) \\ x_3(k) \\ x_4(k) \end{bmatrix} \qquad \mathbf{b}(k) = \begin{bmatrix} a(k) \\ d(k) \\ 0 \\ 0 \end{bmatrix}
$$

and

$$
\mathbf{C} = \begin{bmatrix} c_{11} & 0 & 0 & 0 \\ c_{21} & c_{22} & 0 & 0 \\ 0 & c_{32} & c_{33} & 0 \\ 0 & 0 & c_{43} & c_{44} \end{bmatrix}
$$

Suppose that the initial total enrollment of 1480 consists of 500 freshmen, 400 sophomores, 300 juniors, and 280 seniors. The college wants to study, over a 10-year period, the effects of increasing admissions by 100 each year and transfers by 50 each year until the total enrollment reaches 4000; then admissions and transfers will be held constant. Thus the admissions and transfers for the next 10 years are given by

$$
a(k) = 900 + 100k
$$

$$
d(k) = 150 + 50k
$$

for $k = 1, 2, 3, \ldots$ until the college's total enrollment reaches 4000; then admissions and transfers are held constant at the previous year's levels. We cannot determine when this event will occur without doing a simulation. Table 4.9–1 gives the pseudocode for solving this problem. The enrollment matrix $\mathbf{E}$ is a 4×10 matrix whose columns represent the enrollment in each year.

Because we know the length of the study (10 years), a `for` loop is a natural choice. We use an `if` statement to determine when to switch from the increasing admissions and transfer schedule to the constant schedule. A MATLAB script file to predict the enrollment for the next 10 years appears in Table 4.9–2. Figure 4.9–2 shows the resulting

Table 4.9–1 Pseudocode for Example 4.9–2

Enter the coefficient matrix **C** and the initial enrollment vector **x**.
Enter the initial admissions and transfers, $a(1)$ and $d(1)$.
Set the first column of the enrollment matrix **E** equal to **x**.
Loop over years 2 to 10.
    If the total enrollment is ≤ 4000, increase admissions by 100 and transfers by 50 each year.
    If the total enrollment is > 4000, hold admissions and transfers constant.
    Update the vector **x**, using $\mathbf{x} = \mathbf{Cx} + \mathbf{b}$.
    Update the enrollment matrix **E** by adding another column composed of **x**.
End of the loop over years 2 to 10.
Plot the results.

Table 4.9–2 College enrollment model

```
% Script file enroll1.m. Computes college enrollment.
% Model's coefficients.
C = [0.1,0,0,0;0.75,0.05,0,0;0,0.9,0.05,0;0,0,0.9,0.05];
% Initial enrollment vector.
x = [500;400;300;280];
% Initial admissions and transfers.
a(1) = 1000; d(1) = 200;
% E is the 4 x 10 enrollment matrix.
E(:,1) = x;
% Loop over years 2 to 10.
for k = 2:10
    % The following describes the admissions
    % and transfer policies.
    if sum(x) <= 4000
        % Increase admissions and transfers.
        a(k) = 900+100*k;
        d(k) = 150+50*k;
    else
        % Hold admissions and transfers constant.
        a(k) = a(k-1);
        d(k) = d(k-1);
    end
    % Update enrollment matrix.
    b = [a(k);d(k);0;0];
    x = C*x+b;
    E(:,k) = x;
end
% Plot the results.
plot(E'),hold,plot(E(1,:),'o'),plot(E(2,:),'+'),plot(E(3,:),'*'),...
plot(E(4,:),'x'),xlabel('Year'),ylabel('Number of Students'),...
gtext('Frosh'),gtext('Soph'),gtext('Jr'),gtext('Sr'),...
title('Enrollment as a Function of Time')
```

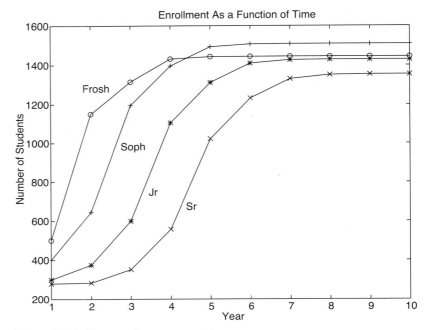

Figure 4.9–2 Class enrollments versus time.

plot. Note that after year 4 there are more sophomores than freshmen. The reason is that the increasing transfer rate eventually overcomes the effect of the increasing admission rate.

In actual practice this program would be run many times to analyze the effects of different admissions and transfer policies and to examine what happens if different values are used for the coefficients in the matrix **C** (indicating different dropout and repeat rates).

Test Your Understanding

T4.9–2 In the program in Table 4.9–2, lines 16 and 17 compute the values of a(k) and d(k). These lines are repeated here:

```
a(k) = 900 + 100 * k
d(k) = 150 + 50 * k;
```

Why does the program contain the line a(1)=1000; d(1)=200;?

Table 4.10–1 Guide to MATLAB commands introduced in Chapter 4

Relational operators	Table 4.2–1
Logical operators	Table 4.3–1
Order of precedence for operator types	Table 4.3–2
Truth table	Table 4.3–3
Logical functions	Table 4.3–4

Miscellaneous commands

Command	Description	Section
break	Terminates the execution of a for or a while loop.	4.5, 4.6
case	Used with switch to direct program execution.	4.7
continue	Passes control to the next iteration of a for or while loop.	4.5, 4.6
double	Converts a logical array to class double.	4.2
else	Delineates an alternate block of statements.	4.4
elseif	Conditionally executes statements.	4.4
end	Terminates for, while, and if statements.	4.4, 4.5, 4.6
for	Repeats statements a specific number of times.	4.5
if	Executes statements conditionally.	4.4
input('s1', 's')	Display the prompt string s1 and stores user input as a string.	4.4
logical	Converts numeric values to logical values.	4.2
nargin	Determines the number of input arguments of a function.	4.4
nargout	Determines the number of output arguments of a function.	4.4
switch	Directs program execution by comparing the input expression with the associated case expressions.	4.7
while	Repeats statements an indefinite number of times.	4.6
xor	Exclusive OR function.	4.3

4.10 Summary

Now that you have finished this chapter, you should be able to write programs that can perform decision-making procedures; that is, the program's operations depend on results of the program's calculations or on input from the user. Sections 4.2, 4.3, and 4.4 covered the necessary functions: the relational operators, the logical operators and functions, and the conditional statements.

You should also be able to use MATLAB loop structures to write programs that repeat calculations a specified number of times or until some condition is satisfied. This feature enables engineers to solve problems of great complexity or requiring numerous calculations. The for loop and while loop structures were covered in Sections 4.5 and 4.6. Section 4.7 covered the switch structure.

Section 4.8 gave an overview and an example of how to debug programs using the Editor/Debugger. Section 4.9 presented an application of these methods to simulation, which enables engineers to study the operation of complicated systems, processes, and organizations.

Tables summarizing the MATLAB commands introduced in this chapter are located throughout the chapter. Table 4.10–1 will help you locate these tables. It also summarizes those commands not found in the other tables.

Key Terms with Page References

Breakpoint, 192
Conditional statement, 164
Flowchart, 150
for loop, 171
Implied loop, 177
Logical operator, 157
Masks, 180
Nested loops, 174
Operations research, 193
Pseudocode, 151

Relational operator, 155
Simulation, 193
State transition diagram, 195
Structure chart, 150
Structured programming, 148
switch structure, 188
Top-down design, 149
Truth table, 159
while loop, 183

Problems

You can find answers to problems marked with an asterisk at the end of the text.

Section 4.1

1. The volume V and surface area A of a sphere of radius r are given by

$$V = \frac{4}{3}\pi r^3 \qquad A = 4\pi r^2$$

   a. Develop a pseudocode description of a program to compute V and A for $0 \le r \le 3$ m and to plot V versus A.

   b. Write and run the program described in part a.

2. The roots of the quadratic equation $ax^2 + bx + c = 0$ are given by

$$x = \frac{-b \pm \sqrt{b^2 - 4ac}}{2a}$$

   a. Develop a pseudocode description of a program to compute both roots given the values of a, b, and c. Be sure to identify the real and imaginary parts.

   b. Write the program described in part a and test it for the following cases:
      1. $a = 2, b = 10, c = 12$
      2. $a = 3, b = 24, c = 48$
      3. $a = 4, b = 24, c = 100$

3. It is desired to compute the sum of the first 10 terms of the series $14k^3 - 20k^2 + 5k, k = 1, 2, 3, \ldots$

   a. Develop a pseudocode description of the required program.

   b. Write and run the program described in part a.

Section 4.2

4.* Suppose that x = 6. Find the results of the following operations by hand and use MATLAB to check your results.

  a. z = (x < 10)
  b. z = (x == 10)
  c. z = (x >= 4)
  d. z = (x ~= 7)

5.* Find the results of the following operations by hand and use MATLAB to check your results.

  a. z = 6 > 3 + 8
  b. z = 6 + 3 > 8
  c. z = 4 > (2 + 9)
  d. z = (4 < 7) + 3
  e. z = 4 < 7 + 3
  f. z = (4<7)*5
  g. z = 4<(7*5)
  h. z = 2/5>=5

6.* Suppose that x = [10, -2, 6, 5, -3] and y = [9, -3, 2, 5, -1]. Find the results of the following operations by hand and use MATLAB to check your results.

  a. z = (x < 6)
  b. z = (x <= y)
  c. z = (x == y)
  d. z = (x ~= y)

7. For the arrays x and y given below, use MATLAB to find all the elements in x that are greater than the corresponding elements in y.

  x = [-3, 0, 0, 2, 6, 8]   y = [-5, -2, 0, 3, 4, 10]

8. The array price given below contains the price in dollars of a certain stock over 10 days. Use MATLAB to determine how many days the price was above $20.

  price = [19, 18, 22, 21, 25, 19, 17, 21, 27, 29]

9. The arrays price_A and price_B given below contain the price in dollars of two stocks over 10 days. Use MATLAB to determine how many days the price of stock A was above the price of stock B.

  price_A = [19, 18, 22, 21, 25, 19, 17, 21, 27, 29]
  price_B = [22, 17, 20, 19, 24, 18, 16, 25, 28, 27]

10. The arrays price_A, price_B, and price_C given below contain the price in dollars of three stocks over 10 days.

  a. Use MATLAB to determine how many days the price of stock A was above both the price of stock B and the price of stock C.

    b. Use MATLAB to determine how many days the price of stock A was above either the price of stock B or the price of stock C.

    c. Use MATLAB to determine how many days the price of stock A was above either the price of stock B or the price of stock C, but not both.

```
price_A = [19, 18, 22, 21, 25, 19, 17, 21, 27, 29]
price_B = [22, 17, 20, 19, 24, 18, 16, 25, 28, 27]
price_C = [17, 13, 22, 23, 19, 17, 20, 21, 24, 28]
```

Section 4.3

11.* Suppose that x = [-3, 0, 0, 2, 5, 8] and y = [-5, -2, 0, 3, 4, 10]. Find the results of the following operations by hand and use MATLAB to check your results.

    a. z = y < ~ x

    b. z = x & y

    c. z = x|y

    d. z = xor(x,y)

12. The height and speed of a projectile (such as a thrown ball) launched with a speed of v_0 at an angle A to the horizontal are given by

$$h(t) = v_0 t \sin A - 0.5 \, gt^2$$
$$v(t) = \sqrt{v_0^2 - 2v_0 gt \sin A + g^2 t^2}$$

where g is the acceleration due to gravity. The projectile will strike the ground when $h(t) = 0$, which gives the time to hit $t_{hit} = 2(v_0/g) \sin A$.

    Suppose that $A = 30°$, $v_0 = 40$ m/s, and $g = 9.81$m/s^2. Use the MATLAB relational and logical operators to find the times when

    a. The height is no less than 15 m.

    b. The height is no less than 15 m and the speed is simultaneously no greater than 36 m/s.

    c. The height is less than 5 m or the speed is greater than 35 m/s.

13.* The price, in dollars, of a certain stock over a 10-day period is given in the following array.

```
price = [19, 18, 22, 21, 25, 19, 17, 21, 27, 29]
```

Suppose you owned 1000 shares at the start of the 10-day period, and you bought 100 shares every day the price was below $20 and sold 100 shares every day the price was above $25. Use MATLAB to compute (*a*) the amount you spent in buying shares, (*b*) the amount you received from the sale of shares, (*c*) the total number of shares you own after the 10th day, and (*d*) the net increase in the worth of your portfolio.

14. Let `e1` and `e2` be logical expressions. DeMorgan's laws for logical expressions state that

NOT(e1 AND e2) implies that (NOT e1) OR (NOT e2)
and
NOT(e1 OR e2) implies that (NOT e1) AND (NOT e2)

Use these laws to find an equivalent expression for each of the following expressions and use MATLAB to verify the equivalence.

a. `~((x < 10) & (x >= 6))`
b. `~((x == 2) | (x > 5))`

15. Are these following expressions equivalent? Use MATLAB to check your answer for specific values of *a*, *b*, *c*, and *d*.

a. 1. `(a == b) & ((b == c)|(a == c))`
    2. `(a == b)|((b == c)&(a == c))`
b. 1. `(a < b) & ((a > c)|(a > d))`
    2. `(a < b) & (a > c)|((a < b)&(a > d))`

16. Write a script file using conditional statements to evaluate the following function, assuming that the scalar variable x has a value. The function is $y = e^{x+1}$ for $x < -1$, $y = 2 + \cos(\pi x)$ for $-1 \le x < 5$, and $y = 10(x - 5) + 1$ for $x \ge 5$. Use your file to evaluate y for $x = -5$, $x = 3$, and $x = 15$, and check the results by hand.

Section 4.4

17. Rewrite the following statements to use only one `if` statement.

```
if x < y
    if z < 10
        w = x*y*z
    end
end
```

18. Write a program that accepts a numerical value x from 0 to 100 as input and computes and displays the corresponding letter grade given by the following table.

A $x \ge 90$
B $80 \le x \le 89$
C $70 \le x \le 79$
D $60 \le x \le 69$
F $x < 60$

a. Use nested `if` statements in your program (do not use `elseif`).
b. Use only `elseif` clauses in your program.

19. Write a program that accepts a year and determines whether the year is a leap year. Use the `mod` function. The output should be the variable `extra_day`, which should be 1 if the year is a leap year and 0 otherwise. The rules for determining leap years in the Gregorian calendar are as follows:

   1. All years evenly divisible by 400 are leap years.
   2. Years evenly divisible by 100 but not by 400 are not leap years.
   3. Years divisible by 4 but not by 100 are leap years.
   4. All other years are not leap years.

   For example, the years 1800, 1900, 2100, 2300, and 2500 are not leap years, but 2400 is a leap year.

20. Figure P20 shows a mass-spring model of the type used to design packaging systems and vehicle suspensions, for example. The springs exert a force that is proportional to their compression, and the proportionality constant is the spring constant k. The two side springs provide additional resistance if the weight W is too heavy for the center spring. When the weight W is gently placed, it moves through a distance x before coming to rest. From statics, the weight force must balance the spring forces at this new position. Thus

$$W = k_1 x \qquad\qquad \text{if } x < d$$
$$W = k_1 x + 2k_2(x - d) \qquad \text{if } x \geq d$$

   These relations can be used to generate the plot of x versus W.

   a. Create a function file that computes the distance x, using the input parameters W, k_1, k_2, and d. Test your function for the following two cases, using the values $k_1 = 10^4$ N/m; $k_2 = 1.5 \times 10^4$ N/m; $d = 0.1$ m.

$$W = 500 \text{ N}$$
$$W = 2000 \text{ N}$$

   b. Use your function to plot x versus W for $0 \leq W \leq 3000$ N for the values of k_1, k_2, and d given in part a.

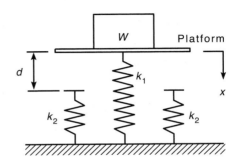

Figure P20

Section 4.5

21. Use a `for` loop to plot the function given in Problem 16 over the interval $-2 \leq x \leq 6$. Properly label the plot. The variable y represents height in kilometers, and the variable x represents time in seconds.

22. Use a `for` loop to determine the sum of the first 10 terms in the series $5k^3$, $k = 1, 2, 3, \ldots, 10$.

23. The (x, y) coordinates of a certain object as a function of time t are given by

$$x(t) = 5t - 10 \qquad y(t) = 25t^2 - 120t + 144$$

for $0 \leq t \leq 4$. Write a program to determine the time at which the object is the closest to the origin at $(0, 0)$. Determine also the minimum distance. Do this in two ways:

   a. By using a `for` loop.
   b. By not using a `for` loop.

24. Consider the array **A**.

$$\mathbf{A} = \begin{bmatrix} 3 & 5 & -4 \\ -8 & -1 & 33 \\ -17 & 6 & -9 \end{bmatrix}$$

Write a program that computes the array **B** by computing the natural logarithm of all the elements of **A** whose value is no less than 1, and adding 20 to each element that is equal to or greater than 1. Do this in two ways:

   a. By using a `for` loop with conditional statements.
   b. By using a logical array as a mask.

25. We want to analyze the mass-spring system discussed in Problem 20 for the case in which the weight W is dropped onto the platform attached to the center spring. If the weight is dropped from a height h above the platform, we can find the maximum spring compression x by equating the weight's gravitational potential energy $W(h + x)$ with the potential energy stored in the springs. Thus

$$W(h + x) = \tfrac{1}{2}k_1 x^2 \qquad \text{if } x < d$$

which can be solved for x as

$$x = \frac{W \pm \sqrt{W^2 + 2k_1 Wh}}{k_1} \qquad \text{if } x < d$$

and

$$W(h + x) = \tfrac{1}{2} k_1 x^2 + \tfrac{1}{2}(2k_2)(x - d)^2 \qquad \text{if } x \geq d$$

which gives the following quadratic equation to solve for x:

$$(k_1 + 2k_2)x^2 - (4k_2d + 2W)x + 2k_2d^2 - 2Wh = 0 \qquad \text{if } x \geq d$$

a. Create a function file that computes the maximum compression x due to the falling weight. The function's input parameters are k_1, k_2, d, W, and h. Test your function for the following two cases, using the values $k_1 = 10^4$ N/m; $k_2 = 1.5 \times 10^4$ N/m; and $d = 0.1$ m.

$$W = 100 \text{ N} \qquad h = 0.5 \text{ m}$$
$$W = 2000 \text{ N} \qquad h = 0.5 \text{ m}$$

b. Use your function file to generate a plot of x versus h for $0 \leq h \leq 2$ m. Use $W = 100$ N and the preceding values for k_1, k_2, and d.

26. Electrical resistors are said to be connected "in series" if the same current passes through each and "in parallel" if the same voltage is applied across each. If in series, they are equivalent to a single resistor whose resistance is given by

$$R = R_1 + R_2 + R_3 + \cdots + R_n$$

If in parallel, their equivalent resistance is given by

$$\frac{1}{R} = \frac{1}{R_1} + \frac{1}{R_2} + \frac{1}{R_3} + \cdots + \frac{1}{R_n}$$

Write an M-file that prompts the user for the type of connection (series or parallel) and the number of resistors n and then computes the equivalent resistance.

27. a. An *ideal* diode blocks the flow of current in the direction opposite that of the diode's arrow symbol. It can be used to make a *half-wave rectifier* as shown in Figure P27a. For the ideal diode, the voltage v_L across the load R_L is given by

$$v_L = \begin{cases} v_S & \text{if } v_S > 0 \\ 0 & \text{if } v_S \leq 0 \end{cases}$$

Suppose the supply voltage is

$$v_S(t) = 3e^{-t/3}\sin(\pi t) \qquad \text{V}$$

where time t is in seconds. Write a MATLAB program to plot the voltage v_L versus t for $0 \leq t \leq 10$.

b. A more accurate model of the diode's behavior is given by the *offset diode* model, which accounts for the offset voltage inherent in semiconductor diodes. The offset model contains an ideal diode and a battery whose voltage equals the offset voltage (which is approximately

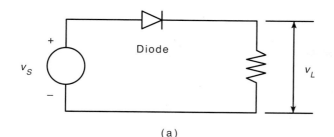

(a)

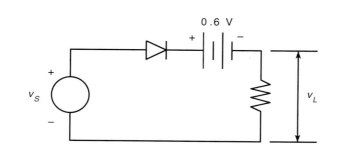

(b)

Figure P27

0.6 V for silicon diodes) [Rizzoni, 2007]. The half-wave rectifier using this model is shown in Figure P27b. For this circuit,

$$v_L = \begin{cases} v_S - 0.6 & \text{if } v_S > 0.6 \\ 0 & \text{if } v_S \le 0.6 \end{cases}$$

Using the same supply voltage given in part a, plot the voltage v_L versus t for $0 \le t \le 10$; then compare the results with the plot obtained in part a.

28.* A company wants to locate a distribution center that will serve six of its major customers in a 30 $\times$ 30 mi area. The locations of the customers relative to the southwest corner of the area are given in the following table in terms of (x, y) coordinates (the x direction is east; the y direction is north) (see Figure P28). Also given is the volume in tons per week that must be delivered from the distribution center to each customer. The weekly delivery cost c_i for customer i depends on the volume V_i and the distance d_i from the distribution center. For simplicity we will assume that this distance is the straight-line distance. (This assumes that the road network is dense.) The weekly cost is given by $c_i = 0.5 d_i V_i$, $i = 1, \ldots, 6$. Find the location of the distribution center (to the nearest mile) that minimizes the total weekly cost to service all six customers.

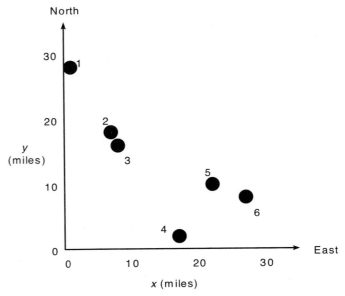

Figure P28

Customer	x location (mi)	y location (mi)	Volume (tons/week)
1	1	28	3
2	7	18	7
3	8	16	4
4	17	2	5
5	22	10	2
6	27	8	6

29. A company has the choice of producing up to four different products with its machinery, which consists of lathes, grinders, and milling machines. The number of hours on each machine required to produce a product is given in the following table, along with the number of hours available per week on each type of machine. Assume that the company can sell everything it produces. The profit per item for each product appears in the last line of the table.

	Product				
	1	**2**	**3**	**4**	**Hours available**
Hours required					
Lathe	1	2	0.5	3	40
Grinder	0	2	4	1	30
Milling	3	1	5	2	45
Unit profit ($)	100	150	90	120	

a. Determine how many units of each product the company should make to maximize its total profit, and then compute this profit. Remember, the company cannot make fractional units, so your answer must be in integers. (*Hint:* First estimate the upper limits on the number of products that can be produced without exceeding the available capacity.)

b. How sensitive is your answer? How much does the profit decrease if you make one more or one less item than the optimum?

30. A certain company makes televisions, stereo units, and speakers. Its parts inventory includes chassis, picture tubes, speaker cones, power supplies, and electronics. The inventory, required components, and profit for each product appear in the following table. Determine how many of each product to make in order to maximize the profit.

	Product			
	Television	**Stereo unit**	**Speaker unit**	**Inventory**
Requirements				
Chassis	1	1	0	450
Picture tube	1	0	0	250
Speaker cone	2	2	1	800
Power supply	1	1	0	450
Electronics	2	2	1	600
Unit profit ($)	80	50	40	

Section 4.6

31. Plot the function $y = 10(1 - e^{-x/4})$ over the interval $0 \le x \le x_{max}$, using a `while` loop to determine the value of x_{max} such that $y(x_{max}) = 9.8$. Properly label the plot. The variable y represents force in newtons, and the variable x represents time in seconds.

32. Use a `while` loop to determine how many terms in the series 2^k, $k = 1$, $2, 3, \ldots$, are required for the sum of the terms to exceed 2000. What is the sum for this number of terms?

33. One bank pays 5.5 percent annual interest, while a second bank pays 4.5 percent annual interest. Determine how much longer it will take to accumulate at least $50 000 in the second bank account if you deposit $1000 initially and $1000 at the end of each year.

34.* Use a loop in MATLAB to determine how long it will take to accumulate $1 000 000 in a bank account if you deposit $10 000 initially and $10 000 at the end of each year; the account pays 6 percent annual interest.

35. A weight W is supported by two cables anchored a distance D apart (see Figure P35). The cable length L_{AB} is given, but the length L_{AC} is to be

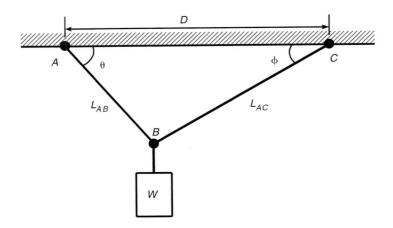

Figure P35

selected. Each cable can support a maximum tension force equal to W. For the weight to remain stationary, the total horizontal force and total vertical force must each be zero. This principle gives the equations

$$-T_{AB} \cos \theta + T_{AC} \cos \phi = 0$$

$$T_{AB} \sin \theta + T_{AC} \sin \phi = W$$

We can solve these equations for the tension forces T_{AB} and T_{AC} if we know the angles θ and ϕ. From the law of cosines

$$\theta = \cos^{-1}\left(\frac{D^2 + L_{AB}^2 - L_{AC}^2}{2DL_{AB}}\right)$$

From the law of sines

$$\phi = \sin^{-1}\left(\frac{L_{AB} \sin \theta}{L_{AC}}\right)$$

For the given values $D = 6$ ft, $L_{AB} = 3$ ft, and $W = 2000$ lb, use a loop in MATLAB to find $L_{AC\text{min}}$, the shortest length L_{AC} we can use without T_{AB} or T_{AC} exceeding 2000 lb. Note that the largest L_{AC} can be is 6.7 ft (which corresponds to $\theta = 90°$). Plot the tension forces T_{AB} and T_{AC} on the same graph versus L_{AC} for $L_{AC\text{min}} \leq L_{AC} \leq 6.7$.

36.* In the structure in Figure P36a, six wires support three beams. Wires 1 and 2 can support no more than 1200 N each, wires 3 and 4 can support no more than 400 N each, and wires 5 and 6 can support no more than 200 N each. Three equal weights W are attached at the points shown. Assuming that the structure is stationary and that the weights of the wires and the beams are very small compared to W, the principles of statics applied to a particular beam state that the sum of vertical forces is zero and

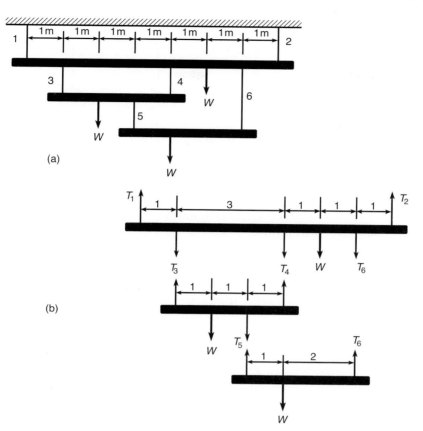

Figure P36

that the sum of moments about any point is also zero. Applying these principles to each beam using the free-body diagrams shown in Figure P36b, we obtain the following equations. Let the tension force in wire i be T_i. For beam 1

$$T_1 + T_2 = T_3 + T_4 + W + T_6$$

$$-T_3 - 4T_4 - 5W - 6T_6 + 7T_2 = 0$$

For beam 2

$$T_3 + T_4 = W + T_5$$

$$-W - 2T_5 + 3T_4 = 0$$

For beam 3

$$T_5 + T_6 = W$$

$$-W + 3T_6 = 0$$

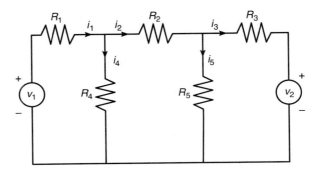

Figure P37

Find the maximum value of the weight W the structure can support. Remember that the wires cannot support compression, so T_i must be nonnegative.

37. The equations describing the circuit shown in Figure P37 are

$$-v_1 + R_1 i_1 + R_4 i_4 = 0$$
$$-R_4 i_4 + R_2 i_2 + R_5 i_5 = 0$$
$$-R_5 i_5 + R_3 i_3 + v_2 = 0$$
$$i_1 = i_2 + i_4$$
$$i_2 = i_3 + i_5$$

a. The given values of the resistances and the voltage v_1 are $R_1 = 5$, $R_2 = 100$, $R_3 = 200$, $R_4 = 150$, $R_5 = 250$ kΩ, and $v_1 = 100$ V. (Note that 1 kΩ = 1000 Ω.) Suppose that each resistance is rated to carry a current of no more than 1 mA ($= 0.001$ A). Determine the allowable range of positive values for the voltage v_2.

b. Suppose we want to investigate how the resistance R_3 limits the allowable range for v_2. Obtain a plot of the allowable limit on v_2 as a function of R_3 for $150 \leq R_3 \leq 250$ kΩ.

38. Many applications require us to know the temperature distribution in an object. For example, this information is important for controlling the material properties, such as hardness, when cooling an object formed from molten metal. In a heat-transfer course, the following description of the temperature distribution in a flat, rectangular metal plate is often derived. The temperature is held constant at T_1 on three sides and at T_2 on the fourth side (see Figure P38). The temperature $T(x, y)$ as a function of the xy coordinates shown is given by

$$T(x, y) = (T_2 - T_1)w(x, y) + T_1$$

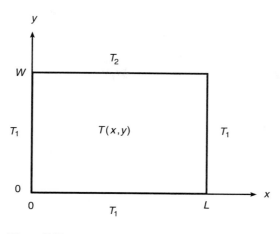

Figure P38

where

$$w(x, y) = \frac{2}{\pi}\sum_{n \text{ odd}}^{\infty} \frac{2}{n} \sin\left(\frac{n\pi x}{L}\right) \frac{\sinh(n\pi y/L)}{\sinh(n\pi W/L)}$$

Use the following data: $T_1 = 70°F$, $T_2 = 200°F$, and $W = L = 2$ ft.

a. The terms in the preceding series become smaller in magnitude as n increases. Write a MATLAB program to verify this fact for $n = 1, \ldots, 19$ for the center of the plate ($x = y = 1$).

b. Using $x = y = 1$, write a MATLAB program to determine how many terms are required in the series to produce a temperature calculation that is accurate to within 1 percent. (That is, for what value of n will the addition of the next term in the series produce a change in T of less than 1 percent?) Use your physical insight to determine whether this answer gives the correct temperature at the center of the plate.

c. Modify the program from part b to compute the temperatures in the plate; use a spacing of 0.2 for both x and y.

39. Consider the following script file. Fill in the lines of the following table with the values that would be displayed immediately after the `while` statement if you ran the script file. Write in the values the variables have each time the `while` statement is executed. You might need more or fewer lines in the table. Then type in the file, and run it to check your answers.

```
k = 1;b = -2;x = -1;y = -2;
while k <= 3
    k, b, x, y
    y = x^2 - 3;
```

```
      if y < b
          b = y;
      end
      x = x + 1;
      k = k + 1;
  end
```

Pass	k	b	x	y
First				
Second				
Third				
Fourth				
Fifth				

40. Assume that the human player makes the first move against the computer in a game of Tic-Tac-Toe, which has a 3 × 3 grid. Write a MATLAB function that lets the computer respond to that move. The function's input argument should be the cell location of the human player's move. The function's output should be the cell location of the computer's first move. Label the cells as 1, 2, 3 across the top row; 4, 5, 6 across the middle row; and 7, 8, 9 across the bottom row.

Section 4.7

41. The following table gives the approximate values of the static coefficient of friction μ for various materials.

Materials	μ
Metal on metal	0.20
Wood on wood	0.35
Metal on wood	0.40
Rubber on concrete	0.70

To start a weight W moving on a horizontal surface, you must push with a force F, where $F = \mu W$. Write a MATLAB program that uses the `switch` structure to compute the force F. The program should accept as input the value of W and the type of materials.

42. The height and speed of a projectile (such as a thrown ball) launched with a speed of v_0 at an angle A to the horizontal are given by

$$h(t) = v_0 t \sin A - 0.5gt^2$$

$$v(t) = \sqrt{v_0^2 - 2v_0 gt \sin A + g^2 t^2}$$

where g is the acceleration due to gravity. The projectile will strike the ground when $h(t) = 0$, which gives the time to hit $t_{hit} = 2(v_0/g) \sin A$.

Use the `switch` structure to write a MATLAB program to compute the maximum height reached by the projectile, the total horizontal distance traveled, or the time to hit. The program should accept as input the user's choice of which quantity to compute and the values of v_0, A, and g. Test the program for the case where $v_0 = 40$ m/s, $A = 30^0$, and $g = 9.81$ m/s^2.

43. Use the `switch` structure to write a MATLAB program to compute the amount of money that accumulates in a savings account in one year. The program should accept the following input: the initial amount of money deposited in the account; the frequency of interest compounding (monthly, quarterly, semiannually, or annually); and the interest rate. Run your program for a $1000 initial deposit for each case; use a 5 percent interest rate. Compare the amounts of money that accumulate for each case.

44. Engineers often need to estimate the pressures and volumes of a gas in a container. The *van der Waals* equation is often used for this purpose. It is

$$P = \frac{RT}{\hat{V} - b} - \frac{a}{\hat{V}^2}$$

where the term b is a correction for the volume of the molecules and the term $a/\hat{V}^2$ is a correction for molecular attractions. The gas constant is R, the *absolute* temperature is T, and the gas specific volume is $\hat{V}$. The value of R is the same for all gases; it is $R = 0.08206$ L-atm/mol-K. The values of a and b depend on the type of gas. Some values are given in the following table. Write a user-defined function using the `switch` structure that computes the pressure P on the basis of the van der Waals equation. The function's input arguments should be T, $\hat{V}$, and a string variable containing the name of a gas listed in the table. Test your function for chlorine (Cl$_2$) for $T = 300$ K and $\hat{V} = 20$ L/mol.

Gas	a (L^2-atm/mol^2)	b (L/mol)
Helium, He	0.0341	0.0237
Hydrogen, H$_2$	0.244	0.0266
Oxygen, O$_2$	1.36	0.0318
Chlorine, Cl$_2$	6.49	0.0562
Carbon dioxide, CO$_2$	3.59	0.0427

45. Using the program developed in Problem 19, write a program that uses the `switch` structure to compute the number of days in a year up to a given date, given the year, the month, and the day of the month.

Section 4.9

46. Consider the college enrollment model discussed in Example 4.9–2. Suppose the college wants to limit freshman admissions to 120 percent of the current sophomore class and limit sophomore transfers to 10 percent of the current freshman class. Rewrite and run the program given in the example to examine the effects of these policies over a 10-year period. Plot the results.

47. Suppose you project that you will be able to deposit the following monthly amounts into a savings account for a period of 5 years. The account initially has no money in it.

Year	1	2	3	4	5
Monthly deposit ($)	300	350	350	350	400

At the end of each year in which the account balance is at least $3000, you withdraw $2000 to buy a certificate of deposit (CD), which pays 6 percent interest compounded annually.

Write a MATLAB program to compute how much money will accumulate in 5 years in the account and in any CDs you buy. Run the program for two different savings interest rates: 4 percent and 5 percent.

48.* A certain company manufactures and sells golf carts. At the end of each week, the company transfers the carts produced that week into storage (inventory). All carts that are sold are taken from the inventory. A simple model of this process is

$$I(k + 1) = P(k) + I(k) - S(k)$$

where

$P(k)$ = number of carts produced in week k

$I(k)$ = number of carts in inventory in week k

$S(k)$ = number of carts sold in week k

The projected weekly sales for 10 weeks are

Week	1	2	3	4	5	6	7	8	9	10
Sales	50	55	60	70	70	75	80	80	90	55

Suppose the weekly production is based on the previous week's sales so that $P(k) = S(k - 1)$. Assume that the first week's production is 50 carts; that is, $P(1) = 50$. Write a MATLAB program to compute and plot the number of carts in inventory for each of the 10 weeks or until the inventory drops below zero. Run the program for two cases: (*a*) an initial inventory of 50 carts so that $I(1) = 50$ and (*b*) an initial inventory of 30 carts so that $I(1) = 30$.

49. Redo Problem 48 with the restriction that the next week's production is set to zero if the inventory exceeds 40 carts.

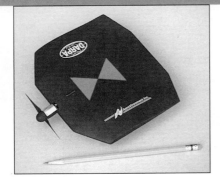

Courtesy of Aero Vironment, Inc.

Engineering in the 21st Century. . .

Low-Speed Aeronautics

Sometimes just when we think a certain technical area is mature and the possibility of further development is unlikely, we are surprised by a novel design. Recent developments in low-speed aeronautics are examples of this phenomenon. Even though engineers have known for years that a human could generate enough power to propel an aircraft, the feat remained impossible until the availability of lightweight materials that enabled the Gossamer Challenger to fly across the English Channel. Solar-powered aircraft that can stay aloft for over a day are other examples.

Another example is the recent appearance of wing-in-ground (WIG) effect vehicles. WIG vehicles make use of an air cushion to create lift. They are a hybrid between an aircraft and a hovercraft, and most are intended for over-water flight only. A hovercraft rides on an air cushion created by fans, but the air cushion of a WIG vehicle is due to the air that is captured under its stubby wings.

Small aircraft with cameras will be useful for search and reconnaissance. An example of such a "micro air vehicle" (MAV) is the 6-in.-long Black Widow produced by Aero Vironment, Inc. It carries a 2-g video camera the size of a sugar cube, and flies at about 65 km/h with a range of 10 km. Proper design of such vehicles requires a systematic methodology to find the optimum combination of airfoil shape, motor type, battery type, and most importantly, the propeller shape.

The MATLAB advanced graphics capabilities make it useful for visualizing flow patterns, and the Optimization toolbox is useful for designing such vehicles. ■

CHAPTER 5

Advanced Plotting

OUTLINE

5.1   *xy* Plotting Functions

5.2   Additional Commands and Plot Types

5.3   Interactive Plotting in MATLAB

5.4   Three-Dimensional Plots

5.5   Summary

Problems

In this chapter you will learn additional features to use to create a variety of two-dimensional plots, which are also called *xy plots,* and three-dimensional plots called *xyz plots,* or *surface plots.* Two-dimensional plots are discussed in Sections 5.1 through 5.3. Section 5.4 discusses three-dimensional plots. These plotting functions are described in the `graph2d` and `graph3d` Help categories, so typing `help graph2d` or `help graph3d` will display a list of the relevant plotting functions.

An important application of plotting is *function discovery,* the technique for using data plots to obtain a mathematical function or "mathematical model" that describes the process that generated the data. This topic is treated in Chapter 6.

5.1 *xy* Plotting Functions

The "anatomy" and nomenclature of a typical *xy* plot is shown in Figure 5.1–1, in which the plot of a data set and a curve generated from an equation appear. A plot can be made from measured data or from an equation. When data are plotted, each data point is plotted with a *data symbol,* or *point marker,* such as the small circles shown in Figure 5.1–1. An exception to this rule would be when there are so many data points that the symbols would be too densely packed.

DATA SYMBOL

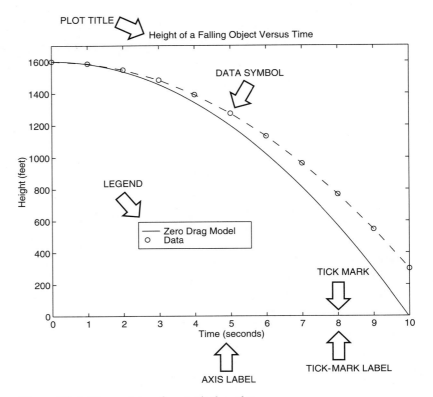

Figure 5.1–1 Nomenclature for a typical *xy* plot.

In that case, the data points should be plotted with a dot. However, when the plot is generated from a function, data symbols must *never* be used! Lines between closely spaced points are always used to plot a function.

The MATLAB basic *xy* plotting function is plot(x,y) as we saw in Chapter 1. If x and y are vectors, a single curve is plotted with the x values on the abscissa and the y values on the ordinate. The xlabel and ylabel commands put labels on the abscissa and the ordinate, respectively. The syntax is xlabel('text'), where text is the text of the label. Note that you must enclose the label's text in single quotes. The syntax for ylabel is the same. The title command puts a title at the top of the plot. Its syntax is title('text'), where text is the title's text.

The plot(x,y) function in MATLAB automatically selects a tick-mark spacing for each axis and places appropriate tick labels. This feature is called *autoscaling*. MATLAB also chooses limits for the *x* and *y* axes. The order of the xlabel, ylabel, and title commands does not matter, but we must place them *after* the plot command, either on separate lines using ellipses or on the same line separated by commas.

After the `plot` command is executed, the plot will appear in the Figure window. You can obtain a hard copy of the plot in one of several ways:

1. Use the menu system. Select **Print** on the **File** menu in the Figure window. Answer **OK** when you are prompted to continue the printing process.
2. Type `print` at the command line. This command sends the current plot directly to the printer.
3. Save the plot to a file to be printed later or imported into another application such as a word processor. You need to know something about graphics file formats to use this file properly. See the subsection **Exporting Figures** later in this section.

Type `help print` to obtain more information.

MATLAB assigns the output of the `plot` command to Figure window number 1. When another `plot` command is executed, MATLAB overwrites the contents of the existing Figure window with the new plot. Although you can keep more than one Figure window active, we do not use this feature in this text.

When you have finished with the plot, close the Figure window by selecting **Close** from the **File** menu in the Figure window. If you do not close the window, it will not reappear when a new `plot` command is executed. However, the figure will still be updated.

Table 5.1–1 lists the requirements essential to producing plots that communicate effectively.

Table 5.1–1 Requirements for a correct plot

1. Each axis must be labeled with the name of the quantity being plotted *and its units!* If two or more quantities having different units are plotted (such as when in a plot of both speed and distance versus time), indicate the units in the axis label, if there is room, or in the legend or labels for each curve.
2. Each axis should have regularly spaced tick marks at convenient intervals—not too sparse, but not too dense—with a spacing that is easy to interpret and interpolate. For example, use 0.1, 0.2, and so on, rather than 0.13, 0.26, and so on.
3. If you are plotting more than one curve or data set, label each on its plot, use different line types, or use a legend to distinguish them.
4. If you are preparing multiple plots of a similar type or if the axes' labels cannot convey enough information, use a title.
5. If you are plotting measured data, plot each data point with a symbol such as a circle, square, or cross (use the same symbol for every point in the same data set). If there are many data points, plot them using the dot symbol.
6. Sometimes data symbols are connected by lines to help the viewer visualize the data, especially if there are few data points. However, connecting the data points, especially with a solid line, might be interpreted to imply knowledge of what occurs between the data points. Thus you should be careful to prevent such misinterpretation.
7. If you are plotting points generated by evaluating a function (as opposed to measured data), do *not* use a symbol to plot the points. Instead, be sure to generate many points, and connect the points with solid lines.

grid and axis Commands

The grid command displays gridlines at the tick marks corresponding to the tick labels. You can use the axis command to override the MATLAB selections

AXIS LIMITS

for the axis limits. The basic syntax is axis([xmin xmax ymin ymax]). This command sets the scaling for the *x* and *y* axes to the minimum and maximum values indicated. Note that, unlike an array, this command does not use commas to separate the values.

Figure 5.1–2 shows a plot in which the command axis([0 10 −2 5]) was used to override the limits chosen by auto scaling (which chose the upper limit of the ordinate to be 4).

The axis command has the following variants:

■ axis square selects the axes' limits so that the plot will be square.
■ axis equal selects the scale factors and tick spacing to be the same on each axis. This variation makes plot(sin(x),cos(x)) look like a circle, instead of an oval.
■ axis auto returns the axis scaling to its default autoscaling mode in which the best axes' limits are computed automatically.

Type help axis to see the full list of variants.

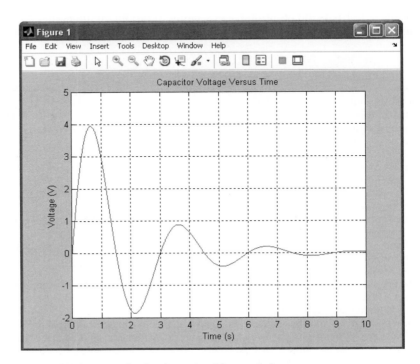

Figure 5.1–2 A sample plot shown in a Figure window.

Plots of Complex Numbers

With only one argument, say, `plot(y)`, the `plot` function will plot the values in the vector `y` versus their indices 1, 2, 3, . . . , and so on. If `y` is complex, `plot(y)` plots the imaginary parts versus the real parts. Thus `plot(y)` in this case is equivalent to `plot(real(y),imag(y))`. This situation is the only time when the `plot` function handles the imaginary parts; in all other variants of the `plot` function, it ignores the imaginary parts. For example, the script file

```
z = 0.1 + 0.9i;
n = 0:0.01:10;
plot(z.^n),xlabel('Real'),ylabel('Imaginary')
```

generates a spiral plot.

The Function Plot Command `fplot`

MATLAB has a "smart" command for plotting functions. The `fplot` command automatically analyzes the function to be plotted and decides how many plotting points to use so that the plot will show all the features of the function. Its syntax is `fplot(function, [xmin xmax])`, where `function` is a function handle to the function to be plotted and `[xmin xmax]` specifies the minimum and maximum values of the independent variable. The range of the dependent variable can also be specified. In this case the syntax is `fplot(function, [xmin xmax ymin ymax])`.

For example, the session

```
>>f = @(x) (cos(tan(x)) - tan(sin(x)));
>>fplot(f,[1 2])
```

produces the plot shown in Figure 5.1–3a. The `fplot` command automatically chooses enough plotting points to display all the variations in the function. We can achieve the same results using the `plot` command, but we need to know how many values to compute to generate the plot. For example, choosing a spacing of 0.01, and using `plot`, we obtain the plot in Figure 5.1–3b. We see that this choice of spacing misses some of the function's behavior.

Another form is `[x,y] = fplot(function, limits)`, where `limits` may be either `[xmin xmax]` or `[xmin xmax ymin ymax]`. With this form the command returns the abscissa and ordinate values in the column vectors `x` and `y`, but no plot is produced. The returned values can then be used for other purposes, such as plotting multiple curves, which is the topic of the next section. Other commands can be used with the `fplot` command to enhance a plot's appearance, for example, the `title`, `xlabel`, and `ylabel` commands and the line type commands to be introduced in the next section.

Plotting Polynomials

We can plot polynomials more easily by using the `polyval` function. The function `polyval(p,x)` evaluates the polynomial p at specified values of its

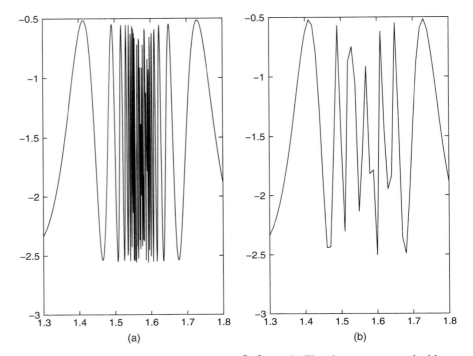

Figure 5.1–3 (a) The plot was generated with `fplot`. (b) The plot was generated with `plot` using 101 points.

independent variable x. For example, to plot the polynomial $3x^5 + 2x^4 - 100x^3 + 2x^2 - 7x + 90$ over the range $-6 \le x \le 6$ with a spacing of 0.01, you type

```
>>x = -6:0.01:6;
>>p = [3,2,-100,2,-7,90];
>>plot(x,polyval(p,x)),xlabel('x'),ylabel('p')
```

Table 5.1–2 summarizes the *xy* plotting commands discussed in this section.

Test Your Understanding

T5.1–1 Plot the equation $y = 0.4\sqrt{1.8x}$ for $0 \le x \le 35$ and $0 \le y \le 3.5$.

T5.1–2 Use the `fplot` command to investigate the function $\tan(\cos x) - \sin(\tan x)$ for $0 \le x \le 2\pi$. How many values of x are needed to obtain the same plot using the `plot` command? (Answer: 292 values.)

T5.1–3 Plot the imaginary part versus the real part of the function $(0.2 + 0.8i)^n$ for $0 \le n \le 20$. Choose enough points to obtain a smooth curve. Label each `axis` and put a title on the plot. Use the `axis` command to change the tick-label spacing.

Table 5.1–2 Basic *xy* plotting commands

Command	Description
`axis([xmin xmax ymin ymax])`	Sets the minimum and maximum limits of the *x* and *y* axes.
`fplot(function,[xmin xmax])`	Performs intelligent plotting of functions, where `function` is a function handle that describes the function to be plotted and `[xmin xmax]` specifies the minimum and maximum values of the independent variable. The range of the dependent variable can also be specified. In this case the syntax is `fplot(function, [xmin xmax ymin ymax])`.
`grid`	Displays gridlines at the tick marks corresponding to the tick labels.
`plot(x,y)`	Generates a plot of the array `y` versus the array `x` on rectilinear axes.
`plot(y)`	Plots the values of `y` versus their indices if `y` is a vector. Plots the imaginary parts of `y` versus the real parts if `y` is a vector having complex values.
`polyval(p,x)`	Evaluates the polynomial `p` at specified values of its independent variable `x`.
`print`	Prints the plot in the Figure window.
`title('text')`	Puts text in a title at the top of a plot.
`xlabel('text')`	Adds a text label to the *x* axis (the abscissa).
`ylabel('text')`	Adds a text label to the *y* axis (the ordinate).

Saving Figures

When you create a plot, the Figure window appears. This window has eight menus, which are discussed in detail in Section 5.3. The **File** menu is used for saving and printing the figure. You can save your figure in a format that can be opened during another MATLAB session or in a format that can be used by other applications.

To save a figure that can be opened in subsequent MATLAB sessions, save it in a figure file with the .fig file name extension. To do this, select **Save** from the Figure window **File** menu or click the **Save** button (the disk icon) on the toolbar. If this is the first time you are saving the file, the **Save As** dialog box appears. Make sure that the type is MATLAB Figure (*.fig). Specify the name you want assigned to the figure file. Click OK. You can also use the `saveas` command.

To open a figure file, select **Open** from the **File** menu or click the **Open** button (the opened folder icon) on the toolbar. Select the figure file you want to open and click OK. The figure file appears in a new figure window.

Exporting Figures

If you want to save the figure in a format that can be used by another application, such as the standard graphics file formats TIFF or EPS, perform these steps.

1. Select **Export Setup** from the **File** menu. This dialog provides options you can specify for the output file, such as the figure size, fonts, line size and style, and output format.
2. Select **Export** from the **Export Setup** dialog. A standard **Save As** dialog appears.

Table 5.1–3 Hints for improving plots

1. Start scales from zero whenever possible. This technique prevents a false impression of the magnitudes of any variations shown on the plot.
2. Use sensible tick-mark spacing. For example, if the quantities are months, choose a spacing of 12 because $1/10$ of a year is not a convenient division. Space tick marks as close as is useful, but no closer.
3. Minimize the number of zeros in the tick labels. For example, use a scale in millions of dollars when appropriate, instead of a scale in dollars with six zeros after every number.
4. Determine the minimum and maximum data values for each axis before plotting the data. Then set the axis limits to cover the entire data range plus an additional amount to allow convenient tick-mark spacing to be selected.

3. Select the format from the list of formats in the **Save As** type menu. This selects the format of the exported file and adds the standard file name extension given to files of that type.
4. Enter the name you want to give the file, less the extension.
5. Click **Save.**

You can also export the figure from the command line, by using the `print` command. See MATLAB Help for more information about exporting figures in different formats.

You can also copy a figure to the clipboard and then paste it into another application:

1. Select **Copy Options** from the **Edit** menu of the Figure window. The **Copying Options** page of the **Preferences** dialog box appears.
2. Complete the fields on the **Copying Options** page and click **OK.**
3. Select **Copy Figure** from the **Edit** menu.

The figure is copied to the Windows clipboard and can be pasted into another application.

MATLAB also enables you to save figures in formats compatible with PowerPoint and MS Word. See the MATLAB Help for more information.

The graphics functions covered in this section and in Section 5.3 can be placed in script files that can be reused to create similar plots. This feature gives them an advantage over the interactive plotting tools discussed in Section 5.3.

When you are creating plots, keep in mind that the actions listed in Table 5.1–3, while not required, can never the less improve the appearance and usefulness of your plots.

5.2 Additional Commands and Plot Types

SUBPLOT

MATLAB can create figures that contain an array of plots, called *subplots*. These are useful when you want to compare the same data plotted with different axis types, for example. The MATLAB `subplot` command creates such figures. We frequently need to plot more than one curve or data set on a single plot. Such a plot is called an *overlay plot*. This section describes these plots and several other types of plots.

Subplots

You can use the `subplot` command to obtain several smaller "subplots" in the same figure. The syntax is `subplot(m,n,p)`. This command divides the Figure window into an array of rectangular panes with *m* rows and *n* columns. The variable `p` tells MATLAB to place the output of the `plot` command following the `subplot` command into the *p*th pane. For example, `subplot(3,2,5)` creates an array of six panes, three panes deep and two panes across, and directs the next plot to appear in the fifth pane (in the bottom left corner). The following script file created Figure 5.2–1, which shows the plots of the functions $y = e^{-1.2x}\sin(10x + 5)$ for $0 \le x \le 5$ and $y = |x^3 - 100|$ for $-6 \le x \le 6$.

```
x = 0:0.01:5;
y = exp(-1.2*x).*sin(10*x+5);
subplot(1,2,1)
plot(x,y),xlabel('x'),ylabel('y'),axis([0 5 -1 1])
x = -6:0.01:6;
y = abs(x.^3-100);
subplot(1,2,2)
plot(x,y),xlabel('x'),ylabel('y'),axis([-6 6 0 350])
```

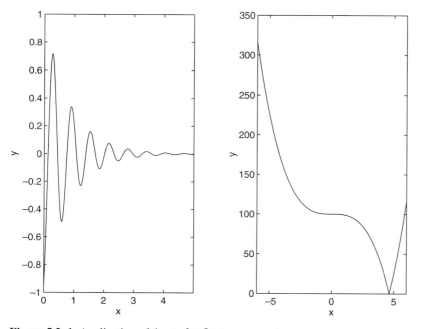

Figure 5.2–1 Application of the `subplot` command.

Test Your Understanding

T5.2–1 Pick a suitable spacing for t and v, and use the subplot command to plot the function $z = e^{-0.5t} \cos(20t - 6)$ for $0 \le t \le 8$ and the function $u = 6 \log_{10}(v^2 + 20)$ for $-8 \le v \le 8$. Label each axis.

Overlay Plots

You can use the following variants of the MATLAB basic plotting functions plot(x,y) and plot(y) to create *overlay plots*:

- plot(A) plots the columns of A versus their indices and generates n curves, where A is a matrix with m rows and n columns.
- plot(x,A) plots the matrix A versus the vector x, where x is either a row vector or a column vector and A is a matrix with m rows and n columns. If the length of x is m, then each *column* of A is plotted versus the vector x. There will be as many curves as there are columns of A. If x has length n, then each *row* of A is plotted versus the vector x. There will be as many curves as there are rows of A.
- plot(A,x) plots the vector x versus the matrix A. If the length of x is m, then x is plotted versus the *columns* of A. There will be as many curves as there are columns of A. If the length of x is n, then x is plotted versus the *rows* of A. There will be as many curves as there are rows of A.
- plot(A,B) plots the columns of the matrix B versus the columns of the matrix A.

Data Markers and Line Types

To plot the vector y versus the vector x and mark each point with a data marker, enclose the symbol for the marker in single quotes in the plot function. Table 5.2–1 shows the symbols for some of the available data markers. For example, to use a small circle, which is represented by the lowercase letter o, type plot(x,y, 'o'). This notation results in a plot like the one on the left

Table 5.2–1 Specifiers for data markers, line types, and colors

Data markers[†]		Line types		Colors	
Dot (·)	·	Solid line	-	Black	k
Asterisk (*)	*	Dashed line	- -	Blue	b
Cross (×)	×	Dash-dotted line	-.	Cyan	c
Circle (o)	o	Dotted line	:	Green	g
Plus sign (+)	+			Magenta	m
Square (□)	s			Red	r
Diamond (◇)	d			White	w
Five-pointed star (★)	p			Yellow	y

[†]Other data markers are available. Search for "markers" in MATLAB Help.

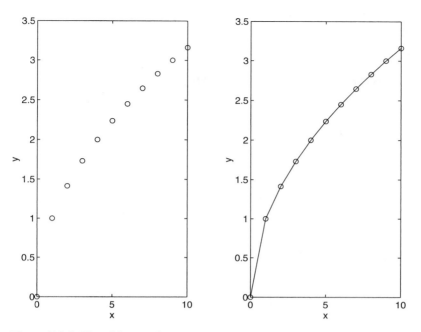

Figure 5.2–2 Use of data markers.

in Figure 5.2–2. To connect each data marker with a straight line, we must plot the data twice, by typing plot(x,y,x,y,'o'). See the plot on the right in Figure 5.2–2.

Suppose we have two curves or data sets stored in the vectors x, y, u, and v. To plot y versus x and v versus u on the same plot, type plot(x,y,u,v). Both sets will be plotted with a solid line, which is the default line style. To distinguish the sets, we can plot them with different line types. To plot y versus x with a solid line and u versus v with a dashed line, type plot(x,y,u,v,'−−'), where the symbols '−−' represent a dashed line. Table 5.2–1 gives the symbols for other line types. To plot y versus x with asterisks (*) connected with a dotted line, you must plot the data twice by typing plot(x,y,'*',x,y,':').

You can obtain symbols and lines of different colors by using the color symbols shown in Table 5.2–1. The color symbol can be combined with the data-marker symbol and the line-type symbol. For example, to plot y versus x with green asterisks (*) connected with a red dashed line, you must plot the data twice by typing plot(x,y,'g*',x,y,'r−−'). (Do not use colors if you are going to print the plot on a black-and-white printer.)

Labeling Curves and Data

When more than one curve or data set is plotted on a graph, we must distinguish between them. If we use different data symbols or different line types, then we must either provide a legend or place a label next to each curve. To create a legend,

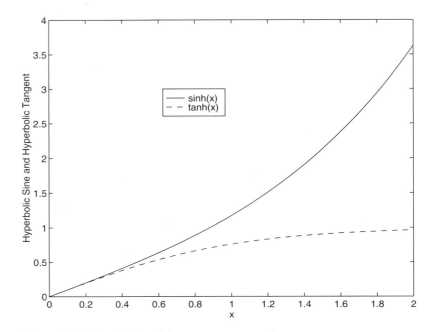

Figure 5.2–3 Application of the `legend` command.

use the `legend` command. The basic form of this command is `legend` (`'string1'`, `'string2'`), where `string1` and `string2` are text strings of your choice. The `legend` command automatically obtains from the plot the line type used for each data set and displays a sample of this line type in the legend box next to the string you selected. The following script file produced the plot in Figure 5.2–3.

```
x = 0:0.01:2;
y = sinh(x);
z = tanh(x);
plot(x,y,x,z,'--'),xlabel('x'),...
    ylabel('Hyperbolic Sine and Hyperbolic Tangent'),...
    legend('sinh(x)','tanh(x)')
```

The `legend` command must be placed somewhere after the `plot` command. When the plot appears in the Figure window, use the mouse to position the legend box. (Hold down the left button on the mouse to move the box.)

Another way to distinguish curves is to place a label next to each. The label can be generated either with the `gtext` command, which lets you place the label by using the mouse, or with the `text` command, which requires you to specify the coordinates of the label. The syntax of the `gtext` command is `gtext('string')`, where `string` is a text string that specifies the label of your choice. When this command is executed, MATLAB waits for a mouse button or a key to be pressed while the mouse pointer is within the Figure window; the label is placed at that position of the mouse pointer. You may use more than one `gtext` command for a given plot. The `text` command `text(x,y,'string')` adds a text string to the plot at the location specified by the coordinates `x,y`. These coordinates are in the same units as the plot's data. Of course, finding the proper coordinates to use with the `text` command usually requires some trial and error.

The `hold` Command

The `hold` command creates a plot that needs two or more `plot` commands. Suppose we wanted to plot $y_2 = 4 + e^{-x} \cos 6x$ versus $y_1 = 3 + e^{-x} \sin 6x$, $-1 \le x \le 1$ on the same plot with the complex function $z = (0.1 + 0.9i)^n$, where $0 \le n \le 10$. The following script file creates the plot in Figure 5.2–4.

```
x = -1:0.01:1;
y1 = 3+exp(-x).*sin(6*x);
y2 = 4+exp(-x).*cos(6*x);
plot((0.1+0.9i).^(0:0.01:10)),hold,plot(y1,y2),...
    gtext('y2 versus y1'),gtext('Imag(z) versus Real(z)')
```

When more than one `plot` command is used, do not place any of the `gtext` commands before any `plot` command. Because the scaling changes as each `plot` command is executed, the label placed by the `gtext` command might end up in the wrong position. Table 5.2–2 summarizes the plot enhancement commands introduced in this section.

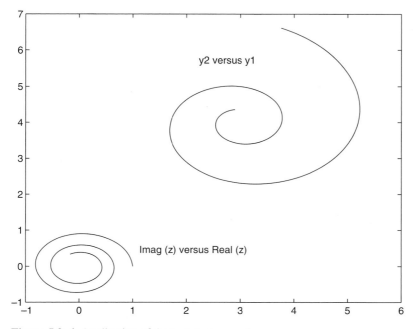

Figure 5.2–4 Application of the `hold` command.

Table 5.2–2 Plot enhancement commands

Command	Description
gtext('text')	Places the string text in the Figure window at a point specified by the mouse.
hold	Freezes the current plot for subsequent graphics commands.
legend('leg1','leg2',...)	Creates a legend using the strings leg1, leg2, and so on and enables its placement with the mouse.
plot(x,y,u,v)	Plots, on rectilinear axes, four arrays: y versus x and v versus u.
plot(x,y,'type')	Plots the array y versus the array x on rectilinear axes, using the line type, data marker, and colors specified in the string type. See Table 5.2–1.
plot(A)	Plots the columns of the $m \times n$ array A versus their indices and generates n curves.
plot(P,Q)	Plots array Q versus array P. See the text for a description of the possible variants involving vectors and/or matrices: plot(x,A), plot(A,x), and plot(A,B).
subplot(m,n,p)	Splits the Figure window into an array of subwindows with m rows and n columns and directs the subsequent plotting commands to the pth subwindow.
text(x,y,'text')	Places the string text in the Figure window at a point specified by coordinates x, y.

Test Your Understanding

T5.2–2 Plot the following two data sets on the same plot. For each set, $x = 0, 1, 2, 3, 4, 5$. Use a different data marker for each set. Connect the markers for the first set with solid lines. Connect the markers for the second set with dashed lines. Use a legend, and label the plot axes appropriately. The first set is $y = 11, 13, 8, 7, 5, 9$. The second set is $y = 2, 4, 5, 3, 2, 4$.

T5.2–3 Plot $y = \cosh x$ and $y = 0.5e^x$ on the same plot for $0 \le x \le 2$. Use different line types and a legend to distinguish the curves. Label the plot axes appropriately.

T5.2–4 Plot $y = \sinh x$ and $y = 0.5e^x$ on the same plot for $0 \le x \le 2$. Use a solid line type for each, the gtext command to label the $\sinh x$ curve, and the text command to label the $0.5e^x$ curve. Label the plot axes appropriately.

T5.2–5 Use the hold command and the plot command twice to plot $y = \sin x$ and $y = x - x^3/3$ on the same plot for $0 \le x \le 1$. Use a solid line type for each, and use the gtext command to label each curve. Label the plot axes appropriately.

Annotating Plots

You can create text, titles, and labels that contain mathematical symbols, Greek letters, and other effects such as italics. The features are based on the T$_E$X typesetting language. For more information, including a list of the available characters, search the online Help for the "Text Properties" page. See also the "Mathematical symbols, Greek Letters, and T$_E$X Characters" page.

You can create a title having the mathematical function $Ae^{-t/\tau}\sin(\omega t)$ by typing

```
>>title('{\it Ae}^{-{\it t/\tau}}\sin({\it \omega t})')
```

The backslash character \ precedes all T$_E$X character sequences. Thus the strings `\tau` and `\omega` represent the Greek letters τ and ω. Superscripts are created by typing ^; subscripts are created by typing _. To set multiple characters as superscripts or subscripts, enclose them in braces. For example, type `x_{13}` to produce x_{13}. In mathematical text variables are usually set in italic, and functions, such as sin, are set in roman type. To set a character, say, x, in italic using the T$_E$X commands, you type `{\it x}`.

Logarithmic Plots

Logarithmic scales—abbreviated log scales—are widely used (1) to represent a data set that covers a wide range of values and (2) to identify certain trends in data. Certain types of functional relationships appear as straight lines when plotted using a log scale. This method makes it easier to identify the function. A *log-log* plot has log scales on both axes. A *semilog* plot has a log scale on only one axis.

Figure 5.2–5 shows a rectilinear plot and a log-log plot of the function

$$y = \sqrt{\frac{100(1 - 0.01x^2)^2 + 0.02x^2}{(1 - x^2)^2 + 0.1x^2}} \qquad 0.1 \le x \le 100 \qquad (5.2\text{–}1)$$

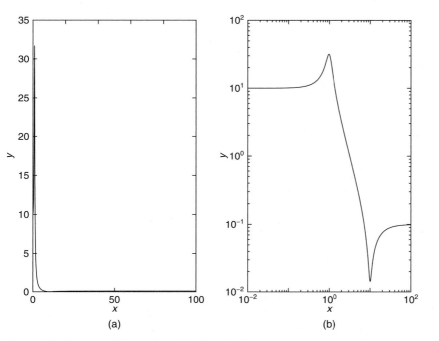

(a)   (b)

Figure 5.2–5 (a) Rectilinear plot of the function in Equation (5.2–1). (b) Log-log plot of the function. Note the wide range of values of both x and y.

Because of the wide range in values on both the abscissa and the ordinate, rectilinear scales do not reveal the important features. The following program produced Figure 5.2–5.

```
% Create the Rectilinear Plot
x1 = 0:0.01:100; u1 = x1.^2;
num1 = 100*(1-0.01*u1).^2 + 0.02*u1;
den1 = (1-u1).^2 + 0.1*u1;
y1 = sqrt(num1./den1);
subplot(1,2,1), plot(x1,y1),xlabel('x'),ylabel('y'),
% Create the Loglog Plot
x2 = logspace(-2, 2, 500); u2 = x2.^2;
num2 = 100*(1-0.01*u2).^2 + 0.02*u2;
den2 = (1-u2).^2 + 0.1*u2;
y2 = sqrt(num2./den2);
subplot(1,2,2), loglog(x2,y2),xlabel('x'),ylabel('y')
```

It is important to remember the following points when using log scales:

1. You cannot plot negative numbers on a log scale, because the logarithm of a negative number is not defined as a real number.
2. You cannot plot the number 0 on a log scale, because $\log_{10} 0 = \ln 0 = -\infty$. You must choose an appropriately small number as the lower limit on the plot.
3. The tick-mark labels on a log scale are the actual values being plotted; they are not the logarithms of the numbers. For example, the range of x values in the plot in Figure 5.2–5b is from $10^{-2} = 0.01$ to $10^2 = 100$, and the range of y values is from 10^{-2} to $10^2 = 100$.

MATLAB has three commands for generating plots having log scales. The appropriate command depends on which axis must have a log scale. Follow these rules:

1. Use the `loglog(x,y)` command to have both scales logarithmic.
2. Use the `semilogx(x,y)` command to have the x scale logarithmic and the y scale rectilinear.
3. Use the `semilogy(x,y)` command to have the y scale logarithmic and the x scale rectilinear.

Table 5.2–3 summarizes these functions. For other two-dimensional plot types, type `help specgraph`. We can plot multiple curves with these commands just as with the `plot` command. In addition, we can use the other commands, such as `grid`, `xlabel`, and `axis`, in the same manner. Figure 5.2–6 shows how these commands are applied. It was created with the following program.

```
x1 = 0:0.01:3; y1 = 25*exp(0.5*x1);
y2 = 40*(1.7.^x1);
x2 = logspace(-1,1,500); y3 = 15*x2.^(0.37);
```

```
subplot(1,2,1),semilogy(x1,y1,x1,y2, '--'),...
    legend ('y = 25e^{0.5x}', 'y = 40(1.7) ^x'),...
    xlabel('x'),ylabel('y'),grid,...
    subplot(1,2,2),loglog(x2,y3),legend('y = 15x^{0.37}'),...
    xlabel('x'),ylabel('y'),grid
```

Table 5.2–3 Specialized plot commands

Command	Description
`bar(x,y)`	Creates a bar chart of y versus x.
`loglog(x,y)`	Produces a log-log plot of y versus x.
`plotyy(x1,y1,x2,y2)`	Produces a plot with two y axes, $y1$ on the left and $y2$ on the right.
`polar(theta,r,'type')`	Produces a polar plot from the polar coordinates `theta` and `r`, using the line type, data marker, and colors specified in the string `type`.
`semilogx(x,y)`	Produces a semilog plot of y versus x with logarithmic abscissa scale.
`semilogy(x,y)`	Produces a semilog plot of y versus x with logarithmic ordinate scale.
`stairs(x,y)`	Produces a stairs plot of y versus x.
`stem(x,y)`	Produces a stem plot of y versus x.

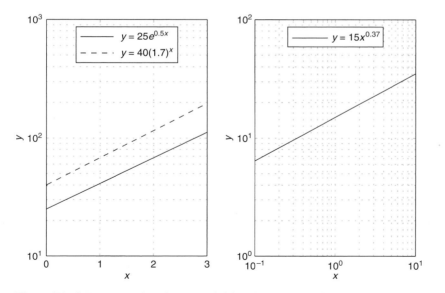

Figure 5.2–6 Two examples of exponential functions plotted with the `semilogy` function (the left-hand plot), and an example of a power function plotted with the `loglog` function (right-hand) plot.

Note that the two *exponential* functions $y = 25e^{0.5x}$ and $y = 40(1.7)^x$ both produce straight lines on a semilog plot with the y axis logarithmic. The *power* function $y = 15x^{0.37}$ produces a straight line on a log-log plot.

Stem, Stairs, and Bar Plots

MATLAB has several other plot types that are related to *xy* plots. These include the stem, stairs, and bar plots. Their syntax is very simple, namely, `stem(x,y)`, `stairs(x,y)`, and `bar(x,y)`. See Table 5.2–3.

Separate *y* Axes

The `plotyy` function generates a graph with two *y* axes. The syntax `plotyy (x1,y1,x2,y2)` plots `y1` versus `x1` with *y* axis labeling on the left, and plots `y2` versus `x2` with *y* axis labeling on the right. The syntax `plotyy(x1,y1,x2, y2,'type1','type2')` generates a `'type1'` plot of `y1` versus `x1` with *y* axis labeling on the left, and generates a `'type2'` plot of `y2` versus `x2` with *y* axis labeling on the right. For example, `plotyy(x1,y1,x2,y2,'plot','stem')` uses `plot(x1,y1)` to generate a plot for the left axis, and `stem(x2,y2)` to generate a plot for the right axis. To see other variations of the `plotyy` function, type `help plotyy`.

Polar Plots

Polar plots are two-dimensional plots made using polar coordinates. If the polar coordinates are (θ, r), where θ is the angular coordinate and r is the radial coordinate of a point, then the command `polar(theta,r)` will produce the polar plot. A grid is automatically overlaid on a polar plot. This grid consists of concentric circles and radial lines every 30°. The `title` and `gtext` commands can be used to place a title and text. The variant command `polar(theta,r,'type')` can be used to specify the line type or data marker, just as with the `plot` command.

EXAMPLE 5.2–1   Plotting Orbits

The equation

$$r = \frac{p}{1 - \epsilon \cos \theta}$$

describes the polar coordinates of an orbit measured from one of the orbit's two focal points. For objects in orbit around the sun, the sun is at one of the focal points. Thus r is the distance of the object from the sun. The parameters p and ϵ determine the size of the orbit and its eccentricity, respectively. Obtain the polar plot that represents an

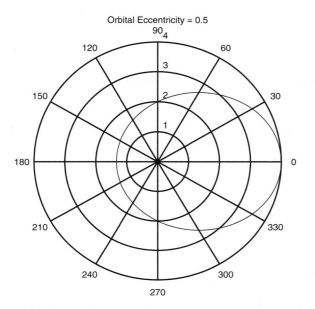

Figure 5.2–7 A polar plot showing an orbit having an eccentricity of 0.5.

orbit having $\epsilon = 0.5$ and $p = 2$ AU (AU stands for "astronomical unit"; 1 AU is the mean distance from the sun to Earth). How far away does the orbiting object get from the sun? How close does it approach Earth's orbit?

■ Solution

Figure 5.2–7 shows the polar plot of the orbit. The plot was generated by the following session.

```
>>theta = 0:pi/90:2*pi;
>>r = 2./(1-0.5*cos(theta));
>>polar(theta,r),title('Orbital Eccentricity = 0.5')
```

The sun is at the origin, and the plot's concentric circular grid enables us to determine that the closest and farthest distances the object is from the sun are approximately 1.3 and 4 AU. Earth's orbit, which is nearly circular, is represented by the innermost circle. Thus the closest the object gets to Earth's orbit is approximately 0.3 AU. The radial grid lines allow us to determine that when $\theta = 90°$ and $270°$, the object is 2 AU from the sun.

Error Bar Plots

Experimental data are often represented with plots containing error bars. The bars show the estimated or calculated errors for each data point. They can also be used to display the error in an approximate formula. For example, keeping two

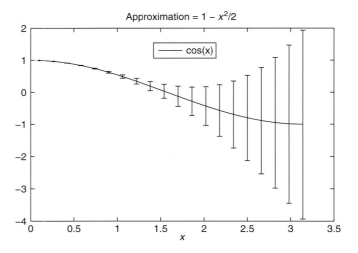

Figure 5.2–8 Error bars for the approximation $\cos x \approx 1 - x^2/2$.

terms in the Taylor series expansion of $\cos x$ about $x = 0$ gives $\cos x \approx 1 - x^2/2$. The following program creates the plot shown in Figure 5.2–8.

```
% errorbar example
x = linspace(0.1, pi, 20);
approx = 1 - x.^2/2;
error = approx - cos(x);
errorbar(x, cos(x), error), legend('cos(x)'), ...
    title('Approximation = 1 - x^2/2')
```

There are more than 20 two-dimensional plot functions available in MATLAB. We have shown the most important ones for engineering applications.

***

Test Your Understanding

T5.2–6 Plot the following functions using axes that will produce a straight-line plot. The power function is $y = 2x^{-0.5}$, and the exponential function is $y = 10^{1-x}$.

T5.2–7 Plot the function $y = 8x^3$ for $-1 \le x \le 1$ with a tick spacing of 0.25 on the x axis and 2 on the y axis.

T5.2–8 The *spiral of Archimedes* is described by the polar coordinates (θ, r), where $r = a\theta$. Obtain a polar plot of this spiral for $0 \le \theta \le 4\pi$, with the parameter $a = 2$.

***

Publishing Reports Containing Graphics

Starting with MATLAB 7, the `publish` function is available for creating reports, which may have embedded graphics. Reports generated by the `publish` function may be exported to a variety of common formats including HTML (Hyper Text Markup Language), which is used for Web-based reports; MS Word; PowerPoint; and LᴬTᴇX. To publish a report, do the following.

1. Open the Editor, type in the M-file that forms the basis of the report, and save it. Use the double percent character (%%) to indicate a section heading in the report. This character marks the beginning of a new cell, which is a group of commands. (Such a cell should not be confused with the cell array data type covered in Section 2.6.) Enter any blank lines you wish to appear in the report. Consider, as a very simple example, the following sample file `polyplot.m`.

```
%% Example of Report Publishing:
% Plotting the cubic y = x^3 - 6x^2 + 10x+4.

%% Create the independent variable.
x = linspace(0, 4, 300); % Use 300 points between 0 and 4.
%% Define the cubic from its coefficients.
p = [1, -6, 10, 4]; % p contains the coefficients.

%% Plot the cubic
plot(x,polyval(p,x)), xlabel('x'),ylabel('y')
```

2. Run the file to check it for errors. (To do this for a larger file, you may use the cell mode of the Debugger to execute its each cell one at a time; see Section 4.7.)

3. Use the `publish` and `open` functions to create the report in the desired format. Using our sample file, we can obtain a report in HTML format by typing

```
>>publish ('polyplot','html')
>>open html/polyplot.html
```

You should see a report like the one shown in Figure 5.2–9.

Instead of using the `publish` and `open` functions, you may select **Publish to HTML** from the **File** menu in the Editor window. To publish to another format, select instead **Publish to** and then choose the desired format from the menu.

Once it is published in HTML, you may click on a section heading in the Contents to go to that section. This is useful for larger reports.

If you want the equation to look professionally typeset, you may edit the resulting report in the appropriate editor (say, MS Word or LᴬTᴇX). For example, to set the cubic polynomial in the resulting LᴬTᴇX file, use the commands presented earlier in this section to replace the equation in the second line of the report with

```
y = {\it x}^3 - 6{\it x}^2 + 10{\it x} + 4
```

Example of Report Publishing:

Plotting the cubic $y = x^3 - 6x^2 + 10x + 4$.

Contents

- Create the independent variable.
- Define the cubic.
- Plot the cubic.

Create the independent variable.

```
x = linspace(0, 4, 300); % Use 300 points between 0 and 4.
```

Define the cubic.

```
p = [1, -6, 10, 4];  % p contains the coefficients.
```

Plot the cubic.

```
plot(x,polyval(p,x)),xlabel('x'),ylabel('y')
```

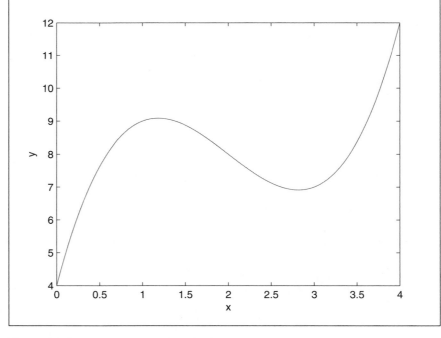

Figure 5.2–9 A sample report published from MATLAB.

5.3 Interactive Plotting in MATLAB

The interactive plotting environment in MATLAB is a set of tools for

- Creating different types of graphs,
- Selecting variables to plot directly from the Workspace Browser,
- Creating and editing subplots,
- Adding annotations such as lines, arrows, text, rectangles, and ellipses, and
- Editing properties of graphics objects, such as their color, line weight, and font.

The Plot Tools interface includes the following three panels associated with a given figure.

- **The Figure Palette:** Use this to create and arrange subplots, to view and plot workspace variables, and to add annotations.
- **The Plot Browser:** Use this to select and control the visibility of the axes or graphics objects plotted in the figure, and to add data for plotting.
- **The Property Editor:** Use this to set basic properties of the selected object and to obtain access to all properties through the Property Inspector.

The Figure Window

When you create a plot, the Figure window appears with the Figure toolbar visible (see Figure 5.3–1). This window has eight menus.

The File Menu   The **File** menu is used for saving and printing the figure. This menu was discussed in Section 5.1 under **Saving Figures** and **Exporting Figures.**

The Edit Menu   You can use the **Edit** menu to cut, copy, and paste items, such as legend or title text, that appear in the figure. Click on **Figure Properties** to open the Property Editor—Figure dialog box to change certain properties of the figure.

Three items on the **Edit** menu are very useful for editing the figure. Clicking the **Axes Properties** item brings up the Property Editor—Axes dialog box. Double-clicking on any axis also brings up this box. You can change the scale type (linear, log, etc.), the labels, and the tick marks by selecting the tab for the desired axis or the font to be edited.

The **Current Object Properties** item enables you to change the properties of an object in the figure. To do this, first click on the object, such as a plotted

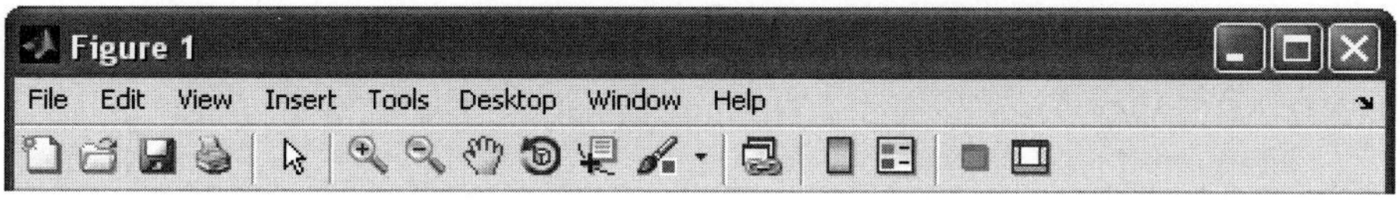

Figure 5.3–1 The Figure toolbar displayed.

line, then click on **Current Object Properties** in the **Edit** menu. You will see the Property Editor—Lineseries dialog box that lets you change properties such as line weight and color, data-marker type, and plot type.

Clicking on any text, such as that placed with the `title`, `xlabel`, `ylabel`, `legend`, or `gtext` commands, and then selecting **Current Object Properties** in the **Edit** menu bring up the Property Editor—Text dialog box, which enables you to edit the text.

The View Menu   The items on the **View** menu are the three toolbars (**Figure Toolbar**, **Plot Edit Toolbar,** and **Camera Toolbar**), the **Figure Palette,** the **Plot Browser,** and the **Property Editor.** These will be discussed later in this section.

The Insert Menu   The **Insert** menu enables you to insert labels, legends, titles, text, and drawing objects, rather than using the relevant commands from the Command window. To insert a label on the *y* axis, for example, click on the **Y Label** item on the menu; a box will appear on the *y* axis. Type the label in this box, and then click outside the box to finish.

The **Insert** menu also enables you to insert arrows, lines, text, rectangles, and ellipses in the figure. To insert an arrow, for example, click on the **Arrow** item; the mouse cursor changes to a crosshair style. Then click the mouse button, and move the cursor to create the arrow. The arrowhead will appear at the point where you release the mouse button. Be sure to add arrows, lines, and other annotations only after you are finished moving or resizing your axes, because these objects are not anchored to the axes. (They can be anchored to the plot by *pinning;* see the MATLAB Help.)

To delete or move a line or arrow, click on it, then press the **Delete** key to delete it, or press the mouse button and move it to the desired location. The **Axes** item lets you use the mouse to place a new set of axes within the existing plot. Click on the new axes, and a box will surround them. Any further plot commands issued from the Command window will direct the output to these axes.

The **Light** item applies to three-dimensional plots.

The Tools Menu   The **Tools** menu includes items for adjusting the view (by zooming and panning) and the alignment of objects on the plot. The **Edit Plot** item starts the plot editing mode, which can also be started by clicking on the northwest-facing arrow on the Figure toolbar. The **Tools** menu also gives access to the **Data Cursor,** which is discussed later in this section. The last two items, **Basic Fitting** and **Data Statistics,** will be discussed in Sections 6.3 and 7.1, respectively.

Other Menus   The **Desktop** menu enables you to dock the Figure window within the desktop. The **Window** menu lets you switch between the Command window and any other Figure windows. The **Help** menu accesses the general MATLAB Help System as well as Help features specific to plotting.

There are three toolbars available in the Figure window: the Figure toolbar, the Plot Edit toolbar, and the Camera toolbar. The **View** menu lets you select which ones you want to appear. We will discuss the Figure toolbar and the Plot

Edit toolbar in this section. The Camera toolbar is useful for three-dimensional plots, which are discussed at the end of this chapter.

The Figure Toolbar

To activate the Figure toolbar, select it from the **View** menu (see Figure 5.3–1). The four leftmost buttons are for opening, saving, and printing the figure. Clicking on the northwest-facing arrow button toggles the plot edit mode on and off.

The **Zoom-in** and **Zoom-out** buttons let you obtain a close-up or faraway view of the figure. The **Pan** and **Rotate 3D** buttons are used for three-dimensional plots.

The **Data Cursor** button enables you to read data directly from a graph by displaying the values of points you select on plotted lines, surfaces, images, and so on.

The **Insert Colorbar** button inserts a color map strip in the graph and is useful for three-dimensional surface plots. The **Insert Legend** button enables you to insert a legend in the plot. The last two buttons hide or show the plot tools and dock the figure if it is undocked.

The Plot Edit Toolbar

Once a plot is in the window, you can display the Plot Edit toolbar from the **View** menu. This toolbar is shown in Figure 5.3–2. You can enable plot editing by clicking on the northwest-facing arrow on the Figure toolbar. Then double-click on an axis, a plotted line, or a label to activate the appropriate property editor. To add text that is not a label, title, or legend, click the button labeled **T,** move the cursor to the desired location for the text, click the mouse button, and type the text. When finished, click outside the text box and note that the nine leftmost buttons become highlighted and available. These enable you to modify the color, font, and other attributes of the text.

To insert arrows, lines, rectangles, and ellipses, click on the appropriate button and follow the instructions given previously for the **Insert** menu.

The Plot Tools

Once a figure has been created, you can display any of or all three Plot Tools (Figure Palette, Plot Browser, and Property Editor) by selecting them from

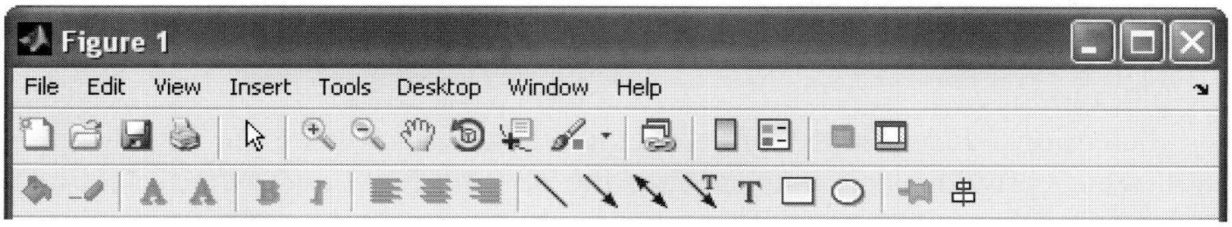

Figure 5.3–2 The Figure and Plot Edit toolbars displayed.

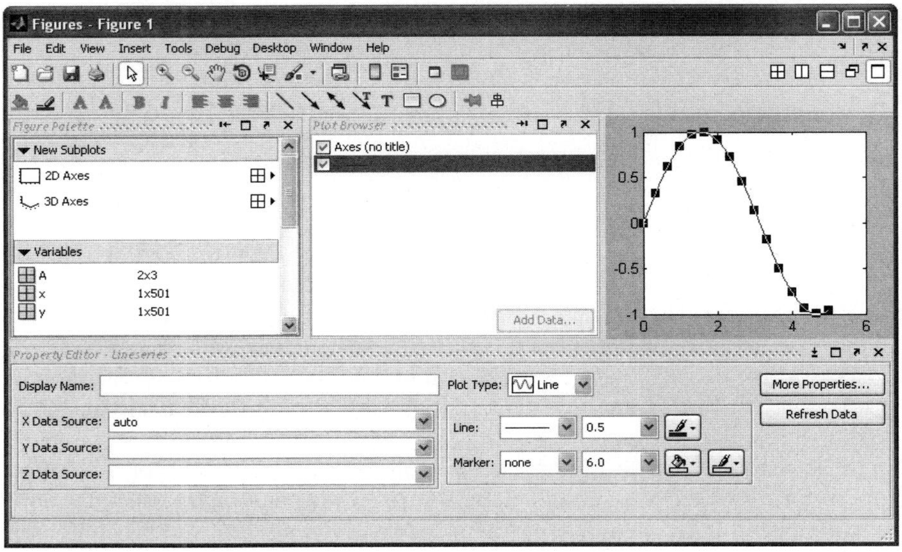

Figure 5.3–3 The Figure window with the Plot Tools activated.

the **View** menu. You can also start the environment by first creating a plot and then clicking on the **Show Plot Tools** icon at the far left of the Figure toolbar (see Figure 5.3–3), or by creating a figure with the plotting tools attached by using the `plottools` command. Remove the tools by clicking on the **Hide Tools** icon, which is second from the left.

Figure 5.3–3 shows the result of clicking on the plotted line after clicking the **Show Plot Tools** icon. The plotting interface then displays the Property Editor—Lineseries.

The Figure Palette

The Figure Palette contains three panels, which are selected and expanded by clicking the appropriate button. Click on the grid icon in the **New Subplots** panel to display the selector grid that enables you to specify the layout of the subplots. In the Variables panel you can select a graphics function to plot the variable by selecting the variable and right-clicking to display the context menu. This menu contains a list of possible plot types based on the type of variable you select. You can also drag the variable into an axes set, and MATLAB will select an appropriate plot type.

Selecting **More Plots** from the context menu activates the Plot Catalog tool, which provides access to most of the plotting functions. After selecting a plot category, and a plot type from that category, you will see its description in the rightmost display. Type the name of one or more variables in the Plotted

Variables field, separated by commas, and they will be passed to the selected plotting function as arguments. You can also type a MATLAB expression that uses any workspace variables shown in the Figure Palette.

Click on the **Annotations** panel to display a menu of objects such as lines, arrows, etc. Click on the desired object, and use the mouse to position and size it.

The Plot Browser

The Plot Browser provides a legend of all the graphs in the figure. For example, if you plot an array with multiple rows and columns, the browser lists each axis and the objects (lines, surfaces, etc.) used to create the graph. To set the properties of an individual line, double-click on the line. Its properties are displayed in the Property Editor—Lineseries box, which opens on the bottom of the figure.

If you select a line in the graph, the corresponding entry in the Plot Browser is highlighted, indicating which column in the variable produced the line. The check box next to each item in the browser controls the object's visibility. For example, if you want to plot only certain columns of data, you can uncheck the columns not wanted. The graph updates as you uncheck each box and rescales the axes as required.

The Property Editor

The Property Editor enables you to access a subset of the selected object's properties. When no object is selected, the Property Editor displays the figure's properties. There are several ways to display the Property Editor.

1. Double-click an object when plot edit mode is enabled.
2. Select an object and right-click to display its context menu, then select **Properties.**
3. Select **Property Editor** from the **View** menu.
4. Use the `propertyeditor` command.

The Property Editor enables you to change the most commonly used object properties. If you want to access all object properties, use the Property Inspector. To display the Property Inspector, click the **Inspector** button on any Property Editor panel. Use of this feature requires detailed knowledge of object properties and handle graphics, and thus will not be covered here.

Recreating Graphs from M-Files

Once your graph is finished, you can generate MATLAB code to reproduce the graph by selecting **Generate M-File** from the **File** menu. MATLAB creates a function that recreates the graph and opens the generated M-File in the editor. This feature is particularly useful for capturing property settings and other modifications made in the plot editor. You can also use the `makemcode` function.

Adding Data to Axes

The Plot Browser provides the mechanism by which you add data to axes. The procedure is as follows:

1. Select a two-dimensional or three-dimensional axis from the New Subplots subpanel.
2. After creating the axis, select it in the Plot Browser panel to enable the **Add Data** button at the bottom of the panel.
3. Click the **Add Data** button to display the Add Data to Axes dialog box. The Add Data to Axes dialog enables you to select a plot type and specify the workspace variables to pass to the plotting function. You can also specify a MATLAB expression, which is evaluated to produce the data to plot.

5.4 Three-Dimensional Plots

MATLAB provides many functions for creating three-dimensional plots. Here we will summarize the basic functions to create three types of plots: line plots, surface plots, and contour plots. Information about the related functions is available in MATLAB Help (category `graph3d`).

Three-Dimensional Line Plots

Lines in three-dimensional space can be plotted with the `plot3` function. Its syntax is `plot3(x,y,z)`. For example, the following equations generate a three-dimensional curve as the parameter t is varied over some range:

$$x = e^{-0.05t} \sin t$$
$$y = e^{-0.05t} \cos t$$
$$z = t$$

If we let t vary from $t = 0$ to $t = 10\pi$, the sine and cosine functions will vary through five cycles, while the absolute values of x and y become smaller as t increases. This process results in the spiral curve shown in Figure 5.4–1, which was produced with the following session.

```
>>t = 0:pi/50:10*pi;
>>plot3(exp(-0.05*t).*sin(t),exp(-0.05*t).*cos(t),t),...
   xlabel('x'),ylabel('y'),zlabel('z'),grid
```

Note that the `grid` and label functions work with the `plot3` function, and that we can label the z axis by using the `zlabel` function, which we have seen for the first time. Similarly, we can use the other plot enhancement functions discussed in Sections 5.1 and 5.2 to add a title and text and to specify line type and color.

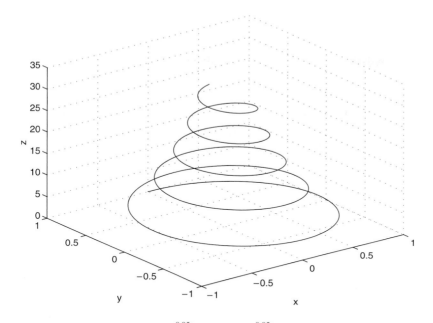

Figure 5.4–1 The curve $x = e^{-0.05t} \sin t$, $y = e^{-0.05t} \cos t$, $z = t$ plotted with the plot3 function.

Surface Mesh Plots

The function $z = f(x, y)$ represents a surface when plotted on xyz axes, and the mesh function provides the means to generate a surface plot. Before you can use this function, you must generate a grid of points in the xy plane and then evaluate the function $f(x, y)$ at these points. The meshgrid function generates the grid. Its syntax is [X,Y] = meshgrid(x,y). If x = xmin:xspacing:xmax and y = ymin:yspacing:ymax, then this function will generate the coordinates of a rectangular grid with one corner at (*xmin, ymin*) and the opposite corner at (*xmax, ymax*). Each rectangular panel in the grid will have a width equal to *xspacing* and a depth equal to *yspacing*. The resulting matrices X and Y contain the coordinate pairs of every point in the grid. These pairs are then used to evaluate the function.

The function [X,Y] = meshgrid(x) is equivalent to [X,Y] = meshgrid(x,x) and can be used if x and y have the same minimum values, the same maximum values, and the same spacing. Using this form, you can type [X,Y] = meshgrid(min:spacing:max), where min and max specify the minimum and maximum values of both x and y and spacing is the desired spacing of the x and y values.

After the grid is computed, you create the surface plot with the mesh function. Its syntax is mesh(x,y,z). The grid, label, and text functions can be used with the mesh function. The following session shows how to generate the

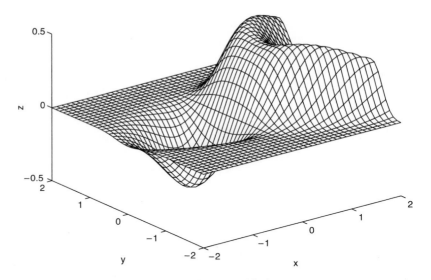

Figure 5.4–2 A plot of the surface $z = xe^{-[(x-y^2)^2+y^2]}$ created with the mesh function.

surface plot of the function $z = xe^{-[(x-y^2)^2+y^2]}$, for $-2 \leq x \leq 2$ and $-2 \leq y \leq 2$, with a spacing of 0.1. This plot appears in Figure 5.4–2.

```
>>[X,Y] = meshgrid(-2:0.1:2);
>>Z = X.*exp(-((X-Y.^2).^2+Y.^2));
>>mesh(X,Y,Z),xlabel('x'),ylabel('y'),zlabel('z')
```

Be careful not to select too small a spacing for the *x* and *y* values for two reasons: (1) Small spacing creates small grid panels, which make the surface difficult to visualize, and (2) the matrices *X* and *Y* can become too large.

The surf and surfc functions are similar to mesh and meshc except that the former create a shaded surface plot. You can use the Camera toolbar and some menu items in the Figure window to change the view and lighting of the figure.

Contour Plots

Topographic plots show the contours of the land by means of constant elevation lines. These lines are also called *contour lines,* and such a plot is called a *contour plot.* If you walk along a contour line, you remain at the same elevation. Contour plots can help you visualize the shape of a function. They can be created with the contour function, whose syntax is contour(X,Y,Z). You use this function the same way you use the mesh function; that is, first use the meshgrid function to generate the grid and then generate the function values. The following session generates the contour plot of the function whose surface plot is shown in

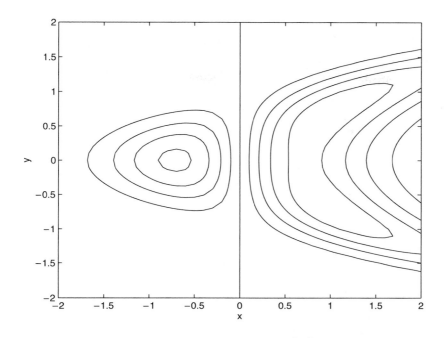

Figure 5.4–3 A contour plot of the surface $z = xe^{-[(x-y^2)^2+y^2]}$ created with the `contour` function.

Figure 5.4–2, namely, $z = xe^{-[(x-y^2)^2+y^2]}$, for $-2 \leq x \leq 2$ and $-2 \leq y \leq 2$, with a spacing of 0.1. This plot appears in Figure 5.4–3.

```
>>[X,Y] = meshgrid(-2:0.1:2);
>>Z = X.*exp(-((X-Y.^2).^2+Y.^2));
>>contour(X,Y,Z),xlabel('x'),ylabel('y')
```

You can add labels to the contour lines. Type `help clabel`.

Contour plots and surface plots can be used together to clarify the function. For example, unless the elevations are labeled on contour lines, you cannot tell whether there is a minimum or a maximum point. However, a glance at the surface plot will make this easy to determine. On the other hand, accurate measurements are not possible on a surface plot; these can be done on the contour plot because no distortion is involved. Thus a useful function is `meshc`, which shows the contour lines beneath the surface plot. The `meshz` function draws a series of vertical lines under the surface plot, while the `waterfall` function draws mesh lines in one direction only. The results of these functions are shown in Figure 5.4–4 for the function $z = xe^{-(x^2+y^2)}$.

Table 5.4–1 summarizes the functions introduced in this section. For other three-dimensional plot types, type `help specgraph`.

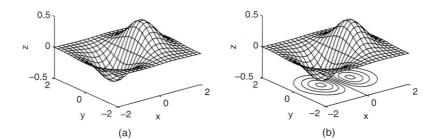

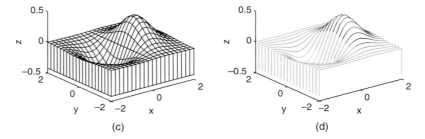

Figure 5.4–4 Plots of the surface $z = xe^{-(x^2+y^2)}$ created with the mesh function and its variant forms: meshc, meshz, and waterfall. (a) mesh, (b) meshc, (c) meshz, (d) waterfall.

Table 5.4–1 Three-dimensional plotting functions

Function	Description
contour(x,y,z)	Creates a contour plot.
mesh(x,y,z)	Creates a three-dimensional mesh surface plot.
meshc(x,y,z)	Same as mesh but draws a contour plot under the surface.
meshz(x,y,z)	Same as mesh but draws a series of vertical reference lines under the surface.
surf(x,y,z)	Creates a shaded three-dimensional mesh surface plot.
surfc(x,y,z)	Same as surf but draws a contour plot under the surface.
[X,Y] = meshgrid(x,y)	Creates the matrices X and Y from the vectors x and y to define a rectangular grid.
[X,Y] = meshgrid(x)	Same as [X,Y] = meshgrid(x,x).
waterfall(x,y,z)	Same as mesh but draws mesh lines in one direction only.

Test Your Understanding

T5.4–1 Create a surface plot and a contour plot of the function $z = (x - 2)^2 + 2xy + y^2$.

5.5 Summary

This chapter explained how to use the powerful MATLAB commands to create effective and pleasing two-dimensional and three-dimensional plots. The following guidelines will help you create plots that effectively convey the desired information.

- Label each axis with the name of the quantity being plotted *and its units!*
- Use regularly spaced tick marks at convenient intervals along each axis.
- If you are plotting more than one curve or data set, label each on its plot or use a legend to distinguish them.
- If you are preparing multiple plots of a similar type or if the axes' labels cannot convey enough information, use a title.
- If you are plotting measured data, plot each data point in a given set with the same symbol, such as a circle, square, or cross.
- If you are plotting points generated by evaluating a function (as opposed to measured data), do *not* use a symbol to plot the points. Instead, connect the points with solid lines.

Key Terms with Page References

Axis limits,   222

Contour plot,   248

Data symbol,   219

Overlay plot,   228

Polar plot,   236

Subplot,   226

Surface mesh plot,   247

Problems

You can find the answers to problems marked with an asterisk at the end of the text.

Sections 5.1, 5.2, and 5.3

1.* *Breakeven analysis* determines the production volume at which the total production cost is equal to the total revenue. At the breakeven point, there is neither profit nor loss. In general, production costs consist of fixed costs and variable costs. Fixed costs include salaries of those not directly involved with production, factory maintenance costs, insurance costs, and so on. Variable costs depend on production volume and include material costs, labor costs, and energy costs. In the following analysis, assume that we produce only what we can sell; thus the production quantity equals the sales. Let the production quantity be Q, in gallons per year.

Consider the following costs for a certain chemical product:
Fixed cost: $3 million per year.
Variable cost: 2.5 cents per gallon of product.

The selling price is 5.5 cents per gallon.

Use these data to plot the total cost and the revenue versus Q, and graphically determine the breakeven point. Fully label the plot and mark the breakeven point. For what range of Q is production profitable? For what value of Q is profit a maximum?

2. Consider the following costs for a certain chemical product:

Fixed cost: $2.045 million/year.

Variable costs:

Material cost: 62 cents per gallon of product.

Energy cost: 24 cents per gallon of product.

Labor cost: 16 cents per gallon of product.

Assume that we produce only what we sell. Let P be the selling price in dollars per gallon. Suppose that the selling price and the sales quantity Q are interrelated as follows: $Q = 6 \times 10^6 - 1.1 \times 10^6 P$. Accordingly, if we raise the price, the product becomes less competitive and sales drop.

Use this information to plot the fixed and total variable costs versus Q, and graphically determine the breakeven point(s). Fully label the plot and mark the breakeven points. For what range of Q is production profitable? For what value of Q is profit a maximum?

3.* *a.* Estimate the roots of the equation

$$x^3 - 3x^2 + 5x \sin\left(\frac{\pi x}{4} - \frac{5\pi}{4}\right) + 3 = 0$$

by plotting the equation.

b. Use the estimates found in part *a* to find the roots more accurately with the `fzero` function.

4. To compute the forces in structures, sometimes we must solve equations similar to the following. Use the `fplot` function to find all the positive roots of this equation:

$$x \tan x = 9$$

5.* Cables are used to suspend bridge decks and other structures. If a heavy uniform cable hangs suspended from its two endpoints, it takes the shape of a *catenary* curve whose equation is

$$y = a \cosh\left(\frac{x}{a}\right)$$

where a is the height of the lowest point on the chain above some horizontal reference line, x is the horizontal coordinate measured to the right from the lowest point, and y is the vertical coordinate measured up from the reference line.

Let $a = 10$ m. Plot the catenary curve for $-20 \leq x \leq 30$ m. How high is each endpoint?

6. Using estimates of rainfall, evaporation, and water consumption, the town engineer developed the following model of the water volume in the reservoir as a function of time

$$V(t) = 10^9 + 10^8(1 - e^{-t/100}) - 10^7 t$$

where V is the water volume in liters and t is time in days. Plot $V(t)$ versus t. Use the plot to estimate how many days it will take before the water volume in the reservoir is 50 percent of its initial volume of 10^9 L.

7. It is known that the following Leibniz series converges to the value $\pi/4$ as $n \to \infty$.

$$S(n) = \sum_{k=0}^{n} (-1)^k \frac{1}{2k + 1}$$

Plot the difference between $\pi/4$ and the sum $S(n)$ versus n for $0 \leq n \leq 200$.

8. A certain fishing vessel is initially located in a horizontal plane at $x = 0$ and $y = 10$ mi. It moves on a path for 10 hr such that $x = t$ and $y = 0.5t^2 + 10$, where t is in hours. An international fishing boundary is described by the line $y = 2x + 6$.

a. Plot and label the path of the vessel and the boundary.

b. The perpendicular distance of the point (x_1, y_1) from the line $Ax + By + C = 0$ is given by

$$d = \frac{Ax_1 + By_1 + C}{\pm \sqrt{A^2 + B^2}}$$

where the sign is chosen to make $d \geq 0$. Use this result to plot the distance of the fishing vessel from the fishing boundary as a function of time for $0 \leq t \leq 10$ hr.

9. Plot columns 2 and 3 of the following matrix $\mathbf{A}$ versus column 1. The data in column 1 are time (seconds). The data in columns 2 and 3 are force (newtons).

$$\mathbf{A} = \begin{bmatrix} 0 & -7 & 6 \\ 5 & -4 & 3 \\ 10 & -1 & 9 \\ 15 & 1 & 0 \\ 20 & 2 & -1 \end{bmatrix}$$

10.* Many applications use the following "small angle" approximation for the sine to obtain a simpler model that is easy to understand and analyze. This approximation states that $\sin x \approx x$, where x must be in radians. Investigate the accuracy of this approximation by creating three plots. For the first, plot $\sin x$ and x versus x for $0 \leq x \leq 1$. For the second, plot the approximation error $\sin x - x$ versus x for $0 \leq x \leq 1$. For the third, plot the relative error $[\sin(x) - x]/\sin(x)$ versus x for $0 \leq x \leq 1$. How small must x be for the approximation to be accurate within 5 percent?

11. You can use trigonometric identities to simplify the equations that appear in many applications. Confirm the identity $\tan(2x) = 2 \tan x / (1 - \tan^2 x)$ by plotting both the left and the right sides versus x over the range $0 \leq x \leq 2\pi$.

12. The complex number identity $e^{ix} = \cos x + i \sin x$ is often used to convert the solutions of equations into a form that is relatively easy to visualize. Confirm this identity by plotting the imaginary part versus the real part for both the left and right sides over the range $0 \leq x \leq 2\pi$.

13. Use a plot over the range $0 \leq x \leq 5$ to confirm that $\sin(ix) = i \sinh x$.

14.* The function $y(t) = 1 - e^{-bt}$, where t is time and $b > 0$, describes many processes, such as the height of liquid in a tank as it is being filled and the temperature of an object being heated. Investigate the effect of the parameter b on $y(t)$. To do this, plot y versus t for several values of b on the same plot. How long will it take for $y(t)$ to reach 98 percent of its steady-state value?

15. The following functions describe the oscillations in electric circuits and the vibrations of machines and structures. Plot these functions on the same plot. Because they are similar, decide how best to plot and label them to avoid confusion.

$$x(t) = 10e^{-0.5t} \sin(3t + 2)$$

$$y(t) = 7e^{-0.4t} \cos(5t - 3)$$

16. In certain kinds of structural vibrations, a periodic force acting on the structure will cause the vibration amplitude to repeatedly increase and decrease with time. This phenomenon, called *beating,* also occurs in musical sounds. A particular structure's displacement is described by

$$y(t) = \frac{1}{f_1^2 - f_2^2}[\cos(f_2 t) - \cos(f_1 t)]$$

where y is the displacement in inches and t is the time in seconds. Plot y versus t over the range $0 \leq t \leq 20$ for $f_1 = 8$ rad/sec and $f_2 = 1$ rad/sec. Be sure to choose enough points to obtain an accurate plot.

17.* The height $h(t)$ and horizontal distance $x(t)$ traveled by a ball thrown at an angle A with a speed v are given by

$$h(t) = vt \sin A - \frac{1}{2}gt^2$$

$$x(t) = vt \cos A$$

At Earth's surface the acceleration due to gravity is $g = 9.81$ m/s².

    a. Suppose the ball is thrown with a velocity $v = 10$ m/s at an angle of 35°. Use MATLAB to compute how high the ball will go, how far it will go, and how long it will take to hit the ground.

    b. Use the values of v and A given in part *a* to plot the ball's *trajectory;* that is, plot h versus x for positive values of h.

    c. Plot the trajectories for $v = 10$ m/s corresponding to five values of the angle A: 20°, 30°, 45°, 60°, and 70°.

    d. Plot the trajectories for $A = 45°$ corresponding to five values of the initial velocity v: 10, 12, 14, 16, and 18 m/s.

18. The perfect gas law relates the pressure p, absolute temperature T, mass m, and volume V of a gas. It states that

$$pV = mRT$$

The constant R is the *gas constant*. The value of R for air is 286.7 $(N \cdot m)/(kg \cdot K)$. Suppose air is contained in a chamber at room temperature ($20°C = 293$ K). Create a plot having three curves of the gas pressure in N/m² versus the container volume V in m³ for $20 \leq V \leq 100$. The three curves correspond to the following masses of air in the container: $m = 1$ kg, $m = 3$ kg, and $m = 7$ kg.

19. Oscillations in mechanical structures and electric circuits can often be described by the function

$$y(t) = e^{-t/\tau} \sin(\omega t + \phi)$$

where t is time and ω is the oscillation frequency in radians per unit time. The oscillations have a period of $2\pi/\omega$, and their amplitudes decay in time at a rate determined by τ, which is called the *time constant*. The smaller τ is, the faster the oscillations die out.

    a. Use these facts to develop a criterion for choosing the spacing of the t values and the upper limit on t to obtain an accurate plot of $y(t)$. (*Hint:* Consider two cases: $4\tau > 2\pi/\omega$ and $4\tau < 2\pi/\omega$.)

    b. Apply your criterion, and plot $y(t)$ for $\tau = 10$, $\omega = \pi$, and $\phi = 2$.

    c. Apply your criterion, and plot $y(t)$ for $\tau = 0.1$, $\omega = 8\pi$, and $\phi = 2$.

20. When a constant voltage was applied to a certain motor initially at rest, its rotational speed $s(t)$ versus time was measured. The data appear in the following table:

Time (sec)	1	2	3	4	5	6	7	8	10
Speed (rpm)	1210	1866	2301	2564	2724	2881	2879	2915	3010

Determine whether the following function can describe the data. If so, find the values of the constants b and c.

$$s(t) = b(1 - e^{ct})$$

21. The following table shows the average temperature for each year in a certain city. Plot the data as a stem plot, a bar plot, and a stairs plot.

Year	2000	2001	2002	2003	2004
Temperature (°C)	21	18	19	20	17

22. A sum of $10 000 invested at 4 percent interest compounded annually will grow according to the formula

$$y(k) = 10^4(1.04)^k$$

where k is the number of years ($k = 0, 1, 2, \ldots$). Plot the amount of money in the account for a 10-year period. Do this problem with four types of plots: the xy plot, the stem plot, the stairs plot, and the bar plot.

23. The volume V and surface area A of a sphere of radius r are given by

$$V = \frac{4}{3}\pi r^3 \qquad A = 4\pi r^2$$

 a. Plot V and A versus r in two subplots, for $0.1 \le r \le 100$ m. Choose axes that will result in straight-line graphs for both V and A.
 b. Plot V and r versus A in two subplots, for $1 \le A \le 10^4$ m^2. Choose axes that will result in straight-line graphs for both V and r.

24. The current amount A of a principal P invested in a savings account paying an annual interest rate r is given by

$$A = P\left(1 + \frac{r}{n}\right)^{nt}$$

where n is the number of times per year the interest is compounded. For continuous compounding, $A = Pe^{rt}$. Suppose $10 000 is initially invested at 3.5 percent ($r = 0.035$).

 a. Plot A versus t for $0 \le t \le 20$ years for four cases: continuous compounding, annual compounding ($n = 1$), quarterly compounding ($n = 4$), and monthly compounding ($n = 12$). Show all four cases on the same subplot and label each curve. On a second subplot, plot the

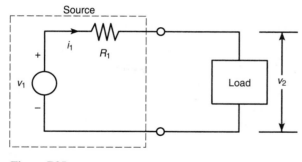

Figure P25

difference between the amount obtained from continuous compounding and the other three cases.

b. Redo part *a*, but plot *A* versus *t* on log-log and semilog plots. Which plot gives a straight line?

25. Figure P25 is a representation of an electrical system with a power supply and a load. The power supply produces the fixed voltage v_1 and supplies the current i_1 required by the load, whose voltage drop is v_2. The current–voltage relationship for a specific load is found from experiments to be

$$i_1 = 0.16(e^{0.12v_2} - 1)$$

Suppose that the supply resistance is $R_1 = 30\ \Omega$ and the supply voltage is $v_1 = 15$ V. To select or design an adequate power supply, we need to determine how much current will be drawn from the power supply when this load is attached. Find the voltage drop v_2 as well.

26. The circuit shown in Figure P26 consists of a resistor and a capacitor and is thus called an *RC* circuit. If we apply a sinusoidal voltage v_i, called the input voltage, to the circuit as shown, then eventually the output voltage v_o will be sinusoidal also, with the same frequency but with a different amplitude and shifted in time relative to the input voltage. Specifically, if $v_i = A_i \sin \omega t$, then $v_o = A_o \sin(\omega t + \phi)$. The frequency–response plot is a plot of A_o/A_i versus frequency ω. It is usually plotted on logarithmic axes.

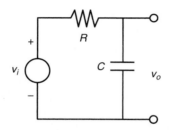

Figure P26

Upper-level engineering courses explain that for the RC circuit shown, this ratio depends on ω and RC as follows:

$$\frac{A_o}{A_i} = \left| \frac{1}{RCs + 1} \right|$$

where $s = \omega i$. For $RC = 0.1$ s, obtain the log-log plot of $|A_o/A_i|$ versus ω and use it to find the range of frequencies for which the output amplitude A_o is less than 70 percent of the input amplitude A_i.

27. An approximation to the function $\sin x$ is $\sin x \approx x - x^3/6$. Plot the $\sin x$ function and 20 evenly spaced error bars representing the error in the approximation.

Section 5.4

28. The popular amusement ride known as the corkscrew has a helical shape. The parametric equations for a circular helix are

$$x = a \cos t$$
$$y = a \sin t$$
$$z = bt$$

where a is the radius of the helical path and b is a constant that determines the "tightness" of the path. In addition, if $b > 0$, the helix has the shape of a right-handed screw; if $b < 0$, the helix is left-handed.

Obtain the three-dimensional plot of the helix for the following three cases and compare their appearance with one another. Use $0 \le t \le 10\pi$ and $a = 1$.

a. $b = 0.1$
b. $b = 0.2$
c. $b = -0.1$

29. A robot rotates about its base at 2 rpm while lowering its arm and extending its hand. It lowers its arm at the rate of 120° per minute and extends its hand at the rate of 5 m/min. The arm is 0.5 m long. The xyz coordinates of the hand are given by

$$x = (0.5 + 5t) \sin\left(\frac{2\pi}{3}t\right) \cos(4\pi t)$$

$$y = (0.5 + 5t) \sin\left(\frac{2\pi}{3}t\right) \sin(4\pi t)$$

$$z = (0.5 + 5t) \cos\left(\frac{2\pi}{3}t\right)$$

where t is time in minutes.

Obtain the three-dimensional plot of the path of the hand for $0 \le t \le 0.2$ min.

30. Obtain the surface and contour plots for the function $z = x^2 - 2xy + 4y^2$, showing the minimum at $x = y = 0$.

31. Obtain the surface and contour plots for the function $z = -x^2 + 2xy + 3y^2$. This surface has the shape of a saddle. At its saddlepoint at $x = y = 0$, the surface has zero slope, but this point does not correspond to either a minimum or a maximum. What type of contour lines corresponds to a saddlepoint?

32. Obtain the surface and contour plots for the function $z = (x - y^2)(x - 3y^2)$. This surface has a singular point at $x = y = 0$, where the surface has zero slope, but this point does not correspond to either a minimum or a maximum. What type of contour lines corresponds to a singular point?

33. A square metal plate is heated to 80°C at the corner corresponding to $x = y = 1$. The temperature distribution in the plate is described by

$$T = 80e^{-(x-1)^2}e^{-3(y-1)^2}$$

Obtain the surface and contour plots for the temperature. Label each axis. What is the temperature at the corner corresponding to $x = y = 0$?

34. The following function describes oscillations in some mechanical structures and electric circuits.

$$z(t) = e^{-t/\tau}\sin(\omega t + \phi)$$

In this function t is time, and ω is the oscillation frequency in radians per unit time. The oscillations have a period of $2\pi/\omega$, and their amplitudes decay in time at a rate determined by τ, which is called the *time constant*. The smaller τ is, the faster the oscillations die out.

Suppose that $\phi = 0$, $\omega = 2$, and τ can have values in the range $0.5 \leq \tau \leq 10$ sec. Then the preceding equation becomes

$$z(t) = e^{-t/\tau}\sin(2t)$$

Obtain a surface plot and a contour plot of this function to help visualize the effect of τ for $0 \leq t \leq 15$ sec. Let the x variable be time t and the y variable be τ.

35. The following equation describes the temperature distribution in a flat rectangular metal plate. The temperature on three sides is held constant at T_1, and at T_2 on the fourth side (see Figure P35). The temperature $T(x, y)$ as a function of the xy coordinates shown is given by

$$T(x,y) = (T_2 - T_1)w(x,y) + T_1$$

where

$$w(x,y) = \frac{2}{\pi}\sum_{n\text{ odd}}^{\infty}\frac{2}{n}\sin\left(\frac{n\pi x}{L}\right)\frac{\sinh(n\pi y/L)}{\sinh(n\pi W/L)}$$

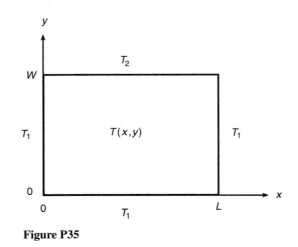

Figure P35

The given data for this problem are $T_1 = 70°F$, $T_2 = 200°F$, and $W = L = 2$ ft.

Using a spacing of 0.2 for both x and y, generate a surface mesh plot and a contour plot of the temperature distribution.

36. The electric potential field V at a point, due to two charged particles, is given by

$$V = \frac{1}{4\pi\epsilon_0}\left(\frac{q_1}{r_1} + \frac{q_2}{r_2}\right)$$

where q_1 and q_2 are the charges of the particles in coulombs (C), r_1 and r_2 are the distances of the charges from the point (in meters), and ϵ_0 is the permittivity of free space, whose value is

$$\epsilon_0 = 8.854 \times 10^{-12}\ C^2/(N \cdot m^2)$$

Suppose the charges are $q_1 = 2 \times 10^{-10}$ C and $q_2 = 4 \times 10^{-10}$ C. Their respective locations in the xy plane are $(0.3, 0)$ and $(-0.3, 0)$ m. Plot the electric potential field on a three-dimensional surface plot with V plotted on the z axis over the ranges $-0.25 \le x \le 0.25$ and $-0.25 \le y \le 0.25$. Create the plot in two ways: (*a*) by using the `surf` function and (*b*) by using the `meshc` function.

37. Refer to Problem 25 of Chapter 4. Use the function file created for that problem to generate a surface mesh plot and a contour plot of x versus h and W for $0 \le W \le 500$ N and for $0 \le h \le 2$ m. Use the values $k_1 = 10^4$ N/m, $k_2 = 1.5 \times 10^4$ N/m, and $d = 0.1$ m.

38. Refer to Problem 28 of Chapter 4. To see how sensitive the cost is to location of the distribution center, obtain a surface plot and a contour plot

of the total cost as a function of the x and y coordinates of the distribution center location. How much would the cost increase if we located the center 1 mi in any direction from the optimal location?

39. Refer to Example 3.2–1. Use a surface plot and a contour plot of the perimeter length L as a function of d and θ over the ranges $1 \le d \le 30$ ft and $0.1 \le \theta \le 1.5$ rad. Are there valleys other than the one corresponding to $d = 7.5984$ and $\theta = 1.0472$? Are there any saddle points?

Engineering in the 21st Century...

Virtual Prototyping

To many people, computer-aided design (CAD) or computer-aided engineering (CAE) means creating engineering drawings. However, it means much more. Engineers can use computers to determine the forces, voltages, currents, and so on that might occur in a proposed design. Then they can use this information to make sure the hardware can withstand the predicted forces or supply the required voltages or currents. Engineers are just beginning to use the full potential of CAE.

The normal stages in the development of a new vehicle, such as an aircraft, formerly consisted of aerodynamic testing a scale model; building a full-size wooden *mock-up* to check for pipe, cable, and structural interferences; and finally building and testing a *prototype*, the first complete vehicle. CAE is changing the traditional development cycle. The Boeing 777 is the first aircraft to be designed and built using CAE, without the extra time and expense of building a mock-up. The design teams responsible for the various subsystems, such as aerodynamics, structures, hydraulics, and electrical systems, all had access to the same computer database that described the aircraft. Thus when one team made a design change, the database was updated, allowing the other teams to see whether the change affected their subsystem. This process of designing and testing with a computer model has been called *virtual prototyping*. Virtual prototyping reduced engineering changes, errors, and rework by 50 percent and greatly enhanced the manufacturability of the airplane. When production began, the parts went together easily.

MATLAB is a powerful tool for many CAE applications. It complements geometric modeling packages because it can do advanced calculations that such packages cannot do. ■

CHAPTER 6

Model Building and Regression

OUTLINE

6.1   Function Discovery

6.2   Regression

6.3   The Basic Fitting Interface

6.4   Summary

Problems

An important application of the plotting techniques covered in Chapter 5 is *function discovery,* the technique for using data plots to obtain a mathematical function or "mathematical model" that describes the process that generated the data. This is the topic of Section 6.1. A systematic way of finding an equation that best fits the data is *regression* (also called the *least-squares method*). Regression is treated in Section 6.2. Section 6.3 introduces the MATLAB Basic Fitting interface, which supports regression.

6.1 Function Discovery

Function discovery is the process of finding, or "discovering," a function that can describe a particular set of data. The following three function types can often describe physical phenomena.

1. The *linear* function: $y(x) = mx + b$. Note that $y(0) = b$.
2. The *power* function: $y(x) = bx^m$. Note that $y(0) = 0$ if $m \geq 0$, and $y(0) = \infty$ if $m < 0$.

3. The *exponential* function: $y(x) = b(10)^{mx}$ or its equivalent form $y = be^{mx}$, where e is the base of the natural logarithm ($\ln e = 1$). Note that $y(0) = b$ for both forms.

Each function gives a straight line when plotted using a specific set of axes:

1. The linear function $y = mx + b$ gives a straight line when plotted on rectilinear axes. Its slope is m and its intercept is b.
2. The power function $y = bx^m$ gives a straight line when plotted on log-log axes.
3. The exponential function $y = b(10)^{mx}$ and its equivalent form $y = be^{mx}$ give a straight line when plotted on a semilog plot whose y axis is logarithmic.

We look for a straight line on the plot because it is relatively easy to recognize, and therefore we can easily tell whether the function will fit the data well.

Use the following procedure to find a function that describes a given set of data. We assume that one of the function types (linear, exponential, or power) can describe the data.

1. Examine the data near the origin. The exponential function can never pass through the origin (unless of course $b = 0$, which is a trivial case). (See Figure 6.1–1 for examples with $b = 1$.) The linear function can pass through the origin only if $b = 0$. The power function can pass through the origin but only if $m > 0$. (See Figure 6.1–2 for examples with $b = 1$.)

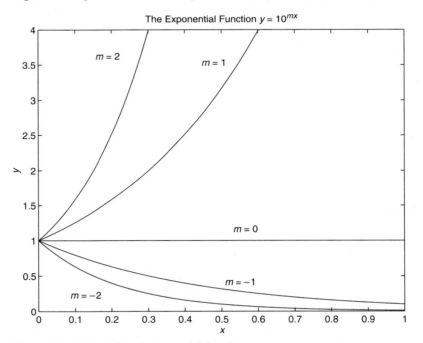

Figure 6.1–1 Examples of exponential functions.

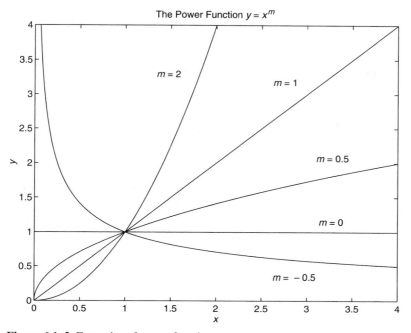

The Power Function $y = x^m$

Figure 6.1–2 Examples of power functions.

2. Plot the data using rectilinear scales. If the data form a straight line, then the data can be represented by the linear function and you are finished. Otherwise, if you have data at $x = 0$, then
   a. If $y(0) = 0$, try the power function.
   b. If $y(0) \neq 0$, try the exponential function.

   If data are not given for $x = 0$, proceed to step 3.
3. If you suspect a power function, plot the data using log-log scales. Only a power function will form a straight line on a log-log plot. If you suspect an exponential function, plot the data using the semilog scales. Only an exponential function will form a straight line on a semilog plot.
4. In function discovery applications, we use the log-log and semilog plots *only* to identify the function type, but not to find the coefficients b and m. The reason is that it is difficult to interpolate on log scales.

We can find the values of b and m with the MATLAB `polyfit` function. This function finds the coefficients of a polynomial of specified degree n that best fits the data, in the so-called least-squares sense. The syntax appears in Table 6.1–1. The mathematical foundation of the least-squares method is presented in Section 6.2.

Because we are assuming that our data will form a straight line on a rectilinear, semilog, or log-log plot, we are interested only in a polynomial that corresponds to a straight line, that is, a first-degree polynomial, which we

Table 6.1–1 The `polyfit` function

Command	Description
`p = polyfit(x,y,n)`	Fits a polynomial of degree n to data described by the vectors x and y, where x is the independent variable. Returns a row vector p of length $n + 1$ that contains the polynomial coefficients in order of descending powers.

will denote as $w = p_1 z + p_2$. Thus, referring to Table 6.1–1, we see that the vector p will be $[p_1, p_2]$ if n is 1. This polynomial has a different interpretation in each of the three cases:

- **The linear function:** $y = mx + b$. In this case the variables w and z in the polynomial $w = p_1 z + p_2$ are the original data variables x and y, and we can find the linear function that fits the data by typing `p = polyfit(x,y,1)`. The first element p_1 of the vector p will be m, and the second element p_2 will be b.

- **The power function:** $y = bx^m$. In this case $\log_{10} y = m \log_{10} x + \log_{10} b$, which has the form $w = p_1 z + p_2$, where the polynomial variables w and z are related to the original data variables x and y by $w = \log_{10} y$ and $z = \log_{10} x$. Thus we can find the power function that fits the data by typing `p = polyfit(log10(x), log10(y),1)`. The first element p_1 of the vector p will be m, and the second element p_2 will be $\log_{10} b$. We can find b from $b = 10^{p_2}$.

- **The exponential function:** $y = b(10)^{mx}$. In this case $\log_{10} y = mx + \log_{10} b$, which has the form $w = p_1 z + p_2$, where the polynomial variables w and z are related to the original data variables x and y by $w = \log_{10} y$ and $z = x$. Thus we can find the exponential function that fits the data by typing `p = polyfit(x, log10(y),1)`. The first element p_1 of the vector p will be m, and the second element p_2 will be $\log_{10} b$. We can find b from $b = 10^{p_2}$.

EXAMPLE 6.1–1	Temperature Dynamics

The temperature of coffee cooling in a porcelain mug at room temperature (68°F) was measured at various times. The data follow.

Time t (sec)	Temperature T (°F)
0	145
620	130
2266	103
3482	90

Develop a model of the coffee's temperature as a function of time, and use the model to estimate how long it took the temperature to reach 120°F.

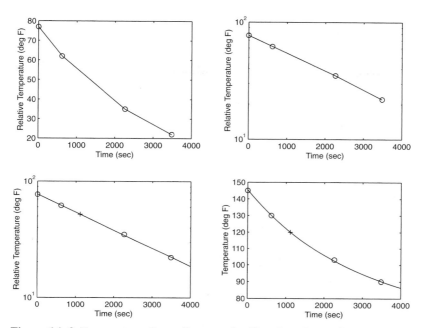

Figure 6.1–3 Temperature of a cooling cup of coffee, plotted on various coordinates.

■ **Solution**

Because $T(0)$ is finite but nonzero, the power function cannot describe these data, so we do not bother to plot the data on log-log axes. Common sense tells us that the coffee will cool and its temperature will eventually equal the room temperature. So we subtract the room temperature from the data and plot the relative temperature, $T - 68$, versus time. If the relative temperature is a linear function of time, the model is $T - 68 = mt + b$. If the relative temperature is an exponential function of time, the model is $T - 68 = b(10)^{mt}$. Figure 6.1–3 shows the plots used to solve the problem. The following MATLAB script file generates the top two plots. The time data are entered in the array `time`, and the temperature data are entered in `temp`.

```
% Enter the data.
time = [0,620,2266,3482];
temp = [145,130,103,90];
% Subtract the room temperature.
temp = temp - 68;
% Plot the data on rectilinear scales.
subplot(2,2,1)
plot(time,temp,time,temp,'o'),xlabel('Time (sec)'),...
   ylabel('Relative Temperature (deg F)')
%
% Plot the data on semilog scales.
```

```
subplot(2,2,2)
semilogy(time,temp,time,temp,'o'),xlabel('Time (sec)'),...
    ylabel('Relative Temperature (deg F)')
```

The data form a straight line on the semilog plot only (the top right plot). Thus the data can be described with the exponential function $T = 68 + b(10)^{mt}$. Using the `polyfit` command, the following lines can be added to the script file.

```
% Fit a straight line to the transformed data.
p = polyfit(time,log10(temp),1);
m = p(1)
b = 10^p(2)
```

The computed values are $m = -1.5557 \times 10^{-4}$ and $b = 77.4469$. Thus our derived model is $T = 68 + b(10)^{mt}$. To estimate how long it will take for the coffee to cool to $120°F$, we must solve the equation $120 = 68 + b(10)^{mt}$ for t. The solution is $t = [\log_{10}(120-68) - \log_{10}(b)]/m$. The MATLAB command for this calculation is shown in the following script file, which is a continuation of the previous script and produces the bottom two subplots shown in Figure 6.1–3.

```
% Compute the time to reach 120 degrees.
t_120 = (log10(120-68)-log10(b))/m
% Show the derived curve and estimated point on semilog scales.
t = 0:10:4000;
T = 68+b*10.^(m*t);
subplot(2,2,3)
semilogy(t,T-68,time,temp,'o',t_120,120-68,'+'),
xlabel('Time (sec)'),...
    ylabel('Relative Temperature (deg F)')
%
% Show the derived curve and estimated point on rectilinear scales.
subplot(2,2,4)
plot(t,T,time,temp+68,'o',t_120,120,'+'),xlabel('Time (sec)'),...
    ylabel('Temperature (deg F)')
```

The computed value of `t_120` is 1112. Thus the time to reach $120°F$ is 1112 sec. The plot of the model, along with the data and the estimated point (1112, 120) marked with a + sign, is shown in the bottom two subplots in Figure 6.1–3. Because the graph of our model lies near the data points, we can treat its prediction of 1112 sec with some confidence.

EXAMPLE 6.1–2   Hydraulic Resistance

A 15-cup coffee pot (see Figure 6.1–4) was placed under a water faucet and filled to the 15-cup line. With the outlet valve open, the faucet's flow rate was adjusted until the water level remained constant at 15 cups, and the time for one cup to flow out of the pot was

Figure 6.1–4 An experiment to verify Torricelli's principle.

measured. This experiment was repeated with the pot filled to the various levels shown in the following table:

Liquid volume V (cups)	Time to fill 1 cup t (sec)
15	6
12	7
9	8
6	9

(a) Use the preceding data to obtain a relation between the flow rate and the number of cups in the pot. (b) The manufacturer wants to make a 36-cup pot using the same outlet valve but is concerned that a cup will fill too quickly, causing spills. Extrapolate the relation developed in part (a) and predict how long it will take to fill one cup when the pot contains 36 cups.

■ **Solution**
(a) Torricelli's principle in hydraulics states that $f = rV^{1/2}$, where f is the flow rate through the outlet valve in cups per second, V is the volume of liquid in the pot in cups, and r is a constant whose value is to be found. We see that this relation is a power function where the exponent is 0.5. Thus if we plot $\log_{10}(f)$ versus $\log_{10}(V)$, we should obtain a straight line. The values for f are obtained from the reciprocals of the given data for t. That is, $f = 1/t$ cups per second.

The MATLAB script file follows. The resulting plots appear in Figure 6.1–5. The volume data are entered in the array `cups`, and the time data are entered in `meas_times`.

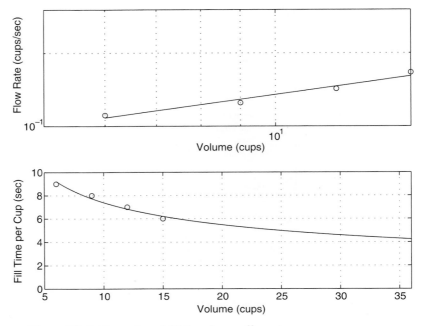

Figure 6.1–5 Flow rate and fill time for a coffee pot.

```
% Data for the problem.
cups = [6,9,12,15];
meas_times = [9,8,7,6];
meas_flow = 1./meas_times;
%
% Fit a straight line to the transformed data.
p = polyfit(log10(cups),log10(meas_flow),1);
coeffs = [p(1),10^p(2)];
m = coeffs(1)
b = coeffs(2)
%
% Plot the data and the fitted line on a loglog plot to see
% how well the line fits the data.
x = 6:0.01:40;
y = b*x.^m;
subplot(2,1,1)
loglog(x,y,cups,meas_flow,'o'),grid,xlabel('Volume (cups)'),...
    ylabel('Flow Rate (cups/sec)'),axis([5 15 0.1 0.3])
```

The computed values are $m = 0.433$ and $b = 0.0499$, and our derived relation is $f = 0.0499V^{0.433}$. Because the exponent is 0.433, not 0.5, our model does not agree exactly with Torricelli's principle, but it is close. Note that the first plot in Figure 6.1–5 shows that the data points do not lie exactly on the fitted straight line. In this application it is difficult to measure the time to fill one cup with an accuracy greater than an integer second, so this inaccuracy could have caused our result to disagree with that predicted by Torricelli.

(b) Note that the fill time is $1/f$, the reciprocal of the flow rate. The remainder of the MATLAB script uses the derived flow rate relation $f = 0.0499V^{0.433}$ to plot the extrapolated fill-time curve $1/f$ versus t.

```
% Plot the fill time curve extrapolated to 36 cups.
subplot(2,1,2)
plot(x,1./y,cups,meas_times,'o'),grid,xlabel('Volume(cups)'),...
    ylabel('Fill Time per Cup (sec)'),axis([5 36 0 10])
%
% Compute the fill time for V = 36 cups.
fill_time = 1/(b*36^m)
```

The predicted fill time for 1 cup is 4.2 sec. The manufacturer must now decide if this time is sufficient for the user to avoid overfilling. (In fact, the manufacturer did construct a 36-cup pot, and the fill time is approximately 4 sec, which agrees with our prediction.)

6.2 Regression

In Section 6.1 we used the MATLAB function `polyfit` to perform regression analysis with functions that are linear or could be converted to linear form by a logarithmic or other transformation. The `polyfit` function is based on the least-squares method, which is also called *regression*. We now show how to use this function to develop polynomial and other types of functions.

The Least-Squares Method

Suppose we have the three data points given in the following table, and we need to determine the coefficients of the straight line $y = mx + b$ that best fit the following data in the least-squares sense.

x	y
0	2
5	6
10	11

According to the least-squares criterion, the line that gives the best fit is the one that minimizes J, the sum of the squares of the vertical differences between the line and the data points. These differences are called the *residuals*. Here there are three data points, and J is given by

$$J = \sum_{i=1}^{3}(mx_i + b - y_i)^2$$

$$= (0m + b - 2)^2 + (5m + b - 6)^2 + (10m + b - 11)^2$$

The values of m and b that minimize J are found by setting the partial derivatives $\partial J/\partial m$ and $\partial J/\partial b$ equal to zero.

$$\frac{\partial J}{\partial m} = 250m + 30b - 280 = 0$$

$$\frac{\partial J}{\partial b} = 30m + 6b - 38 = 0$$

These conditions give two equations that must be solved for the two unknowns m and b. The solution is $m = 0.9$ and $b = 11/6$. The best straight line in the least-squares sense is $y = 0.9x + 11/6$. If we evaluate this equation at the data values $x = 0, 5$, and 10, we obtain the values $y = 1.833, 6.333, 10.8333$. These values are different from the given data values $y = 2, 6, 11$ because the line is not a perfect fit to the data. The value of J is $J = (1.833 - 2)^2 + (6.333 - 6)^2 + (10.8333 - 11)^2 = 0.16656689$. No other straight line will give a lower value of J for these data.

In general, for the polynomial $a_1x^n + a_2x^{n-1} + \cdots + a_nx + a_{n+1}$, the sum of the squares of the residuals for m data points is

$$J = \sum_{i=1}^{m}(a_1x^n + a_2x^{n-1} + \cdots + a_nx + a_{n+1} - y_i)^2$$

The values of the $n + 1$ coefficients a_i that minimize J can be found by solving a set of $n + 1$ linear equations. The `polyfit` function provides this solution. Its syntax is `p = polyfit(x,y,n)`. Table 6.2–1 summarizes the `polyfit` and `polyval` functions.

Consider the data set where $x = 1, 2, 3, \ldots, 9$ and $y = 5, 6, 10, 20, 28, 33, 34, 36, 42$. The following script file computes the coefficients of the first- through fourth-degree polynomials for these data and evaluates J for each polynomial.

```
x = 1:9;
y = [5,6,10,20,28,33,34,36,42];
for k = 1:4
    coeff = polyfit(x,y,k)
    J(k) = sum((polyval(coeff,x)-y).^2)
end
```

The J values are, to two significant figures, 72, 57, 42, and 4.7. Thus the value of J decreases as the polynomial degree is increased, as we would expect. Figure 6.2–1 shows this data and the four polynomials. Note now the fit improves with the higher-degree polynomial.

Table 6.2–1 Functions for polynomial regression

Command	Description
p = polyfit(x,y,n)	Fits a polynomial of degree n to data described by the vectors x and y, where x is the independent variable. Returns a row vector p of length n+1 that contains the polynomial coefficients in order of descending powers.
[p,s,mu] = polyfit(x,y,n)	Fits a polynomial of degree n to data described by the vectors x and y, where x is the independent variable. Returns a row vector p of length n+1 that contains the polynomial coefficients in order of descending powers and a structure s for use with polyval to obtain error estimates for predictions. The optional output variable mu is a two-element vector containing the mean and standard deviation of x.
[y,delta] = polyval(p,x,s,mu)	Uses the optional output structure s generated by [p,s,mu] = polyfit(x,y,n) to generate error estimates. If the errors in the data used with polyfit are independent and normally distributed with constant variance, at least 50 percent of the data will lie within the band y ± delta.

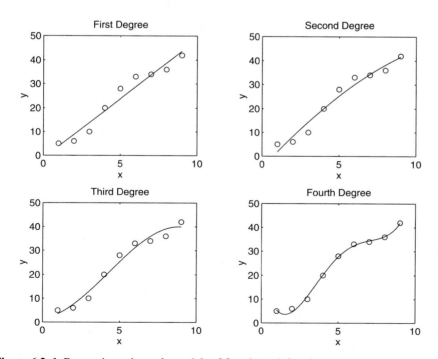

Figure 6.2–1 Regression using polynomials of first through fourth degree.

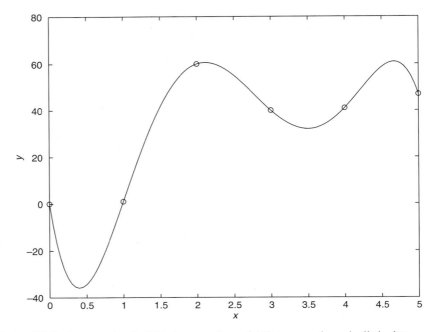

Figure 6.2–2 An example of a fifth-degree polynomial that passes through all six data points but exhibits large excursions between points.

Caution: It is tempting to use a high-degree polynomial to obtain the best possible fit. However, there are two dangers in using high-degree polynomials. High-degree polynomials often exhibit large excursions between the data points and thus should be avoided if possible. Figure 6.2–2 shows an example of this phenomenon. The second danger with using high-degree polynomials is that they can produce large errors if their coefficients are not represented with a large number of significant figures. In some cases it might not be possible to fit the data with a low-degree polynomial. In such cases we might be able to use several cubic polynomials. This method, called *cubic splines*, is covered in Chapter 7.

Test Your Understanding

T6.2–1 Obtain and plot the first- through fourth-degree polynomials for the following data: $x = 0, 1, \ldots, 5$ and $y = 0, 1, 60, 40, 41,$ and 47. Find the coefficients and the J values.

(Answer: The polynomials are $9.5714x + 7.5714$; $-3.6964x^2 + 28.0536x - 4.7500$; $0.3241x^3 - 6.1270x^2 + 32.4934x - 5.7222$; and $2.5208x^4 - 24.8843x^3 + 71.2986x^2 - 39.5304x - 1.4008$. The corresponding J values are 1534, 1024, 1017, and 495, respectively.)

Fitting Other Functions

Given the data (y, z), the logarithmic function $y = m \ln z + b$ can be converted to a first-degree polynomial by transforming the z values into x values by the transformation $x = \ln z$. The resulting function is $y = mx + b$.

Given the data (y, z), the function $y = b(10)^{m/z}$ can be converted to an exponential function by transforming the z values by the transformation $x = 1/z$.

Given the data (v, x), the function $v = 1/(mx + b)$ can be converted to a first-degree polynomial by transforming the v data values with the transformation $y = 1/v$. The resulting function is $y = mx + b$.

To see how to obtain a function $y = kx$ that passes through the origin, see Problem 8.

The Quality of a Curve Fit

The least-squares criterion used to fit a function $f(x)$ is the sum of the squares of the residuals J. It is defined as

$$J = \sum_{i=1}^{m} [f(x_i) - y_i]^2 \qquad (6.2\text{--}1)$$

We can use the J value to compare the quality of the curve fit for two or more functions used to describe the same data. The function that gives the smallest J value gives the best fit.

We denote the sum of the squares of the deviation of the y values from their mean $\bar{y}$ by S, which can be computed from

$$S = \sum_{i=1}^{m} (y_i - \bar{y})^2 \qquad (6.2\text{--}2)$$

This formula can be used to compute another measure of the quality of the curve fit, the *coefficient of determination,* also known as the *r-squared value.* It is defined as

COEFFICIENT OF DETERMINATION

$$r^2 = 1 - \frac{J}{S} \qquad (6.2\text{--}3)$$

For a perfect fit, $J = 0$ and thus $r^2 = 1$. Thus the closer r^2 is to 1, the better the fit. The largest r^2 can be is 1. The value of S indicates how much the data is spread around the mean, and the value of J indicates how much of the data spread is unaccounted for by the model. Thus the ratio J/S indicates the fractional variation unaccounted for by the model. It is possible for J to be larger than S, and thus it is possible for r^2 to be negative. Such cases, however, are indicative of a very poor model that should not be used. As a rule of thumb, a good fit accounts for at least 99 percent of the data variation. This value corresponds to $r^2 \geq 0.99$.

For example, the following table gives the values of J, S, and r^2 for the first- through fourth-degree polynomials used to fit the data $x = 1, 2, 3, \ldots, 9$ and $y = 5, 6, 10, 20, 28, 33, 34, 36, 42$.

Degree n	J	S	r^2
1	72	1562	0.9542
2	57	1562	0.9637
3	42	1562	0.9732
4	4.7	1562	0.9970

Because the fourth-degree polynomial has the largest r^2 value, it represents the data better than the representation from first- through third-degree polynomials, according to the r^2 criterion.

To calculate the values of S and r^2, add the following lines to the end of the script file shown on page 272.

```
mu = mean(y);
for k=1:4
    S(k) = sum((y-mu).^2);
    r2(k) = 1 - J(k)/S(k);
end
S
r2
```

Scaling the Data

The effect of computational errors in computing the coefficients can be lessened by properly scaling the x values. When the function polyfit(x,y,n) is executed, it will issue a warning message if the polynomial degree n is greater than or equal to the number of data points (because there will not be enough equations for MATLAB to solve for the coefficients), or if the vector x has repeated, or nearly repeated, points, or if the vector x needs centering and/or scaling. The alternate syntax [p, s, mu] = polyfit(x,y,n) finds the coefficients p of a polynomial of degree n in terms of the variable

$$\hat{x} = (x - \mu_x)/\sigma_x$$

The output variable mu is a two-element vector $[\mu_x, \sigma_x]$, where μ_x is the mean of x and σ_x is the standard deviation of x (the standard deviation is discussed in Chapter 7).

You can scale the data yourself before using polyfit. Some common scaling methods are

$$\hat{x} = x - x_{min} \quad \text{or} \quad \hat{x} = x - \mu_x$$

if the range of x is small, or

$$\hat{x} = \frac{x}{x_{max}} \quad \text{or} \quad \hat{x} = \frac{x}{x_{mean}}$$

if the range of x is large.

Estimation of Traffic Flow

EXAMPLE 6.2–1

The following data give the number of vehicles (in millions) crossing a bridge each year for 10 years. Fit a cubic polynomial to the data and use the fit to estimate the flow in the year 2010.

Year	2000	2001	2002	2003	2004	2005	2006	2007	2008	2009
Vehicle flow (millions)	2.1	3.4	4.5	5.3	6.2	6.6	6.8	7	7.4	7.8

■ Solution

If we attempt to fit a cubic to these data, as in the following session, we get a warning message.

```
>>Year = 2000:2009;
>>Veh_Flow = [2.1,3.4,4.5,5.3,6.2,6.6,6.8,7,7.4,7.8];
>>p = polyfit(Year,Veh_Flow,3)
Warning: Polynomial is badly conditioned.
```

The problem is caused by the large values of the independent variable `Year`. Because their range is small, we can simply subtract 2000 from each value. Continue the session as follows.

```
>>x = Year-2000; y = Veh_Flow;
>>p = polyfit(x,y,3)
p =
    0.0087    -0.1851   1.5991  2.0362
>>J = sum((polyval(p,x)-y).^2);
>>S = sum((y-mean(y)).^2);
>>r2 = 1 - J/S
r2 =
    0.9972
```

Thus the polynomial fit is good because the coefficient of determination is 0.9972. The corresponding polynomial is

$$f = 0.0087(t - 2000)^3 - 0.1851(t - 2000)^2 + 1.5991(t - 2000) + 2.0362$$

where f is the traffic flow in millions of vehicles and t is the time in years measured from 0. We can use this equation to estimate the flow at the year 2010 by substituting $t = 2010$, or by typing in MATLAB `polyval(p,10)`. Rounded to one decimal place, the answer is 8.2 million vehicles.

Using Residuals

We now show how to use the residuals as a guide to choosing an appropriate function to describe the data. In general, if you see a pattern in the plot of the residuals, it indicates that another function can be found to describe the data better.

EXAMPLE 6.2–2

Modeling Bacteria Growth

The following table gives data on the growth of a certain bacteria population with time. Fit an equation to these data.

Time (min)	Bacteria (ppm)	Time (min)	Bacteria (ppm)
0	6	10	350
1	13	11	440
2	23	12	557
3	33	13	685
4	54	14	815
5	83	15	990
6	118	16	1170
7	156	17	1350
8	210	18	1575
9	282	19	1830

■ Solution

We try three polynomial fits (linear, quadratic, and cubic) and an exponential fit. The script file is given below. Note that we can write the exponential form as $y = b(10)^{mt} = 10^{mt+a}$, where $b = 10^a$.

```
% Time data
x = 0:19;
% Population data
y = [6,13,23,33,54,83,118,156,210,282,...
    350,440,557,685,815,990,1170,1350,1575,1830];
% Linear fit
p1 = polyfit(x,y,1);
% Quadratic fit
p2 = polyfit(x,y,2);
% Cubic fit
p3 = polyfit(x,y,3);
% Exponential fit
p4 = polyfit(x,log10(y),1);
% Residuals
res1 = polyval(p1,x)-y;
res2 = polyval(p2,x)-y;
res3 = polyval(p3,x)-y;
res4 = 10.^polyval(p4,x)-y;
```

You can then plot the residuals as shown in Figure 6.2–3. Note that there is a definite pattern in the residuals of the linear fit. This indicates that the linear function cannot match the curvature of the data. The residuals of the quadratic fit are much smaller, but there is still a pattern, with a random component. This indicates that the quadratic

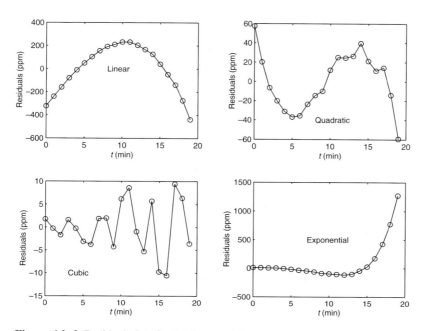

Figure 6.2–3 Residual plots for the four models.

function also cannot match the curvature of the data. The residuals of the cubic fit are even smaller, with no strong pattern and a large random component. This indicates that a polynomial degree higher than 3 will not be able to match the data curvature any better than the cubic. The residuals for the exponential are the largest of all, and indicate a poor fit. Note also how the residuals systematically increase with t, indicating that the exponential cannot describe the data's behavior after a certain time.

Thus the cubic is the best fit of the four models considered. Its coefficient of determination is $r^2 = 0.9999$. The model is

$$y = 0.1916t^3 + 1.2082t^2 + 3.607t + 7.7307$$

where y is the bacteria population in ppm and t is time in minutes.

Multiple Linear Regression

Suppose that y is a linear function of two or more variables $x_1, x_2, \ldots$, for example, $y = a_0 + a_1x_1 + a_2x_2$. To find the coefficient values a_0, a_1, and a_2 to fit a set of data (y, x_1, x_2) in the least-squares sense, we can make use of the fact that the left-division method for solving linear equations uses the least-squares method when the equation set is overdetermined. To use this method, let n be the number of data points and write the linear equation in matrix form as $\mathbf{Xa} = \mathbf{y}$, where

$$\mathbf{a} = \begin{bmatrix} a_0 \\ a_1 \\ a_2 \end{bmatrix} \quad \mathbf{X} = \begin{bmatrix} 1 & x_{11} & x_{21} \\ 1 & x_{12} & x_{22} \\ 1 & x_{13} & x_{23} \\ \cdots & \cdots & \cdots \\ 1 & x_{1n} & x_{2n} \end{bmatrix} \quad \mathbf{y} = \begin{bmatrix} y_1 \\ y_2 \\ y_3 \\ \cdots \\ y_n \end{bmatrix}$$

where x_{1i}, x_{2i}, and y_i are the data, $i = 1, \ldots, n$. The solution for the coefficients is given by `a = X\y`.

EXAMPLE 6.2–3   Breaking Strength and Alloy Composition

We want to predict the strength of metal parts as a function of their alloy composition. The tension force y required to break a steel bar is a function of the percentage x_1 and x_2 of each of two alloying elements present in the metal. The following table gives some pertinent data. Obtain a linear model $y = a_0 + a_1 x_1 + a_2 x_2$ to describe the relationship.

Breaking strength (kN) y	% of element 1 x_1	% of element 2 x_2
7.1	0	5
19.2	1	7
31	2	8
45	3	11

■ Solution
The script file is as follows:

```
x1 = (0:3)';x2 = [5,7,8,11]';
y = [7.1,19.2,31,45]';
X = [ones(size(x1)), x1, x2];
a = X\y
yp = X*a;
Max_Percent_Error = 100*max(abs((yp-y)./y))
```

The vector `yp` is the vector of breaking strength values predicted by the model. The scalar `Max_Percent_Error` is the maximum percent error in the four predictions. The results are `a = [0.8000, 10.2429, 1.2143]'` and `Max_Percent_Error = 3.2193`. Thus the model is $y = 0.8 + 10.2429 x_1 + 1.2143 x_2$. The maximum percent error of the model's predictions, as compared to the given data, is 3.2193 percent.

Linear-in-Parameters Regression

Sometimes we want to fit an expression that is neither a polynomial nor a function that can be converted to linear form by a logarithmic or other transformation. In some cases we can still do a least-squares fit if the function is a linear expression in terms of its parameters. The following example illustrates the method.

Response of a Biomedical Instrument

EXAMPLE 6.2–4

Engineers developing instrumentation often need to obtain a *response* curve that describes how fast the instrument can make measurements. The theory of instrumentation shows that often the response can be described by one of the following equations, where v is the voltage output and t is time. In both models, the voltage reaches a steady-state constant value as $t \rightarrow \infty$, and T is the time required for the voltage to equal 95 percent of the steady-state value.

$$v(t) = a_1 + a_2 e^{-3t/T} \qquad \text{(first-order model)}$$

$$v(t) = a_1 + a_2 e^{-3t/T} + a_3 t e^{-3t/T} \quad \text{(second-order model)}$$

The following data give the output voltage of a certain device as a function of time. Obtain a function that describes these data.

t (s)	0	0.3	0.8	1.1	1.6	2.3	3
v (V)	0	0.6	1.28	1.5	1.7	1.75	1.8

■ Solution

Plotting the data, we estimate that it takes approximately 3 s for the voltage to become constant. Thus we estimate that $T = 3$. The first-order model written for each of the n data points results in n equations, which can be expressed as follows:

$$\begin{bmatrix} 1 & e^{-t_1} \\ 1 & e^{-t_2} \\ \cdots & \cdots \\ 1 & e^{-t_n} \end{bmatrix} \begin{bmatrix} a_1 \\ a_2 \end{bmatrix} = \begin{bmatrix} y_1 \\ y_2 \\ \cdots \\ y_n \end{bmatrix}$$

or, in matrix form,

$$\mathbf{Xa} = \mathbf{y}'$$

which can be solved for the coefficient vector **a** using left division. The following MATLAB script solves the problem.

```
t = [0,0.3,0.8,1.1,1.6,2.3,3];
y = [0,0.6,1.28,1.5,1.7,1.75,1.8];
X = [ones(size(t));exp(-t)]';
a = X\y'
```

The answer is $a_1 = 2.0258$ and $a_2 = -1.9307$.

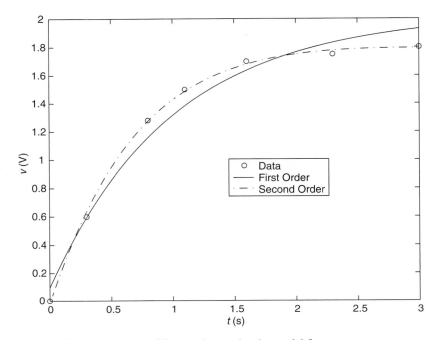

Figure 6.2–4 Comparison of first- and second-order model fits.

A similar procedure can be followed for the second-order model.

$$
\begin{bmatrix}
1 & e^{-t_1} & t_1 e^{-t_1} \\
1 & e^{-t_2} & t_2 e^{-t_2} \\
\cdots & \cdots & \cdots \\
1 & e^{-t_n} & t_n e^{-t_n}
\end{bmatrix}
\begin{bmatrix}
a_1 \\
a_2 \\
a_3
\end{bmatrix}
=
\begin{bmatrix}
y_1 \\
y_2 \\
\cdots \\
y_n
\end{bmatrix}
$$

Continue the previous script as follows.

```
X = [ones(size(t));exp(-t);t.*exp(-t)]';
a = X\y'
```

The answer is $a_1 = 1.7496$, $a_2 = -1.7682$, and $a_3 = 0.8885$. The two models are plotted with the data in Figure 6.2–4. Clearly the second-order model gives the better fit.

6.3 The Basic Fitting Interface

MATLAB supports curve fitting through the Basic Fitting interface. Using this interface, you can quickly perform basic curve-fitting tasks within the same easy-to-use environment. The interface is designed so that you can do the following:

■ Fit data using a cubic spline or a polynomial up to degree 10.
■ Plot multiple fits simultaneously for a given data set.
■ Plot the residuals.

■ Examine the numerical results of a fit.

■ Interpolate or extrapolate a fit.

■ Annotate the plot with the numerical fit results and the norm of residuals.

■ Save the fit and evaluated results to the MATLAB workspace.

Depending on your specific curve-fitting application, you can use the Basic Fitting interface, the command line functions, or both. *Note:* You can use the Basic Fitting interface only with two-dimensional data. However, if you plot multiple data sets as a subplot, and at least one data set is two-dimensional, then the interface is enabled.

Two panes of the Basic Fitting interface are shown in Figure 6.3–1. To reproduce this state:

1. Plot some data.

2. Select **Basic Fitting** from the **Tools** menu of the Figure window.

3. When the first pane of the Basic Fitting interface appears, click the right arrow button once.

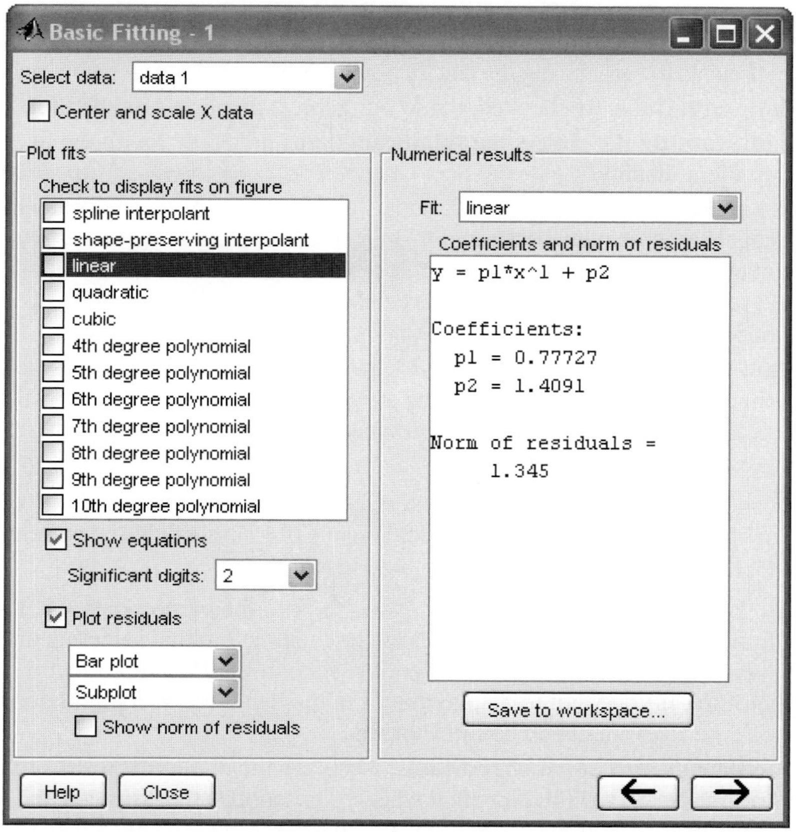

Figure 6.3–1 The Basic Fitting interface.

The third pane is used for interpolating or extrapolating a fit. It appears when you click the right arrow button a second time.

At the top of the first pane is the **Select data** window which contains the names of all the data sets you display in the Figure window associated with the Basic Fitting interface. Use this menu to select the data set to be fit. You can perform multiple fits for the current data set. Use the Plot Editor to change the name of a data set. The remaining items on the first pane are used as follows.

- **Center and scale X data.** If checked, the data are centered at zero mean and scaled to unit standard deviation. You may need to center and scale your data to improve the accuracy of the subsequent numerical computations. As described in the previous section, a warning is returned to the Command window if a fit produces results that might be inaccurate.

- **Plot fits.** This panel allows you to visually explore one or more fits to the current data set.

- **Check to display fits on figure.** Select the fits you want to display for the current data set. You can choose as many fits for a given data set as you want. However, if your data set has n points, then you should use polynomials with at most n coefficients. If you fit using polynomials with more than n coefficients, the interface will automatically set a sufficient number of coefficients to zero during the calculation so that a solution can be obtained.

- **Show equations.** If checked, the fit equation is displayed on the plot.

- **Significant digits.** Select the significant digits associated with the fit coefficient display.

- **Plot residuals.** If checked, the residuals are displayed. You can display the residuals as a bar plot, a scatter plot, a line plot using either the same figure window as the data or using a separate figure window. If you plot multiple data sets as a subplot, then residuals can be plotted only in a separate figure window. See Figure 6.3–2.

- **Show norm of residuals.** If checked, the norm of residuals is displayed. The norm of residuals is a measure of the goodness of fit, where a smaller value indicates a better fit. The norm is the square root of the sum of the squares of the residuals.

The second pane of the Basic Fitting interface is labeled *Numerical Results*. This pane enables you to explore the numerical results of a single fit to the current data set without plotting the fit. It contains three items.

- **Fit.** Use this menu to select an equation to fit to the current data set. The fit results are displayed in the box below the menu. Note that selecting an equation in this menu does not affect the state of the **Plot fits** selection. Therefore, if you want to display the fit in the data plot, you might need to check the relevant check box in **Plot fits.**

- **Coefficients and norm of residuals.** Displays the numerical results for the equation selected in **Fit.** Note that when you first open the **Numerical Results** panel, the results of the last fit you selected in **Plot fits** are displayed.

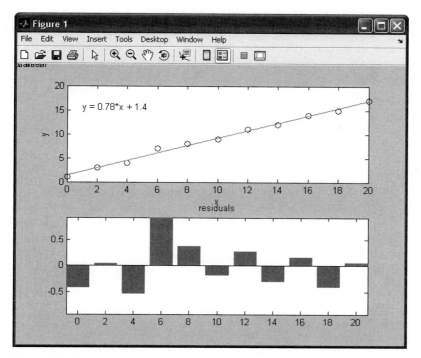

Figure 6.3–2 A figure produced by the Basic Fitting interface.

■ **Save to workspace.** Launches a dialog box that allows you to save the fit results to workspace variables.

The third pane of the Basic Fitting interface contains three items.

■ **Find $Y = f(X)$.** Use this to interpolate or extrapolate the current fit. Enter a scalar or a vector of values corresponding to the independent variable (X). The current fit is evaluated after you click on the **Evaluate** button, and the results are displayed in the associated window. The current fit is displayed in the **Fit** window.

■ **Save to workspace.** Launches a dialog box that allows you to save the evaluated results to workspace variables.

■ **Plot evaluated results.** If checked, the evaluated results are displayed on the data plot.

6.4 Summary

In this chapter you learned an important application of plotting—function discovery—which is the technique for using data plots to obtain a mathematical function that describes the data. Regression can be used to develop a model for cases where there is considerable scatter in the data.

Many physical processes can be modeled with functions that produce a straight line when plotted using a suitable set of axes. In some cases, we can find a transformation that produces a straight line in the transformed variable.

When such a function or transformation cannot be found, we resort to polynomial regression, multiple linear regression, or linear-in-parameters regression to obtain an approximate functional description of the data. The MATLAB Basic Fitting interface is a powerful aid in obtaining regression models.

Key Terms with Page References

Coefficient of determination,   275
Linear-in-parameters,   280
Multiple linear regression,   279
Regression,   271
Residuals,   272

Problems

You can find the answers to problems marked with an asterisk at the end of the text.

Section 6.1

1. The distance a spring stretches from its "free length" is a function of how much tension force is applied to it. The following table gives the spring length y that the given applied force f produced in a particular spring. The spring's free length is 4.7 in. Find a functional relation between f and x, the extension from the free length ($x = y - 4.7$).

Force f (lb)	Spring length y (in.)
0	4.7
0.94	7.2
2.30	10.6
3.28	12.9

2.* In each of the following problems, determine the best function $y(x)$ (linear, exponential, or power function) to describe the data. Plot the function on the same plot with the data. Label and format the plots appropriately.

a.

x	25	30	35	40	45
y	5	260	480	745	1100

b.

x	2.5	3	3.5	4	4.5	5	5.5	6	7	8	9	10
y	1500	1220	1050	915	810	745	690	620	520	480	410	390

c.

x	550	600	650	700	750
y	41.2	18.62	8.62	3.92	1.86

3. The population data for a certain country are as follows:

Year	2004	2005	2006	2007	2008	2009
Population (millions)	10	10.9	11.7	12.6	13.8	14.9

Obtain a function that describes these data. Plot the function and the data on the same plot. Estimate when the population will be double its 2004 size.

4.* The *half-life* of a radioactive substance is the time it takes to decay by one-half. The half-life of carbon 14, which is used for dating previously living things, is 5500 years. When an organism dies, it stops accumulating carbon 14. The carbon 14 present at the time of death decays with time. Let $C(t)/C(0)$ be the fraction of carbon 14 remaining at time t. In radioactive carbon dating, scientists usually assume that the remaining fraction decays exponentially according to the following formula:

$$\frac{C(t)}{C(0)} = e^{-bt}$$

   a. Use the half-life of carbon 14 to find the value of the parameter b, and plot the function.
   b. If 90 percent of the original carbon 14 remains, estimate how long ago the organism died.
   c. Suppose our estimate of b is off by ± 1 percent. How does this error affect the age estimate?

5. *Quenching* is the process of immersing a hot metal object in a bath for a specified time to obtain certain properties such as hardness. A copper sphere 25 mm in diameter, initially at 300°C, is immersed in a bath at 0°C. The following table gives measurements of the sphere's temperature versus time. Find a functional description of these data. Plot the function and the data on the same plot.

Time (s)	0	1	2	3	4	5	6
Temperature (°C)	300	150	75	35	12	5	2

6. The useful life of a machine bearing depends on its operating temperature, as the following data show. Obtain a functional description of these data. Plot the function and the data on the same plot. Estimate a bearing's life if it operates at 150°F.

Temperature (°F)	100	120	140	160	180	200	220
Bearing life (hours $\times 10^3$)	28	21	15	11	8	6	4

7. A certain electric circuit has a resistor and a capacitor. The capacitor is initially charged to 100 V. When the power supply is detached, the capacitor voltage decays with time, as the following data table shows. Find a functional description of the capacitor voltage v as a function of time t. Plot the function and the data on the same plot.

Time (s)	0	0.5	1	1.5	2	2.5	3	3.5	4
Voltage (V)	100	62	38	21	13	7	4	2	3

Sections 6.2 and 6.3

8.* The distance a spring stretches from its free length is a function of how much tension force is applied to it. The following table gives the spring length y that was produced in a particular spring by the given applied force f. The spring's free length is 4.7 in. Find a functional relation between f and x, the extension from the free length ($x = y - 4.7$).

Force f (lb)	Spring length y (in.)
0	4.7
0.94	7.2
2.30	10.6
3.28	12.9

9. The following data give the drying time T of a certain paint as a function of the amount of a certain additive A.
    a. Find the first-, second-, third-, and fourth-degree polynomials that fit the data, and plot each polynomial with the data. Determine the quality of the curve fit for each by computing J, S, and r^2.
    b. Use the polynomial giving the best fit to estimate the amount of additive that minimizes the drying time.

A (oz)	0	1	2	3	4	5	6	7	8	9
T (min)	130	115	110	90	89	89	95	100	110	125

10.* The following data give the stopping distance d as a function of initial speed v, for a certain car model. Find a quadratic polynomial that fits the data. Determine the quality of the curve fit by computing J, S, and r^2.

v (mi/hr)	20	30	40	50	60	70
d (ft)	45	80	130	185	250	330

11.* The number of twists y required to break a certain rod is a function of the percentage x_1 and x_2 of each of two alloying elements present in the rod. The following table gives some pertinent data. Use linear multiple regression to obtain a model $y = a_0 + a_1 x_1 + a_2 x_2$ of the relationship between the number of twists and the alloy percentages. In addition, find the maximum percent error in the predictions.

Number of twists y	Percentage of element 1 x_1	Percentage of element 2 x_2
40	1	1
51	2	1
65	3	1
72	4	1
38	1	2
46	2	2
53	3	2
67	4	2
31	1	3
39	2	3
48	3	3
56	4	3

12. The following represents pressure samples, in pounds per square inch (psi), taken in a fuel line once every second for 10 sec.

Time (sec)	Pressure (psi)	Time (sec)	Pressure (psi)
1	26.1	6	30.6
2	27.0	7	31.1
3	28.2	8	31.3
4	29.0	9	31.0
5	29.8	10	30.5

a. Fit a first-degree polynomial, a second-degree polynomial, and a third-degree polynomial to these data. Plot the curve fits along with the data points.

b. Use the results from part *a* to predict the pressure at $t = 11$ sec. Explain which curve fit gives the most reliable prediction. Consider the coefficients of determination and the residuals for each fit in making your decision.

13. A liquid boils when its vapor pressure equals the external pressure acting on the surface of the liquid. This is why water boils at a lower temperature at higher altitudes. This information is important for people who must design processes utilizing boiling liquids. Data on the vapor pressure P of water as a function of temperature T are given in the following table. From theory we know that $\ln P$ is proportional to $1/T$. Obtain a curve fit for $P(T)$ from these data. Use the fit to estimate the vapor pressure at 285 and 300 K.

T (K)	P (torr)
273	4.579
278	6.543
283	9.209
288	12.788
293	17.535
298	23.756

14. The solubility of salt in water is a function of the water temperature. Let S represent the solubility of NaCl (sodium chloride) as grams of salt in 100 g of water. Let T be temperature in °C. Use the following data to obtain a curve fit for S as a function of T. Use the fit to estimate S when $T = 25°C$.

T (°C)	S (g NaCl/100 g H_2O)
10	35
20	35.6
30	36.25
40	36.9
50	37.5
60	38.1
70	38.8
80	39.4
90	40

15. The solubility of oxygen in water is a function of the water temperature. Let S represent the solubility of O_2 as millimoles of O_2 per liter of water. Let T be temperature in °C. Use the following data to obtain a curve fit for S as a function of T. Use the fit to estimate S when $T = 8°C$ and $T = 50°C$.

T (°C)	S (mmol O_2 /L H_2O)
5	1.95
10	1.7
15	1.55
20	1.40
25	1.30
30	1.15
35	1.05
40	1.00
45	0.95

16. The following function is linear in the parameters a_1 and a_2.

$$y(x) = a_1 + a_2 \ln x$$

Use least-squares regression with the following data to estimate the values of a_1 and a_2. Use the curve fit to estimate the values of y at $x = 2.5$ and at $x = 11$.

x	1	2	3	4	5	6	7	8	9	10
y	10	14	16	18	19	20	21	22	23	23

17. Chemists and engineers must be able to predict the changes in chemical concentration in a reaction. A model used for many single-reactant processes is

$$\text{Rate of change of concentration} = -kC^n$$

where C is the chemical concentration and k is the rate constant. The order of the reaction is the value of the exponent n. Solution methods for differential equations (which are discussed in Chapter 9) can show that the solution for a first-order reaction ($n = 1$) is

$$C(t) = C(0)e^{-kt}$$

The following data describe the reaction

$$(CH_3)_3CBr + H_2O \rightarrow (CH_3)_3COH + HBr$$

Use these data to obtain a least-squares fit to estimate the value of k.

Time t (h)	C (mol of $(CH3)_3$ CBr/L)
0	0.1039
3.15	0.0896
6.20	0.0776
10.0	0.0639
18.3	0.0353
30.8	0.0207
43.8	0.0101

18. Chemists and engineers must be able to predict the changes in chemical concentration in a reaction. A model used for many single-reactant processes is

$$\text{Rate of change of concentration} = -kC^n$$

where C is the chemical concentration and k is the rate constant. The order of the reaction is the value of the exponent n. Solution methods for differential equations (which are discussed in Chapter 9) can show that the solution for a first-order reaction ($n = 1$) is

$$C(t) = C(0)e^{-kt}$$

and the solution for a second-order reaction ($n = 2$) is

$$\frac{1}{C(t)} = \frac{1}{C(0)} + kt$$

The following data (from Brown, 1994) describes the gas-phase decomposition of nitrogen dioxide at 300°C.

$$2NO_2 \rightarrow 2NO + O_2$$

Time t (s)	C (mol NO_2 /L)
0	0.0100
50	0.0079
100	0.0065
200	0.0048
300	0.0038

Determine whether this is a first-order or second-order reaction, and estimate the value of the rate constant k.

19. Chemists and engineers must be able to predict the changes in chemical concentration in a reaction. A model used for many single-reactant processes is

$$\text{Rate of change of concentration} = -kC^n$$

where C is the chemical concentration and k is the rate constant. The order of the reaction is the value of the exponent n. Solution methods for differential equations (which are discussed in Chapter 9) can show that the solution for a first-order reaction ($n = 1$) is

$$C(t) = C(0)e^{-kt}$$

The solution for a second-order reaction ($n = 2$) is

$$\frac{1}{C(t)} = \frac{1}{C(0)} + kt$$

and the solution for a third-order reaction ($n = 3$) is

$$\frac{1}{2C^2(t)} = \frac{1}{2C^2(0)} + kt$$

Time t (min)	C (mol of reactant/L)
5	0.3575
10	0.3010
15	0.2505
20	0.2095
25	0.1800
30	0.1500
35	0.1245
40	0.1070
45	0.0865

The preceding data describe a certain reaction. By examining the residuals, determine whether this is a first-order, second-order, or third-order reaction, and estimate the value of the rate constant k.

Engineering in the 21st Century

Energy-Efficient Transportation

Modern societies have become very dependent on transportation powered by gasoline and diesel fuel. There is some disagreement about how long it will take to exhaust these fuel resources, but it will certainly happen. Novel engineering developments in both personal and mass transportation will be needed to reduce our dependence on such fuels. These developments will be required in a number of areas such as engine design, electric motor and battery technology, lightweight materials, and aerodynamics.

A number of such initiatives are underway. Several projects have the goal of designing a six-passenger car that is one-third lighter and 40 percent more aerodynamic than today's sleekest cars. A hybrid gas-electric vehicle is the most promising at present. An internal combustion engine and an electric motor drive the wheels. A fuel cell or a battery is charged either by a generator driven by the engine or by energy recovered during braking. This is called regenerative braking.

The weight reduction can be achieved with all-aluminum unibody construction and by improved design of the engine, radiator, and brakes to make use of advanced materials such as composites and magnesium. Other manufacturers are investigating plastic bodies made from recycled materials.

There is still much room for improved efficiency, and research and development engineers in this area will remain busy for some time. MATLAB is widely used to assist these efforts with modeling and analysis tools for hybrid vehicle designs. ■

CHAPTER 7

Statistics, Probability, and Interpolation

OUTLINE

7.1   Statistics and Histograms

7.2   The Normal Distribution

7.3   Random Number Generation

7.4   Interpolation

7.5   Summary

Problems

This chapter begins with an introduction to basic statistics in Section 7.1. You will see how to obtain and interpret *histograms,* which are specialized plots for displaying statistical results. The *normal distribution,* commonly called the *bell-shaped curve,* forms the basis of much of probability theory and many statistical methods. It is covered in Section 7.2. In Section 7.3 you will see how to include random processes in your simulation programs. In Section 7.4 you will see how to use interpolation with data tables to estimate values that are not in the table.

When you have finished this chapter, you should be able to use MATLAB to do the following:

■   Solve basic problems in statistics and probability.

■   Create simulations incorporating random processes.

■   Apply interpolation techniques.

7.1 Statistics and Histograms

MEAN

MODE

MEDIAN

BINS

With MATLAB you can compute the *mean* (the average), the *mode* (the most frequently occurring value), and the *median* (the middle value) of a set of data. MATLAB provides the mean(x), mode(x), and median(x) functions to compute the mean, mode, and median of the data values stored in x, if x is a vector. However, if x is a matrix, a row vector is returned containing the mean (or mode or median) value of each column of x. These functions do not require the elements in x to be sorted in ascending or descending order.

The way the data are spread around the mean can be described by a *histogram* plot. A *histogram* is a plot of the frequency of occurrence of data values versus the values themselves. It is a bar plot of the number of data values that occur within each range, with the bar centered in the middle of the range.

To plot a histogram, you must group the data into subranges, called *bins*. The choice of the bin width and bin center can drastically change the shape of the histogram. If the number of data values is relatively small, the bin width cannot be small because some of the bins will contain no data and the resulting histogram might not usefully illustrate the distribution of the data.

To obtain a histogram, first sort the data values if they have has not yet been sorted (you can use the sort function here). Then choose the bin ranges and bin centers and count the number of values in each bin. Use the bar function to plot the number of values in each bin versus the bin centers as a bar chart. The function bar(x,y) creates a bar chart of y versus x.

MATLAB also provides the hist command to generate a histogram. This command has several forms. Its basic form is hist(y), where y is a vector containing the data. This form aggregates the data into 10 bins evenly spaced between the minimum and maximum values in y. The second form is hist(y,n), where n is a user-specified scalar indicating the number of bins. The third form is hist(y,x), where x is a user-specified vector that determines the location of the bin centers; the bin widths are the distances between the centers.

EXAMPLE 7.1–1   Breaking Strength of Thread

To ensure proper quality control, a thread manufacturer selects samples and tests them for breaking strength. Suppose that 20 thread samples are pulled until they break, and the breaking force is measured in newtons rounded off to integer values. The breaking force values recorded were 92, 94, 93, 96, 93, 94, 95, 96, 91, 93, 95, 95, 95, 92, 93, 94, 91, 94, 92, and 93. Plot the histogram of the data.

■ Solution

Store the data in the vector y, which is shown in the following script file. Because there are six outcomes (91, 92, 93, 94, 95, 96 N), we choose six bins. However, if you use hist(y,6), the bins will not be centered at 91, 92, 93, 94, 95, and 96. So use the form

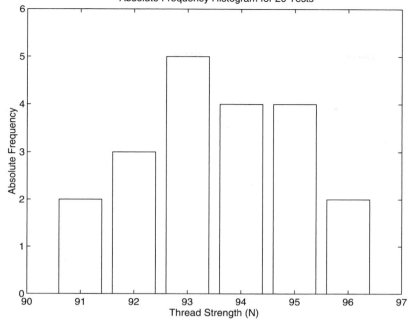

Figure 7.1–1 Histograms for 20 tests of thread strength.

hist(y,x), where x = 91:96. The following script file generates the histogram shown in Figure 7.1–1.

```
% Thread breaking strength data for 20 tests.
y = [92,94,93,96,93,94,95,96,91,93,...
    95,95,95,92,93,94,91,94,92,93];
% The six possible outcomes are 91,92,93,94,95,96.
x = 91:96;
hist(y,x),axis([90 97 0 6]),ylabel('Absolute Frequency'),...
    xlabel('Thread Strength (N)'),...
    title('Absolute Frequency Histogram for 20 Tests')
```

The *absolute frequency* is the number of times a particular outcome occurs. For example, in 20 tests these data show that a 95 occurred 4 times. The absolute frequency is 4, and its *relative frequency* is 4/20, or 20 percent of the time.

When there is a large amount of data, you can avoid typing in every data value by first aggregating the data. The following example shows how this is done using the ones function. The following data were generated by testing 100 thread

ABSOLUTE FREQUENCY

RELATIVE FREQUENCY

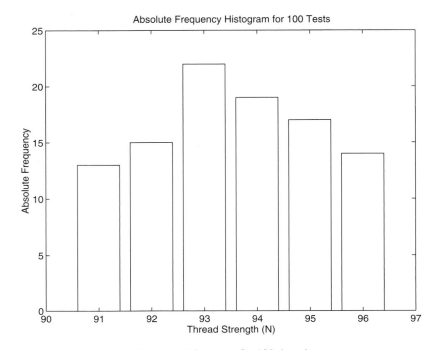

Figure 7.1–2 Absolute frequency histogram for 100 thread tests.

samples. The number of times 91, 92, 93, 94, 95, or 96 N was measured is 13, 15, 22, 19, 17, and 14, respectively.

```
% Thread strength data for 100 tests.
y = [91*ones(1,13),92*ones(1,15),93*ones(1,22),...
    94*ones(1,19),95*ones(1,17),96*ones(1,14)];
x = 91:96;
hist(y,x),ylabel('Absolute Frequency'),...
    xlabel('Thread Strength (N)'),...
    title('Absolute Frequency Histogram for 100 Tests')
```

The result appears in Figure 7.1–2.

The `hist` function is somewhat limited in its ability to produce useful histograms. Unless all the outcome values are the same as the bin centers (as is the case with the thread examples), the graph produced by the `hist` function will not be satisfactory. This case occurs when you want to obtain a *relative* frequency histogram. In such cases you can use the `bar` function to generate the histogram. The following script file generates the relative frequency histogram for the 100 thread tests. Note that if you use the `bar` function, you must aggregate the data first.

```
% Relative frequency histogram using the bar function.
tests = 100;
y = [13,15,22,19,17,14]/tests;
x = 91:96;
bar(x,y),ylabel('Relative Frequency'),...
    xlabel('Thread Strength (N)'),...
    title('Relative Frequency Histogram for 100 Tests')
```

The result appears in Figure 7.1–3.

The fourth, fifth, and sixth forms of the `hist` function do not generate a plot, but are used to compute the frequency counts and bin locations. The `bar` function can then be used to plot the histogram. The syntax of the fourth form is `[z,x] = hist(y)`, where `z` is the returned vector containing the frequency count and `x` is the returned vector containing the bin locations. The fifth and sixth forms are `[z,x] = hist(y,n)` and `[z,x] = hist(y,x)`. In the latter case the returned vector `x` is the same as the user-supplied vector. The following script file shows how the sixth form can be used to generate a relative frequency histogram for the thread example with 100 tests.

```
tests = 100;
y = [91*ones(1,13),92*ones(1,15),93*ones(1,22),...
    94*ones(1,19),95*ones(1,17),96*ones(1,14);
```

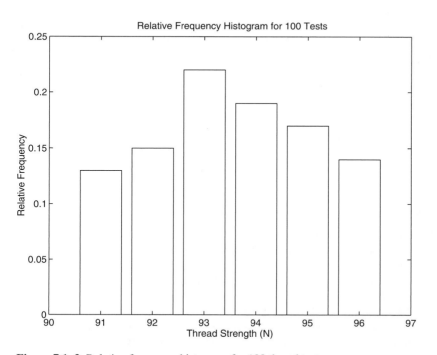

Figure 7.1–3 Relative frequency histogram for 100 thread tests.

Table 7.1–1 Histogram functions

Command	Description
bar(x,y)	Creates a bar chart of y versus x.
hist(y)	Aggregates the data in the vector y into 10 bins evenly spaced between the minimum and maximum values in y.
hist(y,n)	Aggregates the data in the vector y into n bins evenly spaced between the minimum and maximum values in y.
hist(y,x)	Aggregates the data in the vector y into bins whose center locations are specified by the vector x. The bin widths are the distances between the centers.
[z,x] = hist(y)	Same as hist(y) but returns two vectors z and x that contain the frequency count and the bin locations.
[z,x] = hist(y,n)	Same as hist(y,n) but returns two vectors z and x that contain the frequency count and the bin locations.
[z,x] = hist(y,x)	Same as hist(y,x) but returns two vectors z and x that contain the frequency count and the bin locations. The returned vector x is the same as the user-supplied vector x.

```
x = 91:96;
[z,x] = hist(y,x);bar(x,z/tests),...
    ylabel('Relative Frequency'),xlabel('Thread Strength(N)'),...
    title('Relative Frequency Histogram for 100 Tests')
```

The plot generated by this M-file will be identical to that shown in Figure 7.1–3. These commands are summarized in Table 7.1–1.

Test Your Understanding

T7.1–1 In 50 tests of thread, the number of times 91, 92, 93, 94, 95, or 96 N was measured was 7, 8, 10, 6, 12, and 7, respectively. Obtain the absolute and relative frequency histograms.

The Data Statistics Tool

With the Data Statistics tool you can calculate statistics for data and add plots of the statistics to a graph of the data. The tool is accessed from the Figure window after you plot the data. Click on the **Tools** menu, then select **Data Statistics.** The menu appears as shown in Figure 7.1–4. To show the mean of the dependent variable (y) on the plot, click the box in the row labeled mean under the column labeled Y, as shown in the figure. A horizontal line is then placed on the plot at the mean. You can plot other statistics as well; these are shown in the figure. You can save the statistics to the workspace as a structure by clicking on the **Save to Workspace** button. This opens a dialog box that

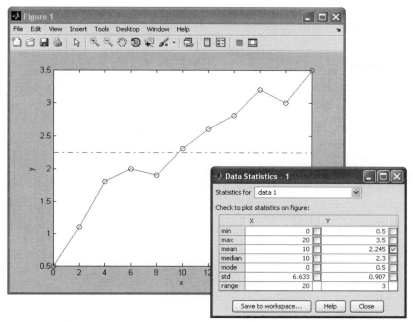

Figure 7.1–4 The Data Statistics tool.

prompts you for a name for the structure containing the x data, and a name for the y data structure.

7.2 The Normal Distribution

Rolling a die is an example of a process whose possible outcomes are a limited set of numbers, namely, the integers from 1 to 6. For such processes the probability is a function of a discrete-valued variable, that is, a variable having a limited number of values. For example, Table 7.2–1 gives the measured heights of 100 men 20 years of age. The heights were recorded to the nearest 1/2 in., so the height variable is discrete-valued.

Scaled Frequency Histogram

You can plot the data as a histogram using either the absolute or relative frequencies. However, another useful histogram uses data scaled so that the total area under the histogram's rectangles is 1. This *scaled frequency histogram* is the absolute frequency histogram divided by the total area of that histogram. The area of each rectangle on the absolute frequency histogram equals the bin width times the absolute frequency for that bin. Because all the rectangles have the same width, the total area is the bin width times the sum of the absolute frequencies. The following M-file produces the scaled histogram shown in Figure 7.2–1.

Table 7.2–1 Height data for men 20 years of age

Height (in.)	Frequency	Height (in.)	Frequency
64	1	70	9
64.5	0	70.5	8
65	0	71	7
65.5	0	71.5	5
66	2	72	4
66.5	4	72.5	4
67	5	73	3
67.5	4	73.5	1
68	8	74	1
68.5	11	74.5	0
69	12	75	1
69.5	10		

```
% Absolute frequency data.
y_abs=[1,0,0,0,2,4,5,4,8,11,12,10,9,8,7,5,4,4,3,1,1,0,1];
binwidth = 0.5;
% Compute scaled frequency data.
area = binwidth*sum(y_abs);
y_scaled = y_abs/area;
```

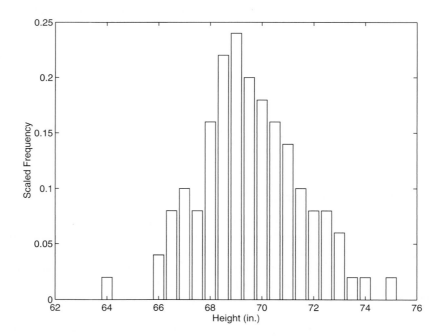

Figure 7.2–1 Scaled histogram of height data.

```
% Define the bins.
bins = 64:binwidth:75;
% Plot the scaled histogram.
bar(bins,y_scaled),...
   ylabel('Scaled Frequency'),xlabel('Height (in.)')
```

Because the total area under the scaled histogram is 1, the fractional area corresponding to a range of heights gives the probability that a randomly selected 20-year-old man will have a height in that range. For example, the heights of the scaled histogram rectangles corresponding to heights of 67 through 69 in. are 0.1, 0.08, 0.16, 0.22, and 0.24. Because the bin width is 0.5, the total area corresponding to these rectangles is $(0.1 + 0.08 + 0.16 + 0.22 + 0.24)(0.5) = 0.4$. Thus 40 percent of the heights lie between 67 and 69 in.

You can use the `cumsum` function to calculate areas under the scaled frequency histogram and therefore to calculate probabilities. If x is a vector, `cumsum(x)` returns a vector the same length as x, whose elements are the sum of the previous elements. For example, if $x = [2, 5, 3, 8]$, `cumsum(x) = [2, 7, 10, 18]`. If A is a matrix, `cumsum(A)` computes the cumulative sum of each row. The result is a matrix the same size as A.

After running the previous script, the last element of `cumsum(y_scaled) * binwidth` is 1, which is the area under the scaled frequency histogram. To compute the probability of a height lying between 67 and 69 in. (that is, above the 6th value up to the 11th value), type

```
>>prob = cumsum(y_scaled)*binwidth;
>>prob67_69 = prob(11)-prob(6)
```

The result is `prob67_69 = 0.4000`, which agrees with our previous calculation of 40 percent.

Continuous Approximation to the Scaled Histogram

For processes having an infinite number of possible outcomes, the probability is a function of a *continuous* variable and is plotted as a curve rather than as rectangles. It is based on the same concept as the scaled histogram; that is, the total area under the curve is 1, and the fractional area gives the probability of occurrence of a specific range of outcomes. A probability function that describes many processes is the *normal* or *Gaussian* function, which is shown in Figure 7.2–2.

This function is also known as the *bell-shaped curve*. Outcomes that can be described by this function are said to be *normally distributed*. The normal probability function is a two-parameter function; one parameter, μ, is the mean of the outcomes, and the other parameter, σ, is the *standard deviation*. The mean μ locates the peak of the curve and is the most likely value to occur. The width, or spread, of the curve is described by the parameter σ. Sometimes the term *variance* is used to describe the spread of the curve. The variance is the square of the standard deviation σ.

NORMAL OR
GAUSSIAN
FUNCTION

NORMALLY
DISTRIBUTED

STANDARD
DEVIATION

VARIANCE

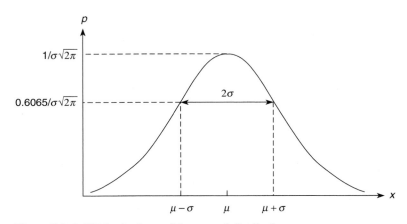

Figure 7.2–2 The basic shape of the normal distribution curve.

The normal probability function is described by the following equation:

$$p(x) = \frac{1}{\sigma\sqrt{2\pi}}\, e^{-(x-\mu)^2/2\sigma^2} \qquad (7.2\text{–}1)$$

It can be shown that approximately 68 percent of the area lies between the limits of $\mu - \sigma \leq x \leq \mu + \sigma$. Consequently, if a variable is normally distributed, there is a 68 percent chance that a randomly selected sample will lie within one standard deviation of the mean. In addition, approximately 96 percent of the area lies between the limits of $\mu - 2\sigma \leq x \leq \mu + 2\sigma$, and 99.7 percent, or practically 100 percent, of the area lies between the limits of $\mu - 3\sigma \leq x \leq \mu + 3\sigma$.

The functions mean(x), var(x), and std(x) compute the mean, variance, and standard deviation of the elements in the vector x.

EXAMPLE 7.2–1   Mean and Standard Deviation of Heights

Statistical analysis of data on human proportions is required in many engineering applications. For example, designers of submarine crew quarters need to know how small they can make bunk lengths without eliminating a large percentage of prospective crew members. Use MATLAB to estimate the mean and standard deviation for the height data given in Table 7.2–1.

■ **Solution**
The script file follows. The data given in Table 7.2–1 are the absolute frequency data and are stored in the vector y_abs. A bin width of 1/2 in. is used because the heights were measured to the nearest 1/2 in. The vector bins contains the heights in 1/2 in. increments.

To compute the mean and standard deviation, reconstruct the original (raw) height data from the absolute frequency data. Note that these data have some zero entries. For example, none of the 100 men had a height of 65 in. Thus to reconstruct the raw data, start with an empty vector y_raw and fill it with the height data obtained from the absolute frequencies. The `for` loop checks to see whether the absolute frequency for a particular bin is nonzero. If it is nonzero, append the appropriate number of data values to the vector `y_raw`. If the particular bin frequency is 0, `y_raw` is left unchanged.

```
% Absolute frequency data.
y_abs = [1,0,0,0,2,4,5,4,8,11,12,10,9,8,7,5,4,4,3,1,1,0,1];
binwidth = 0.5;
% Define the bins.
bins = [64:binwidth:75];
% Fill the vector y_raw with the raw data.
% Start with an empty vector.
y_raw = [];
for i = 1:length(y_abs)
    if y_abs(i)>0
        new = bins(i)*ones(1,y_abs(i));
    else
        new = [];
    end
y_raw = [y_raw,new];
end
% Compute the mean and standard deviation.
mu = mean(y_raw),sigma = std(y_raw)
```

When you run this program, you will find that the mean is $\mu = 69.6$ in. and the standard deviation is $\sigma = 1.96$ in.

If you need to compute probabilities based on the normal distribution, you can use the `erf` function. Typing `erf(x)` returns the area to the left of the value $t = x$ under the curve of the function $2e^{-t^2}/\sqrt{\pi}$. This area, which is a function of x, is known as the *error function* and is written as erf(x). The probability that the random variable x is less than or equal to b is written as $P(x \le b)$ if the outcomes are normally distributed. This probability can be computed from the error function as follows:

ERROR FUNCTION

$$P(x \le b) = \frac{1}{2}\left[1 + \text{erf}\left(\frac{b - \mu}{\sigma\sqrt{2}}\right)\right] \qquad (7.2-2)$$

The probability that the random variable x is no less than a and no greater than b is written as $P(a \le x \le b)$. It can be computed as follows:

$$P(a \le x \le b) = \frac{1}{2}\left[\text{erf}\left(\frac{b - \mu}{\sigma\sqrt{2}}\right) - \text{erf}\left(\frac{a - \mu}{\sigma\sqrt{2}}\right)\right] \qquad (7.2-3)$$

Estimation of Height Distribution

EXAMPLE 7.2–2

Use the results of Example 7.2–1 to estimate how many 20-year-old men are no taller than 68 in. How many are within 3 in. of the mean?

■ **Solution**

In Example 7.2–1 the mean and standard deviation were found to be $\mu = 69.3$ in. and $\sigma = 1.96$ in. In Table 7.2–1, note that few data points are available for heights less than 68 in. However, if you assume that the heights are normally distributed, you can use Equation (7.2–2) to estimate how many men are shorter than 68 in. Use (7.2–2) with $b = 68$, that is,

$$P(x \le 68) = \frac{1}{2}\left[1 + \mathrm{erf}\left(\frac{68 - 69.3}{1.96\sqrt{2}}\right)\right]$$

To determine how many men are within 3 in. of the mean, use Equation (7.2–3) with $a = \mu - 3 = 66.3$ and $b = \mu + 3 = 72.3$, that is,

$$P(66.3 \le x \le 72.3) = \frac{1}{2}\left[\mathrm{erf}\left(\frac{3}{1.96\sqrt{2}}\right) - \mathrm{erf}\left(\frac{-3}{1.96\sqrt{2}}\right)\right]$$

In MATLAB these expressions are computed in a script file as follows:

```
mu = 69.3;
s = 1.96;
% How many are no taller than 68 inches?
b1 = 68;
P1 = (1+erf((b1-mu)/(s*sqrt(2))))/2
% How many are within 3 inches of the mean?
a2 = 66.3;
b2 = 72.3;
P2 = (erf((b2-mu)/(s*sqrt(2)))-erf((a2-mu)/(s*sqrt(2))))/2
```

When you run this program, you obtain the results P1 = 0.2536 and P2 = 0.8741. Thus 25 percent of 20-year-old men are estimated to be 68 in. or less in height, and 87 percent are estimated to be between 66.3 and 72.3 in. tall.

Test Your Understanding

T7.2–1 Suppose that 10 more height measurements are obtained so that the following numbers must be *added* to Table 7.2–1.

Height (in.)	Additional data
64.5	1
65	2
66	1
67.5	2
70	2
73	1
74	1

(a) Plot the scaled frequency histogram. (b) Find the mean and standard deviation. (c) Use the mean and standard deviation to estimate how many

20-year-old men are no taller than 69 in. (d) Estimate how many are between 68 and 72 in. tall.

(Answers: (b) mean = 69.4 in., standard deviation = 2.14 in.; (c) 43 percent; (d) 63 percent.)

Sums and Differences of Random Variables

It can be proved that the mean of the sum (or difference) of two independent normally distributed random variables equals the sum (or difference) of their means, but the variance is always the sum of the two variances. That is, if x and y are normally distributed with means μ_x and μ_y and variances σ_x^2 and σ_y^2, and if $u = x + y$ and $v = x - y$, then

$$\mu_u = \mu_x + \mu_y \tag{7.2-4}$$

$$\mu_v = \mu_x - \mu_y \tag{7.2-5}$$

$$\sigma_u^2 = \sigma_v^2 = \sigma_x^2 + \sigma_y^2 \tag{7.2-6}$$

These properties are applied in some of the homework problems.

7.3 Random Number Generation

We often do not have a simple probability distribution to describe the distribution of outcomes in many engineering applications. For example, the probability that a circuit consisting of many components will fail is a function of the number and the age of the components, but we often cannot obtain a function to describe the failure probability. In such cases we often resort to simulation to make predictions. The simulation program is executed many times, using a random set of numbers to represent the failure of one or more components, and the results are used to estimate the desired probability.

Uniformly Distributed Numbers

In a sequence of *uniformly distributed* random numbers, all values within a given interval are equally likely to occur. The MATLAB function `rand` generates random numbers uniformly distributed over the interval $[0, 1]$. Type `rand` to obtain a single random number in the interval $[0, 1]$. Typing `rand` again generates a different number because the MATLAB algorithm used for the `rand` function requires a "state" to start. MATLAB obtains this state from the computer's CPU clock. Thus every time the `rand` function is used, a different result will be obtained. For example,

```
rand
ans =
    0.6161
rand
ans =
    0.5184
```

Type `rand(n)` to obtain an $n \times n$ matrix of uniformly distributed random numbers in the interval $[0, 1]$. Type `rand(m,n)` to obtain an $m \times n$ matrix of random numbers. For example, to create a 1×100 vector `y` having 100 random values in the interval $[0, 1]$, type `y = rand(1,100)`. Using the `rand` function this way is equivalent to typing `rand` 100 times. Even though there is a single call to the `rand` function, the `rand` function's calculation has the effect of using a different state to obtain each of the 100 numbers so that they will be random.

Use `Y = rand(m,n,p,...)` to generate a multidimensional array `Y` having random elements. Typing `rand(size(A))` produces an array of random entries that is the same size as `A`.

For example, the following script makes a random choice between two equally probable alternatives.

```
if rand < 0.5
   disp('heads')
else
   disp('tails')
end
```

To compare the results of two or more simulations, sometimes you will need to generate the same sequence of random numbers each time the simulation runs. To generate the same sequence, you must use the same state each time. The current state `s` of the uniform number generator can be obtained by typing `s = rand('twister')`. This returns a vector containing the current state of the uniform generator. To set the state of the generator to `s`, type `rand('twister',s)`. Typing `rand('twister',0)` resets the generator to its initial state. Typing `rand('twister',j)`, for integer `j`, resets the generator to state `j`. Typing `rand('twister',sum(100*clock))` resets the generator to a different state each time. Table 7.3–1 summarizes these functions.

The name `'twister'` refers to the specific algorithm used by MATLAB to generate random numbers. In MATLAB Version 4, `'seed'` was used instead of `'twister'`. In Versions 5 through 7.3, `'state'` was used. Use `'twister'` in Version 7.4 and later. The following session shows how to obtain the same sequence every time `rand` is called.

```
>>rand('twister',0)
>>rand
ans =
     0.5488
>>rand
ans =
     0.7152
>>rand('twister',0)
>>rand
```

Table 7.3–1 Random number functions

Command	Description
rand	Generates a single uniformly distributed random number between 0 and 1.
rand(n)	Generates an $n \times n$ matrix containing uniformly distributed random numbers between 0 and 1.
rand(m,n)	Generates an $m \times n$ matrix containing uniformly distributed random numbers between 0 and 1.
s = rand('state')	Returns a vector s containing the current state of the uniformly distributed generator.
rand('twister',s)	Sets the state of the uniformly distributed generator to s.
rand('twister',0)	Resets the uniformly distributed generator to its initial state.
rand('twister',j)	Resets the uniformly distributed generator to state j, for integer j.
rand('twister',sum(100*clock))	Resets the uniformly distributed generator to a different state each time it is executed.
randn	Generates a single normally distributed random number having a mean of 0 and a standard deviation of 1.
randn(n)	Generates an $n \times n$ matrix containing normally distributed random numbers having a mean of 0 and a standard deviation of 1.
randn(m,n)	Generates an $m \times n$ matrix containing normally distributed random numbers having a mean of 0 and a standard deviation of 1.
s = randn('state')	Like rand('state') but for the normally distributed generator.
randn('state',s)	Like rand('state',s) but for the normally distributed generator.
randn('state',0)	Like rand('state',0) but for the normally distributed generator.
randn('state',j)	Like rand('state',j) but for the normally distributed generator.
randn('state',sum(100*clock))	Like rand('state',sum(100*clock)) but for the normally distributed generator.
randperm(n)	Generates a random permutation of the integers from 1 to n.

```
ans =
    0.5488
>>rand
ans =
    0.7152
```

You need not start with the initial state to generate the same sequence. To show this, continue the above session as follows.

```
>>s = rand('twister');
>>rand('twister',s)
>>rand
ans =
    0.6028
>>rand('twister',s)
>>rand
ans =
    0.6028
```

You can use the `rand` function to generate random numbers in an interval other than [0, 1]. For example, to generate values in the interval [2, 10], generate a random number between 0 and 1, multiply it by 8 (the difference between the upper and lower bounds), and add the lower bound (2). The result is a value that is uniformly distributed in the interval [2, 10]. The general formula for generating a uniformly distributed random number y in the interval $[a, b]$ is

$$y = (b - a)x + a \qquad (7.3\text{--}1)$$

where x is a random number uniformly distributed in the interval [0, 1]. For example, to generate a vector y containing 1000 uniformly distributed random numbers in the interval [2, 10], you type `y = 8*rand(1,1000) + 2`. You can check the results with the `mean`, `min`, and `max` functions. You should obtain values close to 6, 2, and 10, respectively.

You can use `rand` to generate random results for games involving dice, for example, but you must use it to create integers. An easier way is to use the `randperm(n)` function, which generates a random permutation of the integers from 1 to n. For example, `randperm(6)` might generate the vector [3 2 6 4 1 5], or some other permutation of the numbers from 1 to 6. Note that `randperm` calls `rand` and therefore changes the state of the generator.

With `twister` you control the internal state of the random number stream used by `rand` and `randn`. Starting with MATLAB Version 7.7, the use of twister is still supported for backwards compatibility but will be discontinued eventually. For Version 7.7 and higher, you can generate the random number stream using `Randstream`, which is an advanced topic. See the MATLAB documentation.

Normally Distributed Random Numbers

In a sequence of normally distributed random numbers, the values near the mean are more likely to occur. We have noted that the outcomes of many processes can be described by the normal distribution. Although a uniformly distributed random variable has definite upper and lower bounds, a normally distributed random variable does not.

The MATLAB function `randn` will generate a single number that is normally distributed with a mean equal to 0 and a standard deviation equal to 1. Type `randn(n)` to obtain an $n \times n$ matrix of such numbers. Type `randn(m,n)` to obtain an $m \times n$ matrix of random numbers.

The functions for retrieving and specifying the state of the normally distributed random number generator are identical to those for the uniformly distributed generator, except that `randn(...)` replaces `rand(...)` in the syntax and `'state'` is used instead of `'twister'`. These functions are summarized in Table 7.3–1.

You can generate a sequence of normally distributed numbers having a mean μ and standard deviation σ from a normally distributed sequence having a mean of 0 and a standard deviation of 1. You do this by multiplying the values by σ and adding μ to each result. Thus if x is a random number with a mean of 0 and a standard deviation of 1, use the following equation to generate a new random number y having a standard deviation of σ and a mean of μ.

$$y = \sigma x + \mu \qquad (7.3\text{--}2)$$

For example, to generate a vector `y` containing 2000 random numbers normally distributed with a mean of 5 and a standard deviation of 3, you type `y = 3*randn(1,2000) + 5`. You can check the results with the `mean` and `std` functions. You should obtain values close to 5 and 3, respectively.

Test Your Understanding

T7.3–1 Use MATLAB to generate a vector `y` containing 1800 random numbers normally distributed with a mean of 7 and a standard deviation of 10. Check your results with the `mean` and `std` functions. Why can't you use the `min` and `max` functions to check your results?

Functions of Random Variables If y and x are linearly related as

$$y = bx + c \qquad (7.3\text{--}3)$$

and if x is normally distributed with a mean μ_x and standard deviation σ_x, it can be shown that the mean and standard deviation of y are given by

$$\mu_y = b\mu_x + c \qquad (7.3\text{--}4)$$

$$\sigma_y = |b|\sigma_x \qquad (7.3\text{--}5)$$

However, it is easy to see that the means and standard deviations do not combine in a straightforward fashion when the variables are related by a nonlinear function. For example, if x is normally distributed with a mean of 0, and if $y = x^2$, it is easy to see that the mean of y is not 0, but is positive. In addition, y is not normally distributed.

Some advanced methods are available for deriving a formula for the mean and variance of $y = f(x)$, but for our purposes, the simplest way is to use random number simulation.

It was noted in the previous section that the mean of the sum (or difference) of two independent normally distributed random variables equals the sum (or difference) of their means, but the variance is always the sum of the two variances. However, if z is a nonlinear function of x and y, then the mean and variance of z cannot be found with a simple formula. In fact, the distribution of z will not even be normal. This outcome is illustrated by the following example.

Statistical Analysis and Manufacturing Tolerances

EXAMPLE 7.3–1

Suppose you must cut a triangular piece off the corner of a square plate by measuring the distances x and y from the corner (see Figure 7.3–1). The desired value of x is 10 in., and

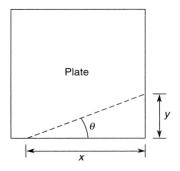

Figure 7.3–1 Dimensions of a triangular cut.

the desired value of θ is $20°$. This requires that $y = 3.64$ in. We are told that measurements of x and y are normally distributed with means of 10 and 3.64, respectively, with a standard deviation equal to 0.05 in. Determine the standard deviation of θ and plot the relative frequency histogram for θ.

■ Solution
From Figure 7.3–1, we see that the angle θ is determined by $\theta = \tan^{-1}(y/x)$. We can find the statistical distribution of θ by creating random variables x and y that have means of 10 and 3.64, respectively, with a standard deviation of 0.05. The random variable θ is then found by calculating $\theta = \tan^{-1}(y/x)$ for each random pair (x, y). The following script file shows this procedure.

```
s = 0.05; % standard deviation of x and y
n = 8000; % number of random simulations
x = 10 + s*randn(1,n);
y = 3.64 + s*randn(1,n);
theta = (180/pi)*atan(y./x);
mean_theta = mean(theta)
sigma_theta = std(theta)
xp = 19:0.1:21;
z = hist(theta,xp);
yp = z/n;
bar(xp,yp),xlabel('Theta (degrees)'),...
    ylabel('Relative Frequency')
```

The choice of 8000 simulations was a compromise between accuracy and the amount of time required to do the calculations. You should try different values of n and compare the results. The results gave a mean of $19.9993°$ for θ with a standard deviation of $0.2730°$. The histogram is shown in Figure 7.3–2. Although the plot resembles the normal distribution, the values of θ are not distributed normally. From the histogram we can calculate that approximately 65 percent of the values of θ lie between 19.8 and 20.2. This range corresponds to a standard deviation of $0.2°$, not $0.273°$ as calculated from the simulation data. Thus the curve is not a normal distribution.

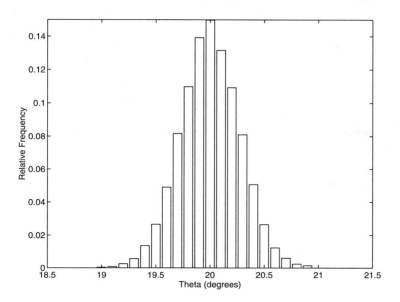

Figure 7.3–2 Scaled histogram of the angle θ.

This example shows that the interaction of two of more normally distributed variables does not produce a result that is normally distributed. In general, the result is normally distributed if and only if the result is a linear combination of the variables.

7.4 Interpolation

Paired data might represent a *cause and effect,* or *input-output relationship,* such as the current produced in a resistor as a result of an applied voltage, or a *time history,* such as the temperature of an object as a function of time. Another type of paired data represents a *profile,* such as a road profile (which shows the height of the road along its length). In some applications we want to estimate a variable's value between the data points. This process is called *interpolation.* In other cases we might need to estimate the variable's value outside the given data range. This process is called *extrapolation.* Interpolation and extrapolation are greatly aided by plotting the data. Such plots, some perhaps using logarithmic axes, often help to discover a functional description of the data.

Suppose we have the following temperature measurements, taken once an hour starting at 7:00 A.M. The measurements at 8 and 10 A.M. are missing for some reason, perhaps because of equipment malfunction.

Time	7 A.M.	9 A.M.	11 A.M.	12 noon
Temperature (°F)	49	57	71	75

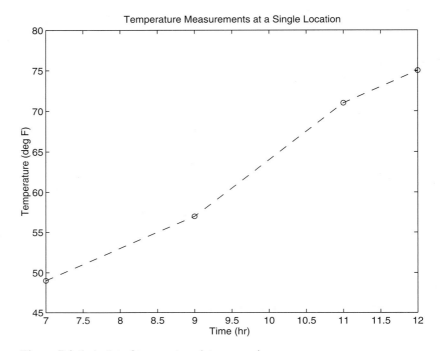

Figure 7.4–1 A plot of temperature data versus time.

A plot of these data is shown in Figure 7.4–1 with the data points connected by dashed lines. If we need to estimate the temperature at 10 A.M., we can read the value from the dashed line that connects the data points at 9 and 11 A.M. From the plot we thus estimate the temperature at 8 A.M. to be 53°F and at 10 A.M. to be 64°F. We have just performed *linear interpolation* on the data to obtain an *estimate* of the missing data. Linear interpolation is so named because it is equivalent to connecting the data points with a linear function (a straight line).

Of course we have no reason to believe that the temperature follows the straight lines shown in the plot, and our estimate of 64°F will most likely be incorrect, but it might be close enough to be useful. Using straight lines to connect the data points is the simplest form of interpolation. Another function could be used if we have a good reason to do so. Later in this section we use polynomial functions to do the interpolation.

Linear interpolation in MATLAB is obtained with the `interp1` and `interp2` functions. Suppose that `x` is a vector containing the independent variable data and that `y` is a vector containing the dependent variable data. If `x_int` is a vector containing the value or values of the independent variable at which we wish to estimate the dependent variable, then typing `interp1(x,y,x_int)` produces a vector the same size as `x_int` containing the interpolated values of y that correspond to x_int. For example, the following session produces an estimate of the temperatures at 8 and 10 A.M. from the preceding data. The vectors x and y contain the times and temperatures, respectively.

```
>>x = [7, 9, 11, 12];
>>y = [49, 57, 71, 75];
>>x_int = [8, 10];
>>interp1(x,y,x_int)
ans =
     53
     64
```

You must keep in mind two restrictions when using the interp1 function. The values of the independent variable in the vector x must be in ascending order, and the values in the interpolation vector x_int must lie within the range of the values in x. Thus we cannot use the interp1 function to estimate the temperature at 6 A.M., for example.

The interp1 function can be used to interpolate in a table of values by defining y to be a matrix instead of a vector. For example, suppose that we now have temperature measurements at three locations and the measurements at 8 and 10 A.M. are missing for all three locations. The data are as follows:

	Temperature (°F)		
Time	**Location 1**	**Location 2**	**Location 3**
7 A.M.	49	52	54
9 A.M.	57	60	61
11 A.M.	71	73	75
12 noon	75	79	81

We define x as before, but now we define y to be a matrix whose three columns contain the second, third, and fourth columns of the preceding table. The following session produces an estimate of the temperatures at 8 and 10 A.M. at each location.

```
>>x = [7, 9, 11, 12]';
>>y(:,1) = [49, 57, 71, 75]';
>>y(:,2) = [52, 60, 73, 79]';
>>y(:,3) = [54, 61, 75, 81]';
>>x_int = [8, 10]';
>>interp1(x,y,x_int)
ans =
     53.0000    56.0000    57.5000
     64.0000    65.5000    68.0000
```

Thus the estimated temperatures at 8 A.M. at each location are 53, 56, and 57.5°F, respectively. At 10 A.M. the estimated temperatures are 64, 65.5, and 68°F. From this example we see that if the first argument x in the `interp1(x,y,x_int)` function is a *vector* and the second argument y is a *matrix,* then the function interpolates between the rows of y and computes a matrix having the same number of columns as y and the same number of rows as the number of values in `x_int`.

Note that we need not define two separate vectors x and y. Rather, we can define a single matrix that contains the entire table. For example, by defining the matrix `temp` to be the preceding table, the session will look like this:

```
>>temp(:,1)  =  [7, 9, 11, 12]';
>>temp(:,2)  =  [49, 57, 71, 75]';
>>temp(:,3)  =  [52, 60, 73, 79]';
>>temp(:,4)  =  [54, 61, 75, 81]';
>>x_int = [8, 10]';
>>interp1(temp(:,1),temp(:,2:4),x_int)
ans =
      53.0000       56.0000       57.5000
      64.0000       65.5000       68.0000
```

Two-Dimensional Interpolation

Now suppose that we have temperature measurements at four locations at 7 A.M. These locations are at the corners of a rectangle 1 mi wide and 2 mi long. Assigning a coordinate system origin (0,0) to the first location, the coordinates of the other locations are (1, 0), (1, 2), and (0, 2); see Figure 7.4–2. The temperature measurements are shown in the figure. The temperature is a function of two

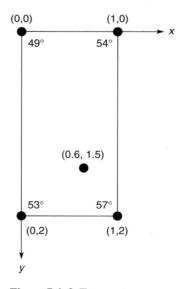

Figure 7.4–2 Temperature measurements at four locations.

variables, the coordinates x and y. MATLAB provides the `interp2` function to interpolate functions of two variables. If the function is written as $z = f(x, y)$ and we wish to estimate the value of z for $x = x_i$ and $y = y_i$, the syntax is `interp2(x,y,z,x_i,y_i)`.

Suppose we want to estimate the temperature at the point whose coordinates are $(0.6, 1.5)$. Put the x coordinates in the vector x and the y coordinates in the vector y. Then put the temperature measurements in a matrix z such that going across a row represents an increase in x and going down a column represents an increase in y. The session to do this is as follows:

```
>>x = [0,1];
>>y = [0,2];
>>z = [49,54;53,57]
z =
    49    54
    53    57
>>interp2(x,y,z,0.6,1.5)
ans =
    54.5500
```

Thus the estimated temperature is 54.55°F.

The syntax of the `interp1` and `interp2` functions is summarized in Table 7.4–1. MATLAB also provides the `interpn` function for interpolating multidimensional arrays.

Cubic Spline Interpolation

High-order polynomials can exhibit undesired behavior between the data points, and this can make them unsuitable for interpolation. A widely used alternative procedure is to fit the data points using a lower-order polynomial between *each pair* of adjacent data points. This method is called *spline* interpolation and is so named for the splines used by illustrators to draw a smooth curve through a set of points.

Spline interpolation obtains an exact fit that is also smooth. The most common procedure uses cubic polynomials, called *cubic splines*, and thus is called

Table 7.4–1 Linear interpolation functions

Command	Description
`y_int=interp1(x,y,x_int)`	Used to linearly interpolate a function of one variable: $y = f(x)$. Returns a linearly interpolated vector `y_int` at the specified value `x_int`, using data stored in `x` and `y`.
`z_int=interp2(x,y,z,x_,y_int)`	Used to linearly interpolate a function of two variables: $y = f(x, y)$. Returns a linearly interpolated vector `z_int` at the specified values `x_int` and `y_int`, using data stored in `x`, `y`, and `z`.

cubic spline interpolation. If the data are given as n pairs of (x, y) values, then $n - 1$ cubic polynomials are used. Each has the form

$$y_i(x) = a_i(x - x_i)^3 + b_i(x - x_i)^2 + c_i(x - x_i) + d_i$$

for $x_i \leq x \leq x_{i+1}$ and $i = 1, 2, \ldots, n - 1$. The coefficients a_i, b_i, c_i, and d_i for each polynomial are determined so that the following three conditions are satisfied for each polynomial:

1. The polynomial must pass through the data points at its endpoints at x_i and x_{i+1}.
2. The slopes of adjacent polynomials must be equal at their common data point.
3. The curvatures of adjacent polynomials must be equal at their common data point.

For example, a set of cubic splines for the temperature data given earlier follows (y represents the temperature values, and x represents the hourly values). The data are repeated here.

x	7	9	11	12
y	49	57	71	75

We will shortly see how to use MATLAB to obtain these polynomials. For $7 \leq x \leq 9$,

$$y_1(x) = -0.35(x - 7)^3 + 2.85(x - 7)^2 - 0.3(x - 7) + 49$$

For $9 \leq x \leq 11$,

$$y_2(x) = -0.35(x - 9)^3 + 0.75(x - 9)^2 + 6.9(x - 9) + 57$$

For $11 \leq x \leq 12$,

$$y_3(x) = -0.35(x - 11)^3 - 1.35(x - 11)^2 + 5.7(x - 11) + 71$$

MATLAB provides the `spline` command to obtain a cubic spline interpolation. Its syntax is `y_int = spline(x,y,x_int)`, where `x` and `y` are vectors containing the data and `x_int` is a vector containing the values of the independent variable x at which we wish to estimate the dependent variable y. The result `y_int` is a vector the same size as `x_int` containing the interpolated values of y that correspond to `x_int`. The spline fit can be plotted by plotting the vectors `x_int` and `y_int`. For example, the following session produces and plots a cubic spline fit to the preceding data, using an increment of 0.01 in the x values.

```
>>x = [7,9,11,12];
>>y = [49,57,71,75];
>>x_int = 7:0.01:12;
>>y_int = spline(x,y,x_int);
>>plot(x,y,'o',x,y,'--',x_int,y_int),...
    xlabel('Time (hr)'),ylabel('Temperature (deg F)'), ...

    title('Measurements at a Single Location'), ...
    axis([7 12 45 80])
```

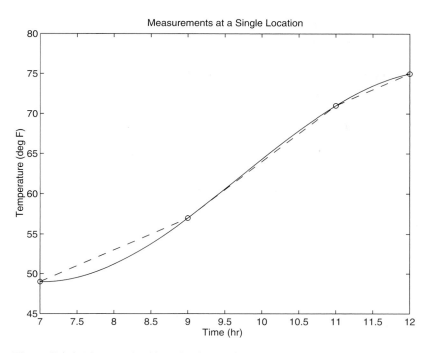

Figure 7.4–3 Linear and cubic-spline interpolation of temperature data.

The plot is shown in Figure 7.4–3. The dashed lines represent linear interpolation, and the solid curve is the cubic spline. If we evaluate the spline polynomial at $x = 8$, we obtain $y(8) = 51.2°F$. This estimate is different from the $53°F$ estimate obtained from linear interpolation. It is impossible to say which estimate is more accurate without having greater understanding of the temperature dynamics.

We can obtain an estimate more quickly by using the following variation of the `interp1` function.

```
y_est = interp1(x,y,x_est,'spline')
```

In this form the function returns a column vector `y_est` that contains the estimated values of y that correspond to the x values specified in the vector `x_est`, using cubic spline interpolation.

In some applications it is helpful to know the polynomial coefficients, but we cannot obtain the spline coefficients from the `interp1` function. However, we can use the form

```
[breaks, coeffs, m, n] = unmkpp(spline(x,y))
```

to obtain the coefficients of the cubic polynomials. The vector `breaks` contains the x values of the data, and the matrix `coeffs` is an $m \times n$ matrix containing the coefficients of the polynomials. The scalars m and n give the dimensions of the matrix coeffs; m is the number of polynomials, and n is the number of coefficients for each polynomial (MATLAB will fit a lower-order polynomial if possible, so there can be fewer than four coefficients). For example, using the same data, the following session produces the coefficients of the polynomials given earlier:

```
>>x = [7,9,11,12];
>>y = [49,57,71,75];
>> [breaks, coeffs, m, n] = unmkpp(spline(x,y))
breaks =
    7    9   11    12
coeffs =
   -0.3500   2.8500  -0.3000  49.0000
   -0.3500   0.7500   6.900   57.0000
   -0.3500  -1.3500   5.7000  71.0000
m =
    3
n =
    4
```

The first row of the matrix coeffs contains the coefficients of the first polynomial, and so on. These functions are summarized in Table 7.4–2. The Basic

Table 7.4–2 Polynomial interpolation functions

Command	Description
y_est = interp1(x,y,x_est, method)	Returns a column vector y_est that contains the estimated values of y that correspond to the x values specified in the vector x_est, using interpolation specified by method. The choices for method are 'nearest', 'linear', 'spline', 'pchip', and 'cubic'.
y_int = spline(x,y,x_int)	Computes a cubic spline interpolation where x and y are vectors containing the data and x_int is a vector containing the values of the independent variable x at which we wish to estimate the dependent variable y. The result y_int is a vector the same size as x_int containing the interpolated values of y that correspond to x_int.
y_int = pchip (x,y,x_int)	Similar to spline but uses piecewise cubic Hermite polynomials for interpolation to preserve shape and respect monotonicity.
[breaks, coeffs, m, n] = unmkpp(spline(x,y))	Computes the coefficients of the cubic spline polynomials for the data in x and y. The vector breaks contains the x values, and the matrix coeffs is an $m \times n$ matrix containing the polynomial coefficients. The scalars m and n give the dimensions of the matrix coeffs; m is the number of polynomials, and n is the number of coefficients for each polynomial.

Fitting interface, which is available on the **Tools** menu of the Figure window, can be used for cubic spline interpolation. See Section 6.3 for instructions for using the interface.

As another example of interpolation, consider 10 evenly spaced data points generated by the function $y = 1/(3 - 3x + x^2)$ over the range $0 \le x \le 4$. The top graph in Figure 7.4–4 shows the results of fitting a cubic polynomial and an eighth-order polynomial to the data. Clearly the cubic is not suitable for interpolation. As we increase the order of the fitted polynomial, we find that the polynomial does not pass through all the data points if the order is less than 7. However, there are two problems with the eighth-order polynomial: we should not use it to interpolate over the interval $0 < x < 0.5$, and its coefficients must be stored with very high accuracy if we use the polynomial to interpolate. The

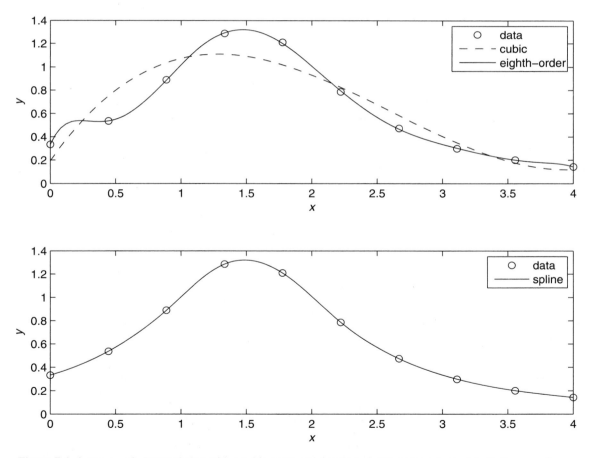

Figure 7.4–4 *Top graph:* Interpolation with a cubic polynomial and an eighth-order polynomial. *Bottom graph:* Interpolation with a cubic spline.

bottom graph in Figure 7.4–4 shows the results of fitting a cubic spline, which is clearly a better choice here.

Interpolation with Hermite Polynomials

The `pchip` function uses piecewise continuous Hermite interpolation polynomials (pchips). Its syntax is identical to that of the `spline` function. With `pchip` the slopes at the data points are computed to preserve the "shape" of the data and to "respect" monotonicity. That is, the fitted function will be monotonic on intervals where the data are monotonic and will have a local extremum on intervals where the data have a local extremum. The differences between the two functions are that

- The second derivatives are continuous with `spline` but may be discontinuous with `pchip`, so `spline` may give a smoother function.
- Therefore, the `spline` function is more accurate if the data are "smoother."
- There are no overshoots and less oscillation in the function produced by `pchip`, even if the data are not smooth.

Consider the data given by $x = [0, 1, 2, 3, 4, 5]$ and $y = [0, -10, 60, 40, 41, 47]$. The top graph in Figure 7.4–5 shows the results of fitting a fifth-order polynomial and a cubic spline to the data. Clearly the fifth-order polynomial is less suitable for interpolation because of the large excursions it makes, especially over the ranges $0 < x < 1$ and $4 < x < 5$. These excursions are often seen with high-order polynomials. Here, the cubic spline is more useful. The bottom graph in Figure 7.4–5 compares the results of a cubic spline fit with a piecewise continuous Hermite polynomial fit (using `pchip`), which is clearly a better choice here.

MATLAB provides a number of other functions to support interpolation for three-dimensional data. See `griddata`, `griddata3`, `griddatan`, `interp3`, and `interpn` in the MATLAB Help.

7.5 Summary

This chapter introduces MATLAB functions that have widespread and important uses in statistics and data analysis. Section 7.1 gives an introduction to basic statistics and probability, including histograms, which are specialized plots for displaying statistical results. The normal distribution that forms the basis of many statistical methods is covered in Section 7.2. Section 7.3 covers random number generators and their use in simulation programs. Section 7.4 covers interpolation methods, including linear and spline interpolation.

Now that you have finished this chapter, you should be able to use MATLAB to

- Solve basic problems in statistics and probability.
- Create simulations incorporating random processes.
- Apply interpolation to data.

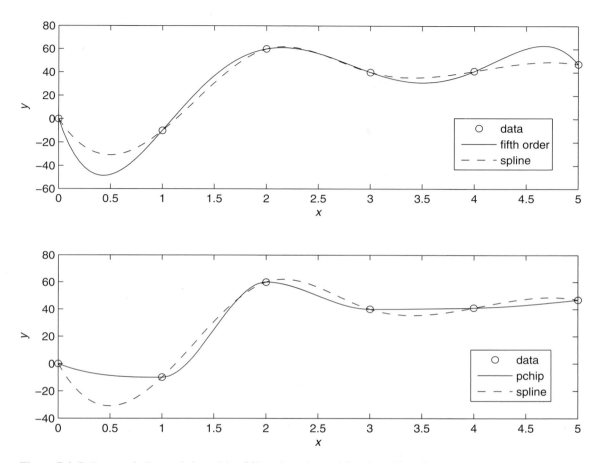

Figure 7.4–5 *Top graph:* Interpolation with a fifth-order polynomial and a cubic spline. *Bottom graph:* Interpolation with piecewise continuous Hermite polynomials (pchip) and a cubic spline.

Key Terms with Page References

Absolute frequency,   297
Bins,   296
Cubic splines,   317
Error function,   305
Gaussian function,   303
Histogram,   296
Interpolation,   313
Mean,   296
Median,   296

Mode,   296
Normally distributed,   303
Normal or Gaussian function,   303
Relative frequency,   297
Scaled frequency histogram,   301
Standard deviation,   303
Uniformly distributed,   307
Variance,   303

Problems

You can find the answers to problems marked with an asterisk at the end of the text.

Section 7.1

1. The following list gives the measured gas mileage in miles per gallon for 22 cars of the same model. Plot the absolute frequency histogram and the relative frequency histogram.

23	25	26	25	27	25	24	22	23	25	26
26	24	24	22	25	26	24	24	24	27	23

2. Thirty pieces of structural timber of the same dimensions were subjected to an increasing lateral force until they broke. The measured force in pounds required to break them is given in the following list. Plot the absolute frequency histogram. Try bin widths of 50, 100, and 200 lb. Which gives the most meaningful histogram? Try to find a better value for the bin width.

243	236	389	628	143	417	205
404	464	605	137	123	372	439
497	500	535	577	441	231	675
132	196	217	660	569	865	725
457	347					

3. The following list gives the measured breaking force in newtons for a sample of 60 pieces of certain type of cord. Plot the absolute frequency histogram. Try bin widths of 10, 30, and 50 N. Which gives the most meaningful histogram? Try to find a better value for the bin width.

311	138	340	199	270	255	332	279	231	296	198	269
257	236	313	281	288	225	216	250	259	323	280	205
279	159	276	354	278	221	192	281	204	361	321	282
254	273	334	172	240	327	261	282	208	213	299	318
356	269	355	232	275	234	267	240	331	222	370	226

Section 7.2

4. For the data given in Problem 1:
   a. Plot the scaled frequency histogram.
   b. Compute the mean and standard deviation and use them to estimate the lower and upper limits of gas mileage corresponding to 68 percent of cars of this model. Compare these limits with those of the data.

5. For the data given in Problem 2:
   a. Plot the scaled frequency histogram.
   b. Compute the mean and standard deviation and use them to estimate the lower and upper limits of strength corresponding to 68 and 96 percent of such timber pieces. Compare these limits with those of the data.

6. For the data given in Problem 3:
   a. Plot the scaled frequency histogram.
   b. Compute the mean and standard deviation, and use them to estimate the lower and upper limits of breaking force corresponding to 68 and 96 percent of cord pieces of this type. Compare these limits with those of the data.

7.* Data analysis of the breaking strength of a certain fabric shows that it is normally distributed with a mean of 300 lb and a variance of 9.
   a. Estimate the percentage of fabric samples that will have a breaking strength no less than 294 lb.
   b. Estimate the percentage of fabric samples that will have a breaking strength no less than 297 lb and no greater than 303 lb.

8. Data from service records show that the time to repair a certain machine is normally distributed with a mean of 65 min and a standard deviation of 5 min. Estimate how often it will take more than 75 min to repair a machine.

9. Measurements of a number of fittings show that the pitch diameter of the thread is normally distributed with a mean of 8.007 mm and a standard deviation of 0.005 mm. The design specifications require that the pitch diameter be 8 ± 0.01 mm. Estimate the percentage of fittings that will be within tolerance.

10. A certain product requires that a shaft be inserted into a bearing. Measurements show that the diameter d_1 of the cylindrical hole in the bearing is normally distributed with a mean of 3 cm and a variance of 0.0064. The diameter d_2 of the shaft is normally distributed with a mean of 2.96 cm and a variance of 0.0036.
    a. Compute the mean and the variance of the clearance $c = d_1 - d_2$.
    b. Find the probability that a given shaft will not fit into the bearing. (*Hint:* Find the probability that the clearance is negative.)

11.* A shipping pallet holds 10 boxes. Each box holds 300 parts of different types. The part weight is normally distributed with a mean of 1 lb and a standard deviation of 0.2 lb.
    a. Compute the mean and standard deviation of the pallet weight.
    b. Compute the probability that the pallet weight will exceed 3015 lb.

12. A certain product is assembled by placing three components end to end. The components' lengths are L_1, L_2, and L_3. Each component is manufactured on a different machine, so the random variations in their lengths are independent of one another. The lengths are normally distributed with means of 1, 2, and 1.5 ft and variances of 0.00014, 0.0002, and 0.0003, respectively.

    a. Compute the mean and variance of the length of the assembled product.

    b. Estimate the percentage of assembled products that will be no less than 4.48 and no more than 4.52 ft in length.

Section 7.3

13. Use a random number generator to produce 1000 uniformly distributed numbers with a mean of 10, a minimum of 2, and a maximum of 18. Obtain the mean and the histogram of these numbers, and discuss whether they appear uniformly distributed with the desired mean.

14. Use a random number generator to produce 1000 normally distributed numbers with a mean of 20 and a variance of 4. Obtain the mean, variance, and histogram of these numbers, and discuss whether they appear normally distributed with the desired mean and variance.

15. The mean of the sum (or difference) of two independent random variables equals the sum (or difference) of their means, but the variance is always the sum of the two variances. Use random number generation to verify this statement for the case where $z = x + y$, where x and y are independent and normally distributed random variables. The mean and variance of x are $\mu_x = 8$ and $\sigma_x^2 = 2$. The mean and variance of y are $\mu_y = 15$ and $\sigma_y^2 = 4$. Find the mean and variance of z by simulation, and compare the results with the theoretical prediction. Do this for 100, 1000, and 5000 trials.

16. Suppose that $z = xy$, where x and y are independent and normally distributed random variables. The mean and variance of x are $\mu_x = 10$ and $\sigma_x^2 = 2$. The mean and variance of y are $\mu_y = 15$ and $\sigma_y^2 = 3$. Find the mean and variance of z by simulation. Does $\mu_z = \mu_x \mu_y$? Does $\sigma_z^2 = \sigma_x^2 \sigma_y^2$? Do this for 100, 1000, and 5000 trials.

17. Suppose that $y = x^2$, where x is a normally distributed random variable with a mean and variance of $\mu_x = 0$ and $\sigma_x^2 = 4$. Find the mean and variance of y by simulation. Does $\mu_y = \mu_x^2$? Does $\sigma_y = \sigma_x^2$? Do this for 100, 1000, and 5000 trials.

18.* Suppose you have analyzed the price behavior of a certain stock by plotting the scaled frequency histogram of the price over a number of months. Suppose that the histogram indicates that the price is normally distributed with a mean $100 and a standard deviation of $5. Write a

MATLAB program to simulate the effects of buying 50 shares of this stock whenever the price is below the $100 mean, and selling all your shares whenever the price is above $105. Analyze the outcome of this strategy over 250 days (the approximate number of business days in a year). Define the profit as the yearly income from selling stock plus the value of the stocks you own at year's end, minus the yearly cost of buying stock. Compute the mean yearly profit you would expect to make, the minimum expected yearly profit, the maximum expected yearly profit, and the standard deviation of the yearly profit. The broker charges 6 cents per share bought or sold with a minimum fee of $40 per transaction. Assume you make only one transaction per day.

19. Suppose that data show that a certain stock price is normally distributed with a mean of $150 and a variance of 100. Create a simulation to compare the results of the following two strategies over 250 days. You start the year with 1000 shares. With the first strategy, every day the price is below $140 you buy 100 shares, and every day the price is above $160 you sell all the shares you own. With the second strategy, every day the price is below $150 you buy 100 shares, and every day the price is above $160 you sell all the shares you own. The broker charges 5 cents per share traded with a minimum of $35 per transaction.

20. Write a script file to simulate 100 plays of a game in which you flip two coins. You win the game if you get two heads, lose if you get two tails, and flip again if you get one head and one tail. Create three user-defined functions to use in the script. Function `flip` simulates the flip of one coin, with the state `s` of the random number generator as the input argument, and the new state `s` and the result of the flip (0 for a tail and 1 for a head) as the outputs. Function `flips` simulates the flipping of two coins and calls `flip`. The input of `flips` is the state `s`, and the outputs are the new state `s` and the result (0 for two tails, 1 for a head and a tail, and 2 for two heads). Function `match` simulates a turn at the game. Its input is the state `s`, and its outputs are the result (1 for win, 0 for lose) and the new state `s`. The script should reset the random number generator to its initial state, compute the state `s`, and pass this state to the user-defined functions.

21. Write a script file to play a simple number guessing game as follows. The script should generate a random integer in the range 1, 2, 3, . . . , 14, 15. It should provide for the player to make repeated guesses of the number, and it should indicate if the player has won or give the player a hint after each wrong guess. The responses and hints are as follows:

- ■ "You won" and then stop the game.
- ■ "Very close" if the guess is within 1 of the correct number.
- ■ "Getting close" if the guess is within 2 or 3 of the correct number.
- ■ "Not close" if the guess is not within 3 of the correct number.

Section 7.4

22.* Interpolation is useful when one or more data points are missing. This situation often occurs with environmental measurements, such as temperature, because of the difficulty of making measurements around the clock. The following table of temperature versus time data is missing readings at 5 and 9 hours. Use linear interpolation with MATLAB to estimate the temperature at those times.

Time (hours, P.M.)	1	2	3	4	5	6	7	8	9	10	11	12
Temperature (°C)	10	9	18	24	?	21	20	18	?	15	13	11

23. The following table gives temperature data in °C as a function of time of day and day of the week at a specific location. Data are missing for the entries marked with a question mark (?). Use linear interpolation with MATLAB to estimate the temperature at the missing points.

	Day				
Hour	Mon	Tues	Wed	Thurs	Fri
1	17	15	12	16	16
2	13	?	8	11	12
3	14	14	9	?	15
4	17	15	14	15	19
5	23	18	17	20	24

24. Computer-controlled machines are used to cut and to form metal and other materials when manufacturing products. These machines often use cubic splines to specify the path to be cut or the contour of the part to be shaped. The following coodinates specify the shape of a certain car's front fender. Fit a series of cubic splines to the coordinates, and plot the splines along with the coordinate points.

x (ft)	0	0.25	0.75	1.25	1.5	1.75	1.875	2	2.125	2.25
y (ft)	1.2	1.18	1.1	1	0.92	0.8	0.7	0.55	0.35	0

25. The following data are the measured temperature T of water flowing from a hot water faucet after it is turned on at time $t = 0$.

t (sec)	T (°F)	t (sec)	T (°F)
0	72.5	6	109.3
1	78.1	7	110.2
2	86.4	8	110.5
3	92.3	9	109.9
4	110.6	10	110.2
5	111.5		

a. Plot the data, connecting them first with straight lines and then with a cubic spline.

b. Estimate the temperature values at the following times, using linear interpolation and then cubic spline interpolation: $t = 0.6, 2.5, 4.7, 8.9$.

c. Use both the linear and cubic spline interpolations to estimate the time it will take for the temperature to equal the following values: $T = 75, 85, 90, 105$.

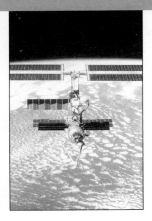

Engineering in the 21st Century...

International Engineering

The end of the Cold War has resulted in increased international cooperation. A vivid example of this is the cooperation between the Russian and American space agencies. In the first joint mission the U.S. space shuttle Atlantis docked with the Russian space station Mir. Now Russia, the United States, and several other countries are collaborating on a more ambitious space project: the International Space Station (ISS). The launch of the first element of the station, a Russian-built cargo block, occurred in 1998. The first permanent crew arrived in 2000. The station will have a crew of seven, will weigh over 1 000 000 lb, and will be 290 ft long and 356 ft wide. To assemble it will require 28 U.S. shuttle flights and 41 Russian flights over a period of 10 years.

The new era of international cooperation has resulted in increased integration of national economies into one global economy. For example, many components for cars assembled in the United States are supplied by companies in other countries. In the 21st century, improved communications and computer networks will enable engineering teams in different countries to work simultaneously on different aspects of the product's design, employing a common design database. To prepare for such a future in international engineering, students need to become familiar with other cultures and languages as well as proficient in computer-aided design.

MATLAB is available in international editions and has been widely used in Europe for many years. Thus engineers familiar with MATLAB will already have a head start in preparing for a career in international engineering. ■

CHAPTER 8

Linear Algebraic Equations

OUTLINE

8.1   Matrix Methods for Linear Equations

8.2   The Left Division Method

8.3   Underdetermined Systems

8.4   Overdetermined Systems

8.5   A General Solution Program

8.6   Summary

Problems

Linear algebraic equations such as

$$5x - 2y = 13$$
$$7x + 3y = 24$$

occur in many engineering applications. For example, electrical engineers use them to predict the power requirements for circuits; civil, mechanical, and aerospace engineers use them to design structures and machines; chemical engineers use them to compute material balances in chemical processes; and industrial engineers apply them to design schedules and operations. The examples and homework problems in this chapter explore some of these applications.

Linear algebraic equations can be solved "by hand" using pencil and paper, by calculator, or with software such as MATLAB. The choice depends on the circumstances. For equations with only two unknown variables, hand solution is easy and adequate. Some calculators can solve equation sets that have many variables. However, the greatest power and flexibility is obtained by using software.

For example, MATLAB can obtain and plot equation solutions as we vary one or more parameters.

Systematic solution methods have been developed for sets of linear equations. In Section 8.1 we introduce some matrix notation that is required for use with MATLAB and that is also useful for expressing solution methods in a compact way. The conditions for the existence and uniqueness of solutions are then introduced. Methods using MATLAB are treated in four sections: Section 8.2 covers the left division method for solving equation sets that have unique solutions. Section 8.3 covers the case where the equation set does not contain enough information to determine all the unknown variables. This is the *underdetermined* case. The *overdetermined* case occurs when the equation set has more independent equations than unknowns (Section 8.4). A general solution program is given in Section 8.5.

8.1 Matrix Methods for Linear Equations

Sets of linear algebraic equations can be expressed as a single equation, using matrix notation. This standard and compact form is useful for expressing solutions and for developing software applications with an arbitrary number of variables. In this application, a vector is taken to be a column vector unless otherwise specified.

Matrix notation enables us to represent multiple equations as a single matrix equation. For example, consider the following set.

$$2x_1 + 9x_2 = 5$$
$$3x_1 - 4x_2 = 7$$

This set can be expressed in vector-matrix form as

$$\begin{bmatrix} 2 & 9 \\ 3 & -4 \end{bmatrix} \begin{bmatrix} x_1 \\ x_2 \end{bmatrix} = \begin{bmatrix} 5 \\ 7 \end{bmatrix}$$

which can be represented in the following compact form

$$\mathbf{Ax} = \mathbf{b} \tag{8.1--1}$$

where we have defined the following matrices and vectors:

$$\mathbf{A} = \begin{bmatrix} 2 & 9 \\ 3 & -4 \end{bmatrix} \qquad \mathbf{x} = \begin{bmatrix} x_1 \\ x_2 \end{bmatrix} \qquad \mathbf{b} = \begin{bmatrix} 5 \\ 7 \end{bmatrix}$$

In general, the set of m equations in n unknowns can be expressed in the form Equation (8.1–1), where $\mathbf{A}$ is $m \times n$, $\mathbf{x}$ is $n \times 1$, and $\mathbf{b}$ is $m \times 1$.

Matrix Inverse

The solution of the scalar equation $ax = b$ is $x = b/a$ if $a \neq 0$. The division operation of scalar algebra has an analogous operation in matrix algebra. For example, to solve the matrix equation (8.1–1) for $\mathbf{x}$, we must somehow "divide" $\mathbf{b}$ by $\mathbf{A}$.

The procedure for doing this is developed from the concept of a *matrix inverse*. The inverse of a matrix $\mathbf{A}$ is denoted by $\mathbf{A}^{-1}$ and has the property that

$$\mathbf{A}^{-1}\mathbf{A} = \mathbf{A}\mathbf{A}^{-1} = \mathbf{I}$$

where $\mathbf{I}$ is the identity matrix. Using this property, we multiply both sides of Equation (8.1–1) from the left by $\mathbf{A}^{-1}$ to obtain $\mathbf{A}^{-1}\mathbf{A}\mathbf{x} = \mathbf{A}^{-1}\mathbf{b}$. Because $\mathbf{A}^{-1}\mathbf{A}\mathbf{x} = \mathbf{I}\mathbf{x} = \mathbf{x}$, we obtain the solution

$$\mathbf{x} = \mathbf{A}^{-1}\mathbf{b} \qquad (8.1–2)$$

The inverse of a matrix $\mathbf{A}$ is defined only if $\mathbf{A}$ is square and nonsingular. A matrix is *singular* if its determinant $|\mathbf{A}|$ is zero. If $\mathbf{A}$ is singular, then a unique solution to Equation (8.1–1) does not exist. The MATLAB functions `inv(A)` and `det(A)` compute the inverse and the determinant of the matrix $\mathbf{A}$. If the `inv(A)` function is applied to a singular matrix, MATLAB will issue a warning to that effect.

SINGULAR MATRIX

An *ill-conditioned* set of equations is a set that is close to being singular. The ill-conditioned status depends on the accuracy with which the solution calculations are made. When internal numerical accuracy used by MATLAB is insufficient to obtain a solution, it prints the message warning that the matrix is close to singular and that the results might be inaccurate.

ILL-CONDITIONED SET

For a 2×2 matrix $\mathbf{A}$,

$$\mathbf{A} = \begin{bmatrix} a & b \\ c & d \end{bmatrix} \qquad \mathbf{A}^{-1} = \frac{1}{ad - bc}\begin{bmatrix} d & -b \\ -c & a \end{bmatrix}$$

where $\det(\mathbf{A}) = ad - bc$. Thus $\mathbf{A}$ is singular if $ad - bc = 0$.

The Matrix Inverse Method

EXAMPLE 8.1-1

Solve the following equations, using the matrix inverse.

$$2x_1 + 9x_2 = 5$$
$$3x_1 - 4x_2 = 7$$

■ Solution
The matrix $\mathbf{A}$ and the vector $\mathbf{b}$ are

$$\mathbf{A} = \begin{bmatrix} 2 & 9 \\ 3 & -4 \end{bmatrix} \qquad \mathbf{b} = \begin{bmatrix} 5 \\ 7 \end{bmatrix}$$

The session is

```
>>A = [2,9;3,-4]; b = [5;7];
>>x = inv(A)*b
x =
   2.3714
   0.0286
```

The solution is $x_1 = 2.3714$ and $x_2 = 0.0286$. MATLAB did not issue a warning, so the solution is unique.

The solution form $\mathbf{x} = \mathbf{A}^{-1}\mathbf{b}$ is rarely applied in practice to obtain numerical solutions to sets of many equations, because calculation of the matrix inverse is likely to introduce greater numerical inaccuracy than the left division method to be introduced.

Test Your Understanding

T8.1–1 For what values of c will the following set (a) have a unique solution and (b) have an infinite number of solutions? Find the relation between x_1 and x_2 for these solutions.

$$6x_1 + cx_2 = 0$$
$$2x_1 + 4x_2 = 0$$

(Answers: (a) $c \neq 12$, $x_1 = x_2 = 0$; (b) $c = 12$, $x_1 = -2x_2$)

T8.1–2 Use the matrix inverse method to solve the following set.

$$3x_1 - 4x_2 = 5$$
$$6x_1 - 10x_2 = 2$$

(Answer: $x_1 = 7$, $x_2 = 4$)

T8.1–3 Use the matrix inverse method to solve the following set.

$$3x_1 - 4x_2 = 5$$
$$6x_1 - 8x_2 = 2$$

(Answer: No solution.)

Existence and Uniqueness of Solutions

The matrix inverse method will warn us if a *unique* solution does not exist, but it does not tell us whether there is no solution or an infinite number of solutions. In addition, the method is limited to cases where the matrix $\mathbf{A}$ is square, that is, cases where the number of equations equals the number of unknowns. For this reason we now introduce a method that allows us to determine easily whether an equation set has a solution and whether it is unique. The method requires the concept of the *rank* of a matrix.

MATRIX RANK

Consider the 3×3 determinant

$$|\mathbf{A}| = \begin{vmatrix} 3 & -4 & 1 \\ 6 & 10 & 2 \\ 9 & -7 & 3 \end{vmatrix} = 0 \qquad (8.1\text{--}3)$$

If we eliminate one row and one column in the determinant, we are left with a 2×2 determinant. Depending on which row and column we choose to eliminate, there are nine possible 2×2 determinants we can obtain. These are called *subdeterminants*. For example, if we eliminate the second row and third column, we obtain

$$\begin{vmatrix} 3 & -4 \\ 9 & -7 \end{vmatrix} = 3(-7) - 9(-4) = 15$$

Subdeterminants are used to define the *rank* of a matrix. The definition of matrix rank is as follows.

Definition of Matrix Rank. An $m \times n$ matrix **A** has a *rank* $r \geq 1$ if and only if |**A**| contains a nonzero $r \times r$ determinant and every square subdeterminant with $r + 1$ or more rows is zero.

For example, the rank of **A** in Equation (8.1–3) is 2 because |**A**| = 0 while |**A**| contains at least one nonzero 2×2 subdeterminant. To determine the rank of a matrix **A** in MATLAB, type rank(A). If **A** is $n \times n$, its rank is n if det(A) $\neq$ 0.

We can use the following test to determine if a solution exists to **Ax** = **b** and whether it is unique. The test requires that we first form the *augmented matrix* [**A b**].

Existence and Uniqueness of Solutions. The set **Ax** = **b** with m equations and n unknowns has solutions if and only if (1) rank(**A**) = rank([**A b**]). Let r = rank(**A**). If condition (1) is satisfied and if $r = n$, then the solution is unique. If condition (1) is satisfied but $r < n$, there are an infinite number of solutions, and r unknown variables can be expressed as linear combinations of the other $n - r$ unknown variables, whose values are arbitrary.

Homogeneous case. The homogeneous set **Ax** = **0** is a special case in which **b** = **0**. For this case, rank(**A**) = rank([**A b**]) always, and thus the set always has the trivial solution **x** = **0**. A nonzero solution, in which at least one unknown is nonzero, exists if and only if rank(**A**) < n. If $m < n$, the homogeneous set always has a nonzero solution.

This test implies that if **A** is square and of dimension $n \times n$, then rank([**A b**]) = rank(**A**), and a unique solution exists for any **b** if rank(**A**) = n.

8.2 The Left Division Method

MATLAB provides the *left division method* for solving the equation set **Ax** = **b.** This method is based on Gauss elimination. To use the left division method to solve for **x,** you type x = A\b. If |**A**| = 0 or if the number of equations does not equal the number of unknowns, then you need to use the other methods to be presented later.

Left Division Method with Three Unknowns

EXAMPLE 8.2–1

Use the left division method to solve the following set.

$$3x_1 + 2x_2 - 9x_3 = -65$$
$$-9x_1 - 5x_2 + 2x_3 = 16$$
$$6x_1 + 7x_2 + 3x_3 = 5$$

■ Solution

The matrices **A** and **b** are

$$\mathbf{A} = \begin{bmatrix} 3 & 2 & -9 \\ -9 & -5 & 2 \\ 6 & 7 & 3 \end{bmatrix} \qquad \mathbf{b} = \begin{bmatrix} -65 \\ 16 \\ 5 \end{bmatrix}$$

The session is

```
>>A = [3,2,-9;-9,-5,2;6,7,3];
>>rank(A)
ans =
   3
```

Because **A** is 3×3 and rank(**A**) = 3, which is the number of unknowns, a unique solution exists. It is obtained by continuing the session as follows.

```
>>b = [-65;16;5];
>>x = A\b
x =
    2.0000
   -4.0000
    7.0000
```

This answer gives the vector **x**, which corresponds to the solution $x_1 = 2$, $x_2 = -4$, $x_3 = 7$.

For the solution $\mathbf{x} = \mathbf{A}^{-1}\mathbf{b}$, vector **x** is proportional to the vector **b**. We can use this linearity property to obtain a more generally useful algebraic solution in cases where the right-hand sides are all multiplied by the same scalar. For example, suppose the matrix equation is $\mathbf{A}\mathbf{y} = \mathbf{b}c$, where c is a scalar. The solution is $\mathbf{y} = \mathbf{A}^{-1}\mathbf{b}c = \mathbf{x}c$. Thus if we obtain the solution to $\mathbf{A}\mathbf{x} = \mathbf{b}$, the solution to $\mathbf{A}\mathbf{y} = \mathbf{b}c$ is given by $\mathbf{y} = \mathbf{x}c$.

EXAMPLE 8.2–2	Calculation of Cable Tension

A mass m is suspended by three cables attached at three points B, C, and D, as shown in Figure 8.2–1. Let T_1, T_2, and T_3 be the tensions in the three cables AB, AC, and AD, respectively. If the mass m is stationary, the sum of the tension components in the x, in the y, and in the z directions must each be zero. This gives the following three equations:

$$\frac{T_1}{\sqrt{35}} - \frac{3T_2}{\sqrt{34}} + \frac{T_3}{\sqrt{42}} = 0$$

$$\frac{3T_1}{\sqrt{35}} - \frac{4T_3}{\sqrt{42}} = 0$$

$$\frac{5T_1}{\sqrt{35}} + \frac{5T_2}{\sqrt{34}} + \frac{5T_3}{\sqrt{42}} - mg = 0$$

Determine T_1, T_2, and T_3 in terms of an unspecified value of the weight mg.

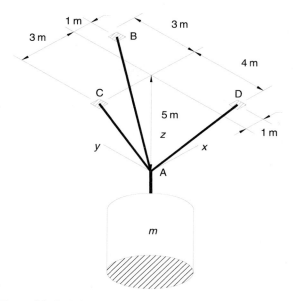

Figure 8.2–1 A mass suspended by three cables.

■ Solution

If we set $mg = 1$, the equations have the form $\mathbf{AT} = \mathbf{b}$ where

$$\mathbf{A} = \begin{bmatrix} \frac{1}{\sqrt{35}} & -\frac{3}{\sqrt{34}} & \frac{1}{\sqrt{42}} \\ \frac{3}{\sqrt{35}} & 0 & -\frac{4}{\sqrt{42}} \\ \frac{5}{\sqrt{35}} & \frac{5}{\sqrt{34}} & \frac{5}{\sqrt{42}} \end{bmatrix} \qquad \mathbf{T} = \begin{bmatrix} T_1 \\ T_2 \\ T_3 \end{bmatrix} \qquad \mathbf{b} = \begin{bmatrix} 0 \\ 0 \\ 1 \end{bmatrix}$$

The script file to solve this system is

```
% File cable.m
s34 = sqrt(34); s35 = sqrt(35); s42 = sqrt(42);
A1 = [1/s35, -3/s34, 1/s42];
A2 = [3/s35, 0, -4/s42];
A3 = [5/s35, 5/s34, 5/s42];
A = [A1; A2; A3];
b = [0; 0; 1];
rank(A)
rank([A, b])
T = A\b
```

When this file is executed by typing `cable`, we find that rank($\mathbf{A}$) = rank ($[\mathbf{A}\ \mathbf{b}]$) = 3 and obtain the values $T_1 = 0.5071$, $T_2 = 0.2915$, and $T_3 = 0.4166$. Because $\mathbf{A}$ is 3×3 and rank($\mathbf{A}$) = 3, which is the number of unknowns, the solution is unique. Using the linearity property, we multiply these results by mg and obtain the general solution $T_1 = 0.5071\ mg$, $T_2 = 0.2915\ mg$, and $T_3 = 0.4166\ mg$.

Linear equations are useful in many engineering fields. Electric circuits are a common source of linear equation models. The circuit designer must be able to solve them to predict the currents that will exist in the circuit. This information is often needed to determine the power supply requirements among other things.

EXAMPLE 8.2–3   An Electric Resistance Network

The circuit shown in Figure 8.2–2 has five resistances and two applied voltages. Assuming that the positive directions of current flow are in the directions shown in the figure, Kirchhoff's voltage law applied to each loop in the circuit gives

$$-v_1 + R_1 i_1 + R_4 i_4 = 0$$
$$-R_4 i_4 + R_2 i_2 + R_5 i_5 = 0$$
$$-R_5 i_5 + R_3 i_3 + v_2 = 0$$

Conservation of charge applied at each node in the circuit gives

$$i_1 = i_2 + i_4$$
$$i_2 = i_3 + i_5$$

You can use these two equations to eliminate i_4 and i_5 from the first three equations. The result is

$$(R_1 + R_4)i_1 - R_4 i_2 = v_1$$
$$-R_4 i_1 + (R_2 + R_4 + R_5)i_2 - R_5 i_3 = 0$$
$$R_5 i_2 - (R_3 + R_5)i_3 = v_2$$

Thus we have three equations in the three unknowns i_1, i_2, and i_3.

Write a MATLAB script file that uses given values of the applied voltages v_1 and v_2 and given values of the five resistances to solve for the currents i_1, i_2, and i_3. Use the

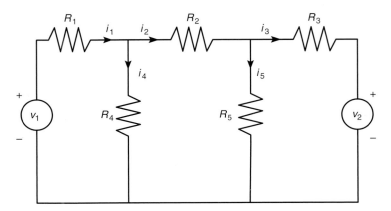

Figure 8.2–2 An electric resistance network.

program to find the currents for the case $R_1 = 5$, $R_2 = 100$, $R_3 = 200$, $R_4 = 150$, and $R_5 = 250$ kΩ and $v_1 = 100$ and $v_2 = 50$ V. (Note that 1 kΩ = 1000 Ω.)

■ **Solution**

Because there are as many unknowns as equations, there will be a unique solution if $|\mathbf{A}| \neq 0$; in addition, the left division method will generate an error message if $|\mathbf{A}| = 0$. The following script file, named `resist.m`, uses the left division method to solve the three equations for i_1, i_2, and i_3.

```
% File resist.m
% Solves for the currents i_1, i_2, i_3
R = [5,100,200,150,250]*1000;
v1 = 100; v2 = 50;
A1 = [R(1) + R(4), -R(4), 0];
A2 = [-R(4), R(2) + R(4) + R(5), -R(5)];
A3 = [0, R(5), -(R(3) + R(5))];
A = [A1; A2; A3];
b=[v1; 0; v2];
current = A\b;
disp('The currents are:')
disp(current)
```

The row vectors `A1`, `A2`, and `A3` were defined to avoid typing the lengthy expression for `A` in one line. This script is executed from the command prompt as follows:

```
>>resist
The currents are:
   1.0e-003*
   0.9544
   0.3195
   0.0664
```

Because MATLAB did not generate an error message, the solution is unique. The currents are $i_1 = 0.9544$, $i_2 = 0.3195$, and $i_3 = 0.0664$ mA, where 1 mA = 1 milli-ampere = 0.001 A.

Ethanol Production

EXAMPLE 8.2–4

Engineers in the food and chemical industries use fermentation in many processes. The following equation describes Baker's yeast fermentation.

$$a(C_6H_{12}O_6) + b(O_2) + c(NH_3)$$
$$\rightarrow C_6H_{10}NO_3 + d(H_2O) + e(CO_2) + f(C_2H_6O)$$

The variables $a, b, \ldots, f$ represent the masses of the products involved in the reaction. In this formula $C_6H_{12}O_6$ represents glucose, $C_6H_{10}NO_3$ represents yeast, and C_2H_6O represents ethanol. This reaction produces ethanol in addition to water and carbon dioxide. We want to determine the amount of ethanol f produced. The number of C, O, N, and H atoms on the left must balance those on the right side of the equation. This gives four equations:

$$6a = 6 + e + 2f$$
$$6a + 2b = 3 + d + 2e + f$$
$$c = 1$$
$$12a + 3c = 10 + 2d + 6f$$

The fermenter is equipped with an oxygen sensor and a carbon dioxide sensor. These enable us to compute the respiratory quotient R:

$$R = \frac{CO_2}{O_2} = \frac{e}{b}$$

Thus the fifth equation is $Rb - e = 0$. The yeast yield Y (grams of yeast produced per gram of glucose consumed) is related to a as follows

$$Y = \frac{144}{180a}$$

where 144 is the molecular weight of yeast and 180 is the molecular weight of glucose. By measuring the yeast yield Y we can compute a as follows: $a = 144/180Y$. This is the sixth equation.

Write a user-defined function that computes f, the amount of ethanol produced, with R and Y as the function's arguments. Test your function for two cases where Y is measured to be 0.5: (a) $R = 1.1$ and (b) $R = 1.05$.

■ Solution

First note that there are only four unknowns because the third equation directly gives $c = 1$, and the sixth equation directly gives $a = 144/180Y$. To write these equations in matrix form, let $x_1 = b$, $x_2 = d$, $x_3 = e$, and $x_4 = f$. Then the equations can be written as

$$-x_3 - 2x_4 = 6 - 6(144/180Y)$$
$$2x_1 - x_2 - 2x_3 - x_4 = 3 - 6(144/180Y)$$
$$-2x_2 - 6x_4 = 7 - 12(144/180Y)$$
$$Rx_1 - x_3 = 0$$

In matrix form these become

$$
\begin{bmatrix}
0 & 0 & -1 & -2 \\
2 & -1 & -2 & -1 \\
0 & -2 & 0 & -6 \\
R & 0 & -1 & 0
\end{bmatrix}
\begin{bmatrix}
x_1 \\ x_2 \\ x_3 \\ x_4
\end{bmatrix}
=
\begin{bmatrix}
6 - 6(144/180Y) \\
3 - 6(144/180Y) \\
7 - 12(144/180Y) \\
0
\end{bmatrix}
$$

The function file is shown below.

```
function E = ethanol(R,Y)
% Computes ethanol produced from yeast reaction.
A = [0,0,-1,-2;2,-1,-2,-1;...
     0,-2,0,-6;R,0,-1,0];
b = [6-6*(144./(180*Y));3-6*(144./(180*Y));...
     7-12*(144./(180*Y));0];
x = A\b;
E = x(4);
```

The session is as follows:

```
>>ethanol(1.1,0.5)
ans =
      0.0654
>>ethanol(1.05,0.5)
ans =
     -0.0717
```

The negative value for E in the second case indicates that ethanol is being consumed rather than produced.

Test Your Understanding

T8.2–1 Use the left division method to solve the following set.

$$5x_1 - 3x_2 = 21$$
$$7x_1 - 2x_2 = 36$$

(Answers: $x_1 = 6$, $x_2 = 3$)

8.3 Underdetermined Systems

An *underdetermined system* does not contain enough information to determine all the unknown variables, usually but not always because it has fewer equations than unknowns. Thus an infinite number of solutions can exist, with one or more of the unknowns dependent on the remaining unknowns. The left division method works for square and nonsquare **A** matrices. However, if **A** is not square, the left division method can give answers that might be misinterpreted. We will show how to interpret MATLAB results correctly.

When there are fewer equations than unknowns, the left division method might give a solution with some of the unknowns set equal to zero, but this is not the general solution. An infinite number of solutions might exist even when the number of equations equals the number of unknowns. This can occur when |**A**| = 0. For such systems the left division method generates an error message warning us that the matrix **A** is singular. In such cases the *pseudoinverse method* $x = \text{pinv(A)} * b$ gives one solution, the *minimum norm solution*. In cases where there are an infinite number of solutions, the `rref` function can be used to express some of the unknowns in terms of the remaining unknowns, whose values are arbitrary.

An equation set can be underdetermined even though it has as many equations as unknowns. This can happen if some of the equations are not independent. Determining by hand whether all the equations are independent might not be easy, especially if the set has many equations, but it is easily done in MATLAB.

An Underdetermined Set with Three Equations and Three Unknowns

Show that the following set does not have a unique solution. How many of the unknowns will be undetermined? Interpret the results given by the left division method.

$$2x_1 - 4x_2 + 5x_3 = -4$$
$$-4x_1 - 2x_2 + 3x_3 = 4$$
$$2x_1 + 6x_2 - 8x_3 = 0$$

■ Solution

A MATLAB session to check the ranks is

```
>>A = [2,-4,5;-4,-2,3;2,6,-8];
>>b = [-4;4;0];
>>rank(A)
ans =
  2
>>rank([A, b])
ans =
  2
>>x = A\b
Warning: Matrix is singular to working precision.
ans =
  NaN
  NaN
  NaN
```

Because the ranks of **A** and [**A b**] are equal, a solution exists. However, because the number of unknowns is 3 and is 1 greater than the rank of **A,** one of the unknowns will be undetermined. An infinite number of solutions exist, and we can solve for only two of the unknowns in terms of the third unknown. The set is underdetermined because there are fewer than three independent equations; the third equation can be obtained from the first two. To see this, add the first and second equations, to obtain $-2x_1 - 6x_2 + 8x_3 = 0$, which is equivalent to the third equation.

Note that we could also tell that the matrix **A** is singular because its rank is less than 3. If we use the left division method, MATLAB returns a message warning that the problem is singular, and it does not produce an answer.

The `pinv` Function and the Euclidean Norm

The `pinv` function (which stands for "pseudoinverse") can be used to obtain a solution of an underdetermined set. To solve the equation set $\mathbf{Ax} = \mathbf{b}$ using the `pinv` function, you type $x = \text{pinv}(A) \ast b$. The `pinv` function gives a solution that gives the minimum value of the *Euclidean norm*, which is the magnitude of the solution vector x. The magnitude of a vector $\mathbf{v}$ in three-dimensional space, having components x, y, z, is $\sqrt{x^2 + y^2 + z^2}$. It can be computed using matrix multiplication and the transpose as follows.

$$\sqrt{\mathbf{v}^T \mathbf{v}} = \sqrt{[x\ y\ z]^T \begin{bmatrix} x \\ y \\ z \end{bmatrix}} = \sqrt{x^2 + y^2 + z^2}$$

The generalization of this formula to an n-dimensional vector $\mathbf{v}$ gives the magnitude of the vector and is the Euclidean norm N. Thus

$$N = \sqrt{\mathbf{v}^T \mathbf{v}} \tag{8.3–1}$$

The MATLAB function `norm(v)` computes the Euclidean norm.

A Statically Indeterminate Problem

EXAMPLE 8.3–2

Determine the forces in the three equally spaced supports that hold up a light fixture. The supports are 5 ft apart. The fixture weighs 400 lb, and its mass center is 4 ft from the right end. Obtain the solution using the MATLAB left-division method and the pseudoinverse method.

■ Solution

Figure 8.3–1 shows the fixture and the free-body diagram, where T_1, T_2, and T_3 are the tension forces in the supports. For the fixture to be in equilibrium, the vertical forces must cancel, and the total moments about an arbitrary fixed point—say, the right endpoint—must be zero. These conditions give the two equations

$$T_1 + T_2 + T_3 - 400 = 0$$
$$400(4) - 10T_1 - 5T_2 = 0$$

or

$$T_1 + T_2 + T_3 = 400 \tag{8.3–2}$$
$$10T_1 + 5T_2 + 0T_3 = 1600 \tag{8.3–3}$$

Because there are more unknowns than equations, the set is underdetermined. Thus we cannot determine a unique set of values for the forces. Such a problem, when the equations of statics do not give enough equations, is called *statically indeterminate*. These

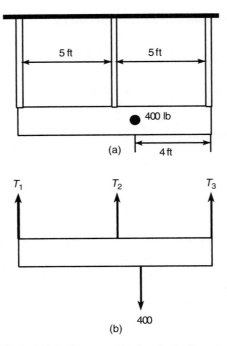

Figure 8.3–1 A light fixture and its free-body diagram.

equations can be written in the matrix form $\mathbf{AT} = \mathbf{b}$ as follows:

$$\begin{bmatrix} 1 & 1 & 1 \\ 10 & 5 & 0 \end{bmatrix} \begin{bmatrix} T_1 \\ T_2 \\ T_3 \end{bmatrix} = \begin{bmatrix} 400 \\ 1600 \end{bmatrix}$$

The MATLAB session is

```
>>A = [1,1,1;10,5,0];
>>b = [400;1600];
>>rank(A)
ans =
   2
>>rank([A, b])
ans =
   2
>>T = A\b
T =
    160.0000
    0
    240.0000
```

```
>>T = pinv(A)*b
T =
   93.3333
  133.3333
  173.3333
```

The left division answer corresponds to $T_1 = 160$, $T_2 = 0$, and $T_3 = 240$. This illustrates how the MATLAB left division operator produces a solution with one or more variables set to zero, for underdetermined sets having more unknowns than equations.

Because the ranks of $\mathbf{A}$ and $[\mathbf{A}\ \mathbf{b}]$ are both 2, a solution exists, but it is not unique. Because the number of unknowns is 3, and is 1 greater than the rank of $\mathbf{A}$, an infinite number of solutions exist, and we can solve for only two of the unknowns in terms of the third.

The pseudoinverse solution gives $T_1 = 93.3333$, $T_2 = 133.3333$, and $T_3 = 173.3333$. This is the minimum norm solution for real values of the variables. The minimum norm solution consists of the real values of T_1, T_2, and T_3 that minimize

$$N = \sqrt{T_1^2 + T_2^2 + T_3^2}$$

To understand what MATLAB is doing, note that we can solve Equations (8.3–2) and (8.3–3) to obtain T_1 and T_2 in terms of T_3 as $T_1 = T_3 - 80$ and $T_2 = 480 - 2T_3$. Then the Euclidean norm can be expressed as

$$N = \sqrt{(T_3 - 80)^2 + (480 - 2T_3)^2 + T_3^2} = \sqrt{6T_3^2 - 2080T_3 + 236{,}800}$$

The real value of T_3 that minimizes N can be found by plotting N versus T_3, or by using calculus. The answer is $T_3 = 173.3333$, the same as the minimum norm solution given by the pseudoinverse method.

Where there are an infinite number of solutions, we must decide whether the solutions given by the left division and the pseudoinverse methods are useful for applications. This must be done in the context of the specific application.

Test Your Understanding

T8.3–1 Find two solutions to the following set.

$$x_1 + 3x_2 + 2x_3 = 2$$

$$x_1 + x_2 + x_3 = 4$$

(Answer: Minimum norm solution: $x_1 = 4.33$, $x_2 = -1.67$, $x_3 = 1.34$. Left division solution: $x_1 = 5$, $x_2 = -1$, $x_3 = 0$.)

The Reduced Row Echelon Form

We can express some of the unknowns in an underdetermined set as functions of the remaining unknowns. In Example 8.3–2, we wrote the solutions for two of the unknowns in terms of the third: $T_1 = T_3 - 80$ and $T_2 = 480 - 2T_3$. These two equations are equivalent to

$$T_1 - T_3 = -80 \qquad T_2 + 2T_3 = 480$$

In matrix form these are

$$\begin{bmatrix} 1 & 0 & -1 \\ 0 & 1 & 2 \end{bmatrix} \begin{bmatrix} T_1 \\ T_2 \\ T_3 \end{bmatrix} = \begin{bmatrix} -80 \\ 480 \end{bmatrix}$$

The augmented matrix [**A b**] for the above set is

$$\begin{bmatrix} 1 & 0 & -1 & -80 \\ 0 & 1 & 2 & 480 \end{bmatrix}$$

Note that the first two columns form a 2×2 identity matrix. This indicates that the corresponding equations can be solved directly for T_1 and T_2 in terms of T_3.

We can always reduce an underdetermined set to such a form by multiplying the set's equations by suitable factors and adding the resulting equations to eliminate an unknown variable. The MATLAB rref function provides a procedure for reducing an equation set to this form, which is called the *reduced row echelon form*. Its syntax is rref([A b]). Its output is the augmented matrix [**C d**] that corresponds to the equation set **Cx = d.** This set is in reduced row echelon form.

| EXAMPLE 8.3–3 | Three Equations in Three Unknowns, *Continued* |

The following underdetermined equation set was analyzed in Example 8.3–1. There it was shown that an infinite number of solutions exist. Use the rref function to obtain the solutions.

$$2x_1 - 4x_2 + 5x_3 = -4$$
$$-4x_1 - 2x_2 + 3x_3 = 4$$
$$2x_1 + 6x_2 - 8x_3 = 0$$

■ **Solution**
The MATLAB session is

```
>>A = [2,-4,5;-4,-2,3;2,6,-8];
>>b = [-4;4;0];
>>rref([A, b])
ans =
    1      0     -0.1    -1.2000
    0      1     -1.3     0.4000
    0      0      0       0
```

The answer corresponds to the augmented matrix [**C d**], where

$$[\mathbf{C} \quad \mathbf{d}] = \begin{bmatrix} 1 & 0 & -0.1 & -1.2 \\ 0 & 1 & -1.3 & 0.4 \\ 0 & 0 & 0 & 0 \end{bmatrix}$$

This matrix corresponds to the matrix equation $\mathbf{Cx} = \mathbf{d}$, or

$$x_1 + 0x_2 - 0.1x_3 = -1.2$$
$$0x_1 + x_2 - 1.3x_3 = 0.4$$
$$0x_1 + 0x_2 - 0x_3 = 0$$

These can be easily solved for x_1 and x_2 in terms of x_3 as follows: $x_1 = 0.1x_3 - 1.2$, $x_2 = 1.3x_3 + 0.4$. This is the general solution to the problem, where x_3 is taken to be the arbitary variable.

Supplementing Underdetermined Systems

Often the linear equations describing the application are underdetermined because not enough information has been specified to determine unique values of the unknowns. In such cases we might be able to include additional information, objectives, or constraints to find a unique solution. We can use the `rref` command to reduce the number of unknown variables in the problem, as illustrated in the next two examples.

Production Planning

EXAMPLE 8.3–4

The following table shows how many hours reactors A and B need to produce 1 ton each of the chemical products 1, 2, and 3. The two reactors are available for 40 and 30 hr per week, respectively. Determine how many tons of each product can be produced each week.

Hours	Product 1	Product 2	Product 3
Reactor A	5	3	3
Reactor B	3	3	4

■ Solution

Let x, y, and z be the number of tons each of products 1, 2, and 3 that can be produced in one week. Using the data for reactor A, the equation for its usage in one week is

$$5x + 3y + 3z = 40$$

The data for reactor B gives

$$3x + 3y + 4z = 30$$

This system is underdetermined. The matrices for the equation $\mathbf{Ax} = \mathbf{b}$ are

$$\mathbf{A} = \begin{bmatrix} 5 & 3 & 3 \\ 3 & 3 & 4 \end{bmatrix} \quad \mathbf{b} = \begin{bmatrix} 40 \\ 30 \end{bmatrix} \quad \mathbf{x} = \begin{bmatrix} x \\ y \\ z \end{bmatrix}$$

Here the rank$(\mathbf{A}) = $ rank$([\mathbf{A} \quad \mathbf{b}]) = 2$, which is less than the number of unknowns. Thus an infinite number of solutions exist, and we can determine two of the variables in terms of the third.

Using the `rref` command `rref([A b])`, where `A = [5,3,3;3,3,4]` and
`b = [40;30]`, we obtain the following reduced echelon augmented matrix:

$$\begin{bmatrix} 1 & 0 & -0.5 & 5 \\ 0 & 1 & 1.8333 & 5 \end{bmatrix}$$

This matrix gives the reduced system

$$x - 0.5z = 5$$
$$y + 1.8333z = 5$$

which can be easily solved as follows:

$$x = 5 + 0.5z \tag{8.3–4}$$
$$y = 5 - 1.8333z \tag{8.3–5}$$

where z is arbitrary. However, z cannot be completely arbitrary if the solution is to be meaningful. For example, negative values of the variables have no meaning here; thus we require that $x \geq 0$, $y \geq 0$, and $z \geq 0$. Equation (8.3–4) shows that $x \geq 0$ if $z \geq -10$. From Equation (8.3–5), $y \geq 0$ implies that $z \leq 5/1.8333 = 2.727$. Thus valid solutions are those given by Equations (8.3–4) and (8.3–5), where $0 \leq z \leq 2.737$ tons. The choice of z within this range must be made on some other basis, such as profit.

For example, suppose we make a profit of \$400, \$600, and \$100 per ton for products 1, 2, and 3, respectively. Then our total profit P is

$$P = 400x + 600y + 100z$$
$$= 400(5 + 0.5z) + 600(5 - 1.8333z) + 100z$$
$$= 5000 - 800z$$

Thus to maximize profit, we should choose z to be the smallest possible value, namely, $z = 0$. This choice gives $x = y = 5$ tons.

However, if the profits for each product were \$3000, \$600, and \$100, the total profit would be $P = 18\,000 + 500z$. Thus we should choose z to be its maximum, namely, $z = 2.727$ tons. From Equations (8.3–4) and (8.3–5), we obtain $x = 6.36$ and $y = 0$ tons.

EXAMPLE 8.3–5   Traffic Engineering

A traffic engineer wants to know if measurements of traffic flow entering and leaving a road network are sufficient to predict the traffic flow on each street in the network. For example, consider the network of one-way streets shown in Figure 8.3–2. The numbers shown are the measured traffic flows in vehicles per hour. Assume that no vehicles park anywhere within the network. If possible, calculate the traffic flows f_1, f_2, f_3, and f_4. If this is not possible, suggest how to obtain the necessary information.

■ Solution

The flow *into* intersection 1 must equal the flow *out* of the intersection. This gives

$$100 + 200 = f_1 + f_4$$

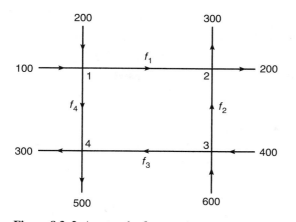

Figure 8.3–2 A network of one-way streets.

Similarly, for the other three intersections, we have

$$f_1 + f_2 = 300 + 200$$
$$600 + 400 = f_2 + f_3$$
$$f_3 + f_4 = 300 + 500$$

Putting these in the matrix form $\mathbf{Ax} = \mathbf{b}$, we obtain

$$\mathbf{A} = \begin{bmatrix} 1 & 0 & 0 & 1 \\ 1 & 1 & 0 & 0 \\ 0 & 1 & 1 & 0 \\ 0 & 0 & 1 & 1 \end{bmatrix} \qquad \mathbf{b} = \begin{bmatrix} 300 \\ 500 \\ 1000 \\ 800 \end{bmatrix} \qquad \mathbf{x} = \begin{bmatrix} f_1 \\ f_2 \\ f_3 \\ f_4 \end{bmatrix}$$

First, check the ranks of $\mathbf{A}$ and $[\mathbf{A}\ \mathbf{b}]$, using the MATLAB `rank` function. Both have a rank of 3, which is 1 less than the number of unknowns, so we can determine three of the unknowns in terms of the fourth. Thus we cannot determine all the traffic flows based on the given measurements.

Using the `rref([A b])` function produces the reduced augmented matrix

$$\begin{bmatrix} 1 & 0 & 0 & 1 & 300 \\ 0 & 1 & 0 & -1 & 200 \\ 0 & 0 & 1 & 1 & 800 \\ 0 & 0 & 0 & 0 & 0 \end{bmatrix}$$

which corresponds to the reduced system

$$f_1 + f_4 = 300$$

$$f_2 - f_4 = 200$$

$$f_3 + f_4 = 800$$

These can be solved easily as follows: $f_1 = 300 - f_4, f_2 = 200 + f_4$, and $f_3 = 800 - f_4$. If we could measure the flow on one of the internal roads, say f_4, then we could compute the other flows. So we recommend that the engineer arrange to have this additional measurement made.

Test Your Understanding

T8.3–2 Use the `rref`, `pinv`, and the left division methods to solve the following set.

$$3x_1 + 5x_2 + 6x_3 = 6$$
$$8x_1 - x_2 + 2x_3 = 1$$
$$5x_1 - 6x_2 - 4x_3 = -5$$

(Answer: There are an infinite number of solutions. The result obtained with the `rref` function is $x_1 = 0.2558 - 0.3721x_3$, $x_2 = 1.0465 - 0.9767x_3$, x_3 arbitrary. The `pinv` function gives $x_1 = 0.0571$, $x_2 = 0.5249$, $x_3 = 0.5340$. The left division method generates an error message.)

T8.3–3 Use the `rref`, `pinv`, and left division methods to solve the following set.

$$3x_1 + 5x_2 + 6x_3 = 4$$
$$x_1 - 2x_2 - 3x_3 = 10$$

(Answer: There are an infinite number of solutions. The result obtained with the `rref` function is $x_1 = 0.2727x_3 + 5.2727$, $x_2 = -1.3636x_3 - 2.2626$, x_3 arbitrary. The solution obtained with left division is $x_1 = 4.8000$, $x_2 = 0$, $x_3 = -1.7333$. The one obtained with the pseudoinverse method is $x_1 = 4.8394$, $x_2 = -0.1972$, $x_3 = -1.5887$.)

8.4 Overdetermined Systems

An *overdetermined system* is a set of equations that has more independent equations than unknowns. Some overdetermined systems have exact solutions, and they can be obtained with the left division method `x = A\b`. For other overdetermined systems, no exact solution exists; in some of these cases, the left division method does not yield an answer, while in other cases the left division method gives an answer that satisfies the equation set only in a "least-squares" sense. We will show what this means in the next example. When MATLAB gives an answer to an overdetermined set, it does not tell us whether the answer is the exact solution. We must determine this information ourselves, and we will now show how to do this.

The Least-Squares Method

EXAMPLE 8.4–1

Suppose we have the following three data points, and we want to find the straight line $y = c_1 x + c_2$ that best fits the data in some sense.

x	y
0	2
5	6
10	11

(a) Find the coefficients c_1 and c_2 using the least-squares criterion. (b) Find the coefficients by using the left division method to solve the three equations (one for each data point) for the two unknowns c_1 and c_2. Compare with the answer from part (a).

■ Solution

(a) Because two points define a straight line, unless we are extremely lucky, our three data points will not lie on the same straight line. A common criterion for obtaining the straight line that best fits the data is the *least-squares* criterion. According to this criterion, the line that minimizes J, the sum of the squares of the vertical differences between the line and the data points, is the "best" fit. Here J is

LEAST-SQUARES METHOD

$$J = \sum_{i=1}^{i=3} (c_1 x_i + c_2 - y_i)^2 = (0c_1 + c_2 - 2)^2 + (5c_1 + c_2 - 6)^2 + (10c_1 + c_2 - 11)^2$$

If you are familiar with calculus, you know that the values of c_1 and c_2 that minimize J are found by setting the partial derivatives $\partial J / \partial c_1$ and $\partial J / \partial c_2$ equal to zero.

$$\frac{\partial J}{\partial c_1} = 250c_1 + 30c_2 - 280 = 0$$

$$\frac{\partial J}{\partial c_2} = 30c_1 + 6c_2 - 38 = 0$$

The solution is $c_1 = 0.9$ and $c_2 = 11/6$. The best straight line in the least-squares sense is $y = 0.9x + 11/6$.

  (b) Evaluating the equation $y = c_1 x + c_2$ at each data point gives the following three equations, which are overdetermined because there are more equations than unknowns.

$$0c_1 + c_2 = 2 \tag{8.4–1}$$

$$5c_1 + c_2 = 6 \tag{8.4–2}$$

$$10c_1 + c_2 = 11 \tag{8.4–3}$$

These equations can be written in the matrix form $\mathbf{Ax} = \mathbf{b}$ as follows.

$$\mathbf{Ax} = \begin{bmatrix} 0 & 1 \\ 5 & 0 \\ 10 & 1 \end{bmatrix} \begin{bmatrix} c_1 \\ c_2 \end{bmatrix} = \begin{bmatrix} 2 \\ 6 \\ 11 \end{bmatrix} = \mathbf{b}$$

where

$$[\mathbf{A}\ \mathbf{b}] = \begin{bmatrix} 0 & 1 & 2 \\ 5 & 1 & 6 \\ 10 & 1 & 11 \end{bmatrix}$$

To use left division, the MATLAB session is

```
>>A = [0,1;5,1;10,1];
>>b = [2;6;11];
>>rank(A)
ans =
  2
>>rank([A, b])
ans =
  3
>>x = A\b
x =
  0.9000
  1.8333
>>A*x
ans =
  1.833
  6.333
  10.8333
```

This result for $\mathbf{x}$ agrees with the least-squares solution obtained previously: $c_1 = 0.9$, $c_2 = 11/6 = 1.8333$. The rank of $\mathbf{A}$ is 2, but the rank of $[\mathbf{A}\ \mathbf{b}]$ is 3, so no exact solution exists for c_1 and c_2. Note that $A*x$ gives the y values generated by the line $y = 0.9x + 1.8333$ at the x data values $x = 0, 5, 10$. These are different from the right-hand sides of the original three equations (8.4–1) through (8.4–3). This is not unexpected, because the least-squares solution is not an exact solution of the equations.

Some overdetermined systems have an exact solution. The left division method sometimes gives an answer for overdetermined systems, but it does not indicate whether the answer is the exact solution. We need to check the ranks of $\mathbf{A}$ and $[\mathbf{A}\ \mathbf{b}]$ to know if the answer is the exact solution. The next example illustrates this situation.

An Overdetermined Set

EXAMPLE 8.4–2

Solve the following equations and discuss the solution for two cases: $c = 9$ and $c = 10$.

$$x_1 + x_2 = 1$$
$$x_1 + 2x_2 = 3$$
$$x_1 + 5x_2 = c$$

■ Solution

The coefficient matrix and the augmented matrix for this problem are

$$\mathbf{A} = \begin{bmatrix} 1 & 1 \\ 1 & 2 \\ 1 & 5 \end{bmatrix} \qquad [\mathbf{A}\ \mathbf{b}] = \begin{bmatrix} 1 & 1 & 1 \\ 1 & 2 & 3 \\ 1 & 5 & c \end{bmatrix}$$

Making the computations in MATLAB, we find that for $c = 9$, rank ($\mathbf{A}$) = rank($[\mathbf{A}\ \mathbf{b}]$) = 2. Thus the system has a solution, and because the number of unknowns (2) equals the rank of $\mathbf{A}$, there is a unique solution. The left division method $\mathtt{A \backslash b}$ gives this solution, which is $x_1 = -1$ and $x_2 = 2$.

For $c = 10$ we find that rank($\mathbf{A}$) = 2, but rank($[\mathbf{A}\ \mathbf{b}]$) = 3. Because rank($\mathbf{A}$) ≠ rank($[\mathbf{A}\ \mathbf{b}]$), there is no solution. However, the left division method $\mathbf{A \backslash b}$ gives $x_1 = -1.3846$ and $x_2 = 2.2692$, which is *not* an exact solution! This can be verified by substituting these values into the original equation set. This answer is the solution to the equation set in a least-squares sense. That is, these values are the values of x_1 and x_2 that minimize J, the sum of the squares of the differences between the equations' left- and right-hand sides.

$$J = (x_1 + x_2 - 1)^2 + (x_1 + 2x_2 - 3)^2 + (x_1 + 5x_2 - 10)^2$$

To interpret MATLAB answers correctly for an overdetermined system, first check the ranks of $\mathbf{A}$ and $[\mathbf{A}\ \mathbf{b}]$ to see if an exact solution exists; if one does not exist, then we know that the left division answer is a least-squares solution. In Section 8.5 we develop a general-purpose program that checks the ranks and solves a general set of linear equations.

Test Your Understanding

T8.4–1 Solve the following set.

$$x_1 - 3x_2 = \quad 2$$
$$3x_1 + 5x_2 = \quad 7$$
$$70x_1 - 28x_2 = 153$$

(Answer: There is a unique solution: $x_1 = 2.2143$, $x_2 = 0.0714$, which is given by the left division method.)

T8.4–2 Show why there is no solution to the following set.

$$x_1 - 3x_2 = 2$$
$$3x_1 + 5x_2 = 7$$
$$5x_1 - 2x_2 = -4$$

8.5 A General Solution Program

In this chapter you saw that the set of linear algebraic equations $\mathbf{Ax} = \mathbf{b}$ with m equations and n unknowns has solutions if and only if (1) rank[$\mathbf{A}$] = rank[$\mathbf{A}$ $\mathbf{b}$]. Let r = rank[$\mathbf{A}$]. If condition (1) is satisfied and if $r = n$, then the solution is unique. If condition (1) is satisfied but $r < n$, an infinite number of solutions exist; in addition, r unknown variables can be expressed as linear combinations of the other $n - r$ unknown variables, whose values are arbitrary. In this case we can use the `rref` command to find the relations between the variables. The pseudocode in Table 8.5–1 can be used to outline an equation solver program before writing it.

A condensed flowchart is shown in Figure 8.5–1. From this chart or the pseudocode, we can develop the script file shown in Table 8.5–2. The program uses the given arrays A and b to check the rank conditions; the left division method to obtain the solution, if one exists; and the `rref` method if there are an infinite number of solutions. Note that the number of unknowns equals the number of columns in A, which is given by `size_A (2)`, the second element in `size_A`. Note also that the rank of $\mathbf{A}$ cannot exceed the number of columns in $\mathbf{A}$.

Test Your Understanding

T8.5–1 Type in the script file `lineq.m` given in Table 8.5–2 and run it for the following cases. Hand-check the answers.
   a. A = [1,-1;1,1], b = [3;5]
   b. A = [1,-1;2,-2], b = [3;6]
   c. A = [1,-1;2,-2], b = [3;5]

Table 8.5–1 Pseudocode for the linear equation solver

If the rank of $\mathbf{A}$ equals the rank of [$\mathbf{A}$ $\mathbf{b}$], then
   determine whether the rank of $\mathbf{A}$ equals the number of unknowns. If so, there is a unique solution, which can be computed using left division. Display the results and stop.
   Otherwise, there are an infinite number of solutions, which can be found from the augmented matrix. Display the results and stop.
Otherwise (if the rank of $\mathbf{A}$ does not equal the rank of [$\mathbf{A}$ $\mathbf{b}$]), there are no solutions. Display this message and stop.

Table 8.5–2 MATLAB program to solve linear equations

```
% Script file lineq.m
% Solves the set Ax = b, given A and b.
% Check the ranks of A and [A b].
if rank(A) == rank([A b])
    % The ranks are equal.
    size_A = size(A);
    % Does the rank of A equal the number of unknowns?
    if rank(A) == size_A(2)
        % Yes. Rank of A equals the number of unknowns.
        disp('There is a unique solution, which is:')
        x = A\b % Solve using left division.
    else
        % Rank of A does not equal the number of unknowns.
        disp('There is an infinite number of solutions.')
        disp('The augmented matrix of the reduced system is:')
        rref([A b]) % Compute the augmented matrix.
    end
else
    % The ranks of A and [A  b] are not equal.
    disp('There are no solutions.')
end
```

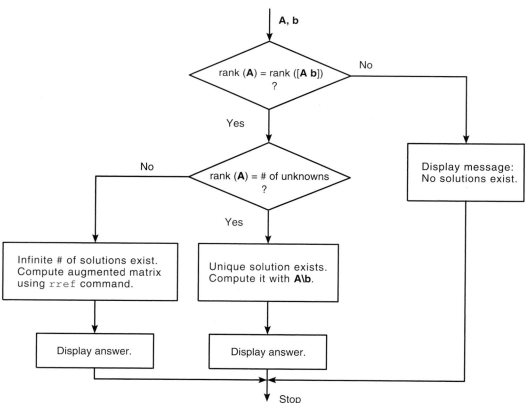

Figure 8.5–1 Flowchart illustrating a program to solve linear equations.

8.6 Summary

If the number of equations in the set *equals* the number of unknown variables, MATLAB provides two ways of solving the equation set $\mathbf{Ax} = \mathbf{b}$: the matrix inverse method, $x = \text{inv}(A)*b$, and the matrix left division method, $x = A\backslash b$. If MATLAB does not generate an error message when you use one of these methods, then the set has a unique solution. You can always check the solution for x by typing Ax to see if the result is the same as b. If you receive an error message, the set is underdetermined (even though it may have an equal number of equations and unknowns), and either it does not have a solution, or it has more than one solution.

For underdetermined sets, MATLAB provides three ways of dealing with the equation set $\mathbf{Ax} = \mathbf{b}$ (note that the matrix inverse method will never work with such sets):

1. The matrix left division method (which gives one specific solution, but not the general solution).
2. The pseudoinverse method. Solve for x by typing $x = \text{pinv}(A)*b$. This gives the minimum norm solution.
3. The reduced row echelon form (RREF) method. This method uses the MATLAB command rref to obtain a general solution for some of the unknowns in terms of the other unknowns.

The four methods are summarized in Table 8.6–1. You should be able to determine whether a unique solution, an infinite number of solutions, or no solution exists. You can do this by applying the existence and uniqueness test given on page 335.

Some overdetermined systems have exact solutions, and they can be obtained with the left division method, but the method does not indicate that the solution is exact. To determine this, first check the ranks of $\mathbf{A}$ and $[\mathbf{A} \quad \mathbf{b}]$ to see if a solution exists; if one does not exist, then we know that the left division solution is a least-squares answer.

Table 8.6–1 Matrix functions and commands for solving linear equations

Function	Description
det(A)	Computes the determinant of the array **A**.
inv(A)	Computes the inverse of the matrix **A**.
pinv(A)	Computes the pseudoinverse of the matrix **A**.
rank(A)	Computes the rank of the matrix **A**.
rref([A b])	Computes the reduced row echelon form corresponding to the augmented matrix [**A b**].
x = inv(A)*b	Solves the matrix equation $\mathbf{Ax} = \mathbf{b}$ using the matrix inverse.
x = A\b	Solves the matrix equation $\mathbf{Ax} = \mathbf{b}$ using left division.

Key Terms with Page References

Augmented matrix, 335
Euclidean norm, 343
Gauss elimination, 335
Homogeneous equations, 335
Ill-conditioned equations, 333
Least-squares method, 351
Left division method, 335
Matrix inverse, 332
Matrix rank, 334

Minimum norm solution, 342
Overdetermined system, 350
Pseudoinverse method, 342
Reduced row echelon form, 345
Singular matrix, 333
Statically indeterminate, 343
Subdeterminant, 335
Underdetermined system, 341

Problems

You can find the answers to problems marked with an asterisk at the end of the text.

Section 8.1

1. Solve the following problems using matrix inversion. Check your solutions by computing $A^{-1}A$.

   a. $2x + y = 5$
   $3x - 9y = 7$

   b. $-8x - 5y = 4$
   $-2x + 7y = 10$

   c. $12x - 5y = 11$
   $-3x + 4y + 7x_3 = -3$
   $6x + 2y + 3x_3 = 22$

   d. $6x - 3y + 4x_3 = 41$
   $12x + 5y - 7x_3 = -26$
   $-5x + 2y + 6x_3 = 16$

2.* a. Solve the following matrix equation for the matrix C.

$$A(BC + A) = B$$

   b. Evaluate the solution obtained in part a for the case

$$A = \begin{bmatrix} 7 & 9 \\ -2 & 4 \end{bmatrix} \qquad B = \begin{bmatrix} 4 & -3 \\ 7 & 6 \end{bmatrix}$$

3. Use MATLAB to solve the following problems.

a. $-2x + y = -5$
   $-2x + y = 3$

b. $-2x + y = 3$
   $-8x + 4y = 12$

c. $-2x + y = -5$
   $-2x + y = -5.00001$

d. $x_1 + 5x_2 - x_3 + 6x_4 = 19$
   $2x_1 - x_2 + x_3 - 2x_4 = 7$
   $-x_1 + 4x_2 - x_3 + 3x_4 = 30$
   $3x_1 - 7x_2 - 2x_3 + x_4 = -75$

Section 8.2

4. The circuit shown in Figure P4 has five resistances and one applied voltage. Kirchhoff's voltage law applied to each loop in the circuit shown gives

$$v - R_2 i_2 - R_4 i_4 = 0$$
$$-R_2 i_2 + R_1 i_1 + R_3 i_3 = 0$$
$$-R_4 i_4 - R_3 i_3 + R_5 i_5 = 0$$

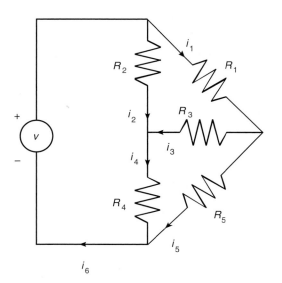

Figure P4

Conservation of charge applied at each node in the circuit gives

$$i_6 = i_1 + i_2$$
$$i_2 + i_3 = i_4$$
$$i_1 = i_3 + i_5$$
$$i_4 + i_5 = i_6$$

a. Write a MATLAB script file that uses given values of the applied voltage v and the values of the five resistances and solves for the six currents.

b. Use the program developed in part a to find the currents for the case where $R_1 = 1$, $R_2 = 5$, $R_3 = 2$, $R_4 = 10$, $R_5 = 5$ kΩ, and $v = 100$ V. (1 k$\Omega = 1000 \ \Omega$.)

5.*   a. Use MATLAB to solve the following equations for x, y, and z as functions of the parameter c.

$$x - 5y - 2z = 11c$$
$$6x + 3y + z = 13c$$
$$7x + 3y - 5z = 10c$$

b. Plot the solutions for x, y, and z versus c on the same plot, for $-10 \le c \le 10$.

6.   Fluid flows in pipe networks can be analyzed in a manner similar to that used for electric resistance networks. Figure P6 shows a network with three pipes. The volume flow rates in the pipes are q_1, q_2, and q_3. The pressures at the pipe ends are p_a, p_b, and p_c. The pressure at the junction is p_1. Under certain conditions, the pressure–flow rate relation in a pipe has the same form as the voltage-current relation in a resistor. Thus, for the three pipes, we have

$$q_1 = \frac{1}{R_1}(p_a - p_1)$$

$$q_2 = \frac{1}{R_2}(p_1 - p_b)$$

$$q_3 = \frac{1}{R_3}(p_1 - p_c)$$

where the R_i are the pipe resistances. From conservation of mass, $q_1 = q_2 + q_3$.

a. Set up these equations in a matrix form $\mathbf{Ax} = \mathbf{b}$ suitable for solving for the three flow rates q_1, q_2, and q_3 and the pressure p_1, given the values of pressures p_a, p_b, and p_c and the values of resistances R_1, R_2, and R_3. Find the expressions for $\mathbf{A}$ and $\mathbf{b}$.

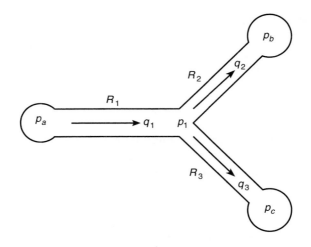

(a)

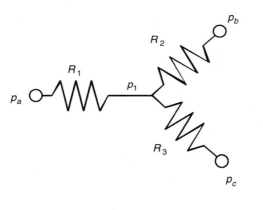

(b)

Figure P6

b. Use MATLAB to solve the matrix equations obtained in part a for the case where $p_a = 4320$ lb/ft^2, $p_b = 3600$ lb/ft^2, and $p_c = 2880$ lb/ft^2. These correspond to 30, 25, and 20 psi, respectively (1 psi = 1 lb/in^2, and atmospheric pressure is 14.7 psi). Use the resistance values $R_1 = 10\ 000$, $R_2 = 14\ 000$ lb sec/ft^5. These values correspond to fuel oil flowing through pipes 2 ft long, with 2- and 1.4-in. diameters, respectively. The units of the answers are ft^3/sec for the flow rates and lb/ft^2 for pressure.

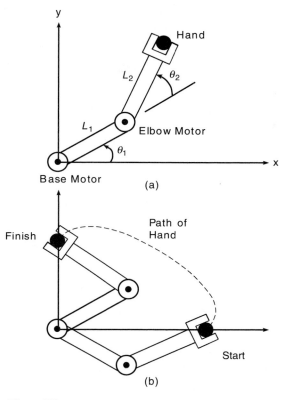

Figure P7

7. Figure P7 illustrates a robot arm that has two "links" connected by two "joints"—a shoulder or base joint and an elbow joint. There is a motor at each joint. The joint angles are θ_1 and θ_2. The (x, y) coordinates of the hand at the end of the arm are given by

$$x = L_1 \cos \theta_1 + L_2 \cos (\theta_1 + \theta_2)$$
$$y = L_1 \sin \theta_1 + L_2 \sin (\theta_1 + \theta_2)$$

where L_1 and L_2 are the lengths of the links.

Polynomials are used for controlling the motion of robots. If we start the arm from rest with zero velocity and acceleration, the following polynomials are used to generate commands to be sent to the joint motor controllers

$$\theta_1(t) = \theta_1(0) + a_1 t^3 + a_2 t^4 + a_3 t^5$$
$$\theta_2(t) = \theta_2(0) + b_1 t^3 + b_2 t^4 + b_3 t^5$$

where $\theta_1(0)$ and $\theta_2(0)$ are the starting values at time $t = 0$. The angles $\theta_1(t_f)$ and $\theta_2(t_f)$ are the joint angles corresponding to the desired destination of the arm at time t_f. The values of $\theta_1(0)$, $\theta_2(0)$, $\theta_1(t_f)$, and $\theta_2(t_f)$ can be found from trigonometry, if the starting and ending (x, y) coordinates of the hand are specified.

   a. Set up a matrix equation to be solved for the coefficients a_1, a_2, and a_3, given values for $\theta_1(0)$, $\theta_1(t_f)$, and t_f. Obtain a similar equation for the coefficients b_1, b_2, and b_3.

   b. Use MATLAB to solve for the polynomial coefficients given the values $t_f = 2$ sec, $\theta_1(0) = -19°$, $\theta_2(0) = 44°$, $\theta_1(t_f) = 43°$, and $\theta_2(t_f) = 151°$. (These values correspond to a starting hand location of $x = 6.5$, $y = 0$ ft and a destination location of $x = 0$, $y = 2$ ft for $L_1 = 4$ and $L_2 = 3$ ft.)

   c. Use the results of part b to plot the path of the hand.

8.* Engineers must be able to predict the rate of heat loss through a building wall to determine the heating system requirements. They do this by using the concept of *thermal resistance R*, which relates the heat flow rate q through a material to the temperature difference ΔT across the material: $q = \Delta T/R$. This relation is like the voltage-current relation for an electric resistor: $i = v/R$. So the heat flow rate plays the role of electric current, and the temperature difference plays the role of the voltage difference. The SI unit for q is the *watt* (W), which is 1 joule/second (J/s).

The wall shown in Figure P8 consists of four layers: an inner layer of plaster/lathe 10 mm thick, a layer of fiberglass insulation 125 mm thick, a

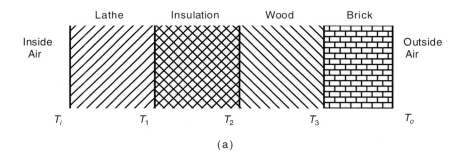

(a)

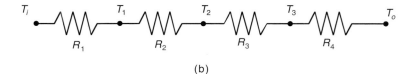

(b)

Figure P8

layer of wood 60 mm thick, and an outer layer of brick 50 mm thick. If we assume that the inner and outer temperatures T_i and T_o have remained constant for some time, then the heat energy stored in the layers is constant, and thus the heat flow rate through each layer is the same. Applying conservation of energy gives the following equations.

$$q = \frac{1}{R_1}(T_i - T_1) = \frac{1}{R_2}(T_1 - T_2) = \frac{1}{R_3}(T_2 - T_3) = \frac{1}{R_4}(T_3 - T_o)$$

The thermal resistance of a solid material is given by $R = D/k$, where D is the material thickness and k is the material's *thermal conductivity.* For the given materials, the resistances for a wall area of 1 m^2 are $R_1 = 0.036$, $R_2 = 4.01$, $R_3 = 0.408$, and $R_4 = 0.038$ K/W.

Suppose that $T_i = 20°C$ and $T_o = -10°C$. Find the other three temperatures and the heat loss rate q, in watts. Compute the heat loss rate if the wall's area is 10 m^2.

9.   The concept of thermal resistance described in Problem 8 can be used to find the temperature distribution in the flat square plate shown in Figure P9(a).

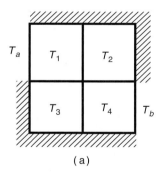

(a)

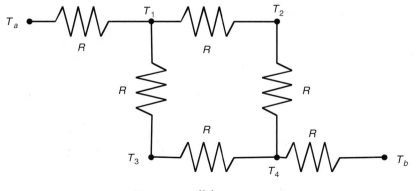

(b)

Figure P9

The plate's edges are insulated so that no heat can escape, except at two points where the edge temperature is heated to T_a and T_b, respectively. The temperature varies through the plate, so no single point can describe the plate's temperature. One way to estimate the temperature distribution is to imagine that the plate consists of four subsquares and to compute the temperature in each subsquare. Let R be the thermal resistance of the material between the centers of adjacent subsquares. Then we can think of the problem as a network of electric resistors, as shown in part (b) of the figure. Let q_{ij} be the heat flow rate between the points whose temperatures are T_i and T_j. If T_a and T_b remain constant for some time, then the heat energy stored in each subsquare is constant also, and the heat flow rate between each subsquare is constant. Under these conditions, conservation of energy says that the heat flow into a subsquare equals the heat flow out. Applying this principle to each subsquare gives the following equations.

$$q_{a1} = q_{12} + q_{13}$$
$$q_{12} = q_{24}$$
$$q_{13} = q_{34}$$
$$q_{34} + q_{24} = q_{4b}$$

Substituting $q = (T_i - T_j)/R$, we find that R can be canceled out of every equation, and they can be rearranged as follows:

$$T_1 = \frac{1}{3}(T_a + T_2 + T_3)$$

$$T_2 = \frac{1}{2}(T_1 + T_4)$$

$$T_3 = \frac{1}{2}(T_1 + T_4)$$

$$T_4 = \frac{1}{3}(T_2 + T_3 + T_5)$$

These equations tell us that the temperature of each subsquare is the average of the temperatures in the adjacent subsquares!

Solve these equations for the case where $T_a = 150°C$ and $T_b = 20°C$.

10. Use the averaging principle developed in Problem 9 to find the temperature distribution of the plate shown in Figure P10, using the 3×3 grid and the given values $T_a = 150°C$ and $T_b = 20°C$.

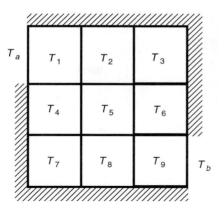

Figure P10

Section 8.3

11.* Solve the following equations:

$$7x + 9y - 9z = 22$$

$$3x + 2y - 4z = 12$$

$$x + 5y - z = -2$$

12. The following table shows how many hours in process reactors A and B are required to produce 1 ton each of chemical products 1, 2, and 3. The two reactors are available for 35 and 40 hrs per week, respectively.

Hours	Product 1	Product 2	Product 3
Reactor A	6	2	10
Reactor B	3	5	2

Let x, y, and z be the number of tons each of products 1, 2, and 3 that can be produced in one week.

a. Use the data in the table to write two equations in terms of x, y, and z. Determine whether a unique solution exists. If not, use MATLAB to find the relations between x, y, and z.

b. Note that negative values x, y, and z have no meaning here. Find the allowable ranges for x, y, and z.

c. Suppose the profits for each product are $200, $300, and $100 for products 1, 2, and 3, respectively. Find the values of x, y, and z to maximize the profit.

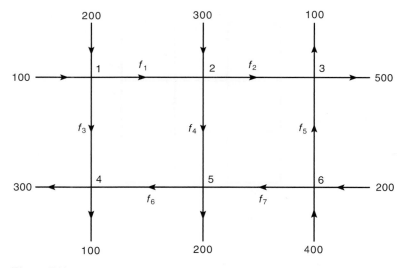

Figure P13

    d. Suppose the profits for each product are \$200, \$500, and \$100 for products 1, 2, and 3, respectively. Find the values of *x, y,* and *z* to maximize the profit.

13. See Figure P13. Assume that no vehicles stop within the network. A traffic engineer wants to know if the traffic flows $f_1, f_2, \ldots, f_7$ (in vehicles per hour) can be computed given the measured flows shown in the figure. If not, then determine how many more traffic sensors need to be installed, and obtain the expressions for the other traffic flows in terms of the measured quantities.

Section 8.4

14.* Use MATLAB to solve the following problem:

$$x - 3y = 2$$
$$x + 5y = 18$$
$$4x - 6y = 20$$

15.* Use MATLAB to solve the following problem:

$$x - 3y = 2$$
$$x + 5y = 18$$
$$4x - 6y = 10$$

16. *a.* Use MATLAB to find the coefficients of the quadratic polynomial $y = ax^2 + bx + c$ that passes through the three points $(x, y) = (1, 4)$, $(4, 73)$, $(5, 120)$.

  b. Use MATLAB to find the coefficients of the cubic polynomial $y = ax^3 + bx^2 + cx + d$ that passes through the three points given in part *a*.

17. Use the MATLAB program given in Table 8.5–2 to solve the following problems:

  a. Problem 3d

  b. Problem 11

  c. Problem 14

  d. Problem 15

Engineering in the 21st Century. . .

Rebuilding the Infrastructure

During the Great Depression, many public works projects that improved the nation's infrastructure were undertaken to stimulate the economy and provide employment. These projects included highways, bridges, water supply systems, sewer systems, and electrical power distribution networks. Following World War II another burst of such activity culminated in the construction of the interstate highway system. As we enter the 21st century, much of the infrastructure is 30 to 70 years old and is literally crumbling or not up to date. One survey showed that more than 25 percent of the nation's bridges are substandard. These need to be repaired or replaced.

Rebuilding the infrastructure requires engineering methods different from those in the past because labor and material costs are now higher and environmental and social issues have greater importance than before. Infrastructure engineers must take advantage of new materials, inspection technology, construction techniques, and labor-saving machines.

Also, some infrastructure components, such as communications networks, need to be replaced because they are outdated and do not have sufficient capacity or ability to take advantage of new technology. An example is the *information infrastructure,* which includes physical facilities to transmit, store, process, and display voice, data, and images. Better communications and computer networking technology will be needed for such improvements. Many of the MATLAB toolboxes provide advanced support for such work, including the Financial, Communications, Image Processing, Signal Processing, PDE, and Wavelet toolboxes. ■

CHAPTER 9

Numerical Methods for Calculus and Differential Equations

OUTLINE

9.1 Numerical Integration

9.2 Numerical Differentiation

9.3 First-Order Differential Equations

9.4 Higher-Order Differential Equations

9.5 Special Methods for Linear Equations

9.6 Summary

Problems

This chapter covers numerical methods for computing integrals and derivatives and for solving ordinary differential equations. Some integrals cannot be evaluated analytically, and we need to compute them numerically with an approximate method (Section 9.1). In addition, it is often necessary to use data to estimate rates of change, and this requires a numerical estimate of the derivative (Section 9.2). Finally, many differential equations cannot be solved analytically, and so we must be able to solve them by using appropriate numerical techniques. Section 9.3 covers first-order differential equations, and Section 9.4 extends the methods to higher-order equations. More powerful methods are available for linear equations. Section 9.5 treats these methods.

When you have finished this chapter, you should be able to

■ Use MATLAB to numerically evaluate integrals.

■ Use numerical methods with MATLAB to estimate derivatives.

■ Use the MATLAB numerical differential equation solvers to obtain solutions.

9.1 Numerical Integration

The integral of a function $f(x)$ for $a \leq x \leq b$ can be interpreted as the area between the $f(x)$ curve and the x axis, bounded by the limits $x = a$ and $x = b$. If we denote this area by A, then we can write A as

$$A = \int_a^b f(x)\,dx \tag{9.1--1}$$

DEFINITE
INTEGRAL

INDEFINITE
INTEGRAL

IMPROPER
INTEGRAL

SINGULARITIES

An integral is called a *definite* integral if it has specified limits of integration. *Indefinite* integrals have no specified limits. *Improper* integrals can have infinite values, depending on their integration limits. For example, the following integral can be found in most integral tables:

$$\int \frac{1}{x-1}\,dx = \ln|x-1|$$

However, it is an improper integral if the integration limits include the point $x = 1$. So, even though an integral can be found in an integral table, you should examine the integrand to check for *singularities,* which are points at which the integrand is undefined. The same warning applies when you are using numerical methods to evaluate integrals.

Trapezoidal Integration

The simplest way to find the area under a curve is to split the area into rectangles (Figure 9.1–1a). If the widths of the rectangles are small enough, the sum of their areas gives the approximate value of the integral. A more sophisticated method is to use trapezoidal elements (Figure 9.1–1b). Each trapezoid is called a *panel.* It is not necessary to use panels of the same width; to increase the method's

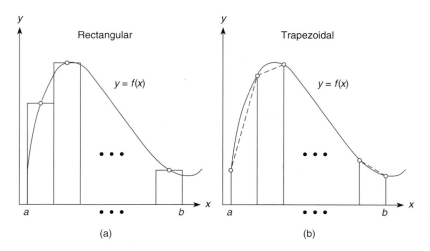

Figure 9.1–1 Illustration of (a) rectangular and (b) trapezoidal numerical integration.

accuracy, you can use narrow panels where the function is changing rapidly. When the widths are adjusted according to the function's behavior, the method is said to be *adaptive*. MATLAB implements trapezoidal integration with the `trapz` function. Its syntax is `trapz(x,y)`, where the array `y` contains the function values at the points contained in the array `x`. If you want the integral of a single function, then `y` is a vector. To integrate more than one function, place their values in a matrix `y`; `trapz(x,y)` will compute the integral of each column of `y`.

You cannot directly specify a function to integrate with the `trapz` function; you must first compute and store the function's values ahead of time in an array. Later we will discuss two other integration functions, the `quad` and `quadl` functions, that can accept functions directly. However, they cannot handle arrays of values. So the functions complement each other. The `trapz` function is summarized in Table 9.1–1.

As a simple example of the use of the `trapz` function, let us compute the integral

$$A = \int_0^{\pi} \sin x \, dx \qquad (9.1-2)$$

whose exact answer is $A = 2$. To investigate the effect of panel width, let us first use 10 panels with equal widths of $\pi/10$. The script file is

```
x = linspace(0,pi,10);
y = sin(x);
A = trapz(x,y)
```

Table 9.1–1 Basic syntax of numerical integration functions

Command	Description
`dblquad(fun,a,b,c,d)`	Computes the double integral of the function $f(x, y)$ between the limits $a \leq x \leq b$ and $c \leq y \leq d$. The input `fun` specifies the function that computes the integrand. It must accept a vector argument x and scalar y, and it must return a vector result.
`polyint(p,C)`	Computes the integral of the polynomial p using an optional user-specified constant of integration C.
`quad(fun,a,b)`	Uses an adaptive Simpson rule to compute the integral of the function `fun` between the limits a and b. The input `fun`, which represents the integrand $f(x)$, is a function handle for the integrand function. It must accept a vector argument x and return the vector result y.
`quadl(fun,a,b)`	Uses Lobatto integration. The syntax is identical to `quad`.
`trapz(x,y)`	Uses trapezoidal integration to compute the integral of y with respect to x, where the array y contains the function values at the points contained in the array x.
`triplequad(fun,a,b,c,d,e,f)`	Computes the triple integral of the function $f(x, y, z)$ between the limits $a \leq x \leq b$, $c \leq y \leq d$, and $e \leq y \leq f$. The input `fun` specifies the function that computes the integrand. It must accept a vector argument x, scalar y, and scalar z, and it must return a vector result.

The answer is $A = 1.9797$, which gives a relative error of $100(2 - 1.9797)/2 = 1$ percent. Now try 100 panels of equal width; replace the array x with x = linspace(0,pi,100). The answer is $A = 1.9998$ for a relative error of $100(2 - 1.9998)/2 = 0.01$ percent. If we examine the plot of the integrand $\sin x$, we see that the function is changing faster near $x = 0$ and $x = \pi$ than near $x = \pi/2$. Thus we could achieve the same accuracy by using fewer panels if narrower panels are used near $x = 0$ and $x = \pi$.

We normally use the trapz function when the integrand is given as a table of values. Otherwise, if the integrand is given as a function, use the quad or quadl functions, to be introduced shortly.

<table>
<tr><td>**EXAMPLE 9.1–1**</td><td style="text-align:right">Velocity from an Accelerometer</td></tr>
</table>

An *accelerometer* is used in aircraft, rockets, and other vehicles to estimate the vehicle's velocity and displacement. The accelerometer integrates the acceleration signal to produce an estimate of the velocity, and it integrates the velocity estimate to produce an estimate of displacement. Suppose the vehicle starts from rest at time $t = 0$, and its measured acceleration is given in the following table.

Time (s)	0	1	2	3	4	5	6	7	8	9	10
Acceleration (m/s^2)	0	2	4	7	11	17	24	32	41	48	51

(a) Estimate the velocity v after 10 s.
(b) Estimate the velocity at times $t = 1, 2, \ldots, 10$ s.

■ **Solution**
(a) The initial velocity is zero, so $v(0) = 0$. The relation between the velocity and acceleration $a(t)$ is

$$v(10) = \int_0^{10} a(t)\, dt + v(0) = \int_0^{10} a(t)\, dt$$

The script file is shown below.

```
t = 0:10;
a = [0,2,4,7,11,17,24,32,41,48,51];
v10 = trapz(t,a)
```

The answer for the velocity after 10 s is v10, and it is 211.5 m/s.

(b) The following script file uses the fact that the velocity can be expressed as

$$v(t_{k+1}) = \int_{t_k}^{t_{k+1}} a(t)\, dt + v(t_k) \qquad k = 1, 2, \ldots, 10$$

where $v(t_1) = 0$.

```
t = 0:10;
a = [0,2,4,7,11,17,24,32,41,48,51];
```

```
v(1) = 0;
for k = 1:10
   v(k+1) = trapz(t(k:k+1), a(k:k+1))+v(k);
end
disp([t',v'])
```

The answers are given in the following table.

Time (s)	0	1	2	3	4	5	6	7	8	9	10
Velocity (m/s)	0	1	4	9.5	18.5	32.5	53	81	117	162	211.5

Test Your Understanding

T9.1–1 Modify the script file given in part (b) of Example 9.1–1 to estimate the displacement at times $t = 1, 2, \ldots, 10$ s. (Partial answer: The displacement after 10 s is 584.25 m.)

Quadrature Functions

Another approach to numerical integration is *Simpson's rule*, which divides the integration range $b - a$ into an even number of sections and uses a different quadratic function to represent the integrand for each panel. A quadratic function has three parameters, and Simpson's rule computes these parameters by requiring that the quadratic pass through the function's three points corresponding to the two adjacent panels. To obtain greater accuracy, we can use polynomials of degree higher than 2.

The MATLAB function `quad` implements an adaptive version of Simpson's rule. The `quadl` function is based on an adaptive Lobatto integration method, where the letter "l" in `quadl` stands for Lobatto. The term *quad* is an abbreviation of *quadrature,* which is an old term for the process of measuring areas. Some writers distinguish between the terms *quadrature* and *integration* and reserve *integration* to mean numerical integration of ordinary differential equations. We will not make that distinction.

The function `quad(fun,a,b)` computes the integral of the function `fun` between the limits a and b. The input `fun`, which represents the integrand $f(x)$, is either a function handle of the integrand function (the preferred method) or the name of the function as a character string (i.e., placed in single quotes). The function $y = f(x)$ must accept a vector argument x and must return the vector result y. The basic syntax of `quadl` is identical and is summarized in Table 9.1–1.

To illustrate, let us compute the integral given in Equation (9.1–2). The session consists of one command: `A = quad(@sin, 0, pi)` or `A = quad('sin', 0, pi)`. The answer given by MATLAB is $A = 2.0000$, which is correct to four decimal places. We use `quadl` the same way.

Because the `quad` and `quadl` functions call the integrand function using vector arguments, you must always use array operations when defining the function. The following example shows how this is done.

EXAMPLE 9.1–2   Evaluation for Fresnel's Cosine Integral

Some simple-looking integrals cannot be evaluated in closed form. An example is Fresnel's cosine integral

$$A = \int_0^b \cos x^2 \, dx \qquad (9.1\text{--}3)$$

(a) Demonstrate two ways to compute the integral when the upper limit is $b = \sqrt{2\pi}$.

(b) Demonstrate the use of a nested function to compute the more general integral

$$A = \int_0^b \cos x^n \, dx \qquad (9.1\text{--}4)$$

for $n = 2$ and for $n = 3$.

Solution

(a) The integrand $\cos x^2$ obviously does not contain any singularities that might cause problems for the integration function. We demonstrate two ways to use the `quad` function.

1. With a function file: Define the integrand with a user-defined function as shown by the following function file.

```
function c2 = cossq(x)
c2 = cos(x.^2);
```

The `quad` function is called as follows: `A = quad(@cossq,0,sqrt(2*pi))`. The result is $A = 0.6119$.

2. With an anonymous function (anonymous functions are discussed in Section 3.3): The session is

```
>>cossq = @(x)cos(x.^2);
>>A = quad(cossq,0,sqrt(2*pi))
  A =
     0.6119
```

The two lines can be combined into one as follows:

```
A = quad(@(x)cos(x.^2),0,sqrt(2*pi))
```

The advantage of using an anonymous function is that you need not create and save a function file. However, for complicated integrand functions, using a function file is preferable.

(b) Because `quad` requires that the integrand function have only one argument, the following code will not work.

```
>>cossq = @(x)cos(x.^n);
>>n = 2;
>>A = quad(cossq,0,sqrt(2*pi))
??? Undefined function or variable 'n'.
```

Instead we will use parameter passing with a nested function (nested functions are discussed in Section 3.3). First create and save the following function.

```
function A = integral_n(n)
A = quad(@cossq_n,0,sqrt(2*pi));

% Nested function
   function integrand = cossq_n(x)
      integrand = cos(x.^n);
   end
end
```

The session for $n = 2$ and $n = 3$ is as follows.

```
>>A = integral_n(2)
   A =
      0.6119
>>A = integral_n(3)
   A =
      0.7734
```

The quad functions have some optional arguments for analyzing and adjusting the algorithm's efficiency and accuracy. Type help quad for details.

Test Your Understanding

T9.1–2 Use both the quad and quadl functions to compute the integral

$$A = \int_2^5 \frac{1}{x}\, dx$$

and compare the answers with that obtained from the closed-form solution, which is $A = 0.9163$.

Polynomial Integration

MATLAB provides the polyint function to compute the integral of a polynomial. The syntax q = polyint(p, C) returns a polynomial q representing the integral of polynomial p with a user-specified scalar constant of integration C. The elements of the vector p are the coefficients of the polynomial, arranged in descending powers. The syntax polyint(p) assumes the constant of integration C is zero.

For example, the integral of $12x^3 + 9x^2 + 8x + 5$ is obtained from q = polyint([12,9,8,5], 10). The answer is q = [3, 3, 4, 5, 10], which corresponds to $3x^4 + 3x^3 + 4x^2 + 5x + 10$. Because polynomial integrals can be obtained from a symbolic formula, the polyint function is not a numerical integration operation.

Double Integrals

The function `dblquad` computes double integrals. Consider the integral

$$A = \int_c^d \int_a^b f(x, y)\,dx\,dy$$

The basic syntax is

```
A = dblquad(fun, a, b, c, d)
```

where `fun` is the handle to a user-defined function that defines the integrand $f(x, y)$. The function must accept a vector x and a scalar y, and it must return a vector result, so the appropriate array operations must be used. The extended syntax enables the user to adjust the accuracy and to use `quadl` or a user-defined quadrature routine. See the MATLAB Help for details.

For example, using an anonymous function to compute the integral

$$A = \int_0^1 \int_1^3 xy^2\,dx\,dy$$

you type

```
>>fun = @(x,y)x.*y^2;
>>A = dblquad(fun, 1, 3, 0, 1)
```

The answer is $A = 1.3333$.

The preceding integral is carried out over the rectangular region specified by $1 \le x \le 3, 0 \le y \le 1$. Some double integrals are specified over a nonrectangular region. These problems can be handled by a transformation of variables. You can also use a rectangular region that encloses the nonrectangular region and force the integrand to be zero outside of the nonrectangular region, by using the MATLAB relational operators, for example. See Problem 15. The following example illustrates the former approach.

EXAMPLE 9.1–3   Double Integral over a Nonrectangular Region

Compute the integral

$$A = \int \int_R (x - y)^4 (2x + y)^2\,dx\,dy$$

over the region R bounded by the lines

$$x - y = \pm 1 \qquad 2x + y = \pm 2$$

Solution

We must convert the integral into one that is specified over a rectangular region. To do this, let $u = x - y$ and $v = 2x + y$. Thus, using the Jacobian, we obtain

$$dx\,dy = \begin{vmatrix} \partial x/\partial u & \partial x/\partial v \\ \partial y/\partial u & \partial y/\partial v \end{vmatrix} du\,dv = \begin{vmatrix} 1/3 & 1/3 \\ -2/3 & 1/3 \end{vmatrix} du\,dv = \frac{1}{3} du\,dv$$

Then the region R is specified as a rectangular region in terms of u and v. Its boundaries are given by $u = \pm 1$ and $v = \pm 2$, and the integral becomes

$$A = \frac{1}{3} \int_{-2}^{2} \int_{-1}^{1} u^4 v^2 \, du \, dv$$

and the MATLAB session is

```
>>fun = @(u,v)u.^4*v^2;
>>A = (1/3)*dblquad(fun, -1, 1, -2, 2)
```

The answer is $A = 0.7111$.

Triple Integrals

The function `triplequad` computes triple integrals. Consider the integral

$$A = \int_{e}^{f} \int_{c}^{d} \int_{a}^{b} f(x, y, z) \, dx \, dy \, dz$$

The basic syntax is

```
A = triplequad(fun, a, b, c, d, e, f)
```

where `fun` is the handle to a user-defined function that defines the integrand $f(x, y, z)$. The function must accept a vector x, a scalar y, and a scalar z, and it must return a vector result, so the appropriate array operations must be used. The extended syntax enables the user to adjust the accuracy and to use `quadl` or a user-defined quadrature routine. See the MATLAB Help for details. For example, to compute the integral

$$A = \int_{1}^{2} \int_{0}^{2} \int_{1}^{3} \left(\frac{xy - y^2}{z} \right) dx \, dy \, dz$$

You type

```
>>fun = @(x,y,z)(x*y-y^2)/z;
>>A = triplequad(fun, 1, 3, 0, 2, 1, 2)
```

The answer is $A = 1.8484$.

9.2 Numerical Differentiation

The derivative of a function can be interpreted graphically as the slope of the function. This interpretation leads to various methods for computing the derivative of a set of data. Figure 9.2–1 shows three data points that represent a function $y(x)$. Recall that the definition of the derivative is

$$\frac{dy}{dx} = \lim_{\Delta x \to 0} \frac{\Delta y}{\Delta x} \qquad (9.2\text{–}1)$$

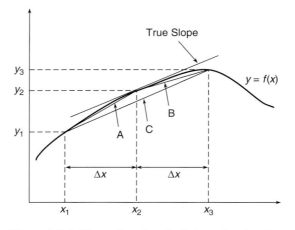

Figure 9.2–1 Illustration of methods for estimating the derivative dy/dx.

The success of numerical differentiation depends heavily on two factors: the spacing of the data points and the scatter present in the data due to measurement error. The greater the spacing, the more difficult it is to estimate the derivative. We assume here that the spacing between the measurements is regular; that is, $x_3 - x_2 = x_2 - x_1 = \Delta x$. Suppose we want to estimate the derivative dy/dx at the point x_2. The correct answer is the slope of the straight line passing through the point (x_2, y_2); but we do not have a second point on that line, so we cannot find its slope. Therefore, we must estimate the slope by using nearby data points. One estimate can be obtained from the straight line labeled A in the figure. Its slope is

$$m_A = \frac{y_2 - y_1}{x_2 - x_1} = \frac{y_2 - y_1}{\Delta x} \tag{9.2–2}$$

BACKWARD DIFFERENCE

This estimate of the derivative is called the *backward difference* estimate, and it is actually a better estimate of the derivative at $x = x_1 + (\Delta x)/2$ than at $x = x_2$. Another estimate can be obtained from the straight line labeled B. Its slope is

$$m_B = \frac{y_3 - y_2}{x_3 - x_2} = \frac{y_3 - y_2}{\Delta x} \tag{9.2–3}$$

FORWARD DIFFERENCE

This estimate is called the *forward difference* estimate, and it is a better estimate of the derivative at $x = x_2 + (\Delta x)/2$ than at $x = x_2$. Examining the plot, you might think that the average of these two slopes would provide a better estimate of the derivative at $x = x_2$, because the average tends to cancel out the effects of measurement error. The average of m_A and m_B is

$$m_C = \frac{m_A + m_B}{2} = \frac{1}{2}\left(\frac{y_2 - y_1}{\Delta x} + \frac{y_3 - y_2}{\Delta x}\right) = \frac{y_3 - y_1}{2\,\Delta x} \tag{9.2–4}$$

This is the slope of the line labeled *C*, which connects the first and third data points. This estimate of the derivative is called the *central difference* estimate.

The diff Function

MATLAB provides the diff function to use for computing derivative estimates. Its syntax is d = diff(x), where x is a vector of values, and the result is a vector d containing the differences between adjacent elements in x. That is, if x has *n* elements, d will have *n* − 1 elements, where $d = [x(2) - x(1), x(3) - x(2), \ldots, x(n) - x(n - 1)]$. For example, if x = [5, 7, 12, -20], then diff(x) returns the vector [2, 5, -32]. The derivative dy/dx can be estimated from diff(y)./diff(x).

The following script file implements the backward difference and central difference methods for artificial data generated from a sinusoidal signal that is measured 51 times during one half-period. The measurement error is uniformly distributed between –0.025 and 0.025.

```
x = 0:pi/50:pi;
n = length(x);
% Data-generation function with +/-0.025 random error.
y = sin(x)+.05*(rand(1,51)-0.5);
% Backward difference estimate of dy/dx.
d1 = diff(y)./diff(x);
subplot(2,1,1)
plot(x(2:n),d1,x(2:n),d1,'o')
% Central difference estimate of dy/dx.
d2 = (y(3:n)-y(1:n-2))./(x(3:n)-x(1:n-2));
subplot(2,1,2)
plot(x(2:n-1),d2,x(2:n-1),d2,'o')
```

Test Your Understanding

T9.2–1 Modify the previous program to use the forward difference method to estimate the derivative. Plot the results, and compare with the results from the backward and central difference methods.

Polynomial Derivatives

MATLAB provides the polyder function to compute the derivative of a polynomial. Its syntax has several forms. The basic form is d = polyder(p), where p is a vector whose elements are the coefficients of the polynomial, arranged in descending powers. The output d is a vector containing the coefficients of the derivative polynomial.

The second syntax form is d = polyder(p1,p2). This form computes the derivative of the *product* of the two polynomials p1 and p2. The third form is [num, den] = polyder(p2,p1). This form computes the derivative of the *quotient* p2/p1. The vector of coefficients of the numerator of the derivative is given by num. The denominator is given by den.

Here are some examples of the use of polyder. Let $p_1 = 5x + 2$ and $p_2 = 10x^2 + 4x - 3$. Then

$$\frac{dp_2}{dx} = 20x + 4$$

$$p_1 p_2 = 50x^3 + 40x^2 - 7x - 6$$

$$\frac{d(p_1 p_2)}{dx} = 150x^2 + 80x - 7$$

$$\frac{d(p_2/p_1)}{dx} = \frac{50x^2 + 40x + 23}{25x^2 + 20x + 4}$$

These results can be obtained with the following program.

```
p1 = [5, 2];p2 = [10, 4, -3];
% Derivative of p2.
der2 = polyder(p2)
% Derivative of p1*p2.
prod = polyder(p1,p2)
% Derivative of p2/p1.
[num, den] = polyder(p2,p1)
```

The results are der2 = [20, 4], prod = [150, 80, -7], num = [50, 40, 23], and den = [25, 20, 4].

Because polynomial derivatives can be obtained from a symbolic formula, the polyder function is not a numerical differentiation operation.

Gradients

The *gradient* ∇f of a function $f(x, y)$ is a vector pointing in the direction of increasing values of $f(x, y)$. It is defined by

$$\nabla f = \frac{\partial f}{\partial x}\mathbf{i} + \frac{\partial f}{\partial y}\mathbf{j}$$

where $\mathbf{i}$ and $\mathbf{j}$ are the unit vectors in the x and y directions, respectively. The concept can be extended to functions of three or more variables.

In MATLAB the gradient of a set of data representing a two-dimensional function $f(x, y)$ can be computed with the gradient function. Its syntax is [df_dx, df_dy] = gradient (f, dx, dy), where df_dx and df_dy represent $\partial f/\partial x$ and $\partial f/\partial y$ and dx and dy are the spacing in the x and y values

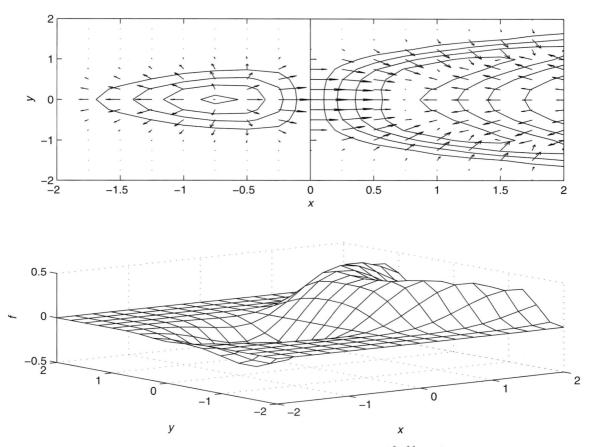

Figure 9.2–2 Gradient, contour, and surface plots of the function $f(x, y) = xe^{-(x^2+y^2)^2} + y^2$.

associated with the numerical values of f. The syntax can be extended to include functions of three or more variables.

The following program plots the contour plot and the gradient (shown by arrows) for the function

$$f(x, y) = xe^{-(x^2+y^2)^2} + y^2$$

The plots are shown in Figure 9.2–2. The arrows point in the direction of increasing f.

```
[x,y] = meshgrid(-2:0.25:2);
f = x.*exp(-((x-y.^2).^2+y.^2));
dx = x(1,2) - x(1,1); dy = y(2,1) - y(1,1);
[df_dx, df_dy] = gradient(f, dx, dy);
subplot(2,1,1)
contour(x,y,f), xlabel('x'), ylabel('y'), . . .
```

Table 9.2–1 Numerical differentiation functions

Command	Description
`d = diff(x)`	Returns a vector `d` containing the differences between adjacent elements in the vector `x`.
`[df_dx,df_dy] = gradient(f,dx,dy)`	Computes the gradient of the function $f(x, y)$, where `df_dx` and `df_dy` represent $\partial f / \partial x$ and $\partial f / \partial y$, and `dx` and `dy` are the spacing in the x and y values associated with the numerical values of f.
`d = polyder(p)`	Returns a vector `d` containing the coefficients of the derivative of the polynomial represented by the vector `p`.
`d = polyder(p1,p2)`	Returns a vector `d` containing the coefficients of the polynomial that is the derivative of the product of the polynomials represented by `p1` and `p2`.
`[num, den] = polyder(p2,p1)`	Returns the vectors `num` and `den` containing the coefficients of the numerator and denominator polynomials of the derivative of the quotient p_2/p_1, where `p1` and `p2` are polynomials.

```
hold on, quiver(x,y,df_dx, df_dy), hold off
subplot(2,1,2)
mesh(x,y,f),xlabel('x'),ylabel('y'),zlabel('f')
```

LAPLACIAN

The curvature is given by the second-order derivative expression called the *Laplacian*.

$$\nabla^2 f(x, y) = \frac{\partial^2 f}{\partial x^2} + \frac{\partial^2 f}{\partial y^2}$$

It can be computed with the `del2` function. See the MATLAB Help for details.

The MATLAB differentiation functions discussed here are summarized in Table 9.2–1.

9.3 First-Order Differential Equations

In this section, we introduce numerical methods for solving first-order differential equations. In Section 9.4 we show how to extend the techniques to higher-order equations.

ORDINARY DIFFERENTIAL EQUATION (ODE)

An *ordinary differential equation* (ODE) is an equation containing ordinary derivatives of the dependent variable. An equation containing partial derivatives with respect to two or more independent variables is a *partial differential equation* (PDE). Solution methods for PDEs are an advanced topic, and we will not treat them in this text. In this chapter we limit ourselves to *initial-value problems* (IVPs). These are problems where the ODE must be solved for a given set of

INITIAL-VALUE PROBLEM

values specified at some initial time, which is usually taken to be $t = 0$. Other types of ODE problems are discussed at the end of Section 9.6.

It will be convenient to use the following abbreviated "dot" notation for derivatives.

$$\dot{y}(t) = \frac{dy}{dt} \qquad \ddot{y}(t) = \frac{d^2y}{dt^2}$$

The *free response* of a differential equation, sometimes called the homogeneous solution or the initial response, is the solution for the case where there is no forcing function. The free response depends on the initial conditions. The *forced response* is the solution due to the forcing function when the initial conditions are zero. For linear differential equations, the complete or total response is the sum of the free and the forced responses. Nonlinear ODEs can be recognized by the fact that the dependent variable or its derivatives appear raised to a power or in a transcendental function. For example, the equations $\dot{y} = y^2$ and $\dot{y} = \cos y$ are nonlinear.

FREE RESPONSE

FORCED RESPONSE

The essence of a numerical method is to convert the differential equation into a difference equation that can be programmed. Numerical algorithms differ partly as a result of the specific procedure used to obtain the difference equations. It is important to understand the concept of "step size" and its effects on solution accuracy. To provide a simple introduction to these issues, we consider the simplest numerical methods, the *Euler method* and the *predictor-corrector method*.

The Euler Method

The *Euler method* is the simplest algorithm for numerical solution of a differential equation. Consider the equations

EULER METHOD

$$\frac{dy}{dt} = f(t,y) \qquad y(0) = y_0 \tag{9.3–1}$$

where $f(t, y)$ is a known function and y_0 is the initial condition, which is the given value of $y(t)$ at $t = 0$. From the definition of the derivative,

$$\frac{dy}{dt} = \lim_{\Delta t \to 0} \frac{y(t + \Delta t) - y(t)}{\Delta t}$$

If the time increment Δt is chosen small enough, the derivative can be replaced by the approximate expression

$$\frac{dy}{dt} \approx \frac{y(t + \Delta t) - y(t)}{\Delta t} \tag{9.3–2}$$

Assume that the function $f(t, y)$ in Equation (9.3–1) remains constant over the time interval $(t, t + \Delta t)$, and replace Equation (9.3–1) by the following approximation:

$$\frac{y(t + \Delta t) - y(t)}{\Delta t} = f(t, y)$$

or

$$y(t + \Delta t) = y(t) + f(t, y)\Delta t \qquad (9.3\text{--}3)$$

The smaller Δt is, the more accurate are our two assumptions leading to Equation (9.3–3). This technique for replacing a differential equation with a difference equation is the *Euler method*. The increment Δt is called the *step size*.

STEP SIZE

Equation (9.3–3) can be written in more convenient form as

$$y(t_{k+1}) = y(t_k) + \Delta t f[t_k, y(t_k)] \qquad (9.3\text{--}4)$$

where $t_{k+1} = t_k + \Delta t$. This equation can be applied successively at the times t_k by putting it in a `for` loop. The accuracy of the Euler method can be improved sometimes by using a smaller step size. However, very small step sizes require longer run times and can result in a large accumulated error due to roundoff effects.

The Predictor-Corrector Method

The Euler method can have a serious deficiency in problems where the variables are rapidly changing, because the method assumes the variables are constant over the time interval Δt. One way of improving the method is to use a better approximation to the right-hand side of Equation (9.3–1). Suppose instead of the Euler approximation (9.3–4) we use the average of the right-hand side of Equation (9.3–1) on the interval (t_k, t_{k+1}). This gives

$$y(t_{k+1}) = y(t_k) + \frac{\Delta t}{2}(f_k + f_{k+1}) \qquad (9.3\text{--}5)$$

where

$$f_k = f[t_k, y(t_k)] \qquad (9.3\text{--}6)$$

with a similar definition for f_{k+1}. Equation (9.3–5) is equivalent to integrating Equation (9.3–1) with the trapezoidal rule.

The difficulty with Equation (9.3–5) is that f_{k+1} cannot be evaluated until $y(t_{k+1})$ is known, but this is precisely the quantity being sought. A way out of this difficulty is to use the Euler formula (9.3–4) to obtain a preliminary estimate of $y(t_{k+1})$. This estimate is then used to compute f_{k+1} for use in Equation (9.3–5) to obtain the required value of $y(t_{k+1})$.

The notation can be changed to clarify the method. Let $h = \Delta t$ and $y_k = y(t_k)$, and let x_{k+1} be the estimate of $y(t_{k+1})$ obtained from the Euler formula (9.3–4). Then, by omitting the t_k notation from the other equations, we obtain the following description of the predictor-corrector process.

Euler predictor $x_{k+1} = y_k + hf(t_k, y_k)$ $(9.3\text{--}7)$

Trapezoidal corrector $y_{k+1} = y_k + \dfrac{h}{2}[f(t_k, y_k) + f(t_{k+1}, x_{k+1})]$ $(9.3\text{--}8)$

MODIFIED EULER
METHOD

This algorithm is sometimes called the *modified Euler method*. However, note that any algorithm can be tried as a predictor or a corrector. Thus many other methods can be classified as predictor-corrector.

Runge-Kutta Methods

The Taylor series representation forms the basis of several methods of solving differential equations, including the Runge-Kutta methods. The Taylor series may be used to represent the solution $y(t + h)$ in terms of $y(t)$ and its derivatives as follows.

$$y(t + h) = y(t) + h\dot{y}(t) + \frac{1}{2}h^2\ddot{y}(t) + \cdots \qquad (9.3\text{–}9)$$

The number of terms kept in the series determines its accuracy. The required derivatives are calculated from the differential equation. If these derivatives can be found, Equation (9.3–9) can be used to march forward in time. In practice, the high-order derivatives can be difficult to calculate, and the series (9.3–9) is truncated at some term. The Runge-Kutta methods were developed because of the difficulty in computing the derivatives. These methods use several evaluations of the function $f(t, y)$ in a way that approximates the Taylor series. The number of terms in the series that is duplicated determines the order of the Runge-Kutta method. Thus, a fourth-order Runge-Kutta algorithm duplicates the Taylor series through the term involving h^4.

MATLAB ODE Solvers

In addition to the many variations of the predictor-corrector and Runge-Kutta algorithms that have been developed, there are more-advanced algorithms that use a variable step size. These "adaptive" algorithms use larger step sizes when the solution is changing more slowly. MATLAB provides several functions, called *solvers*, that implement the Runge-Kutta and other methods with variable step size. Two of these are the ode45 and ode15s functions. The ode45 function uses a combination of fourth- and fifth-order Runge-Kutta methods. It is a general-purpose solver whereas ode15s is suitable for more-difficult equations called "stiff" equations. These solvers are more than sufficient to solve the problems in this text. It is recommended that you try ode45 first. If the equation proves difficult to solve (as indicated by a lengthy solution time or by a warning or error message), then use ode15s.

In this section we limit our coverage to first-order equations. Solution of higher-order equations is covered in Section 9.4. When used to solve the equation $\dot{y} = f(t, y)$, the basic syntax is (using ode45 as the example)

```
[t,y] = ode45(@ydot, tspan, y0)
```

where @ydot is the handle of the function file whose inputs must be t and y, and whose output must be a column vector representing dy/dt, that is, $f(t, y)$. The number of rows in this column vector must equal the order of the equation. The syntax for ode15s is identical. The function file ydot may also be specified by a character string (i.e., its name placed in single quotes), but use of the function handle is now the preferred approach.

The vector tspan contains the starting and ending values of the independent variable t, and optionally any intermediate values of t where the solution is desired. For example, if no intermediate values are specified, `tspan is [t0, tfinal]`, where `t0` and `tfinal` are the desired starting and ending values of the independent parameter t. As another example, using `tspan = [0, 5, 10]` tells MATLAB to find the solution at $t = 5$ and at $t = 10$. You can solve equation backward in time by specifying `t0` to be greater than `tfinal`.

The parameter `y0` is the initial value y(0). The function file must have its first two input arguments as `t` and `y` in that order, even for equations where $f(t, y)$ is not a function of t. You need not use array operations in the function file because the ODE solvers call the file with scalar values for the arguments. The solvers may have an additional argument, `options`, which is discussed at the end of this section.

First consider an equation whose solution is known in closed form, so that we can make sure we are using the method correctly.

EXAMPLE 9.3–1   Response of an *RC* Circuit

The model of the *RC* circuit shown in Figure 9.3–1 can be found from Kirchhoff's voltage law and conservation of charge. It is $RC\dot{y} + y = v(t)$. Suppose the value of *RC* is 0.1 s. Use a numerical method to find the free response for the case where the applied voltage y is zero and the initial capacitor voltage is y(0) = 2 V. Compare the results with the analytical solution, which is $y(t) = 2e^{-10t}$.

Solution
The equation for the circuit becomes $0.1\dot{y} + y = 0$. First solve this for y: $\dot{y} = -10y$. Next define and save the following function file. Note that the order of the input arguments must be t and y even though t does not appear on the right-hand side of the equation.

```
function ydot = RC_circuit(t,y)
% Model of an RC circuit with no applied voltage.
ydot = -10*y;
```

The initial time is $t = 0$, so set `t0` to be 0. Here we know from the analytical solution that $y(t)$ will be close to 0 for $t \geq 0.5$ s, so we choose `tfinal` to be 0.5 s. In other problems we generally do not have a good guess for `tfinal`, so we must try several increasing values of `tfinal` until we see enough of the response on the plot.

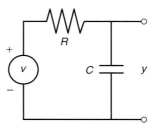

Figure 9.3–1 An *RC* circuit.

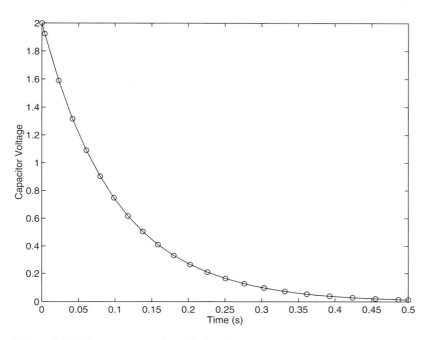

Figure 9.3–2 Free response of an *RC* circuit.

The function `ode45` is called as follows, and the solution plotted along with the analytical solution `y_true`.

```
[t, y] = ode45(@RC_circuit, [0, 0.5], 2);
y_true = 2*exp(-10*t);
plot(t,y,'o',t,y_true), xlabel('Time(s)'),...
   ylabel('Capacitor Voltage')
```

Note that we need not generate the array `t` to evaluate `y_true` because `t` is generated by the `ode45` function. The plot is shown in Figure 9.3–2. The numerical solution is marked by the circles, and the analytical solution is indicated by the solid line. Clearly the numerical solution gives an accurate answer. Note that the step size has been automatically selected by the `ode45` function.

Earlier versions of MATLAB required that the function name, here `RC_circuit`, be enclosed within single quotes, but this might not be allowed in future versions. The use of function handles is now preferred, such as `@RC_circuit`. As we will see, additional capabilities are available with function handles.

Test Your Understanding

T9.3–1 Use MATLAB to compute and plot the solution of the following equation.

$$10\frac{dy}{dt} + y = 20 + 7\sin 2t \qquad y(0) = 15$$

When the differential equation is nonlinear, we often have no analytical solution to use for checking our numerical results. In such cases we can use our physical insight to guard against grossly incorrect results. We can also check the equation for singularities that might affect the numerical procedure. Finally, we can sometimes use an approximation to replace the nonlinear equation with a linear one that can be solved analytically. Although the linear approximation does not give the exact answer, it can be used to see if our numerical answer is "in the ballpark." The following example illustrates this approach.

Liquid Height in a Spherical Tank

Figure 9.3–3 shows a spherical tank for storing water. The tank is filled through a hole in the top and drained through a hole in the bottom. If the tank's radius is r, you can use integration to show that the volume of water in the tank as a function of its height h is given by

$$V(h) = \pi r h^2 - \pi \frac{h^3}{3} \qquad (9.3–10)$$

Torricelli's principle states that the liquid flow rate through the hole is proportional to the square root of the height h. Further studies in fluid mechanics have identified the relation more precisely, and the result is that the volume flow rate through the hole is given by

$$q = C_d A \sqrt{2gh} \qquad (9.3–11)$$

where A is the area of the hole, g is the acceleration due to gravity, and C_d is an experimentally determined value that depends partly on the type of liquid. For water, $C_d = 0.6$ is a common value. We can use the principle of conservation of mass to obtain a differential equation for the height h. Applied to this tank, the principle says that the rate of change of liquid volume in the tank must equal the flow rate out of the tank; that is,

$$\frac{dV}{dt} = -q \qquad (9.3–12)$$

From Equation (9.3–10),

$$\frac{dV}{dt} = 2\pi r h \frac{dh}{dt} - \pi h^2 \frac{dh}{dt} = \pi h (2r - h) \frac{dh}{dt}$$

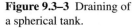

Figure 9.3–3 Draining of
a spherical tank.

Substituting this and Equation (9.3–11) into Equation (9.3–12) gives the required equation for h.

$$\pi(2rh - h^2)\frac{dh}{dt} = -C_d A\sqrt{2gh} \tag{9.3–13}$$

Use MATLAB to solve this equation to determine how long it will take for the tank to empty if the initial height is 9 ft. The tank has a radius of $r = 5$ ft and has a 1-in.-diameter hole in the bottom. Use $g = 32.2$ ft/sec^2. Discuss how to check the solution.

Solution:
With $C_d = 0.6$, $r = 5$, $g = 32.2$, and $A = \pi(1/24)^2$, Equation (9.3–13) becomes

$$\frac{dh}{dt} = -\frac{0.0334\sqrt{h}}{10h - h^2} \tag{9.3–14}$$

We can first check the above expression for dh/dt for singularities. The denominator does not become zero unless $h = 0$ or $h = 10$, which correspond to a completely empty and a completely full tank. So we will avoid singularities if $0 < h < 10$.

Finally, we can use the following approximation to estimate the time to empty. Replace h on the right side of Equation (9.3–14) with its average value, namely, $(9 - 0)/2 = 4.5$ ft. This gives $dh/dt = -0.00286$, whose solution is $h(t) = h(0) - 0.00286t = 9 - 0.00286t$. According to this equation, the tank will be empty at $t = 9/0.00286 = 3147$ sec, or 52 min. We will use this value as a "reality check" on our answer.

The function file based on Equation (9.3–14) is

```
function hdot = height(t,h)
hdot = -(0.0334*sqrt(h))/(10*h-h^2);
```

The file is called as follows, using the `ode45` solver.

```
[t, h]=ode45(@height, [0, 2475], 9);
plot(t,h),xlabel('Time (sec)'), ylabel('Height (ft)')
```

The resulting plot is shown in Figure 9.3–4. Note how the height changes more rapidly when the tank is nearly full or nearly empty. This is to be expected because of the effects of the tank's curvature. The tank empties in 2475 sec, or 41 min. This value is not grossly different from our rough estimate of 52 min, so we should feel comfortable accepting the numerical results. The value of the final time of 2475 sec was found by increasing the final time until the plot showed that the height became 0.

9.4 Higher-Order Differential Equations

To use the ODE solvers to solve an equation higher than order 1, you must first write the equation as a set of first-order equations. This is easily done. Consider the second-order equation

$$5\ddot{y} + 7\dot{y} + 4y = f(t) \tag{9.4–1}$$

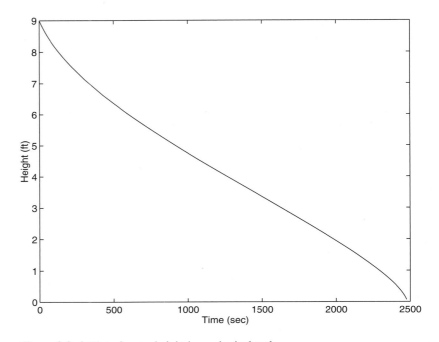

Figure 9.3–4 Plot of water height in a spherical tank.

Solve it for the highest derivative:

$$\ddot{y} = \frac{1}{5}f(t) - \frac{4}{5}y - \frac{7}{5}\dot{y} \qquad (9.4\text{–}2)$$

Define two new variables x_1 and x_2 to be y and its derivative $\dot{y}$. That is, define $x_1 = y$ and $x_2 = \dot{y}$. This implies that

$$\dot{x}_1 = x_2$$

$$\dot{x}_2 = \frac{1}{5}f(t) - \frac{4}{5}x_1 - \frac{7}{5}x_2$$

CAUCHY OR STATE-VARIABLE FORM

This form is sometimes called the *Cauchy form* or the *state-variable form*.

Now write a function file that computes the values of $\dot{x}_1$ and $\dot{x}_2$ and stores them in a *column* vector. To do this, we must first have a function specified for $f(t)$. Suppose that $f(t) = \sin t$. Then the required file is

```
function xdot = example_1(t,x)
% Computes derivatives of two equations
xdot(1) = x(2);
xdot(2) = (1/5)*(sin(t)-4*x(1)-7*x(2));
xdot = [xdot(1); xdot(2)];
```

Note that `xdot(1)` represents $\dot{x}_1$, `xdot(2)` represents $\dot{x}_2$, `x(1)` represents x_1, and `x(2)` represents x_2. Once you become familiar with the notation for the state-variable form, you will see that the previous code could be replaced with the following shorter form.

```
function xdot = example_1(t,x)
% Computes derivatives of two equations
xdot = [x(2); (1/5)*(sin(t)-4*x(1)-7*x(2))];
```

Suppose we want to solve Equation (9.4–1) for $0 \le t \le 6$ with the initial conditions $x(0) = 3$, $\dot{x}(0) = 9$. Then the initial condition for the *vector* x is `[3, 9]`. To use `ode45`, you type

```
[t, x] = ode45(@example_1, [0, 6], [3, 9]);
```

Each row in the vector x corresponds to a time returned in the column vector `t`. If you type `plot(t,x)`, you will obtain a plot of both x_1 and x_2 versus t. Note x is a matrix with two columns. The first column contains the values of x_1 at the various times generated by the solver; the second column contains the values of x_2. Thus, to plot only x_1, type `plot(t,x(:,1))`. To plot only x_2, type `plot(t,x(:,2))`.

When we are solving nonlinear equations, sometimes it is possible to check the numerical results by using an approximation that reduces the equation to a linear one. The following example illustrates such an approach with a second-order equation.

A Nonlinear Pendulum Model

EXAMPLE 9.4–1

The pendulum shown in Figure 9.4–1 consists of a concentrated mass m attached to a rod whose mass is small compared to m. The rod's length is L. The equation of motion for this pendulum is

$$\ddot{\theta} + \frac{g}{L}\sin\theta = 0 \qquad (9.4\text{–}3)$$

Suppose that $L = 1$ m and $g = 9.81$ m/s^2. Use MATLAB to solve this equation for $\theta(t)$ for two cases: $\theta(0) = 0.5$ rad and $\theta(0) = 0.8\pi$ rad. In both cases $\dot{\theta}(0) = 0$. Discuss how to check the accuracy of the results.

Solution
If we use the small-angle approximation $\sin \approx \theta$, the equation becomes

$$\ddot{\theta} + \frac{g}{L}\theta = 0 \qquad (9.4\text{–}4)$$

which is linear and has the solution

$$\theta(t) = \theta(0)\cos\sqrt{\frac{g}{L}}\,t \qquad (9.4\text{–}5)$$

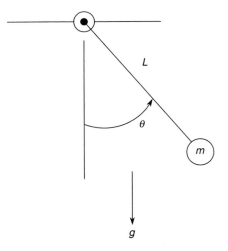

Figure 9.4–1 A pendulum.

if $\dot{\theta}(0) = 0$. Thus the amplitude of oscillation is $\theta(0)$, and the period is $P = 2\pi\sqrt{L/g} = 2.006$ s. We can use this information to select a final time and to check our numerical results.

First rewrite the pendulum equation (9.4–3) as two first-order equations. To do this, let $x_1 = \theta$ and $x_2 = \dot{\theta}$. Thus

$$\dot{x}_1 = \dot{\theta} = x_2$$
$$\dot{x}_2 = \ddot{\theta} = -\frac{g}{L}\sin x_1$$

The following function file is based on the last two equations. Remember that the output xdot must be a *column* vector.

```
function xdot = pendulum(t,x)
g = 9.81; L = 1;
xdot = [x(2); -(g/L)*sin(x(1))];
```

This file is called as follows. The vectors ta and xa contain the results for the case where $\theta(0) = 0.5$. In both cases, $\dot{\theta}(0) = 0$. The vectors tb and xb contain the results for $\theta(0) = 0.8\pi$.

```
[ta, xa] = ode45(@pendulum, [0,5], [0.5, 0];
[tb, xb] = ode45(@pendulum, [0,5], [0.8*pi, 0];
plot(ta, xa(:,1), tb,xb(:,1)), xlabel ('Time (s)'), ...
    ylabel('Angle (rad)'), gtext('Case 1'), gtext('Case 2')
```

The results are shown in Figure 9.4–2. The amplitude remains constant, as predicted by the small-angle analysis, and the period for the case where $\theta(0) = 0.5$ is a little larger than 2 s, the value predicted by the small-angle analysis. So we can place some confidence in the numerical procedure. For the case where $\theta(0) = 0.8\pi$, the period of the

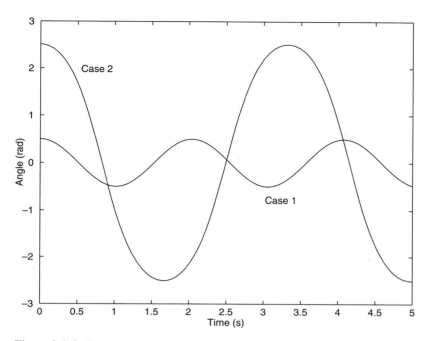

Figure 9.4–2 The pendulum angle as a function of time for two starting positions.

numerical solution is about 3.3 s. This illustrates an important property of nonlinear differential equations. The free response of a linear equation has the same period for any initial conditions; however, the form and therefore the period of the free response of a nonlinear equation often depend on the particular values of the initial conditions.

In this example, the values of g and L were encoded in the function `pendulum(t,x)`. Now suppose you want to obtain the pendulum response for different lengths L or different gravitational accelerations g. You could use the `global` command to declare g and L as global variables, or you could pass parameter values through an argument list in the `ode45` function; but starting with MATLAB 7, the preferred method is to use a nested function. Nested functions are discussed in Section 3.3. The following program shows how this is done.

```
function pendula
g = 9.81; L = 0.75; % First case.
tF = 6*pi*sqrt(L/g); % Approximately 3 periods.
[t1, x1] = ode45(@pendulum, [0,tF], [0.4, 0];
%
g = 1.63; L = 2.5; % Second case.
tF = 6*pi*sqrt(L/g); % Approximately 3 periods.
[t2, x2] = ode45(@pendulum, [0,tF], [0.2, 0];
plot(t1, x1(:,1), t2, x2(:,1)), ...
```

```
xlabel ('time (s)'), ylabel ('\theta (rad)')
% Nested function.
    function xdot = pendulum(t,x)
    xdot = [x(2);-(g/L)*sin(x(1))];
end
end
```

Advanced Solver Capabilities

The complete preferred ODE solver syntax in MATLAB 7, using `ode45` as an example, is

```
[t, y] = ode45(@ydot, tspan, y0, options)
```

where the `options` argument is created with the `odeset` function.

The `odeset` Function   The `odeset` function creates an options structure to be supplied to the solver. Its syntax is

```
options = odeset('name1', 'value1' 'name2', 'value2',...)
```

where `name` is the name of a *property* and `value` is the value to be assigned to the property.

A simple example will clarify things. The `Refine` property is used to increase the number of output points from the solver by an integer factor n. For `ode45` the default value of n is 4 because of the solver's large step sizes. Suppose we want to solve the $\dot{y} = \sin^2 t$ for $0 \le t \le 4\pi$ with $y(0) = 0$. Define the following function file.

```
function ydot = sinefn(t,y)
ydot = sin(t)^2;
```

Then use the `odeset` function to set the value of `Refine` to $n = 8$, and call the `ode45` solver, as shown in the following code. This will produce twice as many points to plot to obtain a smoother curve.

```
options = odeset('Refine',8);
[t, y] = ode45(@sinefn, [0, 4*pi], 0, options);
```

Another property is the `Events` property, which has two possible values: `on` and `off`. It can be used to locate transitions to, from, or through zeros of a user-defined function. This can be used to detect in the ODE solution when a variable makes a transition to, from, or through a certain value, such as zero. This feature can be used, for example, to simulate a dropped ball bouncing up from the floor. See the MATLAB Help for other examples.

There are many properties that can be set with the `odeset` function. To see a list of these, type `odeset`. Table 9.4–1 summarizes the syntax of the ODE solvers using `ode45` as an example.

Table 9.4–1 Syntax of the ODE solver `ode45`

Command	Description
`[t, y] = ode45(@ydot, tspan, y0, options)`	Solves the vector differential equation $\dot{\mathbf{y}} = \mathbf{f}(t, \mathbf{y})$ specified by the function file whose handle is `@ydot` and whose inputs must be t and $\mathbf{y}$, and whose output must be a *column* vector representing $d\mathbf{y}/dt$; that is, $\mathbf{f}(t, \mathbf{y})$. The number of rows in this column vector must equal the order of the equation. The vector `tspan` contains the starting and ending values of the independent variable t, and optionally any intermediate values of t where the solution is desired. The vector `y0` contains the initial values. The function file must have two input arguments, `t` and `y`, even for equations where $f(t, y)$ is not a function of t. The `options` argument is created with the `odeset` function. The syntax is identical for the solver `ode15s`.
`options = odeset ('name1', 'value1' 'name2', 'value2', . . .)`	Creates an integrator options structure `options` to be used with the ODE solver, in which the named properties have the specified values, where `name` is the name of a *property* and `value` is the value to be assigned to the property. Any unspecified properties have default values. Typing `odeset` with no input arguments displays all property names and their possible values.

9.5 Special Methods for Linear Equations

MATLAB provides some convenient tools to use if the differential equation model is linear. Even though there are general methods available for finding the analytical solutions of linear differential equations, it is sometimes more convenient to use a numerical method to find the solution. Examples of such situations occur when the forcing function is a complicated function or when the order of the differential equation is higher than 2. In such cases the algebra involved in obtaining the analytical solution might not be worth the effort, especially if the main objective is to obtain a plot of the solution.

Matrix Methods

We can use matrix operations to reduce the number of lines to be typed in the derivative function file. For example, the following equation describes the motion of a mass connected to a spring, with viscous friction acting between the mass and the surface. Another force $u(t)$ also acts on the mass.

$$m\ddot{y} + c\dot{y} + ky = u(t) \tag{9.5-1}$$

This can be put into Cauchy form by letting $x_1 = y$ and $x_2 = \dot{y}$. This gives

$$\dot{x}_1 = x_2$$

$$\dot{x}_2 = \frac{1}{m}u(t) - \frac{k}{m}x_1 - \frac{c}{m}x_2$$

This can be written as one matrix equation as follows.

$$\begin{bmatrix} \dot{x}_1 \\ \dot{x}_2 \end{bmatrix} = \begin{bmatrix} 0 & 1 \\ -\dfrac{k}{m} & -\dfrac{c}{m} \end{bmatrix} \begin{bmatrix} x_1 \\ x_2 \end{bmatrix} + \begin{bmatrix} 0 \\ \dfrac{1}{m} \end{bmatrix} u(t)$$

In compact form this is

$$\dot{\mathbf{x}} = \mathbf{A}\mathbf{x} + \mathbf{B}u(t) \tag{9.5–2}$$

where

$$\mathbf{A} = \begin{bmatrix} 0 & 1 \\ -\dfrac{k}{m} & -\dfrac{c}{m} \end{bmatrix} \qquad \mathbf{B} = \begin{bmatrix} 0 \\ \dfrac{1}{m} \end{bmatrix} \qquad \mathbf{x} = \begin{bmatrix} x_1 \\ x_2 \end{bmatrix}$$

The following function file shows how to use matrix operations. In this example, $m = 1$, $c = 2$, $k = 5$, and the applied force is $u(t) = 10$.

```
function xdot = msd(t,x)
% Function file for mass with spring and damping.
% Position is first variable, velocity is second variable.
u = 10;
m = 1;c = 2;k = 5;
A = [0, 1;-k/m, -c/m];
B = [0; 1/m];
xdot = A*x+B*u;
```

Note that the output xdot will be a column vector because of the definition of matrix-vector multiplication. We try different values of the final time until we see the entire response. Using a final time of 5 and the initial conditions $x_1(0) = 0$ and $x_2(0) = 0$, we call the solver and plot the solution as follows:

```
[t, x] = ode45(@msd, [0,5], [0,0];
plot(t,x(:,1),t,x(:,2))
```

Figure 9.5–1 shows the edited plot. Note that we could have avoided embedding the values of the parameters m, c, k, and u by making msd a nested function as was done with the functions pendulum and pendula in Section 9.4.

Test Your Understanding

T9.5–1 Plot the position and velocity of a mass with a spring and damping, having the parameter values $m = 2$, $c = 3$, and $k = 7$. The applied force is $u = 35$, the initial position is $y(0) = 2$, and the initial velocity is $\dot{y}(0) = -3$.

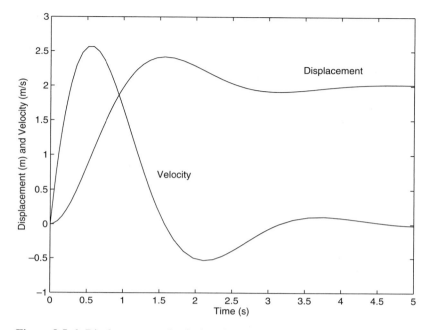

Figure 9.5–1 Displacement and velocity of the mass as a function of time.

Characteristic Roots from the `eig` Function

The characteristic roots of a linear differential equation give information about the speed of response and the oscillation frequency, if any.

MATLAB provides the `eig` function to compute the characteristic roots when the model is given in the state-variable form (9.5–2). Its syntax is `eig(A)`, where `A` is the matrix that appears in Equation (9.5–2). (The function's name is an abbreviation of *eigenvalue*, which is another name for characteristic root.) For example, consider the equations

EIGENVALUE

$$\dot{x}_1 = -3x_1 + x_2 \qquad (9.5–3)$$

$$\dot{x}_2 = -x_1 - 7x_2 \qquad (9.5–4)$$

The matrix **A** for these equations is

$$\mathbf{A} = \begin{bmatrix} -3 & 1 \\ -1 & -7 \end{bmatrix}$$

To find the characteristic roots, type

```
>>A = [-3, 1;-1, -7];
>>r = eig(A)
```

The answer so obtained is `r = [-6.7321, -3.2679]`. To find the time constants, which are the negative reciprocals of the real parts of the roots, you type `tau = -1./real(r)`. The time constants are 0.1485 and 0.3060. Four times the dominant time constant, or 4(0.3060) = 1.224, gives the time it takes for the free response to become approximately zero.

ODE Solvers in the Control System Toolbox

Many of the functions from the Control System toolbox are available in the Student Edition of MATLAB. Some of these can be used to solve linear, time-invariant (constant-coefficient) differential equations. They are sometimes more convenient to use and more powerful than the ODE solvers discussed thus far, because general solutions can be found for linear, time-invariant equations. Here we discuss several of these functions. These are summarized in Table 9.5–1. The other features of the Control System toolbox require advanced methods, and will not be covered here. See [Palm, 2005] for coverage of these methods.

LTI OBJECT

An *LTI object* describes a linear, time-invariant equation, or sets of equations, here referred to as the *system*. An LTI object can be created from different descriptions of the system, it can be analyzed with several functions, and it can be accessed to provide alternate descriptions of the system. For example, the equation

$$2\ddot{x} + 3\dot{x} + 5x = u(t) \tag{9.5–5}$$

is one description of a particular system. This description is called the *reduced form*. The following is a state-model description of the same system

$$\dot{\mathbf{x}} = \mathbf{A}\mathbf{x} + \mathbf{B}u \tag{9.5–6}$$

where $x_1 = x$, $x_2 = \dot{x}$, and

$$\mathbf{A} = \begin{bmatrix} 0 & 1 \\ -\frac{5}{2} & -\frac{3}{2} \end{bmatrix} \qquad \mathbf{B} = \begin{bmatrix} 0 \\ \frac{1}{2} \end{bmatrix} \qquad \mathbf{x} = \begin{bmatrix} x_1 \\ x_2 \end{bmatrix} \tag{9.5–7}$$

Both model forms contain the same information. However, each form has its own advantages, depending on the purpose of the analysis.

Because there are two or more state variables in a state model, we need to be able to specify which state variable, or combination of variables, constitutes the output of the simulation. For example, models (9.5–6) and (9.5–7) can represent the motion of a mass, with x_1 the position and x_2 the velocity of the mass. We need to be able to specify whether we want to see a plot of the position, or the velocity, or both. This specification of the output, denoted by the vector $\mathbf{y}$, is done in general with the matrices $\mathbf{C}$ and $\mathbf{D}$, which must be compatible with the equation

$$\mathbf{y} = \mathbf{C}\mathbf{x} + \mathbf{D}u(t) \tag{9.5–8}$$

where the vector $\mathbf{u}(t)$ allows for multiple inputs. To continue the previous example, if we want the output to be the position $x = x_1$, then $y = x_1$, and we would select $\mathbf{C} = [1, 0]$ and $\mathbf{D} = 0$. Thus, in this case, Equation (9.5–8) reduces to $y = x_1$.

To create an LTI object from the reduced form (9.5–5), use the `tf(right, left)` function, and type

```
>>sys1 = tf(1, [2, 3, 5]);
```

where the vector `right` is the vector of coefficients of the right-hand side of the equation, arranged in descending derivative order, and `left` is the vector of coefficients of the left-hand side of the equation, also arranged in descending derivative order. The result, `sys1`, is the LTI object that describes the system in reduced form, also called the *transfer function form*. (The name of the function, `tf`, stands for *transfer function*, which is an equivalent way of describing the coefficients on the left- and right-hand sides of the equation.)

The LTI object `sys2` in transfer function form for the equation

$$6\frac{d^3x}{dt^3} - 4\frac{d^2x}{dt^2} + 7\frac{dx}{dt} + 5x = 3\frac{d^2u}{dt^2} + 9\frac{du}{dt} + 2u \tag{9.5–9}$$

is created by typing

```
>>sys2 = tf([3, 9, 2], [6, -4, 7, 5]);
```

To create an LTI object from a state model, you use the `ss(A, B, C, D)` function, where `ss` stands for *state space*. For example, to create an LTI object in state-model form for the system described by Equations (9.5–6) through (9.5–8), you type

```
>>A = [0, 1; -5/2, -3/2]; B = [0; 1/2];
>>C = [1, 0]; D = 0;
>>sys3 = ss(A,B,C,D);
```

An LTI object defined using the `tf` function can be used to obtain an equivalent state-model description of the system. To create a state model for the system described by the LTI object `sys1` created previously in transfer function form, you type `ss(sys1)`. You will then see the resulting **A**, **B**, **C**, and **D** matrices on the screen. To extract and save the matrices, use the `ssdata` function as follows.

```
>>[A1, B1, C1, D1] = ssdata(sys1);
```

The results are

$$\mathbf{A1} = \begin{bmatrix} -1.5 & -1.25 \\ 2 & 0 \end{bmatrix} \qquad \mathbf{B1} = \begin{bmatrix} 0.5 \\ 0 \end{bmatrix} \qquad \mathbf{C1} = [0 \quad 0.5] \qquad \mathbf{D1} = [0]$$

When using `ssdata` to convert a transfer function form to a state model, note that the output y will be a scalar that is identical to the solution variable of the reduced form; in this case the solution variable of Equation (9.5–1) is the variable y. To interpret the state model, we need to relate its state variables x_1 and x_2 to y. The values of the matrices **C1** and **D1** tell us that the output variable $y = 0.5x_2$. Thus we see that $x_2 = 2y$. The other state variable x_1 is related to x_2 by $\dot{x}_2 = 2x_1$. Thus $x_1 = \dot{y}$.

Table 9.5–1 LTI object functions

Command	Description
sys = ss(A, B, C, D)	Creates an LTI object in state-space form, where the matrices A, B, C, and D correspond to those in the model $\dot{\mathbf{x}} = \mathbf{Ax} + \mathbf{Bu}, \mathbf{y} = \mathbf{Cx} + \mathbf{Du}$.
[A, B, C, D] = ssdata(sys)	Extracts the matrices A, B, C and D corresponding to those in the model $\dot{\mathbf{x}} = \mathbf{Ax} + \mathbf{Bu}, \mathbf{y} = \mathbf{Cx} + \mathbf{Du}$.
sys = tf(right,left)	Creates an LTI object in transfer function form, where the vector right is the vector of coefficients of the right-hand side of the equation, arranged in descending derivative order, and left is the vector of coefficients of the left-hand side of the equation, also arranged in descending derivative order.
sys2 = tf(sys1)	Creates the transfer function model sys2 from the state model sys1.
sys1 = ss(sys2)	Creates the state model sys1 from the transfer function model sys2.
[right, left] = tfdata(sys,'v')	Extracts the coefficients on the right- and left-hand sides of the reduced-form model specified in the transfer function model sys. When the optional parameter 'v' is used, the coefficients are returned as vectors rather than as cell arrays.

To create a transfer function description of the system sys3, previously created from the state model, you type tfsys3 = tf(sys3). To extract and save the coefficients of the reduced form, use the tfdata function as follows:

```
[right, left] = tfdata(sys3, 'v')
```

For this example, the vectors returned are right = 1 and left = [1, 1.5, 2.5]. The optional parameter 'v' tells MATLAB to return the coefficients as vectors; otherwise, they are returned as cell arrays. These functions are summarized in Table 9.5–1.

Test Your Understanding

T9.5–2 Obtain the state model for the reduced-form model

$$5\ddot{x} + 7\dot{x} + 4x = u(t)$$

Then convert the state model back to reduced form, and see if you get the original reduced-form model.

Linear ODE Solvers

The Control System toolbox provides several solvers for linear models. These solvers are categorized by the type of input function they can accept: zero input, impulse input, step input, and a general input function. These are summarized in Table 9.5–2.

The initial Function The initial function computes and plots the free response of a state model. This is sometimes called the *initial-condition response* or the *undriven response* in the MATLAB documentation. The basic syntax is initial(sys,x0), where sys is the LTI object in state-model form and x0

Table 9.5–2 Basic syntax of the LTI ODE solvers

Command	Description
impulse(sys)	Computes and plots the impulse response of the LTI object sys.
initial(sys,x0)	Computes and plots the free response of the LTI object sys given in state-model form, for the initial conditions specified in the vector x0.
lsim(sys,u,t)	Computes and plots the response of the LTI object sys to the input specified by the vector u, at the times specified by the vector t.
step(sys)	Computes and plots the step response of the LTI object sys.

See the text for description of extended syntax.

is the initial-condition vector. The time span and number of solution points are chosen automatically. For example, to find the free response of the state model (9.5–5) through (9.5–8), for $x_1(0) = 5$ and $x_2(0) = -2$, first define it in state-model form. This was done previously to obtain the system sys3. Then use the initial function as follows.

```
>>initial(sys3, [5, -2])
```

The plot shown in Figure 9.5–2 will be displayed on the screen. Note that MATLAB automatically labels the plot, computes the steady-state response, and displays it with a dotted line.

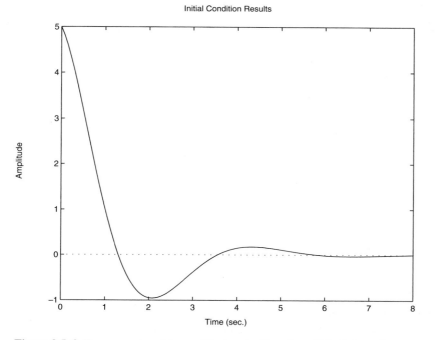

Figure 9.5–2 Free response of the model given by Equations (9.5–5) through (9.5–8) for $x_1(0) = 5$ and $x_2(0) = -2$.

To specify the final time `tF`, use the syntax `initial(sys,x0,tF)`. To specify a vector of times of the form `t = 0:dt:tF`, at which to obtain the solution, use the syntax `initial(sys,x0,t)`.

When called with left-hand arguments, as `[y, t, x] = initial(sys,x0,...)`, the function returns the output response `y`, the time vector `t` used for the simulation, and the state vector `x` evaluated at those times. The columns of the matrices `y` and `x` are the outputs and the states, respectively. The number of rows in `y` and `x` equals `length(t)`. No plot is drawn. The syntax `initial(sys1, sys2, ...,x0,t)` plots the free response of multiple LTI systems on a single plot. The time vector `t` is optional. You can specify line color, line style, and marker for each system, for example, `initial(sys1,'r', sys2,'y--',sys3,'gx',x0)`.

The `impulse` Function   The `impulse` function plots the unit-impulse response for each input-output pair of the system, assuming that the initial conditions are zero. (The unit impulse is also called the Dirac delta function.) The basic syntax is `impulse(sys)`, where `sys` is the LTI object. Unlike the `initial` function, the `impulse` function can be used with either a state model or a transfer function model. The time span and number of solution points are chosen automatically. For example, the impulse response of Equation (9.5–5) is found as follows:

```
>>sys1 = tf(1, [2, 3, 5]);
>>impulse(sys1)
```

The extended syntax of the `impulse` function is similar to that of the `initial` function.

The `step` Function   The `step` function plots the unit-step response for each input-output pair of the system, assuming that the initial conditions are zero. [The unit step function $u(t)$ is 0 for $t < 0$ and 1 for $t > 0$.] The basic syntax is `step(sys)`, where `sys` is the LTI object. The `step` function can be used with either a state model or a transfer function model. The time span and number of solution points are chosen automatically. The extended syntax of the `step` function is similar to that of the `initial` and the `impulse` functions.

To find the unit-step response, for zero initial conditions, of the state model (9.5–6) through (9.5–8), and the reduced-form model

$$5\ddot{x} + 7\dot{x} + 5x = 5\dot{f} + f \tag{9.5–10}$$

the session is (assuming `sys3` is still available in the workspace)

```
>>sys4 = tf([5, 1], [5, 7, 5]);
>>step(sys3,'b',sys4,'--')
```

The result is shown in Figure 9.5–3. The steady-state response is indicated by the horizontal dotted line. Note how the steady-state response and the time to reach that state are automatically determined.

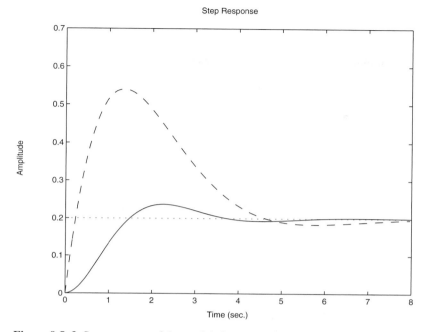

Figure 9.5–3 Step response of the model given by Equations (9.5–6) through (9.5–8) and the model (9.5–10), for zero initial conditions.

Step response can be characterized by the following parameters.

■ *Steady-state value*: The limit of the response as $t \rightarrow \infty$.

■ *Settling time*: The time for the response to reach and stay within a certain percentage (usually 2 percent) of its steady-state value.

■ *Rise time*: The time required for the response to rise from 10 to 90 percent of its steady-state value.

■ *Peak response*: The largest value of the response.

■ *Peak time*: The time at which the peak response occurs.

When the `step(sys)` function puts a plot on the screen, you may use the plot to calculate these parameters by right-clicking anywhere within the plot area. This brings up a menu. Choose **Characteristics** to obtain a submenu that contains the response characteristics. When you select a specific characteristic, for example, "peak response," MATLAB puts a large dot on the peak and displays dashed lines indicating the value of the peak response and the peak time. Move the cursor over this dot to see a display of the values. You can use the other solvers in the same way, although the menu choices may be different. For example, peak response and settling time are available when you use the `impulse(sys)` function, but not the rise time. If instead of choosing **Characteristics** you choose **Properties** and select the **Options** tab, you can change the defaults for the settling time and rise time, which are 2 percent and 10 to 90 percent.

Using this method, we find that the solid curve in Figure 9.5–3 has the following characteristics:

■ Steady-state value: 0.2
■ 2 percent settling time: 5.22
■ 10 to 90 percent rise time: 1.01
■ Peak response: 0.237
■ Peak time: 2.26

You can also read values off any part of the curve by placing the cursor on the curve at the desired point. You can move the cursor along the curve and read the values as they change. Using this method, we find that the solid curve in Figure 9.5–3 crosses the steady-state value of 0.2 for the second time at $t = 3.74$.

You can suppress the plot generated by `step` and create your own plot as follows, assuming `sys3` is still available in the workspace.

```
[x,t] = step(sys3);
plot(t,x)
```

You can then use the Plot Editor tools to edit the plot. However, with this approach, right-clicking on the plot will no longer give you information about the step response characteristics.

Suppose the step input is not a *unit* step but instead is 0 for $t < 0$ and 10 for $t > 0$. There are two ways to obtain the solution with the factor 10. Using `sys3` as the example, these are `step(10*sys3)` and

```
[x,t] = step(sys3);
plot(t,10*x)
```

The `lsim` Function   The `lsim` function plots the response of the system to an arbitrary input. The basic syntax for zero initial conditions is `lsim(sys,u,t)`, where `sys` is the LTI object, `t` is a time vector having regular spacing, as `t = 0:dt:tF`, and `u` is a matrix with as many columns as inputs, and whose *i*th row specifies the value of the input at time `t(i)`. To specify nonzero initial conditions for a state-space model, use the syntax `lsim(sys,u,t,x0)`. This computes and plots the total response (the free plus forced response). Right-clicking on the plot brings up the menu containing the **Characteristics** choice, although the only characteristic available is the peak response.

When called with left-hand arguments, as `[y, t] = lsim(sys, u, . . .)`, the function returns the output response `y` and the time vector `t` used for the simulation. The columns of the matrix `y` are the outputs, and the number of its rows equals `length(t)`. No plot is drawn. To obtain the state vector solution for state-space models, use the syntax `[y, t, x] = lsim(sys,u,. . .)`. The syntax `lsim(sys1,sys2,. . .,u,t,x0)` plots the responses of multiple LTI systems on a single plot. The initial-condition vector `xo` is needed only if the initial conditions are nonzero. You can specify line color, line style, and marker for each system, for example, `lsim(sys1,'r',sys2, 'y--',sys3,'gx',u,t)`.

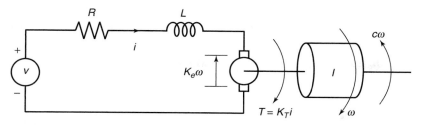

Figure 9.5–4 An armature-controlled dc motor.

We will see an example of the `lsim` function shortly.

Programming Detailed Forcing Functions

As a final example of higher-order equations, we now show how to program a detailed forcing function for use with the `lsim` function. We use a dc motor as the application. The equations for an armature-controlled dc motor (such as a permanent-magnet motor) shown in Figure 9.5–4 are the following. They result from Kirchhoff's voltage law and Newton's law applied to a rotating inertia. The motor's current is i and its rotational velocity is ω.

$$L\frac{di}{dt} = -Ri - K_e\omega + v(t) \tag{9.5–11}$$

$$I\frac{d\omega}{dt} = K_T i - c\omega \tag{9.5–12}$$

where L, R, and I are the motor's inductance, resistance, and inertia; K_T and K_e are the torque constant and back emf constant; c is a viscous damping constant; and $v(t)$ is the applied voltage. These equations can be put into matrix form as follows, where $x_1 = i$ and $x_2 = \omega$.

$$\begin{bmatrix} \dot{x}_1 \\ \dot{x}_2 \end{bmatrix} = \begin{bmatrix} -\frac{R}{L} & -\frac{K_e}{L} \\ \frac{K_T}{I} & -\frac{c}{I} \end{bmatrix} \begin{bmatrix} x_1 \\ x_2 \end{bmatrix} + \begin{bmatrix} \frac{1}{L} \\ 0 \end{bmatrix} v(t)$$

Trapezoidal Profile for a DC Motor

EXAMPLE 9.5–1

In many applications we want to accelerate the motor to a desired speed and allow it to run at that speed for some time before decelerating to a stop. Investigate whether an applied voltage having a trapezoidal profile will accomplish this. Use the values $R = 0.6\ \Omega$, $L = 0.002$ H, $K_T = 0.04$ N · m/A, $K_e = 0.04$ V· s/rad, $c = 0$, and $I = 6 \times 10^{-5}$ kg · m². The applied voltage in volts is given by

$$y(t) = \begin{cases} 100t & 0 \le t < 0.1 \\ 10 & 0.1 \le t \le 0.4 \\ -100(t - 0.4) + 10 & 0.4 < t \le 0.5 \\ 0 & t > 0.5 \end{cases}$$

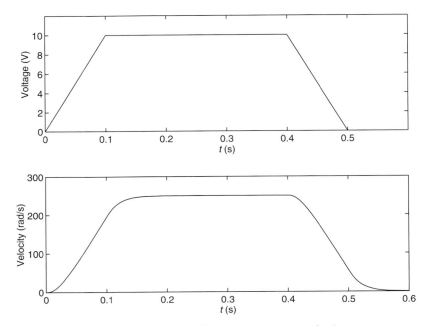

Figure 9.5–5 Voltage input and resulting velocity response of a dc motor.

This is shown in the top graph in Figure 9.5–5.

■ Solution

The following program first creates the model sys from the matrices **A**, **B**, **C**, and **D**. We choose **C** and **D** to obtain the speed x_2 as the only output. (To obtain both the speed and the current as outputs, we would choose C = [1, 0; 0, 1] and D = [0; 0].) The program computes the time constants using the eig function and then creates time, the array of time values to be used by lsim. We choose the time increment 0.0001 to be a very small fraction of the total time, 0.6 s.

The trapezoidal voltage function is then created with a for loop. This is perhaps the easiest way because the if-elseif-else structure mimics the equations that define $v(t)$. The initial conditions $x_1(0)$ and $x_2(0)$ are assumed to be zero, and so they need not be specified in the lsim function.

```
% File dcmotor.m
R = 0.6; L = 0.002; c = 0;
K_T = 0.04; K_e = 0.04; I = 6e-5;
A = [-R/L, -K_e/L; K_T/I, -c/I];
B = [1/L; 0]; C = [0,1]; D = [0];
sys = ss(A,B,C,D);
Time_constants = -1./real(eig(A))
time = 0:0.0001:0.6;
k = 0;
for t = 0:0.0001:0.6
```

```
    k = k + 1;
    if t < 0.1
        v(k) = 100*t;
    elseif t <= 0.4
        v(k) = 10;
    elseif t <= 0.5
        v(k) = -100*(t-0.4) + 10;
    else
        v(k) = 0;
    end
end
[y,t] = lsim(sys, v, time);
subplot(2,1,1), plot(time,v)
subplot(2,1,2), plot(time,y)
```

The time constants are computed to be 0.0041 and 0.0184 s. The largest time constant indicates that the motor's response time is approximately $4(0.0184) = 0.0736$ s. Because this time is less than the time needed for the applied voltage to reach 10 V, the motor should be able to follow the desired trapezoidal profile reasonably well. To know for certain, we must solve the motor's differential equations. The results are plotted in the bottom graph of Figure 9.5–5. The motor's velocity follows a trapezoidal profile as expected, although there is some slight deviation because of its electric resistance and mechanical inertia.

LTI Viewer   The Control System toolbox contains the LTI Viewer, which assists in the analysis of LTI systems. It provides an interactive user interface that allows you to switch between different types of response plots and between the analysis of different systems. The viewer is invoked by typing `ltiview`. See the MATLAB Help for more information.

Predefined Input Functions

You can always create any complicated input function to use with the ODE solver `ode45` or `lsim` by defining a vector containing the input function's values at specified times, as was done in Example 9.5–1 for the trapezoidal profile. However, MATLAB provides the `gensig` function that makes it easy to construct periodic input functions.

The syntax `[u, t] = gensig(type, period)` generates a periodic input of a specified type `type`, having a period `period`. The following types are available: sine wave (`type` = 'sin'), square wave (`type` = 'square'), and narrow-width periodic pulse (`type` = 'pulse'). The vector `t` contains the times, and the vector `u` contains the input values at those times. All generated inputs have unit amplitudes. The syntax `[u, t] = gensig(type, period, tF, dt)` specifies the time duration `tF` of the input and the spacing `dt` between the time instants.

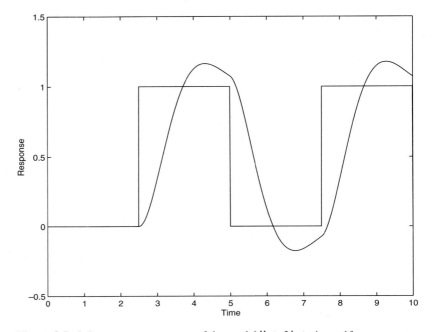

Figure 9.5–6 Square wave response of the model $\ddot{x} + 2\dot{x} + 4x = 4f$.

For example, suppose a square wave with period 5 is applied to the following reduced-form model.

$$\ddot{x} + 2\dot{x} + 4x = 4f \qquad\qquad (9.5–13)$$

To find the response for zero initial conditions, over the interval $0 \leq t \leq 10$, using a step size of 0.01, the session is

```
>>sys5 = tf(4,[1,2,4]);
>>[u, t] = gensig('square',5,10,0.01);
>>[y, t] = lsim (sys5,u,t);plot(t,y,u), . . .
    axis([0 10 -0.5 1.5]), . . .
    xlabel('Time'),ylabel('Response')
```

The result is shown in Figure 9.5–6.

9.6 Summary

This chapter covered numerical methods for computing integrals and derivatives, and for solving ordinary differential equations. Now that you have finished this chapter, you should be able to do the following.

■   Numerically evaluate single, double, and triple integrals whose integrands are given functions.

- Numerically evaluate single integrals whose integrands are given as numerical values.
- Numerically estimate the derivative of a set of data.
- Compute the gradient and Laplacian of a given function.
- Obtain in closed form the integral and derivative of a polynomial function.
- Use the MATLAB ODE solvers to solve single first-order ordinary differential equations whose initial conditions are specified.
- Convert higher-order ordinary differential equations into a set of first-order equations.
- Use the MATLAB ODE solvers to solve sets of higher-order ordinary differential equations whose initial conditions are specified.
- Use MATLAB to convert a model from transfer function form to state-variable form, and vice versa.
- Use the MATLAB linear solvers to solve linear differential equations to obtain the free response and the step response for arbitrary forcing functions.

We have not covered all the differential equation solvers provided in MATLAB, but limited our coverage to ordinary differential equations whose initial conditions are specified. MATLAB provides algorithms for solving *boundary-value problems* (BVPs) such as

$$\ddot{x} + 7\dot{x} + 10x = 0 \qquad x(0) = 2 \qquad x(5) = 8 \qquad 0 \le t \le 5$$

See the Help for the function `bvp4c`. Some differential equations are specified *implicitly* as $f(t, y, \dot{y}) = 0$. The solver `ode15i` can be used for such problems. MATLAB can also solve *delay-differential equations* (DDEs) such as

$$\ddot{x} + 7\dot{x} + 10x + 5x(t - 3) = 0$$

See the help for the functions `dde23`, `ddesd`, and `deval`. The function `pdepe` can solve *partial* differential equations. See also `pdeval`. In addition, MATLAB provides support for analyzing and plotting the solver's output. See the functions `odeplot`, `odephas2`, `odephas3`, and `odeprint`.

Key Terms with Page References

Backward difference, 378
Cauchy form, 390
Central difference, 379
Definite integral, 370
Eigenvalue, 397
Euler method, 383
Forced response, 383
Forward difference, 378
Free response, 383
Improper integral, 370
Indefinite integral, 370

Initial-value problem (IVP), 382
Laplacian, 382
LTI object, 398
Modified Euler method, 384
Ordinary differential equation, 382
ODE, 382
Predictor-corrector method, 384
Quadrature, 373
Singularities, 370
State-variable form, 390
Step size, 384

Problems

You can find answers to problems marked with an asterisk at the end of the text.

Section 9.1

1.* An object moves at a velocity $v(t) = 5 + 7t^2$ m/s starting from the position $x(2) = 5$ m at $t = 2$ s. Determine its position at $t = 10$ s.

2. The total distance traveled by an object moving at velocity $v(t)$ from the time $t = a$ to the time $t = b$ is

$$x(b) = \int_a^b |v(t)|\, dt + x(a)$$

The absolute value $|v(t)|$ is used to account for the possibility that $v(t)$ might be negative. Suppose an object starts at time $t = 0$ and moves with a velocity of $v(t) = \cos(\pi t)$ m. Find the object's location at $t = 1$ s if $x(0) = 2$ m.

3. An object starts with an initial velocity of 3 m/s at $t = 0$, and it accelerates with an acceleration of $a(t) = 7t$ m/s^2. Find the total distance the object travels in 4 s.

4. The equation for the voltage $v(t)$ across a capacitor as a function of time is

$$v(t) = \frac{1}{C}\left[\int_0^t i(t)\, dt + Q_0 \right]$$

where $i(t)$ is the applied current and Q_0 is the initial charge. A certain capacitor initially holds no charge. Its capacitance is $C = 10^{-7}$ F. If a current $i(t) = 0.2[1 + \sin(0.2t)]$ A is applied to the capacitor, compute the voltage $v(t)$ at $t = 1.2$ s if its initial velocity is zero.

5. A certain object's acceleration is given by $a(t) = 7t \sin 5t$ m/s^2. Compute its velocity at $t = 10$ s if its initial velocity is zero.

6. A certain object moves with the velocity $v(t)$ given in the table below. Determine the object's position $x(t)$ at $t = 10$ s if $x(0) = 3$.

Time (s)	0	1	2	3	4	5	6	7	8	9	10
Velocity (m/s)	0	2	5	7	9	12	15	18	22	20	17

7.* A tank having vertical sides and a bottom area of 100 ft^2 is used to store water. The tank is initially empty. To fill the tank, water is pumped into the top at the rate given in the following table. Determine the water height $h(t)$ at $t = 10$ min.

Time (min)	0	1	2	3	4	5	6	7	8	9	10
Flow rate (ft^3/min)	0	80	130	150	150	160	165	170	160	140	120

8. A cone-shaped paper drinking cup (like the kind supplied at water fountains) has a radius R and a height H. If the water height in the cup is h, the
water volume is given by

$$V = \frac{1}{3}\pi\left(\frac{R}{H}\right)^2 h^3$$

Suppose that the cup's dimensions are $R = 1.5$ in. and $H = 4$ in.

a. If the flow rate from the fountain into the cup is 2 in.^{3}/s, how long will
it take to fill the cup to the brim?

b. If the flow rate from the fountain into the cup is given by $2(1 - e^{-2t})$
in.^{3}/s, how long will it take to fill the cup to the brim?

9. A certain object has a mass of 100 kg and is acted on by a force
$f(t) = 500[2 - e^{-t}\sin(5\pi t)]$ N. The mass is at rest at $t = 0$. Determine the
object's velocity at $t = 5$ s.

10.* A rocket's mass decreases as it burns fuel. The equation of motion for a
rocket in vertical flight can be obtained from Newton's law, and it is

$$m(t)\frac{dv}{dt} = T - m(t)g$$

where T is the rocket's thrust and its mass as a function of time is given
by $m(t) = m_0(1 - rt/b)$. The rocket's initial mass is m_0, the burn time is b,
and r is the fraction of the total mass accounted for by the fuel.

Use the values $T = 48{,}000$ N, $m_0 = 2200$ kg, $r = 0.8$, $g = 9.81$ m/s^2, and
$b = 40$ s. Determine the rocket's velocity at burnout.

11. The equation for the voltage $v(t)$ across a capacitor as a function of time is

$$v(t) = \frac{1}{C}\left[\int_0^t i(t)\,dt + Q_0\right]$$

where $i(t)$ is the applied current and Q_0 is the initial charge. Suppose that
$C = 10^{-7}$ F and that $Q_0 = 0$. Suppose the applied current is $i(t) = 0.3 +
0.1e^{-5t}\sin(25\pi t)$ A. Plot the voltage $v(t)$ for $0 \le t \le 7$ s.

12. Compute the indefinite integral of $p(x) = 5x^2 - 9x + 8$.

13. Compute the double integral

$$A = \int_0^3 \int_1^3 (x^2 + 3xy)\,dx\,dy$$

14. Compute the double integral

$$A = \int_0^4 \int_0^\pi x^2 \sin y\,dx\,dy$$

15.   Compute the double integral

$$A = \int_0^1 \int_y^3 x^2(x + y)\,dx\,dy$$

Note that the region of integration lies to the right of the line $y = x$. Use this fact and a MATLAB relational operator to eliminate values for which $y > x$.

16.   Compute the triple integral

$$A = \int_1^2 \int_0^1 \int_1^3 xe^{yz}\,dx\,dy\,dz$$

Section 9.2

17.   Plot the estimate of the derivative dy/dx from the following data. Do this by using forward, backward, and central differences. Compare the results.

x	0	1	2	3	4	5	6	7	8	9	10
y	0	2	5	7	9	12	15	18	22	20	17

18.   At a relative maximum of a curve $y(x)$, the slope dy/dx is zero. Use the following data to estimate the values of x and y that correspond to a maximum point.

x	0	1	2	3	4	5	6	7	8	9	10
y	0	2	5	7	9	10	8	7	6	8	10

19.   Compare the performance of the forward, backward, and central difference methods for estimating the derivative of $y(x) = e^{-x}\sin(3x)$. Use 101 points from $x = 0$ to $x = 4$. Use a random additive error of ± 0.01.

20.   Compute the expressions for dp_2/dx, $d(p_1 p_2)/dx$, and $d(p_2/p_1)/dx$ for $p_1 = 5x^2 + 7$ and $p_2 = 5x^2 - 6x + 7$.

21.   Plot the contour plot and the gradient (shown by arrows) for the function

$$f(x, y) = -x^2 + 2xy + 3y^2$$

Section 9.3

22.   Plot the solution of the equation

$$6\dot{y} + y = f(t)$$

if $f(t) = 0$ for $t < 0$ and $f(t) = 15$ for $t \geq 0$. The initial condition is $y(0) = 7$.

23. The equation for the voltage y across the capacitor of an RC circuit is

$$RC\frac{dy}{dt} + y = v(t)$$

where $v(t)$ is the applied voltage. Suppose that $RC = 0.2$ s and that the capacitor voltage is initially 2 V. Suppose also that the applied voltage goes from 0 to 10 V at $t = 0$. Plot the voltage $y(t)$ for $0 \leq t \leq 1$ s.

24. The following equation describes the temperature $T(t)$ of a certain object immersed in a liquid bath of constant temperature T_b.

$$10\frac{dT}{dt} + T = T_b$$

Suppose the object's temperature is initially $T(0) = 70°F$ and the bath temperature is $T_b = 170°F$.
a. How long will it take for the object's temperature T to reach the bath temperature?
b. How long will it take for the object's temperature T to reach 168°F?
c. Plot the object's temperature $T(t)$ as a function of time.

25.* The equation of motion of a rocket-propelled sled is, from Newton's law,

$$m\dot{v} = f - cv$$

where m is the sled mass, f is the rocket thrust, and c is an air resistance coefficient. Suppose that $m = 1000$ kg and $c = 500$ N · s/m. Suppose also that $v(0) = 0$ and $f = 75{,}000$ N for $t \geq 0$. Determine the speed of the sled at $t = 10$ s.

26. The following equation describes the motion of a mass connected to a spring, with viscous friction on the surface.

$$m\ddot{y} + c\dot{y} + ky = 0$$

Plot $y(t)$ for $y(0) = 10$, $\dot{y}(0) = 5$ if
a. $m = 3$, $c = 18$, and $k = 102$
b. $m = 3$, $c = 39$ and $k = 120$

27. The equation for the voltage y across the capacitor of an RC circuit is

$$RC\frac{dy}{dt} + y = v(t)$$

where $v(t)$ is the applied voltage. Suppose that $RC = 0.2$ s and that the capacitor voltage is initially 2 V. Suppose also that the applied voltage is $v(t) = 10[2 - e^{-t}\sin(5\pi t)]$ V. Plot the voltage $y(t)$ for $0 \leq t \leq 5$ s.

28. The equation describing the water height h in a spherical tank with a drain at the bottom is

$$\pi(2rh - h^2)\frac{dh}{dt} = -C_d A\sqrt{2gh}$$

Suppose the tank's radius is $r = 3$ m and the circular drain hole has a radius of 2 cm. Assume that $C_d = 0.5$ and that the initial water height is $h(0) = 5$ m. Use $g = 9.81$ m/s^2.

a. Use an approximation to estimate how long it takes for the tank to empty.

b. Plot the water height as a function of time until $h(t) = 0$.

29. The following equation describes a certain dilution process, where $y(t)$ is the concentration of salt in a tank of freshwater to which salt brine is being added.

$$\frac{dy}{dt} + \frac{5}{10 + 2t}y = 4$$

Suppose that $y(0) = 0$. Plot $y(t)$ for $0 \le t \le 10$.

Section 9.4

30. The following equation describes the motion of a certain mass connected to a spring, with viscous friction on the surface

$$3\ddot{y} + 18\dot{y} + 102y = f(t)$$

where $f(t)$ is an applied force. Suppose that $f(t) = 0$ for $t < 0$ and $f(t) = 10$ for $t \ge 0$.

a. Plot $y(t)$ for $y(0) = \dot{y}(0) = 0$.

b. Plot $y(t)$ for $y(0) = 0$ and $\dot{y}(0) = 10$. Discuss the effect of the nonzero initial velocity.

31. The following equation describes the motion of a certain mass connected to a spring, with viscous friction on the surface

$$3\ddot{y} + 39\dot{y} + 120y = f(t)$$

where $f(t)$ is an applied force. Suppose that $f(t) = 0$ for $t < 0$ and $f(t) = 10$ for $t \ge 0$.

a. Plot $y(t)$ for $y(0) = \dot{y}(0) = 0$.

b. Plot $y(t)$ for $y(0) = 0$ and $\dot{y}(0) = 10$. Discuss the effect of the nonzero initial velocity.

32. The following equation describes the motion of a certain mass connected to a spring, with no friction

$$3\ddot{y} + 75y = f(t)$$

where $f(t)$ is an applied force. Suppose the applied force is sinusoidal with a frequency of ω rad/s and an amplitude of 10 N: $f(t) = 10\sin(\omega t)$. Suppose that the initial conditions are $y(0) = \dot{y}(0) = 0$. Plot $y(t)$ for $0 \leq t \leq 20$ s. Do this for the following three cases. Compare the results of each case.

a. $\omega = 1$ rad/s
b. $\omega = 5$ rad/s
c. $\omega = 10$ rad/s

33. Van der Pol's equation has been used to describe many oscillatory processes. It is

$$\ddot{y} - \mu(1 - y^2)\dot{y} + y = 0$$

Plot $y(t)$ for $\mu = 1$ and $0 \leq t \leq 20$, using the initial conditions $y(0) = 5$, $\dot{y}(0) = 0$.

34. The equation of motion for a pendulum whose base is accelerating horizontally with an acceleration $a(t)$ is

$$L\ddot{\theta} + g\sin\theta = a(t)\cos\theta$$

Suppose that $g = 9.81$ m/s^2, $L = 1$ m, and $\dot{\theta}(0) = 0$. Plot $\theta(t)$ for $0 \leq t \leq 10$ s for the following three cases.

a. The acceleration is constant: $a = 5$ m/s^2, and $\theta(0) = 0.5$ rad.
b. The acceleration is constant: $a = 5$ m/s^2, and $\theta(0) = 3$ rad.
c. The acceleration is linear with time: $a = 0.5t$ m/s^2, and $\theta(0) = 3$ rad.

35. Van der Pol's equation is

$$\ddot{y} - \mu(1 - y^2)\dot{y} + y = 0$$

This equation is stiff for large values of the parameter μ. Compare the performance of `ode45` and `ode15s` for this equation. Use $\mu = 1000$ and $0 \leq t \leq 3000$, with the initial conditions $y(0) = 2$, $\dot{y}(0) = 0$. Plot $y(t)$ versus t.

Section 9.5

36. The equations for an armature-controlled dc motor are the following. The motor's current is i and its rotational velocity is ω.

$$L\frac{di}{dt} = -Ri - K_e\omega + v(t) \tag{9.6--1}$$

$$I\frac{d\omega}{dt} = K_Ti - c\omega \tag{9.6--2}$$

where L, R, and I are the motor's inductance, resistance, and inertia; K_T and K_e are the torque constant and back emf constant; c is a viscous damping constant; and $v(t)$ is the applied voltage.

Use the values $R = 0.8\ \Omega$, $L = 0.003\ H$, $K_T = 0.05\ N \cdot m/A$, $K_e = 0.05\ V \cdot s/rad$, $c = 0$, and $I = 8 \times 10^{-5}\ kg \cdot m^2$.

a. Suppose the applied voltage is 20 V. Plot the motor's speed and current versus time. Choose a final time large enough to show the motor's speed becoming constant.

b. Suppose the applied voltage is trapezoidal as given below.

$$v(t) = \begin{cases} 400t & 0 \le t < 0.05 \\ 20 & 0.05 \le t \le 0.2 \\ -400(t - 0.2) + 20 & 0.2 < t \le 0.25 \\ 0 & t > 0.25 \end{cases}$$

Plot the motor's speed versus time for $0 \le t \le 0.3$ s. Also plot the applied voltage versus time. How well does the motor speed follow a trapezoidal profile?

37. Compute and plot the unit-impulse response of the following model.

$$10\ddot{y} + 3\dot{y} + 7y = f(t)$$

38. Compute and plot the unit-step response of the following model.

$$10\ddot{y} + 6\dot{y} + 2y = f + 7\dot{f}$$

39.* Find the reduced form of the following state model.

$$\begin{bmatrix} \dot{x}_1 \\ \dot{x}_2 \end{bmatrix} = \begin{bmatrix} -4 & -1 \\ 2 & -3 \end{bmatrix} \begin{bmatrix} x_1 \\ x_2 \end{bmatrix} + \begin{bmatrix} 2 \\ 5 \end{bmatrix} u(t)$$

40. The following state model describes the motion of a certain mass connected to a spring, with viscous friction on the surface, where $m = 1$, $c = 2$, and $k = 5$.

$$\begin{bmatrix} \dot{x}_1 \\ \dot{x}_2 \end{bmatrix} = \begin{bmatrix} 0 & 1 \\ -5 & -2 \end{bmatrix} \begin{bmatrix} x_1 \\ x_2 \end{bmatrix} + \begin{bmatrix} 0 \\ 1 \end{bmatrix} f(t)$$

a. Use the `initial` function to plot the position x_1 of the mass, if the initial position is 5 and the initial velocity is 3.

b. Use the `step` function to plot the step response of the position and velocity for zero initial conditions, where the magnitude of the step input is 10. Compare your plot with that shown in Figure 9.5–1.

41. Consider the following equation.

$$5\ddot{y} + 2\dot{y} + 10y = f(t)$$

a. Plot the free response for the initial conditions $y(0) = 10$, $\dot{y}(0) = -5$.

b. Plot the unit-step response (for zero initial conditions).

c. The *total response* to a step input is the sum of the free response and the step response. Demonstrate this fact for this equation by plotting the sum of the solutions found in parts *a* and *b* and comparing the plot with that generated by solving for the total response with $y(0) = 10$, $\dot{y}(0) = -5$.

42. The model for the *RC* circuit shown in Figure P42 is

$$RC\frac{dv_o}{dt} + v_o = v_i$$

For $RC = 0.2$ s, plot the voltage response $v_o(t)$ for the case where the applied voltage is a single square pulse of height 10 V and duration 0.4 s, starting at $t = 0$. The initial capacitor voltage is zero.

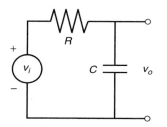

Figure P42

Nick Koudis/Getty Images/RF

Engineering in the 21st Century. . .

Embedded Control Systems

An embedded control system is a microprocessor and sensor suite designed to be an integral part of a product. The aerospace and automotive industries have used embedded controllers for some time, but the decreasing cost of components now makes embedded controllers feasible for more consumer and biomedical applications.

For example, embedded controllers can greatly increase the performance of orthopedic devices. One model of an artificial leg now uses sensors to measure in real time the walking speed, the knee joint angle, and the loading due to the foot and ankle. These measurements are used by the controller to adjust the hydraulic resistance of a piston to produce a more stable, natural, and efficient gait. The controller algorithms are adaptive in that they can be tuned to an individual's characteristics and their settings changed to accommodate different physical activities.

Engines incorporate embedded controllers to improve efficiency. Embedded controllers in new active suspensions use actuators to improve on the performance of traditional passive systems consisting only of springs and dampers. One design phase of such systems is *hardware-in-the-loop testing,* in which the controlled object (the engine or vehicle suspension) is replaced with a real-time simulation of its behavior. This enables the embedded system hardware and software to be tested faster and less expensively than with the physical prototype, and perhaps even before the prototype is available.

Simulink is often used to create the simulation model for hardware-in-the-loop testing. The Control Systems and the Signal Processing toolboxes, and the DSP and Fixed Point block sets, are also useful for such applications. ■

CHAPTER 10

Simulink

OUTLINE

10.1   Simulation Diagrams

10.2   Introduction to Simulink

10.3   Linear State-Variable Models

10.4   Piecewise-Linear Models

10.5   Transfer-Function Models

10.6   Nonlinear State-Variable Models

10.7   Subsystems

10.8   Dead Time in Models

10.9   Simulation of a Nonlinear Vehicle Suspension Model

10.10   Summary

Problems

Simulink is built on top of MATLAB, so you must have MATLAB to use Simulink. It is included in the Student Edition of MATLAB and is also available separately from The MathWorks, Inc. Simulink is widely used in industry to model complex systems and processes that are difficult to model with a simple set of differential equations.

Simulink provides a graphical user interface that uses various types of elements called *blocks* to create a simulation of a dynamic system, that is, a system that can be modeled with differential or difference equations whose independent variable is time. For example, one block type is a multiplier, another performs a sum, and still another is an integrator. The Simulink graphical interface enables you to position the blocks, resize them, label them, specify block parameters, and interconnect the blocks to describe complicated systems for simulation.

This chapter starts with simulations of simple systems that require few blocks. Gradually, through a series of examples, more block types are introduced. The chosen applications require only a basic knowledge of physics and thus can be appreciated by readers from any engineering or scientific discipline. By the time you have finished this chapter, you will have seen the block types needed to simulate a large variety of common applications.

10.1 Simulation Diagrams

BLOCK DIAGRAM

You develop Simulink models by constructing a diagram that shows the elements of the problem to be solved. Such diagrams are called *simulation diagrams* or *block diagrams.* Consider the equation $\dot{y} = 10f(t)$. Its solution can be represented symbolically as

$$y(t) = \int 10f(t)\,dt$$

which can be thought of as two steps, using an intermediate variable x:

$$x(t) = 10f(t) \qquad \text{and} \qquad y(t) = \int x(t)\,dt$$

This solution can be represented graphically by the simulation diagram shown in Figure 10.1–1a. The arrows represent the variables y, x, and f. The blocks represent cause-and-effect processes. Thus, the block containing the number 10 represents the process $x(t) = 10f(t)$, where $f(t)$ is the cause (the *input*) and $x(t)$ represents the effect (the *output*). This type of block is called a *multiplier* or *gain block.*

GAIN BLOCK

The block containing the integral sign $\int$ represents the integration process $y(t) = \int x(t)\,dt$, where $x(t)$ is the cause (the *input*) and $y(t)$ represents the effect (the *output*). This type of block is called an *integrator block.*

INTEGRATOR BLOCK

There is some variation in the notation and symbols used in simulation diagrams. Figure 10.1–1b shows one variation. Instead of being represented by a box, the multiplication process is now represented by a triangle like that used to represent an electrical amplifier, hence the name *gain block.*

In addition, the integration symbol in the integrator block has been replaced by the operator symbol $1/s$, which derives from the notation used for the Laplace transform (see Section 11.7 for a discussion of this transform). Thus the equation $\dot{y} = 10f(t)$ is represented by $sy = 10f$, and the solution is represented as

$$y = \frac{10f}{s}$$

or as the two equations

$$x = 10f \qquad \text{and} \qquad y = \frac{1}{s}x$$

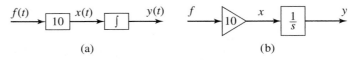

(a)   (b)

Figure 10.1–1 Simulation diagrams for $\dot{y} = 10f(t)$.

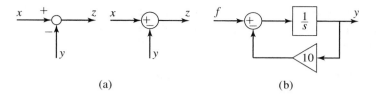

Figure 10.1–2 (a) The summer element. (b) Simulation diagram for $\dot{y} = f(t) - 10y$.

Another element used in simulation diagrams is the *summer* that, despite its name, is used to subtract as well as to sum variables. Two versions of its symbol are shown in Figure 10.1–2a. In each case the symbol represents the equation $z = x - y$. Note that a plus or minus sign is required for each input arrow.

The summer symbol can be used to represent the equation $\dot{y} = f(t) - 10y$, which can be expressed as

$$y(t) = \int [f(t) - 10y]\, dt$$

or as

$$y = \frac{1}{s}(f - 10y)$$

You should study the simulation diagram shown in Figure 10.1–2b to confirm that it represents this equation. This figure forms the basis for developing a Simulink model to solve the equation.

10.2 Introduction to Simulink

Type `simulink` in the MATLAB Command window to start Simulink. The Simulink Library Browser window opens. See Figure 10.2–1. The Simulink blocks are located in "libraries." These libraries are displayed under the Simulink heading in Figure 10.2–1. Depending on what other MathWorks products are installed, you might see additional items in this window, such as the Control System Toolbox and Stateflow. These provide additional Simulink blocks, which can be displayed by clicking on the plus sign to the left of the item. As Simulink evolves through new versions, some libraries are renamed and some blocks are moved to different libraries, so the library we specify here might change in later releases. The best way to locate a block, given its name, is to type its name in the search pane at the top of the Simulink Library Browser. When you press Enter, Simulink will take you to the block location.

To create a new model, click on the icon that resembles a clean sheet of paper, or select **New** from the **File** menu in the browser. A new **Untitled** window opens for you to create the model. To select a block from the Library Browser, double-click on the appropriate library, and a list of blocks within that library then appears as shown in Figure 10.2–1. This figure shows the result of clicking on the Continuous library, then clicking on the Integrator block.

SUMMER

LIBRARY BROWSER

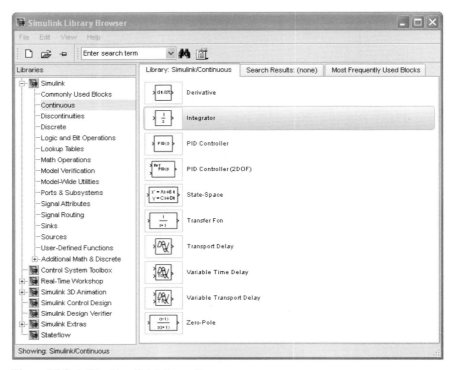

Figure 10.2–1 The Simulink Library Browser.

Click on the block name or icon, hold the mouse button down, drag the block to the new model window, and release the button. You can access help for that block by right-clicking on its name or icon and selecting **Help** from the drop-down menu.

Simulink model files have the extension .mdl. Use the **File** menu in the model window to Open, Close, and Save model files. To print the block diagram of the model, select **Print** on the **File** menu. Use the **Edit** menu to copy, cut, and paste blocks. You can also use the mouse for these operations. For example, to delete a block, click on it and press the **Delete** key.

Getting started with Simulink is best done through examples, which we now present.

EXAMPLE 10.2–1   Simulink Solution of $\dot{y} = 10 \sin t$

Use Simulink to solve the following problem for $0 \le t \le 13$.

$$\frac{dy}{dt} = 10 \sin t \qquad y(0) = 0$$

The exact solution is $y(t) = 10(1 - \cos t)$.

■ **Solution**

To construct the simulation, do the following steps. Refer to Figure 10.2–2. Figure 10.2–3 shows the Model window after completing the following steps.

1. Start Simulink and open a new model window as described previously.
2. Select and place in the new window the Sine Wave block from the Sources library. Double-click on it to open the Block Parameters window, and make sure the Amplitude is set to 1, the Bias to 0, the Frequency to 1, the Phase to 0, and the Sample time to 0. Then click **OK.**
3. Select and place the Gain block from the Math Operations library, double-click on it, and set the Gain value to 10 in the Block Parameters window. Then click **OK.** Note that the value 10 then appears in the triangle. To make the number more visible, click on the block, and drag one of the corners to expand the block so that all the text is visible.
4. Select and place the Integrator block from the Continuous library, double-click on it to obtain the Block Parameters window, and set the Initial condition to 0 [because $y(0) = 0$]. Then click **OK.**
5. Select and place the Scope block from the Sinks library.
6. Once the blocks have been placed as shown in Figure 10.2–2, connect the input port on each block to the outport port on the preceding block. To do this, move the cursor to an input port or an output port; the cursor will change to a cross. Hold the mouse button down, and drag the cursor to a port on another block. When you release the mouse button, Simulink will connect them with an arrow pointing at the input port. Your model should now look like that shown in Figure 10.2–2.
7. Enter 13 for the Stop time in the window to the right of the **Start Simulation** icon (the black triangle). See Figure 10.2–3. The default value is 10, which can be deleted and replaced with 13.

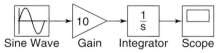

Figure 10.2–2 Simulink model for $\dot{y} = 10 \sin t$.

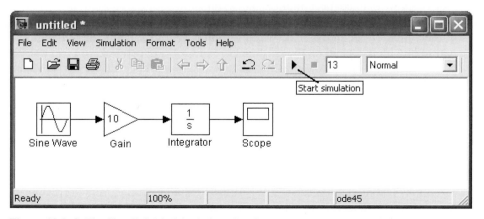

Figure 10.2–3 The Simulink Model window showing the model created in Example 10.2–1.

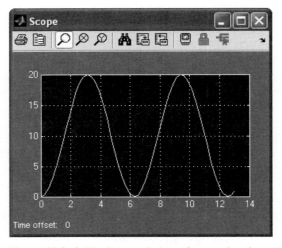

Figure 10.2–4 The Scope window after running the model in Example 10.2–1.

8. Run the simulation by clicking on the **Start Simulation** icon on the toolbar.

9. You will hear a bell sound when the simulation is finished. Then double-click on the Scope block and click on the binoculars icon in the Scope display to enable autoscaling. You should see an oscillating curve with an amplitude of 10 and a period of 2π (Figure 10.2–4). The independent variable in the Scope block is time t; the input to the block is the dependent variable y. This completes the simulation.

In the **Configuration Parameters** submenu under the **Simulation** menu, you can select the ODE solver to use by clicking on the Solver tab. The default is ode45, as indicated in the lower right-hand corner of the Model window.

To have Simulink automatically connect two blocks, select the Source block, hold down the **Ctrl** key, and left-click on the Destination block. Simulink also provides easy ways to connect multiple blocks and lines; see the Help for information.

Note that blocks have a Block Parameters window that opens when you double-click on the block. This window contains several items, the number and nature of which depend on the specific type of block. In general, you can use the default values of these parameters, except where we have explicitly indicated that they should be changed. You can always click on **Help** within the Block Parameters window to obtain more information.

When you click on **Apply,** any changes immediately take effect and the window remains open. If you click on **OK,** the changes take effect but the window closes.

Note that most blocks have default labels. You can edit text associated with a block by clicking on the text and making the changes. You can save the Simulink model as an .mdl file by selecting **Save** from the **File** menu in Simulink. The model file can then be reloaded at a later time. You can also print the diagram by selecting **Print** on the **File** menu.

The Scope block is useful for examining the solution, but if you want to obtain a labeled and printed plot, you can use the To Workspace block, which is described in the next example.

Exporting to the MATLAB Workspace

EXAMPLE 10.2–2

We now demonstrate how to export the results of the simulation to the MATLAB workspace, where they can be plotted or analyzed with any of the MATLAB functions.

■ Solution

Modify the Simulink model constructed in Example 10.2–1 as follows. Refer to Figure 10.2–5.

1. Delete the arrow connecting the Scope block by clicking on it and pressing the **Delete** key. Delete the Scope block in the same way.
2. Select and place the To Workspace block from the Sinks library and the Clock block from the Sources library.
3. Select and place the Mux block from the Signal Routing library, double-click on it, and set the Number of inputs to 2. Click **OK.** (The name *Mux* is an abbreviation for multiplexer, which is an electrical device for combining several signals.)
4. Connect the top input port of the Mux block to the output port of the Integrator block. Then use the same technique to connect the bottom input port of the Mux block to the outport port of the Clock block. Your model should now look like that shown in Figure 10.2–5.
5. Double-click on the To Workspace block. You can specify any variable name you want as the output; the default is `simout`. Change its name to `y`. The output variable `y` will have as many rows as there are simulation time steps, and as many columns as there are inputs to the block. The second column in our simulation will be time, because of the way we have connected the Clock to the second input port of the Mux. Specify the Save format as Array. Use the default values for the other parameters (these should be `inf`, `1`, and `-1` for Limit data points to last: Decimation, and Sample time, respectively). Click on **OK.**

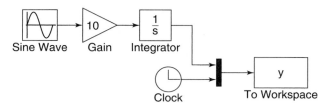

Figure 10.2–5 Simulink model using the Clock and To Workspace blocks.

6.   After running the simulation, you can use the MATLAB plotting commands from the Command window to plot the columns of y (or simout in general). To plot $y(t)$, type in the MATLAB Command window

```
>>plot(y(:,2),y(:,1)),xlabel('t'),ylabel('y')
```

Simulink can be configured to put the time variable tout into the MATLAB workspace automatically when you are using the To Workspace block. This is done with the **Data Import/Export** menu item under **Configuration Parameters** on the **Simulation** menu. The alternative is to use the Clock block to put tout into the workspace. The Clock block has one parameter, Decimation. If this parameter is set to 1, the Clock block will output the time every time step; if set to 10, for example, the block will output every 10 time steps, and so on.

Simulink Model for $\dot{y} = -10y + f(t)$

Construct a Simulink model to solve

$$\dot{y} = -10y + f(t) \qquad y(0) = 1$$

where $f(t) = 2 \sin 4t$, for $0 \le t \le 3$.

■ **Solution**

To construct the simulation, do the following steps.

1.   You can use the model shown in Figure 10.2–2 by rearranging the blocks as shown in Figure 10.2–6. You will need to add a Sum block.

2.   Select the Sum block from the Math Operations library and place it as shown in the simulation diagram. Its default setting adds two input signals. To change this, double-click on the block, and in the List of Signs window, type |+– . The signs are ordered counterclockwise from the top. The symbol | is a spacer indicating here that the top port is to be empty.

3.   To reverse the direction of the Gain block, right-click on the block, select **Format** from the pop-up menu, and select **Flip Block.**

4.   When you connect the negative input port of the Sum block to the output port of the Gain block, Simulink will attempt to draw the shortest line. To obtain the more

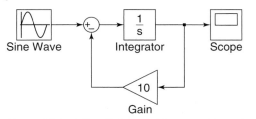

Figure 10.2–6 Simulink model for $\dot{y} = -10y + f(t)$.

standard appearance shown in Figure 10.2–6, first extend the line vertically down from the Sum input port. Release the mouse button, and then click on the end of the line and attach it to the Gain block. The result will be a line with a right angle. Do the same to connect the input of the Gain block to the arrow connecting the Integrator and the Scope blocks. A small dot appears to indicate that the lines have been successfully connected. This point is called a *takeoff point* because it takes the value of the variable represented by the arrow (here, the variable y) and makes that value available to another block.

5. Set the Stop time to 3.
6. Run the simulation as before and observe the results in the Scope.

10.3 Linear State-Variable Models

State-variable models, unlike transfer-function models, can have more than one input and more than one output. Simulink has the State-Space block that represents the linear state-variable model $\dot{\mathbf{x}} = \mathbf{A}\mathbf{x} + \mathbf{B}\mathbf{u}, \mathbf{y} = \mathbf{C}\mathbf{x} + \mathbf{D}\mathbf{u}$. (See Section 9.5 for discussion of this model form.) The vector $\mathbf{u}$ represents the inputs, and the vector $\mathbf{y}$ represents the outputs. Thus, when you are connecting inputs to the State-Space block, care must be taken to connect them in the proper order. Similar care must be taken when connecting the block's outputs to another block. The following example illustrates how this is done.

Simulink Model of a Two-Mass Suspension System | **EXAMPLE 10.3–1**

The following are the equations of motion of the two-mass suspension model shown in Figure 10.3–1.

$$m_1\ddot{x}_1 = k_1(x_2 - x_1) + c_1(\dot{x}_2 - \dot{x}_1)$$

$$m_2\ddot{x}_2 = -k_1(x_2 - x_1) - c_1(\dot{x}_2 - \dot{x}_1) + k_2(y - x_2)$$

Develop a Simulink model of this system to obtain the plots of x_1 and x_2. The input $y(t)$ is a unit step function, and the initial conditions are zero. Use the following values: $m_1 = 250$ kg, $m_2 = 40$ kg, $k_1 = 1.5 \times 10^4$ N/m, $k_2 = 1.5 \times 10^5$ N/m, and $c_1 = 1917$ N $\cdot$ s/m.

■ **Solution**

The equations of motion can be expressed in state-variable form by letting $z_1 = x_1$, $z_2 = \dot{x}_1$, $z_3 = x_2$, $z_4 = \dot{x}_2$. The equations of motion become

$$\dot{z}_1 = z_2 \qquad \dot{z}_2 = \frac{1}{m_1}(-k_1z_1 - c_1z_2 + k_1z_3 + c_1z_4)$$

$$\dot{z}_3 = z_4 \qquad \dot{z}_4 = \frac{1}{m_2}[k_1z_1 + c_1z_2 - (k_1 + k_2)z_3 - c_1z_4 + k_2y]$$

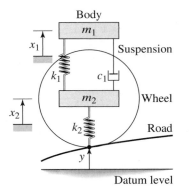

Figure 10.3–1 Two-mass suspension model.

These equations are expressed in vector-matrix form as

$$\dot{\mathbf{z}} = \mathbf{A}\mathbf{z} + \mathbf{B}y(t)$$

where

$$\mathbf{A} = \begin{bmatrix} 0 & 1 & 0 & 0 \\ -\dfrac{k_1}{m_1} & -\dfrac{c_1}{m_1} & \dfrac{k_1}{m_1} & \dfrac{c_1}{m_1} \\ 0 & 0 & 0 & 1 \\ \dfrac{k_1}{m_2} & \dfrac{c_1}{m_2} & -\dfrac{k_1+k_2}{m_2} & -\dfrac{c_1}{m_2} \end{bmatrix} \qquad \mathbf{B} = \begin{bmatrix} 0 \\ 0 \\ 0 \\ \dfrac{k_2}{m_2} \end{bmatrix}$$

and

$$\mathbf{z} = \begin{bmatrix} z_1 \\ z_2 \\ z_3 \\ z_4 \end{bmatrix} = \begin{bmatrix} x_1 \\ \dot{x}_1 \\ x_2 \\ \dot{x}_2 \end{bmatrix}$$

To simplify the notation, let $a_1 = k_1/m_1$, $a_2 = c_1/m_1$, $a_3 = k_1/m_2$, $a_4 = c_1/m_2$, $a_5 = k_2/m_2$, and $a_6 = a_3 + a_5$. The matrices **A** and **B** become

$$\mathbf{A} = \begin{bmatrix} 0 & 1 & 0 & 0 \\ -a_1 & -a_2 & a_1 & a_2 \\ 0 & 0 & 0 & 1 \\ a_3 & a_4 & -a_6 & -a_4 \end{bmatrix} \qquad \mathbf{B} = \begin{bmatrix} 0 \\ 0 \\ 0 \\ a_5 \end{bmatrix}$$

Next select appropriate values for the matrices in the output equation $\mathbf{y} = \mathbf{C}\mathbf{z} + \mathbf{B}y(t)$. Because we want to plot x_1 and x_2, which are z_1 and z_3, we must use the following matrices for **C** and **D**.

$$\mathbf{C} = \begin{bmatrix} 1 & 0 & 0 & 0 \\ 0 & 0 & 1 & 0 \end{bmatrix} \qquad \mathbf{D} = \begin{bmatrix} 0 \\ 0 \end{bmatrix}$$

Note that the dimensions of **B** tell Simulink that there is one input. The dimensions of **C** and **D** tell Simulink that there are two outputs.

Open a new model window, and then do the following to create the model shown in Figure 10.3–2.

1. Select and place the Step block from the Sources library. Double-click on it to open the Block Parameters window, and set the Step Time to 0, the Initial Value to 0, and the Final Value to 1. Do not change the default value of any other parameters in this window. Click **OK**. The Step Time is the time at which the step input begins.
2. Select and place the State-Space block from the Continuous library. Open its Block Parameters window and enter the following values for the matrices **A, B, C,** and **D.** For **A** enter

   `[0, 1, 0, 0; -a1,-a2, a1, a2; 0, 0, 0, 1; a3, a4, -a6, -a4]`

   For **B** enter `[0; 0; 0; a5]`. For **C** enter `[1, 0, 0, 0; 0, 0, 1, 0]`, and for **D** enter `[0; 0]`. Then enter `[0; 0; 0; 0]` for the initial conditions. Click **OK.**

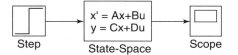

Figure 10.3–2 Simulink model containing the State-Space block and Step block.

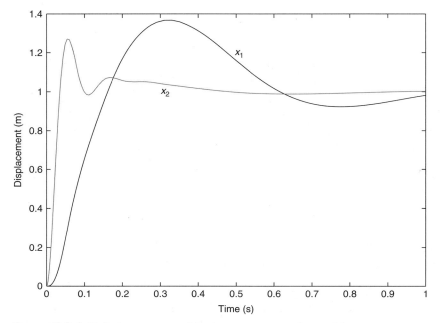

Figure 10.3–3 Unit-step response of the two-mass suspension model.

3. Select and place the Scope block from the Sinks library.
4. Connect the input and output ports as shown in Figure 10.3–2, and save the model.
5. In the Workspace window enter the parameter values and compute the a_i constants as shown in the following session.

```
>>m1 = 250; m2 = 40; k1 = 1.5e+4;
>>k2 = 1.5e+5; c1 =1917;
>>a1 = k1/m1; a2 = c1/m1; a3 = k1/m2;
>>a4 = c1/m2; a5 = k2/m2; a6 = a3 + a5;
```

6. Experiment with different values of Stop Time until the Scope shows that steady state has been reached. Using this method a Stop Time of 1 s was found to be satisfactory. The plots of both x_1 and x_2 will appear in the Scope. The default `ode45` solver gave jagged lines for the x_2 plot, so the `ode15s` solver was used instead. A To Workspace block can be added to obtain the plot in MATLAB. Figure 10.3–3 was created in this way.

10.4 Piecewise-Linear Models

Unlike linear models, closed-form solutions are not available for most nonlinear differential equations, and we must therefore solve such equations numerically. A nonlinear ordinary differential equation can be recognized by the fact that the dependent variable or its derivatives appear raised to a power or in a transcendental function. For example, the following equations are nonlinear.

$$y\ddot{y} + 5\dot{y} + y = 0 \qquad \dot{y} + \sin y = 0 \qquad \dot{y} + \sqrt{y} = 0$$

Piecewise-linear models are actually nonlinear, although they may appear to be linear. They are composed of linear models that take effect when certain conditions are satisfied. The effect of switching back and forth between these linear models makes the overall model nonlinear. An example of such a model is a mass attached to a spring and sliding on a horizontal surface with Coulomb friction. The model is

$$m\ddot{x} + kx = \begin{cases} f(t) - \mu mg & \text{if } \dot{x} \geq 0 \\ f(t) + \mu mg & \text{if } \dot{x} < 0 \end{cases}$$

These two linear equations can be expressed as the single, nonlinear equation

$$m\ddot{x} + kx = f(t) - \mu mg \, \text{sign}(\dot{x}) \qquad \text{where} \qquad \text{sign}(\dot{x}) = \begin{cases} +1 & \text{if } \dot{x} \geq 0 \\ -1 & \text{if } \dot{x} < 0 \end{cases}$$

Solutions of models that contain piecewise-linear functions are very tedious to program. However, Simulink has built-in blocks that represent many of the commonly found functions such as Coulomb friction. Therefore Simulink is especially useful for such applications. One such block is the Saturation block in the Discontinuities library. The block implements the saturation function shown in Figure 10.4–1.

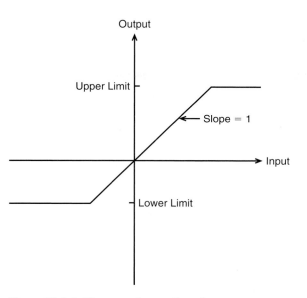

Figure 10.4–1 The saturation nonlinearity.

Simulink Model of a Rocket-Propelled Sled

EXAMPLE 10.4–1

A rocket-propelled sled on a track is represented in Figure 10.4–2 as a mass m with an applied force f that represents the rocket thrust. The rocket thrust initially is horizontal, but the engine accidentally pivots during firing and rotates with an angular acceleration of $\ddot{\theta} = \pi/50$ rad/s. Compute the sled's velocity v for $0 \leq t \leq 6$ if $v(0) = 0$. The rocket thrust is 4000 N and the sled mass is 450 kg.

The sled's equation of motion is

$$450\dot{v} = 4000 \cos \theta(t)$$

To obtain $\theta(t)$, note that

$$\dot{\theta} = \int_0^t \ddot{\theta} \, dt = \frac{\pi}{50} t$$

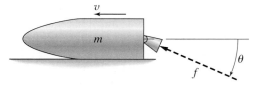

Figure 10.4–2 A rocket-propelled sled.

and

$$\theta = \int_0^t \dot{\theta}\, dt = \int_0^t \frac{\pi}{50} t\, dt = \frac{\pi}{100} t^2$$

Thus the equation of motion becomes

$$450\dot{v} = 4000 \cos\left(\frac{\pi}{100} t^2\right)$$

or

$$\dot{v} = \frac{80}{9} \cos\left(\frac{\pi}{100} t^2\right)$$

The solution is formally given by

$$v(t) = \frac{80}{9} \int_0^t \cos\left(\frac{\pi}{100} t^2\right) dt$$

Unfortunately, no closed-form solution is available for the integral, which is called *Fresnel's cosine integral*. The value of the integral has been tabulated numerically, but we will use Simulink to obtain the solution.

(a) Create a Simulink model to solve this problem for $0 \le t \le 10\,\text{s}$.

(b) Now suppose that the engine angle is limited by a mechanical stop to $60°$, which is $60\pi/180$ rad. Create a Simulink model to solve the problem.

■ **Solution**

(a) There are several ways to create the input function $\theta = (\pi/100)t^2$. Here we note that $\ddot{\theta} = \pi/50$ rad/s and that

$$\dot{\theta} = \int_0^t \ddot{\theta}\, dt$$

and

$$\theta = \int_0^t \dot{\theta}\, dt = \frac{\pi}{100} t^2$$

Thus we can create $\theta(t)$ by integrating the constant $\ddot{\theta} = \pi/50$ twice. The simulation diagram is shown in Figure 10.4–3. This diagram is used to create the corresponding Simulink model shown in Figure 10.4–4.

There are two new blocks in this model. The Constant block is in the Sources library. After placing it, double-click on it and type `pi/50` in its Constant Value window.

The Trigonometric block is in the Math Operations library. After placing it, double-click on it and select `cos` in its Function window.

Figure 10.4–3 Simulation diagram for $v = (80/9)\cos(\pi t^2/100)$.

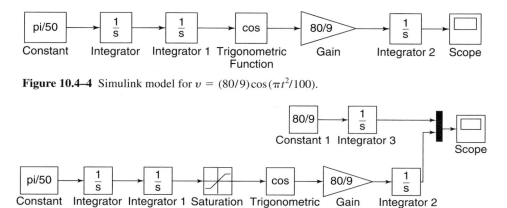

Figure 10.4–4 Simulink model for $v = (80/9)\cos(\pi t^2/100)$.

Figure 10.4–5 Simulink model for $v = (80/9)\cos(\pi t^2/100)$ with a Saturation block.

Set the Stop Time to 10, run the simulation, and examine the results in the Scope.

(b) Modify the model in Figure 10.4–4 as follows to obtain the model shown in Figure 10.4–5. We use the Saturation block in the Discontinuities library to limit the range of θ to $60\pi/180$ rad. After placing the block as shown in Figure 10.4–5, double-click on it and type `60*pi/180` in its Upper Limit window. Then type 0 in its Lower Limit window.

Enter and connect the remaining elements as shown, and run the simulation. The upper Constant block and Integrator block are used to generate the solution when the engine angle is $\theta = 0$, as a check on our results. [The equation of motion for $\theta = 0$ is $\dot{v} = 80/9$, which gives $v(t) = 80t/9$.]

If you prefer, you can substitute a To Workspace block for the Scope. Then you can plot the results in MATLAB. The resulting plot is shown in Figure 10.4–6.

The Relay Block

The Simulink Relay block is an example of something that is tedious to program in MATLAB but is easy to implement in Simulink. Figure 10.4–7a is a graph of the logic of a relay. The relay switches the output between two specified values, named *On* and *Off* in the figure. Simulink calls these values "Output when on" and "Output when off." When the relay output is *On,* it remains *On* until the input drops below the value of the Switch-off point parameter, named *SwOff* in the figure. When the relay output is *Off,* it remains *Off* until the input exceeds the value of the Switch-on point parameter, named *SwOn* in the figure.

The Switch-on point parameter value must be greater than or equal to the Switch-off point value. Note that the value of *Off* need not be zero. Note also that the value of *Off* need not be less than the value of *On.* The case where *Off* > *On* is

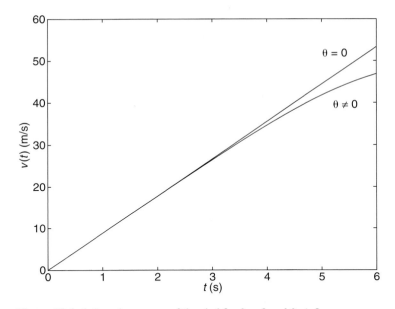

Figure 10.4–6 Speed response of the sled for $\theta = 0$ and $\theta \neq 0$.

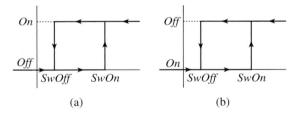

Figure 10.4–7 The relay function. (a) The case where
On > Off. (b) The case where *On < Off*.

shown in Figure 10.4–7b. As we will see in the following example, it is sometimes necessary to use this case.

EXAMPLE 10.4–2   Model of a Relay-Controlled Motor

The model of an armature-controlled dc motor was discussed in Section 9.5. See Figure 10.4–8. The model is

$$L\frac{di}{dt} = -Ri - K_e\omega + v(t)$$

$$I\frac{d\omega}{dt} = K_T i - c\omega - T_d(t)$$

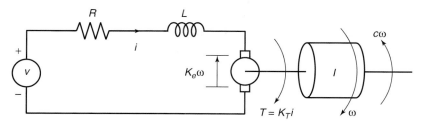

Figure 10.4–8 An armature-controlled dc motor.

where the model now includes a torque $T_d(t)$ acting on the motor shaft due, for example, to some unwanted source such as Coulomb friction or wind gusts. Control system engineers call this a *disturbance*. These equations can be put into matrix form as follows, where $x_1 = i$ and $x_2 = \omega$.

$$\begin{bmatrix} \dot{x}_1 \\ \dot{x}_2 \end{bmatrix} = \begin{bmatrix} -\dfrac{R}{L} & -\dfrac{K_e}{L} \\ \dfrac{K_T}{I} & -\dfrac{c}{I} \end{bmatrix} \begin{bmatrix} x_1 \\ x_2 \end{bmatrix} + \begin{bmatrix} \dfrac{1}{L} & 0 \\ 0 & -\dfrac{1}{I} \end{bmatrix} \begin{bmatrix} v(t) \\ T_d(t) \end{bmatrix}$$

Use the values $R = 0.6\ \Gamma$, $L = 0.002$ H, $K_T = 0.04$ N·m/A, $K_e = 0.04$ V·s/rad, $c = 0.01$ N·m·s/rad, and $I = 6 \times 10^{-5}$ kg·m^2.

Suppose we have a sensor that measures the motor speed, and we use the sensor's signal to activate a relay to switch the applied voltage $v(t)$ between 0 and 100 V to keep the speed between 250 and 350 rad/s. This corresponds to the relay logic shown in Figure 10.4–7b, with *SwOff* = 250, *SwOn* = 350, *Off* = 100, and *On* = 0. Investigate how well this scheme will work if the disturbance torque is a step function that increases from 0 to 3 N·m, starting at $t = 0.05$ s. Assume that the system starts from rest with $\omega(0) = 0$ and $i(0) = 0$.

■ Solution

For the given parameter values,

$$\mathbf{A} = \begin{bmatrix} -300 & -20 \\ 666.7 & -166.7 \end{bmatrix} \qquad \mathbf{B} = \begin{bmatrix} 500 & 0 \\ 0 & -16\,667 \end{bmatrix}$$

To examine the speed ω as output, we choose $\mathbf{C} = [0, 1]$ and $\mathbf{D} = [0, 0]$. To create this simulation, first obtain a new model window. Then do the following.

1. Select and place in the new window the Step block from the Sources library. Label it Disturbance Step as shown in Figure 10.4–9. Double-click on it to obtain the Block Parameters window, and set the Step Time to 0.05, the Initial and Final values to 0 and 3, and the Sample time to 0. Click **OK.**

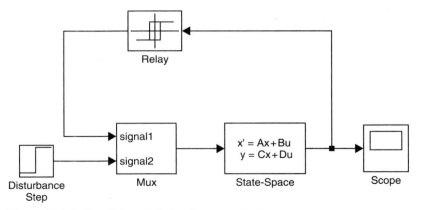

Figure 10.4–9 Simulink model of a relay-controlled motor.

2. Select and place the Relay block from the Discontinuities library. Double-click on it, and set the Switch-on and Switch-off points to 350 and 250, and set the Output when on and Output when off to 0 and 100. Click **OK.**
3. Select and place the Mux block from the Signal Routing library. The Mux block combines two or more signals into a vector signal. Double-click on it, and set the Display option to signals. Click **OK.** Then click on the Mux icon in the model window, and drag one of the corners to expand the box so that all the text is visible.
4. Select and place the State-Space block from the Continuous library. Double-click on it, and enter [-300, -20; 666.7, -166.7] for **A,** [500, 0; 0, -16667] for **B,** [0, 1] for **C,** and [0, 0] for **D.** Then enter [0; 0] for the initial conditions. Click **OK.** Note that the dimension of the matrix **B** tells Simulink that there are two inputs. The dimensions of the matrices **C** and **D** tell Simulink that there is one output.
5. Select and place the Scope block from the Sinks library.
6. Once the blocks have been placed, connect the input port on each block to the outport port on the preceding block as shown in the figure. It is important to connect the top port of the Mux block [which corresponds to the first input, $v(t)$] to the output of the Relay block, and to connect the bottom port of the Mux block [(which corresponds to the second input, $T_d(t)$] to the output of the Disturbance Step block.
7. Set the Stop time to 0.1 (which is simply an estimate of how long is needed to see the complete response), run the simulation, and examine the plot of $\omega(t)$ in the Scope. You should see something like Figure 10.4–10. If you want to examine the current $i(t)$, change the matrix **C** to [1, 0] and run the simulation again.

The results show that the relay logic control scheme keeps the speed within the desired limits of 250 and 350 before the disturbance torque starts to act. The speed oscillates because when the applied voltage is zero, the speed decreases as a result of the back-emf and the viscous damping. The speed drops below 250 when the disturbance torque starts to act, because the applied voltage is 0 at that time. As soon as the speed drops below 250,

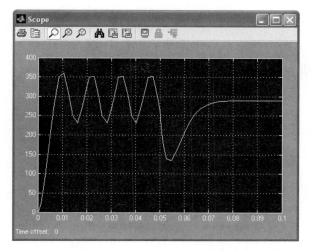

Figure 10.4–10 Scope display of the speed response of a relay-controlled motor.

the relay controller switches the voltage to 100, but it now takes longer for the speed to increase because the motor torque must now work against the disturbance.

Note that the speed becomes constant, instead of oscillating. This is so because with $v = 100$, the system achieves a steady-state condition in which the motor torque equals the sum of the disturbance torque and the viscous damping torque. Thus the acceleration is zero.

One practical use of this simulation is to determine how long the speed is below the limit of 250. The simulation shows that this time is approximately 0.013 s. Other uses of the simulation include finding the period of the speed's oscillation (about 0.013 s) and the maximum value of the disturbance torque that can be tolerated by the relay controller (it is about 3.7 N $\cdot$ m).

10.5 Transfer-Function Models

The equation of motion of a mass-spring-damper system is

$$m\ddot{y} + c\dot{y} + ky = f(t) \qquad (10.5-1)$$

As with the Control System toolbox, Simulink can accept a system description in transfer-function form and in state-variable form. (See Section 9.5 for a discussion of these forms.) If the mass-spring system is subjected to a sinusoidal forcing function $f(t)$, it is easy to use the MATLAB commands presented thus far to solve and plot the response $y(t)$. However, suppose that the force $f(t)$ is created by applying a sinusoidal input voltage to a hydraulic piston that has a *dead-zone* nonlinearity. This means that the piston does not generate a force until the input

DEAD ZONE

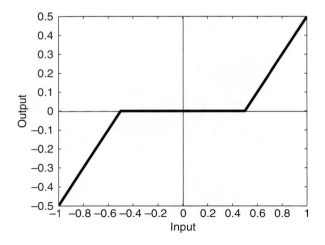

Figure 10.5–1 A dead-zone nonlinearity

voltage exceeds a certain magnitude, and thus the system model is piecewise linear.

A graph of a particular dead-zone nonlinearity is shown in Figure 10.5–1. When the input (the independent variable on the graph) is between -0.5 and 0.5, the output is zero. When the input is greater than or equal to the upper limit of 0.5, the output is the input minus the upper limit. When the input is less than or equal to the lower limit of -0.5, the output is the input minus the lower limit. In this example, the dead zone is symmetric about 0, but it need not be in general.

Simulations with dead-zone nonlinearities are somewhat tedious to program in MATLAB, but are easily done in Simulink. The following example illustrates how it is done.

EXAMPLE 10.5–1   Response with a Dead Zone

Create and run a Simulink simulation of a mass-spring-damper model (Equation 10.5–1) using the parameter values $m = 1$, $c = 2$, and $k = 4$. The forcing function is the function $f(t) = \sin 1.4t$. The system has the dead-zone nonlinearity shown in Figure 10.5–1.

■ **Solution**

To construct the simulation, do the following steps.

1. Start Simulink and open a new Model window as described previously.
2. Select and place in the new window the Sine Wave block from the Sources library. Double-click on it, and set the Amplitude to 1, the Frequency to 1.4, the Phase to 0, and the Sample time to 0. Click **OK.**

3. Select and place the Dead Zone block from the Discontinuities library, double-click on it, and set the Start of dead zone to -0.5 and the End of dead zone to 0.5. Click **OK.**

4. Select and place the Transfer Fcn block from the Continuous library, double-click on it, and set the Numerator to `[1]` and the Denominator to `[1, 2, 4]`. Click **OK.**

5. Select and place the Scope block from the Sinks library.

6. Once the blocks have been placed, connect the input port on each block to the outport port on the preceding block. Your model should now look like Figure 10.5–2.

7. Set the Stop time to 10.

8. Run the simulation. You should see an oscillating curve in the Scope display.

It is informative to plot both the input and the output of the Transfer Fcn block versus time on the same graph. To do this:

1. Delete the arrow connecting the Scope block to the Transfer Fcn block. Do this by clicking on the arrow line and then pressing the **Delete** key.

2. Select and place the Mux block from the Signal Routing library, double-click on it, and set the Number of inputs to 2. Click **OK.**

3. Connect the top input port of the Mux block to the output port of the Transfer Fcn block. Then use the same technique to connect the bottom input port of the Mux block to the arrow from the outport port of the Dead Zone block. Just remember to start with the input port. Simulink will sense the arrow automatically and make the connection. Your model should now look like Figure 10.5–3.

4. Set the Stop time to 10, run the simulation as before, and bring up the Scope display. You should see what is shown in Figure 10.5–4. This plot shows the effect of the dead zone on the sine wave.

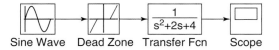

Sine Wave   Dead Zone   Transfer Fcn   Scope

Figure 10.5–2 The Simulink model of dead-zone response.

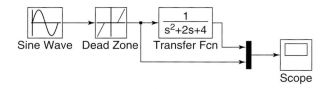

Sine Wave   Dead Zone   Transfer Fcn   Scope

Figure 10.5–3 Modification of the dead-zone model to include a Mux block.

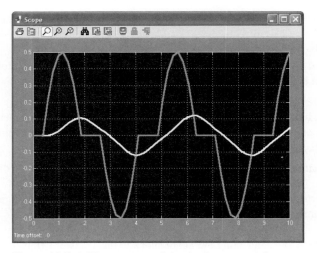

Figure 10.5–4 The response of the dead-zone model.

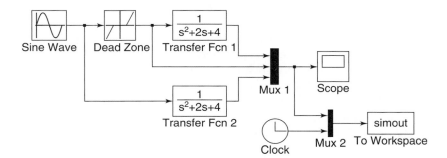

Figure 10.5–5 Modification of the dead-zone model to export variables to the MATLAB workspace.

You can bring the simulation results into the MATLAB workspace by using the To Workspace block. For example, suppose we want to examine the effects of the dead zone by comparing the response of the system with and without a dead zone. We can do this with the model shown in Figure 10.5–5. To create this model:

1. Copy the Transfer Fcn block by right-clicking on it, holding down the mouse button, and dragging the block copy to a new location. Then release the button. Copy the Mux block in the same way.

2. Double-click on the first Mux block and change the number of its inputs to 3.

3. In the usual way, select and place the To Workspace block from the Sinks library and optionally the Clock box from the Sources library. Double-click on the To Workspace block. You can specify any variable name you want as the output; the default is simout. Change its name to y. The output variable y will have as many

rows as there are simulation time steps, and as many columns as there are inputs to the block. The fourth column in our simulation will be time, because of the way we have connected the Clock to the second Mux. Specify the Save format as Matrix. Use the default values for the other parameters (these should be `inf`, `1`, and `-1` for Maximum number of rows, Decimation, and Sample time, respectively). Click on **OK.**

4. Connect the blocks as shown, and run the simulation.

5. You can use the MATLAB plotting commands from the Command window to plot the columns of `y`; for example, to plot the response of the two systems and the output of the Dead Zone block versus time, type

```
>>plot(y(:,4),y(:,1),y(:,4),y(:,2),y(:,4),y(:,3))
```

10.6 Nonlinear State-Variable Models

Nonlinear models cannot be put into transfer-function form or the state-variable form $\dot{\mathbf{x}} = \mathbf{A}\mathbf{x} + \mathbf{B}\mathbf{u}$. However, they can be simulated in Simulink. The following example shows how this can be done.

Model of a Nonlinear Pendulum

The pendulum shown in Figure 10.6–1 has the following nonlinear equation of motion, if there is viscous friction in the pivot and if there is an applied moment $M(t)$ about the pivot

$$I\ddot{\theta} + c\dot{\theta} + mgL \sin \theta = M(t)$$

where I is the mass moment of inertia about the pivot. Create a Simulink model for this system for the case where $I = 4$, $mgL = 10$, $c = 0.8$, and $M(t)$ is a square wave with an amplitude of 3 and a frequency of 0.5 Hz. Assume that the initial conditions are $\theta(0) = \pi/4$ rad and $\dot{\theta}(0) = 0$.

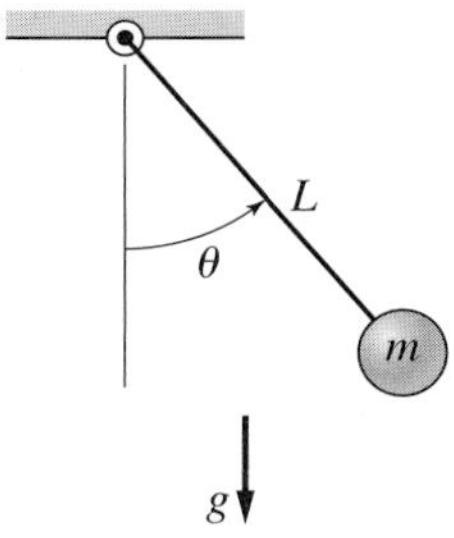

Figure 10.6–1
A pendulum.

■ **Solution**

To simulate this model in Simulink, define a set of variables that lets you rewrite the equation as two first-order equations. Thus let $\omega = \dot{\theta}$. Then the model can be written as

$$\dot{\theta} = \omega$$

$$\dot{\omega} = \frac{1}{I}[-c\omega - mgL \sin \theta + M(t)] = 0.25[-0.8\omega - 10 \sin \theta + M(t)]$$

Integrate both sides of each equation over time to obtain

$$\theta = \int \omega \, dt$$

$$\omega = 0.25 \int [-0.8\omega - 10 \sin \theta + M(t)] \, dt$$

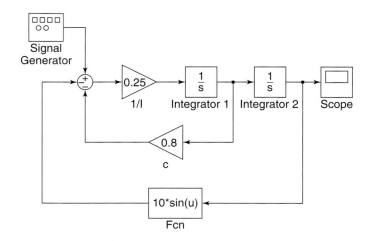

Figure 10.6–2 Simulink model of nonlinear pendulum dynamics.

We will introduce four new blocks to create this simulation. Obtain a new Model window and do the following.

1. Select and place in the new window the Integrator block from the Continuous library, and change its label to Integrator 1 as shown in Figure 10.6–2. You can edit text associated with a block by clicking on the text and making the changes. Double-click on the block to obtain the Block Parameters window, and set the Initial condition to 0 [this is the initial condition $\dot{\theta}(0) = 0$]. Click **OK.**

2. Copy the Integrator block to the location shown and change its label to Integrator 2. Set its initial condition to $\pi/4$ by typing `pi/4` in the Block Parameters window. This is the initial condition $\theta(0) = \pi/4$.

3. Select and place a Gain block from the Math Operations library, double-click on it, and set the Gain value to 0.25. Click **OK.** Change its label to `1/I`. Then click on the block, and drag one of the corners to expand the box so that all the text is visible.

4. Copy the Gain box, change its label to c, and place it as shown in Figure 10.6–2. Double-click on it, and set the Gain value to 0.8. Click **OK.** To flip the box left to right, right-click on it, select **Format,** and select **Flip.**

5. Select and place the Scope block from the Sinks library.

6. For the term $10 \sin \theta$, we cannot use the Trig function block in the Math library because we need to multiply the $\sin \theta$ by 10. So we use the Fcn block under the User-Defined Functions library (Fcn stands for function). Select and place this block as shown. Double-click on it, and type `10*sin(u)` in the expression window. This block uses the variable u to represent the input to the block. Click **OK.** Then flip the block.

7. Select and place the Sum block from the Math Operations library. Double-click on it, and select round for the Icon shape. In the List of Signs window, type `+--`. Click **OK.**

8. Select and place the Signal Generator block from the Sources library. Double-click on it, select square wave for the Wave form, 3 for the Amplitude, 0.5 for the Frequency, and Hertz for the Units. Click **OK.**

9. Once the blocks have been placed, connect arrows as shown in the figure.

10. Set the Stop time to 10, run the simulation, and examine the plot of $\theta(t)$ in the Scope. This completes the simulation.

10.7 Subsystems

One potential disadvantage of a graphical interface such as Simulink is that, to simulate a complex system, the diagram can become rather large and therefore somewhat cumbersome. Simulink, however, provides for the creation of *subsystem blocks,* which play a role analogous to that of subprograms in a programming language. A subsystem block is actually a Simulink program represented by a single block. A subsystem block, once created, can be used in other Simulink programs. We also introduce some other blocks in this section.

To illustrate subsystem blocks, we will use a simple hydraulic system whose model is based on the conservation of mass principle familiar to engineers. Because the governing equations are similar to other engineering applications, such as electric circuits and devices, the lessons learned from this example will enable you to use Simulink for other applications.

A Hydraulic System

The working fluid in a *hydraulic* system is an incompressible fluid such as water or a silicon-based oil. (*Pneumatic* systems operate with compressible fluids, such as air.) Consider a hydraulic system composed of a tank of liquid of mass density ρ (Figure 10.7–1). The tank shown in cross section in the figure is cylindrical with a bottom area A. A flow source dumps liquid into the tank at the mass flow rate $q_{mi}(t)$. The total mass in the tank is $m = \rho A h$, and from conservation of mass we have

$$\frac{dm}{dt} = \rho A \frac{dh}{dt} = q_{mi} - q_{mo} \qquad (10.7\text{–}1)$$

since ρ and A are constants.

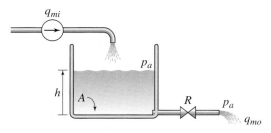

Figure 10.7–1 A hydraulic system with a flow source.

If the outlet is a pipe that discharges to atmospheric pressure p_a and provides a resistance to flow that is proportional to the pressure difference across its ends, then the outlet flow rate is

$$q_{mo} = \frac{1}{R}[(\rho g h + p_a) - p_a] = \frac{\rho g h}{R}$$

where R is called the *fluid resistance*. Substituting this expression into Equation (10.7–1) gives the model

$$\rho A \frac{dh}{dt} = q_{mi}(t) - \frac{\rho g}{R}h \qquad (10.7\text{–}2)$$

The transfer function is

$$\frac{H(s)}{Q_{mi}(s)} = \frac{1}{\rho A s + \rho g/R}$$

On the other hand, the outlet may be a valve or other restriction that provides nonlinear resistance to the flow. In such cases, a common model is the signed-square-root (SSR) relation

$$q_{mo} = \frac{1}{R}\text{SSR}(\Delta p)$$

where q_{mo} is the outlet mass flow rate, R is the resistance, Δp is the pressure difference across the resistance, and

$$\text{SSR}(\Delta p) = \begin{cases} \sqrt{\Delta p} & \text{if } \Delta p \geq 0 \\ -\sqrt{|\Delta p|} & \text{if } \Delta p < 0 \end{cases}$$

Note that we may express the SSR(u) function in MATLAB as follows: `sgn(u)*sqrt(abs(u))`.

Consider the slightly different system shown in Figure 10.7–2, which has a flow source q and two pumps that supply liquid at the pressures p_l and p_r. Suppose

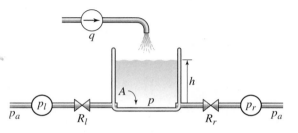

Figure 10.7–2 A hydraulic system with a flow source and two pumps.

the resistances are nonlinear and obey the signed-square-root relation. Then the model of the system is

$$\rho A \frac{dh}{dt} = q + \frac{1}{R_l}\text{SSR}(p_l - p) - \frac{1}{R_r}\text{SSR}(p - p_r)$$

where A is the bottom area and $p = \rho gh$. The pressures p_l and p_r are the *gauge* pressures at the left- and right-hand sides. Gauge pressure is the difference between the absolute pressure and atmospheric pressure. Note that the atmospheric pressure p_a cancels out of the model because of the use of gauge pressure.

We will use this application to introduce the following Simulink elements:

■ Subsystem blocks
■ Input and output ports

You can create a subsystem block in one of two ways: by dragging the Subsystem block from the library to the Model window or by first creating a Simulink model and then "encapsulating" it within a bounding box. We will illustrate the latter method.

We will create a subsystem block for the liquid-level system shown in Figure 10.7–2. First construct the Simulink model shown in Figure 10.7–3. The oval blocks are Input and Output Ports (In 1 and Out 1), which are available in the Ports and Subsystems library. Note that you can use MATLAB variables and expressions when entering the gains in each of the four Gain blocks.

Before running the program, we will assign values to these variables in the MATLAB Command window. Enter the gains for the four Gain blocks using the expressions shown in the block. You may also use a variable as the Initial condition of the Integrator block. Name this variable h0.

The SSR blocks are examples of the Fcn block, which is in the User-Defined Functions library. Double-click on the block and enter the MATLAB expression

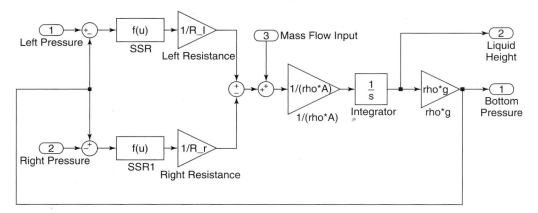

Figure 10.7–3 Simulink model of the system shown in Figure 10.7–2.

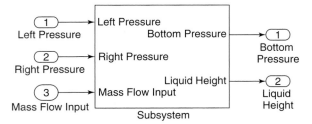

Figure 10.7–4 The Subsystem block.

`sgn(u)*sqrt(abs(u))`. Note that the Fcn block requires you to use the variable u. The output of the Fcn block must be a scalar, as is the case here, and you cannot perform matrix operations in the Fcn block, but these are not needed here. (An alternative to the Fcn block is the MATLAB Fcn block to be discussed in Section 10.9.) Save the model and give it a name, such as Tank.

Now create a "bounding box" surrounding the diagram. Do this by placing the mouse cursor in the upper left, holding the mouse button down, and dragging the expanding box to the lower right to enclose the entire diagram. Then choose **Create Subsystem** from the **Edit** menu. Simulink will then replace the diagram with a single block having as many input and output ports as required and will assign default names. You can resize the block to make the labels readable. You may view or edit the subsystem by double-clicking on it. The result is shown in Figure 10.7–4.

Connecting Subsystem Blocks

We now create a simulation of the system shown in Figure 10.7–5, where the mass inflow rate q is a step function. To do this, create the Simulink model shown in Figure 10.7–6. The square blocks are Constant blocks from the Sources library. These give constant inputs (which are not the same as step function inputs).

The larger rectangular blocks are two subsystem blocks of the type just created. To insert them into the model, open the Tank subsystem model, select **Copy** from the **Edit** menu, then paste it twice into the new model window. Connect the

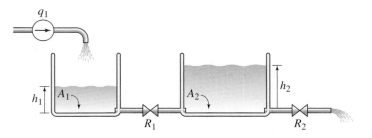

Figure 10.7–5 A hydraulic system with two tanks.

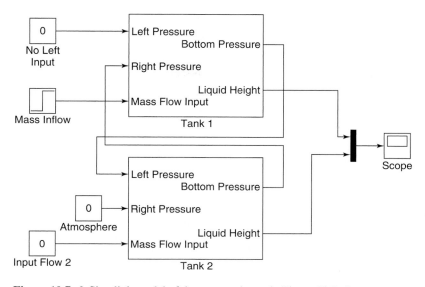

Figure 10.7–6 Simulink model of the system shown in Figure 10.7–5.

input and output ports and edit the labels as shown. Then double-click on the Tank 1 subsystem block, set the left-side gain $1/R\_l$ equal to 0, the right-side gain $1/R\_r$ equal to $1/R\_1$, and the gain $1/rho*A$ equal to $1/rho*A\_1$. Set the Initial condition of the integrator to $h10$. Note that setting the gain $1/R\_l$ equal to 0 is equivalent to $R\_l = \infty$, which indicates no inlet on the left-hand side.

Then double-click on the Tank 2 subsystem block, set the left-side gain $1/R\_l$ equal to $1/R\_1$, the right-side gain $1/R\_r$ equal to $1/R\_2$, and the gain $1/rho*A$ equal to $1/rho*A\_2$. Set the Initial condition of the integrator to $h20$. For the Step block, set the Step time to 0, the Initial value to 0, the Final value to the variable q_1, and the Sample time to 0. Save the model using a name other than Tank.

Before you run the model, in the Command window assign numerical values to the variables. As an example, you may type the following values for water, in U.S. Customary units, in the Command window.

```
>>A_1 = 2;A_2 = 5;rho = 1.94;g = 32.2;
>>R_1 = 20;R_2 = 50;q_1 = 0.3;h10 = 1;h20 = 10;
```

After selecting a simulation Stop time, you may run the simulation. The Scope will display the plots of the heights h_1 and h_2 versus time.

Figures 10.7–7, 10.7–8, and 10.7–9 illustrate some electrical and mechanical systems that are likely candidates for application of subsystem blocks. In Figure 10.7–7, the basic element for the subsystem block is an *RC* circuit. In Figure 10.7–8, the basic element for the subsystem block is a mass connected to two elastic elements.

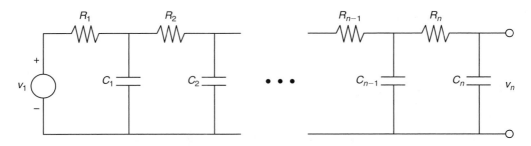

Figure 10.7–7 A network of *RC* loops.

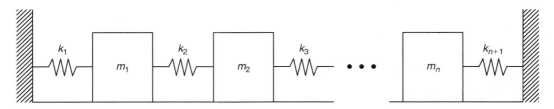

Figure 10.7–8 A vibrating system.

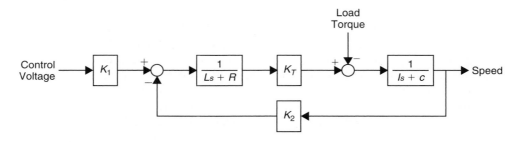

Figure 10.7–9 An armature-controlled dc motor.

Figure 10.7–9 is the block diagram of an armature-controlled dc motor, which may be converted to a subsystem block. The inputs for the block would be the voltage from a controller and a load torque, and the output would be the motor speed. Such a block would be useful in simulating systems containing several motors, such as a robot arm.

10.8 Dead Time in Models

TRANSPORT DELAY

Dead time, also called *transport delay,* is a time delay between an action and its effect. It occurs, for example, when a fluid flows through a conduit. If the fluid velocity v is constant and the conduit length is L, it takes a time $T = L/v$ for the fluid to move from one end to the other. The time T is the dead time.

Let $\theta_1(t)$ denote the incoming fluid temperature and $\theta_2(t)$ the temperature of the fluid leaving the conduit. If no heat energy is lost, then $\theta_2(t) = \theta_1(t - T)$.

From the shifting property of the Laplace transform,

$$\Theta_2(s) = e^{-Ts}\Theta_1(s)$$

So the transfer function for a dead-time process is e^{-Ts}.

Dead time may be described as a "pure" time delay, in which no response at all occurs for a time T, as opposed to the time lag associated with the time constant of a response, for which $\theta_2(t) = (1 - e^{-t/\tau})\theta_1(t)$.

Some systems have an unavoidable time delay in the interaction between components. The delay often results from the physical separation of the components and typically occurs as a delay between a change in the actuator signal and its effect on the system being controlled, or as a delay in the measurement of the output.

Another, perhaps unexpected, source of dead time is the computation time required for a digital control computer to calculate the control algorithm. This can be a significant dead time in systems using inexpensive and slower microprocessors.

The presence of dead time means the system does not have a characteristic equation of finite order. In fact, there are an infinite number of characteristic roots for a system with dead time. This can be seen by noting that the term e^{-Ts} can be expanded in an infinite series as

$$e^{-Ts} = \frac{1}{e^{Ts}} = \frac{1}{1 + Ts + T^2s^2/2 + \cdots}$$

The fact that there are an infinite number of characteristic roots means that the analysis of dead-time processes is difficult, and often simulation is the only practical way to study such processes.

Systems having dead-time elements are easily simulated in Simulink. The block implementing the dead-time transfer function e^{-Ts} is called the *Transport Delay* block.

Consider the model of the height h of liquid in a tank, such as that shown in Figure 10.7–1, whose input is a mass flow rate q_i. Suppose that it takes a time T for the change in input flow to reach the tank following a change in the valve opening. Thus, T is a dead time. For specific parameter values, the transfer function has the form

$$\frac{H(s)}{Q_i(s)} = e^{-Ts}\frac{2}{5s + 1}$$

Figure 10.8–1 shows a Simulink model for this system. After placing the Transport Delay block, set the delay to 1.25. Set the Step time to 0 in the Step Function block. We will now discuss the other blocks in the model.

Specifying Initial Conditions with Transfer Functions

The "Transfer Fcn (with initial outputs)" block, so-called to distinguish it from the Transfer Fcn block, enables us to set the initial value of the block output. In our model, this corresponds to the initial liquid height in the tank. This feature thus provides a useful improvement over traditional transfer-function analysis, in which initial conditions are assumed to be zero.

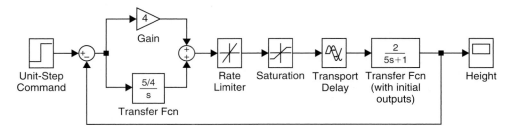

Figure 10.8–1 Simulink model of a hydraulic system with dead time.

The Transfer Fcn (with initial outputs) block is equivalent to adding the free response to the block output, with all the block's state variables set to zero except for the output variable. The block also lets you assign an initial value to the block input, but we will not use this feature and so will leave the Initial input set to 0 in the Block Parameters window. Set the Initial output to 0.2 to simulate an initial liquid height of 0.2.

The Saturation and Rate Limiter Blocks

Suppose that the minimum and maximum flow rates available from the input flow valve are 0 and 2. These limits can be simulated with the Saturation block, discussed in Section 10.4. After placing the block as shown in Figure 10.8–1, double-click on it and type 2 in its Upper limit window and 0 in the Lower limit window.

In addition to being limited by saturation, some actuators have limits on how fast they can react. This limitation might be due to deliberate restrictions placed on the unit by its manufacturer to avoid damage to the unit. An example is a flow control valve whose rate of opening and closing is controlled by a *rate limiter*. Simulink has such a block, and it can be used in series with the Saturation block to model the valve behavior. Place the Rate Limiter block as shown in Figure 10.8–1. Set the Rising slew rate to 1 and the Falling slew rate to −1.

A Control System

PI CONTROLLER

The Simulink model shown in Figure 10.8–1 is for a specific type of control system called a *PI controller,* whose response $f(t)$ to the error signal $e(t)$ is the sum of a term proportional to the error signal and a term proportional to the integral of the error signal. That is,

$$f(t) = K_P e(t) + K_I \int_0^t e(t)\, dt$$

where K_P and K_I are called the proportional and integral gains. Here the error signal $e(t)$ is the difference between the unit-step command representing the desired height and the actual height. In transform notation this expression becomes

$$F(s) = K_P E(s) + \frac{K_I}{s} E(s) = \left(K_p + \frac{K_I}{s} \right) E(s)$$

In Figure 10.8–1, we used the values $K_P = 4$ and $K_I = 5/4$. These values are computed using the methods of control theory. (For a discussion of control systems, see, for example, [Palm, 2010].) The simulation is now ready to be run. Set the Stop time to 30 and observe the behavior of the liquid height $h(t)$ in the Scope. Does it reach the desired height of 1?

10.9 Simulation of a Nonlinear Vehicle Suspension Model

Linear or linearized models are useful for predicting the behavior of dynamic systems because powerful analytical techniques are available for such models, especially when the inputs are relatively simple functions such as the impulse, step, ramp, and sine. Often in the design of an engineering system, however, we must eventually deal with nonlinearities in the system and with more complicated inputs such as trapezoidal functions, and this must often be done with simulation.

In this section we introduce four additional Simulink elements that enable us to model a wide range of nonlinearities and input functions:

■ Derivative block
■ Signal Builder block
■ Look-Up Table block
■ MATLAB Fcn block

As our example, we will use the single-mass suspension model shown in Figure 10.9–1, where the spring and damper forces f_s and f_d have the nonlinear models shown in Figures 10.9–2 and 10.9–3. The damper model is unsymmetric and represents a damper whose force during rebound is higher than during jounce (in order to minimize the force transmitted to the passenger compartment when the vehicle strikes a bump). The bump is represented by the trapezoidal function $y(t)$ shown in Figure 10.9–4. This function corresponds approximately to a vehicle traveling at 30 mi/h over a road surface elevation 0.2 m high and 48 m long.

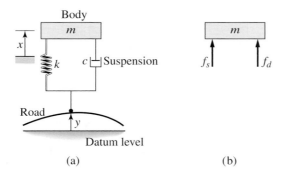

(a)   (b)

Figure 10.9–1 Single-mass model of a vehicle suspension.

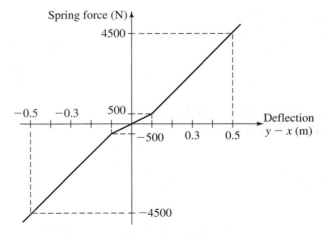

Figure 10.9–2 Nonlinear spring function.

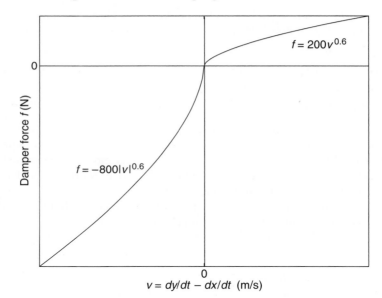

Figure 10.9–3 Nonlinear damping function.

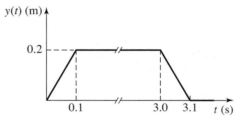

Figure 10.9–4 Road surface profile.

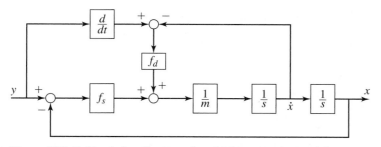

Figure 10.9–5 Simulation diagram of a vehicle suspension model.

The system model from Newton's law is

$$m\ddot{x} = f_s(y - x) + f_d(\dot{y} - \dot{x})$$

where $m = 400$ kg, $f_s(y-x)$ is the nonlinear spring function shown in Figure 10.9–2, and $f_d(\dot{y} - \dot{x})$ is the nonlinear damper function shown in Figure 10.9–3. The corresponding simulation diagram is shown in Figure 10.9–5.

The Derivative and Signal Builder Blocks

The simulation diagram shows that we need to compute $\dot{y}$. Because Simulink uses numerical and not analytical methods, it computes derivatives only approximately, using the Derivative block. We must keep this in mind when using rapidly changing or discontinuous inputs. The Derivative block has no settings, so merely place it in the Simulink diagram as shown in Figure 10.9–6.

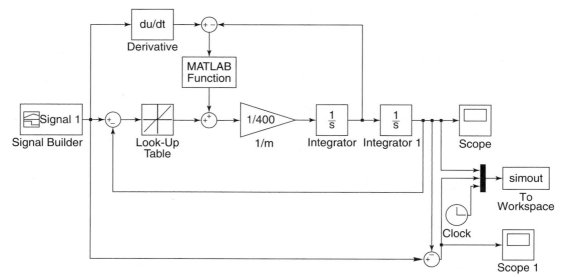

Figure 10.9–6 Simulink model of a vehicle suspension system.

Next, place the Signal-Builder block, then double-click on it. A plot window appears that enables you to place points to define the input function. Follow the directions in the window to create the function shown in Figure 10.9–4.

The Look-Up Table Block

The spring function f_s is created with the Look-Up Table block. After placing it as shown, double-click on it and enter $[-0.5, -0.1, 0, 0.1, 0.5]$ for the Vector of input values and $[-4500, -500, 0, 500, 4500]$ for the Vector of output values. Use the default settings for the remaining parameters.

Place the two integrators as shown, and make sure the initial values are set to 0. Then place the Gain block and set its gain to $1/400$. The To Workspace block and the Clock will enable you to plot $x(t)$ and $y(t)-x(t)$ versus t in the MATLAB Command window.

The MATLAB Fcn Block

In Section 10.7 we used the Fcn block to implement the signed-square-root function. We cannot use this block for the damper function shown in Figure 10.9–3 because we must write a user-defined function to describe it. This function is as follows.

```
function f = damper(v)
if v <= 0
    f = -800*(abs(v)).^(0.6);
else
    f = 200*v.^(0.6);
end
```

Create and save this function file. After placing the MATLAB Fcn block, double-click on it and enter its name `damper`. Make sure Output dimensions is set to -1 and the Output signal type is set to auto.

The Fcn, MATLAB Fcn, Math Function, and S-Function blocks can be used to implement functions, but each has its advantages and limitations. The Fcn block can contain an expression, but its output must be a scalar, and it cannot call a function file. The MATLAB Fcn block is slower than the Fcn block, but its output can be an array, and it can call a function file.

The Math Function block can produce an array output, but it is limited to a single MATLAB function and cannot use an expression or call a file. The S-Function block provides more advanced features, such as the ability to use C language code.

The Simulink model when completed should look like Figure 10.9–6. You can plot the response $x(t)$ in the Command window as follows:

```
>>x = simout(:,1);
>>t = simout(:,3);
>>plot(t,x),grid,xlabel('t (s)'),ylabel('x (m)')
```

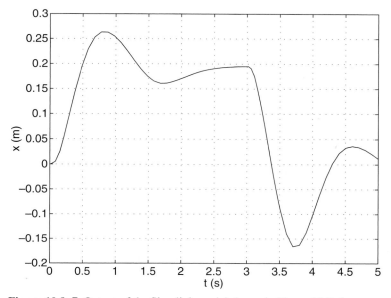

Figure 10.9–7 Output of the Simulink model shown in Figure 10.9–6.

The result is shown in Figure 10.9–7. The maximum overshoot is seen to be $0.26 - 0.2 = 0.06$ m, but the maximum *undershoot* is seen to be much greater, -0.168 m.

10.10 Summary

The Simulink model window contains menu items we have not discussed. However, the ones we have discussed are the most important ones for getting started. We have introduced just a few of the blocks available within Simulink. Some of the blocks not discussed deal with discrete-time systems (ones modeled with difference, rather than differential, equations), digital logic systems, and other types of mathematical operations. In addition, some blocks have additional properties that we have not mentioned. However, the examples given here will help you get started in exploring the other features of Simulink. Consult the online help for information about these items.

Key Terms with Page References

Block diagram,   420
Dead time,   448
Dead zone,   437
Derivative block,   453
Fcn block,   454
Gain block,   420

Integrator block,   420
Library Browser,   421
Look-Up Table block,   454
PI controller,   450
Piecewise-linear models,   430
Rate Limiter block,   450

Relay block,   433
Saturation block,   450
Signal Builder block,   453
Simulation diagrams,   420
State-variable models,   441

Subsystems,   443
Summer,   421
Transfer-function models,   437
Transport delay,   448

Problems

Section 10.1

1. Draw a simulation diagram for the following equation.

$$\dot{y} = 5f(t) - 7y$$

2. Draw a simulation diagram for the following equation.

$$5\ddot{y} = 3\dot{y} + 7y = f(t)$$

3. Draw a simulation diagram for the following equation.

$$3\dot{y} + 5\sin y = f(t)$$

Section 10.2

4. Create a Simulink model to plot the solution of the following equation for $0 \le t \le 6$.

$$10\ddot{y} = 7\sin 4t + 5\cos 3t \quad y(0) = 3 \quad \dot{y}(0) = 2$$

5. A projectile is launched with a velocity of 100 m/s at an angle of 30° above the horizontal. Create a Simulink model to solve the projectile's equations of motion where x and y are the horizontal and vertical displacements of the projectile.

$$\ddot{x} = 0 \quad x(0) = 0 \quad \dot{x}(0) = 100\cos 30°$$
$$\ddot{y} = -g \quad y(0) = 0 \quad \dot{y}(0) = 100\sin 30°$$

Use the model to plot the projectile's trajectory y versus x for $0 \le t \le 10\ s$.

6. The following equation has no analytical solution even though it is linear.

$$\dot{x} + x = \tan t \quad x(0) = 0$$

The approximate solution, which is less accurate for large values of t, is

$$x(t) = \frac{1}{3}t^3 - t^2 + 3t - 3 + 3e^{-t}$$

Create a Simulink model to solve this problem, and compare its solution with the approximate solution over the range $0 \le t \le 1$.

7. Construct a Simulink model to plot the solution of the following equation for $0 \le t \le 10$

$$15\dot{x} + 5x = 4u_s(t) - 4u_s(t-2) \quad x(0) = 5$$

where $u_s(t)$ is a unit-step function (in the Block Parameters window of the Step block, set the Step time to 0, the Initial value to 0, and the Final value to 1).

8. A tank having vertical sides and a bottom area of 100 ft^2 is used to store water. To fill the tank, water is pumped into the top at the rate given in the following table. Use Simulink to solve for and plot the water height $h(t)$ for $0 \le t \le 10$ min.

Time (min)	0	1	2	3	4	5	6	7	8	9	10
Flow Rate (ft^3/min)	0	80	130	150	150	160	165	170	160	140	120

Section 10.3

9. Construct a Simulink model to plot the solution of the following equations for $0 \le t \le 2$

$$\dot{x}_1 = -6x_1 + 4x_2$$

$$\dot{x}_2 = 5x_1 - 7x_2 + f(t)$$

where $f(t) = 3t$. Use the Ramp block in the Sources library.

10. Construct a Simulink model to plot the solution of the following equations for $0 \le t \le 3$

$$\dot{x}_1 = -6x_1 + 4x_2 + f_1(t)$$

$$\dot{x}_2 = 5x_1 - 7x_2 + f_2(t)$$

where $f_1(t)$ is a step function of height 3 starting at $t = 0$ and $f_2(t)$ is a step function of height -3 starting at $t = 1$.

Section 10.4

11. Use the Saturation block to create a Simulink model to plot the solution of the following equation for $0 \le t \le 6$.

$$3\dot{y} + y = f(t) \qquad y(0) = 3$$

where

$$f(t) = \begin{cases} 8 & \text{if } 10 \sin 3t > 8 \\ -8 & \text{if } 10 \sin 3t < -8 \\ 10 \sin 3t & \text{otherwise} \end{cases}$$

12.   Construct a Simulink model of the following problem.
$$5\dot{x} + \sin x = f(t) \qquad x(0) = 0$$

The forcing function is

$$f(t) = \begin{cases} -5 & \text{if } g(t) \leq -5 \\ g(t) & \text{if } -5 \leq g(t) \leq 5 \\ 5 & \text{if } g(t) \geq 5 \end{cases}$$

where $g(t) = 10 \sin 4t$.

13.   If a mass-spring system has Coulomb friction on the surface rather than viscous friction, its equation of motion is

$$m\ddot{y} = \begin{cases} -ky + f(t) - \mu mg & \text{if } \dot{y} \geq 0 \\ -ky + f(t) + \mu mg & \text{if } \dot{y} < 0 \end{cases}$$

where μ is the coefficient of friction. Develop a Simulink model for the case where $m = 1$kg, $k = 5$ N/m, $\mu = 0.4$, and $g = 9.8$ m/s^2. Run the simulation for two cases: (a) the applied force $f(t)$ is a step function with a magnitude of 10 N and (b) the applied force is sinusoidal: $f(t) = 10 \sin 2.5t$. Either the Sign block in the Math Operations library or the Coulomb and Viscous Friction block in the Discontinuities library can be used, but since there is no viscous friction in this problem, the Sign block is easier to use.

14.   A certain mass, $m = 2$ kg, moves on a surface inclined at an angle $\phi = 30°$ above the horizontal. Its initial velocity is $v(0) = 3$ m/s up the incline. An external force of $f_1 = 5$ N acts on it parallel to and up the incline. The coefficient of Coulomb friction is $\mu = 0.5$. Use the Sign block and create a Simulink model to solve for the velocity of the mass until the mass comes to rest. Use the model to determine the time at which the mass comes to rest.

15.   *a.* Develop a Simulink model of a thermostatic control system in which the temperature model is

$$RC\frac{dT}{dt} + T = Rq + T_a(t)$$

where T is the room air temperature in °F, T_a is the ambient (outside) air temperature in °F, time t is measured in hours, q is the input from the heating system in lb · ft/hr, R is the thermal resistance, and C is the thermal capacitance. The thermostat switches q on at the value q_{max} whenever the temperature drops below 69°F and switches q to $q = 0$ whenever the temperature is above 71°F. The value of q_{max} indicates the heat output of the heating system.

Run the simulation for the case where $T(0) = 70°F$, and $T_a(t) = 50 + 10 \sin(\pi t/12)$. Use the values $R = 5 \times 10^{-5} °F \cdot hr/lb \cdot ft$ and $C = 4 \times 10^4 lb \cdot ft/°F$. Plot the temperatures T and T_a versus t on the same graph, for $0 \le t \le 24$ hr. Do this for two cases: $q_{max} = 4 \times 10^5$ and $q_{max} = 8 \times 10^5$ lb · ft/hr. Investigate the effectiveness of each case.

b. The integral of q over time is the energy used. Plot $\int q \, dt$ versus t and determine how much energy is used in 24 hr for the case where $q_{max} = 8 \times 10^5$.

16. Refer to Problem 15. Use the simulation with $q_{max} = 8 \times 10^5$ to compare the energy consumption and the thermostat cycling frequency for the two temperature bands (69°, 71°) and (68°, 72°).

17. Consider the liquid-level system shown in Figure 10.7–1. The governing equation based on conservation of mass is Equation (10.7–2). Suppose that the height h is controlled by using a relay to switch the input flow rate between the values 0 and 50 kg/s. The flow rate is switched on when the height is less than 4.5 m and is switched off when the height reaches 5.5 m. Create a Simulink model for this application using the values $A = 2 \, m^2$, $R = 400 \, m^{-1} \cdot s^{-1}$, $\rho = 1000 \, kg/m^3$, and $h(0) = 1$ m. Obtain a plot of $h(t)$.

Section 10.5

18. Use the Transfer Function block to construct a Simulink model to plot the solution of the following equation for $0 \le t \le 4$.

$$2\ddot{x} + 12\dot{x} + 10x = 5u_s(t) - 5u_s(t-2) \quad x(0) = \dot{x}(0) = 0$$

19. Use Transfer Function blocks to construct a Simulink model to plot the solution of the following equations for $0 \le t \le 2$.

$$3\ddot{x} + 15\dot{x} + 18x = f(t) \quad x(0) = \dot{x}(0) = 0$$
$$2\ddot{y} + 16\dot{y} + 50y = x(t) \quad y(0) = \dot{y}(0) = 0$$

where $f(t) = 75u_s(t)$.

20. Use Transfer Function blocks to construct a Simulink model to plot the solution of the following equations for $0 \le t \le 2$

$$3\ddot{x} + 15\dot{x} + 18x = f(t) \quad x(0) = \dot{x}(0) = 0$$
$$2\ddot{y} + 16\dot{y} + 50y = x(t) \quad y(0) = \dot{y}(0) = 0$$

where $f(t) = 50u_s(t)$. At the output of the first block there is a dead zone for $-1 \le x \le 1$. This limits the input to the second block.

21. Use Transfer Function blocks to construct a Simulink model to plot the solution of the following equations for $0 \le t \le 2$

$$3\ddot{x} + 15\dot{x} + 18x = f(t) \quad x(0) = \dot{x}(0) = 0$$
$$2\ddot{y} + 16\dot{y} + 50y = x(t) \quad y(0) = \dot{y}(0) = 0$$

where $f(t) = 50u_s(t)$. At the output of the first block there is a saturation that limits x to be $|x| \leq 1$. This limits the input to the second block.

Section 10.6

22. Construct a Simulink model to plot the solution of the following equation for $0 \leq t \leq 4$.
$$2\ddot{x} + 12\dot{x} + 10x^2 = 8 \sin 0.8t \quad x(0) = \dot{x}(0) = 0$$

23. Create a Simulink model to plot the solution of the following equation for $0 \leq t \leq 3$.
$$\dot{x} + 10x^2 = 5 \sin 3t \quad x(0) = 1$$

24. Construct a Simulink model of the following problem.
$$10\dot{x} + \sin x = f(t) \quad x(0) = 0$$

The forcing function is $f(t) = \sin 2t$. The system has the dead-zone non-linearity shown in Figure 10.5–1.

25. The following model describes a mass supported by a nonlinear, hardening spring. The units are SI. Use $g = 9.81$ m/s^2.
$$5\ddot{y} = 5g - (900y + 1700y^3) \quad y(0) = 0.5 \quad \dot{y}(0) = 0$$

Create a Simulink model to plot the solution for $0 \leq t \leq 2$.

26. Consider the system for lifting a mast shown in Figure P26. The 70-ft-long mast weighs 500 lb. The winch applies a force $f = 380$ lb to the cable. The mast is supported initially at an angle of 30°, and the cable at A is initially horizontal. The equation of motion of the mast is
$$25\,400\,\ddot{\theta} = -17\,500 \cos \theta + \frac{626\,000}{Q} \sin(1.33 + \theta)$$

where
$$Q = \sqrt{2020 + 1650 \cos(1.33 + \theta)}$$

Create and run a Simulink model to solve for and plot $\theta(t)$ for $\theta(t) \leq \pi/2$ rad.

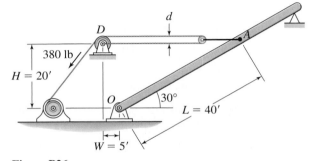

Figure P26

27. The equation describing the water height h in a spherical tank with a drain at the bottom is

$$\pi(2rh - h^2)\frac{dh}{dt} = -C_d A\sqrt{2gh}$$

Suppose that the tank's radius is $r = 3$ m and the circular drain hole of area A has a radius of 2 cm. Assume that $C_d = 0.5$ and that the initial water height is $h(0) = 5$ m. Use $g = 9.81$ m/s². Use Simulink to solve the nonlinear equation, and plot the water height as a function of time until $h(t) = 0$.

28. A cone-shaped paper drinking cup (like the kind used at water fountains) has a radius R and a height H. If the water height in the cup is h, the water volume is given by

$$V = \frac{1}{3}\pi\left(\frac{R}{H}\right)^2 h^3$$

Suppose that the cup's dimensions are $R = 1.5$ in. and $H = 4$ in.

a. If the flow rate from the fountain into the cup is 2 in.³/sec, use Simulink to determine how long will it take to fill the cup to the brim.

b. If the flow rate from the fountain into the cup is given by $2(1 - e^{-2t})$ in.³/sec, use Simulink to determine how long will it take to fill the cup to the brim.

Section 10.7

29. Refer to Figure 10.7–2. Assume that the resistances obey the linear relation, so that the mass flow q_l through the left-hand resistance is $q_l = (p_l - p)/R_l$, with a similar linear relation for the right-hand resistance.

a. Create a Simulink subsystem block for this element.

b. Use the subsystem block to create a Simulink model of the system shown in Figure 10.7–5. Assume that the mass inflow rate is a step function.

c. Use the Simulink model to obtain plots of $h_1(t)$ and $h_2(t)$ for the following parameter values: $A_1 = 2$ m², $A_2 = 5$ m², $R_1 = 400$ m⁻¹ · s⁻¹ $R_2 = 600$ m⁻¹ · s⁻¹, $\rho = 1000$ kg/m³, $q_{mi} = 50$ kg/s, $h_1(0) = 1.5$ m, and $h_2(0) = 0.5$ m.

30. a. Use the subsystem block developed in Section 10.7 to construct a Simulink model of the system shown in Figure P30. The mass inflow rate is a step function.

b. Use the Simulink model to obtain plots of $h_1(t)$ and $h_2(t)$ for the following parameter values: $A_1 = 3$ ft², $A_2 = 5$ ft², $R_1 = 30$ ft⁻¹ · sec⁻¹, $R_2 = 40$ ft⁻¹ · sec⁻¹, $\rho = 1.94$ slug/ft³, $q_{mi} = 0.5$ slug/sec, $h_1(0) = 2$ ft, and $h_2(0) = 5$ ft.

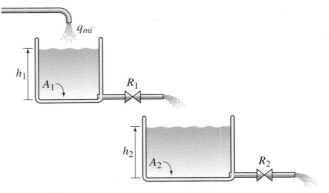

Figure P30

31. Consider Figure 10.7–7 for the case where there are three RC loops with the values $R_1 = R_3 = 10^4\,\Omega$, $R_2 = 5 \times 10^4\,\Omega$, $C_1 = C_3 = 10^{-6}\,$F, and $C_2 = 4 \times 10^{-6}\,$F.

   a. Develop a subsystem block for one RC loop.
   b. Use the subsystem block to construct a Simulink model of the entire system of three loops. Plot $v_3(t)$ over $0 \le t \le 3$ for $v_1(t) = 12 \sin 10t$ V.

32. Consider Figure 10.7–8 for the case where there are three masses. Use the values $m_1 = m_3 = 10$ kg, $m_2 = 30$ kg, $k_1 = k_4 = 10^4$ N/m, and $k_2 = k_3 = 2 \times 10^4$ N/m.

   a. Develop a subsystem block for one mass.
   b. Use the subsystem block to construct a Simulink model of the entire system of three masses. Plot the displacements of the masses over $0 \le t \le 2$ s for if the initial displacement of m_1 is 0.1 m.

Section 10.8

33. Refer to Figure P30. Suppose there is a dead time of 10 sec between the outflow of the top tank and the lower tank. Use the subsystem block developed in Section 10.7 to create a Simulink model of this system. Using the parameters given in Problem 30, plot the heights h_1 and h_2 versus time.

Section 10.9

34. Redo the Simulink suspension model developed in Section 10.9, using the spring relation and input function shown in Figure P34 and the following damper relation.

$$f_d(v) = \begin{cases} -500|v|^{1.2} & v \le 0 \\ 50v^{1.2} & v > 0 \end{cases}$$

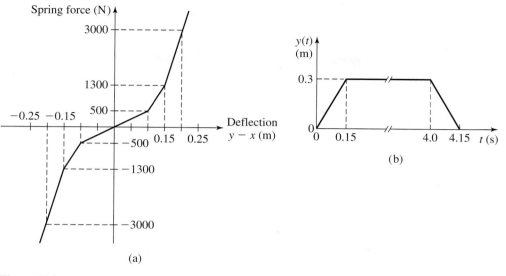

(a)

(b)

Figure P34

Use the simulation to plot the response. Evaluate the overshoot and undershoot.

35. Consider the system shown in Figure P35. The equations of motion are

$$m_1\ddot{x}_1 + (c_1 + c_2)\dot{x}_1 + (k_1 + k_2)x_1 - c_2\dot{x}_2 - k_2x_2 = 0$$
$$m_2\ddot{x}_2 + c_2\dot{x}_2 + k_2x_2 - c_2\dot{x}_1 - k_2x_1 = f(t)$$

Suppose that $m_1 = m_2 = 1$, $c_1 = 3$, $c_2 = 1$, $k_1 = 1$, and $k_2 = 4$.

a. Develop a Simulink model of this system. In doing this, consider whether to use a state-variable representation or a transfer-function representation of the model.

b. Use the Simulink model to plot the response $x_1(t)$ for the following input. The initial conditions are zero.

$$f(t) = \begin{cases} t & 0 \le t \le 1 \\ 2 - t & 1 < t < 2 \\ 0 & t \ge 2 \end{cases}$$

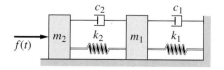

Figure P35

Engineering in the 21st Century...

Developing Alternative Sources of Energy

It now appears that the United States and much of the rest of the world have recognized the need to reduce their dependence on nonrenewable energy sources such as natural gas, oil, coal, and perhaps even uranium. Supplies of these fuels will eventually be exhausted, they have harmful environmental effects, and when imported, they cause huge trade imbalances that hurt the economy. One of the major engineering challenges of the 21st century will be to develop renewable energy sources.

Renewable energy sources include solar energy (both solar-thermal and solar-electric), geothermal power, tidal and wave power, wind power, as well as crops that can be converted to alcohol. In solar-thermal applications, energy from the sun is used to heat a fluid, which can be used to heat a building or power an electrical generator such as a steam turbine. In solar-electric applications, sunlight is directly converted to electricity.

Geothermal power is obtained from ground heat or steam vents. With tidal power the tidal currents are used to drive a turbine to generate electricity. Wave power uses the change in water surface level due to waves to drive water through a turbine or other device. Wind power uses a wind turbine to drive a generator.

The difficulty with most renewable energy sources is that they are diffuse, so the energy must be concentrated somehow, and they are intermittent, which requires a storage method. At present most renewable energy systems are not very efficient, and so the engineering challenge of the future is to improve their efficiency. ■

CHAPTER 11

MuPAD

OUTLINE

11.1 Introduction to MuPAD

11.2 Symbolic Expressions and Algebra

11.3 Algebraic and Transcendental Equations

11.4 Linear Algebra

11.5 Calculus

11.6 Ordinary Differential Equations

11.7 Laplace Transforms

11.8 Special Functions

11.9 Summary

Problems

MuPAD is a very large program with many capabilities. In this chapter we cover a subset of those capabilities, emphasizing the ones of greatest use to beginning students in engineering and science. Specifically we treat the following:

- MuPAD basics and the Help system.
- Symbolic algebra.
- Methods for solving algebraic and transcendental equations.
- Selected topics in linear algebra.
- Symbolic methods for solving ordinary differential equations.
- Symbolic calculus.
- Laplace transforms.
- The special functions of mathematics.

When you have finished this chapter, you should be able to use MATLAB to do the following:

- Create symbolic expressions and manipulate them algebraically.
- Obtain symbolic and numeric solutions to algebraic and transcendental equations.
- Perform symbolic linear algebra operations, including obtaining expressions for determinants, matrix inverses, eigenvectors, and eigenvalues.
- Perform symbolic differentiation and integration.
- Evaluate limits and series symbolically.
- Obtain symbolic solutions to ordinary differential equations.
- Obtain and apply Laplace transforms.
- Solve ordinary differential equations in terms of special functions or series.

11.1 Introduction to MuPAD

MuPAD is a big package with a lot of capabilities. In this chapter we introduce a subset of those capabilities, ones that are most useful to engineers and scientists, and only the basic syntax of those commands. Most commands in MuPAD have an extended syntax that is documented in the MuPAD Help system. You will need to refer frequently to this Help system, because space limitations make it impossible to cover MuPAD in detail in one chapter.

MuPAD comes with the Symbolic Math toolbox. The toolbox itself has its own syntax, different somewhat from that of MuPAD, and its commands are entered in the Command window or in M-files. The MuPAD interface, however, is a notebook. Prior to MATLAB Release 2008b (Version 5.1 of the toolbox), the toolbox used a licensed version of Maple as its engine. Now the Maple engine has been replaced by the MuPAD engine, but the toolbox syntax remains the same. However, the MuPAD engine behaves somewhat differently from the Maple engine. One example of the differing behavior is the solution of transcendental equations. Whereas the Maple engine sometimes correctly gave more than one solution, the MuPAD engine often gives only one of the possible solutions. Thus previous users of the toolbox will find that their old program may behave differently now.

In this chapter we cover MuPAD exclusively, because of page length limitations and because MuPAD appears to be the future of the Symbolic Math toolbox.

The MuPAD Welcome Screen

In MuPAD you perform operations in "notebooks." Thus you can organize your work by topics. To start MuPAD, first start MATLAB; then to open a new notebook, type `mupadwelcome` at the MATLAB prompt. This opens the screen shown in Figure 11.1–1. It enables you to get help under the subwindow **First**

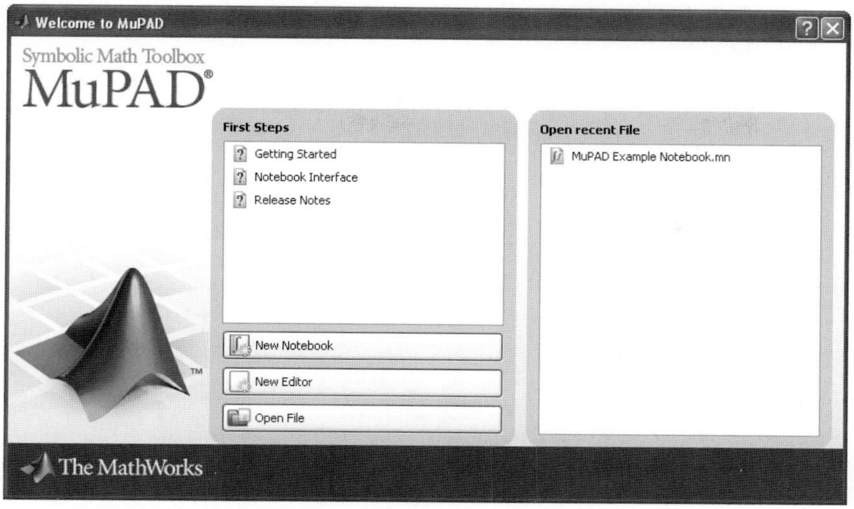

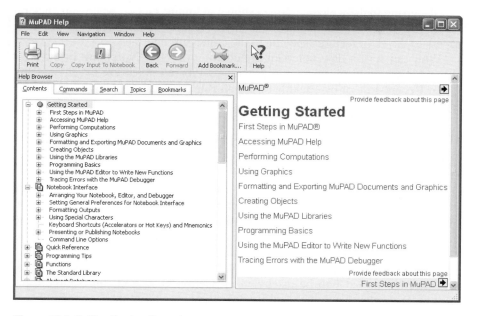

Figure 11.1–1 The MuPAD welcome screen.

Figure 11.1–2 The Getting Started screen.

Steps. Clicking on **Getting Started** opens the screen shown in Figure 11.1–2, which has a variety of Help topics. The left-hand pane has a list of topics that can be expanded or collapsed by clicking on the + or − signs. The right-hand pane contains a shorter list of topics.

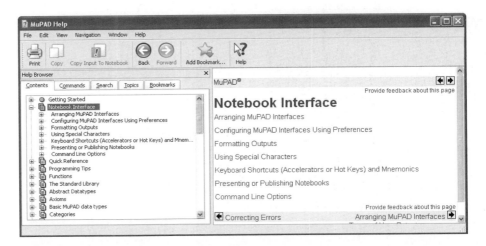

Figure 11.1–3 The Notebook Interface Help screen.

Referring to the welcome screen shown in Figure 11.1–1, when you click on **Notebook Interface**, the screen shown in Figure 11.1–3 appears. It provides access to the Help features specific to the Notebook Interface.

You can start a new notebook by clicking on **New Notebook** in the welcome screen, or you can open a previously created notebook by clicking on **Open Recent File** or by clicking on **Open File** to access files not in the default directory.

You can avoid the welcome screen by typing `mupad` at the MATLAB prompt. This opens a new notebook. Once MuPAD opens, you can select a previously created notebook if you wish.

The Menu Bar

Figure 11.1–4 shows a previously created notebook containing commands, outputs (the results of executing the commands), and comments. Commands and comments are placed in input and text regions, respectively. Results appear in output regions. The results can be mathematical expressions, numbers, or graphs.

INPUT, TEXT, AND OUTPUT REGIONS

In the default desktop arrangement, the Command bar appears to the right of the screen. This bar will be discussed in Section 11.2. You can configure the desktop to display the Search and Replace bar instead, but the Command bar is more useful in general.

The Menu bar is at the top of the interface. It contains nine menus, some of which you will not need to do basic computations. The **File**, **Edit**, and **Help** menus are similar to those in MATLAB except that the latter accesses the MuPAD **First Steps** screen. The **View** menu lets you configure the screen by adding or removing toolbars. The **Navigation** menu enables you to go back and forth in the notebook.

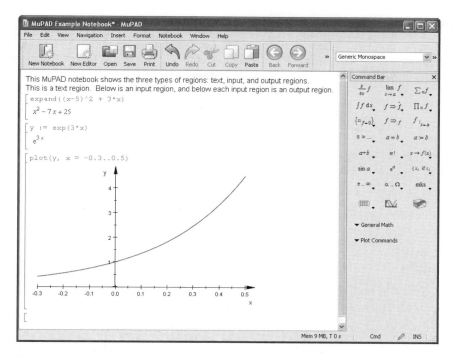

Figure 11.1–4 The MuPAD Notebook Interface.

Figure 11.1–5 The Standard bar.

The **Insert** menu is very useful because it enables you to insert calculations and comments (text paragraphs) at any place in the notebook. By clicking on one of the **Evaluate** options in the **Notebook** menu, the results of any newly inserted calculations will be propagated throughout the notebook.

The **Format** menu lets you change font sizes and page formatting. The **Window** menu lets you switch between notebooks if more than one is open.

You quit a MuPAD notebook by selecting **Close** on the **File** menu. Selecting **Exit** on the **File** menu will close MuPAD and return you to MATLAB.

The Standard Toolbar

Below the Menu bar is the Standard toolbar (Figure 11.1–5). This contains short-cut buttons for commonly used operations whose functions are obvious from the name of the button. The most useful are the **Calculation** button, which inserts a new input region below the current cursor position; the **Text Paragraph** button, which inserts a text paragraph; and the **Evaluate** button, which evaluates the cal-culation in the input region containing the cursor.

Entering Commands

You enter commands into input regions in the notebook. Figure 11.1–4 shows a notebook with some comments, commands, and responses displayed. Input and output regions have a bracket at the left; text regions do not. After you type a command in a blank input region, press **Enter** and MuPAD will evaluate the command. The result will be displayed in the next bracket, just below the input region. An example session that computes cos(3.14) is the following. The command is **cos(3.14)** and the output is –0.9999987317. In this chapter we display commands in boldface type, outputs and comments in italic and roman type.

[> **cos(3.14)**
[−0.9999987317

Note that the answer is not exactly −1 because 3.14 is not exactly π. For exact answers use **PI,** which is the MuPAD symbol for π.

[> **cos(PI)**
[−1

Some commonly used reserved symbols in MuPAD are **PI** for π, **I** for the complex number $i = \sqrt{-1}$, and **E** for e, the base of the natural logarithm. You should not use these as identifiers to represent something other than their reserved meaning.

RESERVED SYMBOLS

You can edit previous inputs by clicking on them and entering or deleting text, using the keyboard editing keys. Obtain an updated result by pressing **Enter**. You can cut and paste from the input region, but output cannot be edited or copied because it is a graphic. Terminate the command with a colon to suppress the output display. Enter more than one command on a line by separating them with either a semicolon or a colon.

MuPAD uses the standard arithmetic operators $(+ - * /)$ and follows the same precedence rules as MATLAB. MuPAD uses function names that are for the most part identical to those used by MATLAB, for example, **sin, tan, exp.** Two notable exceptions are the natural logarithm ln, which is **ln** in MuPAD, and the base-10 logarithm, $\log_{10}$, which is **log** in MuPAD. Answers are exact (not rounded floating-point numbers) when computing with integers and rational numbers. For example,

[> **1 + 3*7/2**
$$\frac{23}{2}$$

[> **(3 + (7/2*5))/(1/5 + 2/9)^2**
$$\frac{83025}{722}$$

[> **sqrt(52)**
[$2\sqrt{13}$

Use **E** or **exp(1)** for the base of the natural logarithm, although MuPAD displays the results using the lowercase **e**. Use **ln(x)** for $\ln x$ and **log(x)** for $\log x$.

[> 2*ln(E^3/2)

$$\left[2 \ln \left(\frac{e^3}{2} \right) \right.$$

[> float(%)
[4.613705639

The **float** function returns a floating-point result. You can use % to refer to the previous result. MuPAD displays results according to the value of **DIGITS**, which sets the number of decimal digits to display. The default value is 10.

Warning! It is easy to make mistakes using the **float** function with the % symbol, which represents the last *computed* result, *not* necessarily the result just above the **float(%)** command. For example, if you go back several steps in the notebook and recompute a command, then typing **float(%)** later on will cause it to evaluate the results of that command, even though you type **float(%)** as the last command in the notebook.

Approximate computation is used if at least one of the numbers involved is a floating-point value. For example, changing the 3 in the earlier example to 3.2 gives a floating-point result.

[> (3.2 + (7/2*5))/(1/5 + 2/9)^2
[116.1149584

Most of the common mathematical functions will return an exact value if their argument is exact, and a floating-point value if their argument is floating-point. For example,

[> 0.8*cos(2)
[0.8 cos(2)

[> 2/5*cos(2.1)
[−0.2019384418

Complex number calculations are easily done using **I** for $\sqrt{-1}$.

[> (1/3 + 2*I)*(0.1 + I/2)^3
[0.1953333333 − 0.1846666666*i*

The functions **Re(p)** and **Im(p)** return the real and the imaginary part of a complex number p. The functions **conjugate(p)**, **abs(p)**, and **arg(p)** compute the complex conjugate, the absolute value, and the polar angle of the number p. The function **rectform** converts a number to rectangular form.

[> conjugate(1/(sqrt(3) + I))

$$\left[\frac{1}{\sqrt{3} - i} \right.$$

[> **rectform(conjugate(1/(sqrt(3) + I)))**

$$\frac{\sqrt{3}}{4} + \frac{i}{4}$$

[> **abs(1/(sqrt(3) + I))**

$$\frac{1}{2}$$

[> **arg(1/sqrt(3) + I))**

$$-\frac{\pi}{6}$$

Inserting Text and Calculations

You can insert text for documentation purposes. An example is shown in Figure 11.1–4. To insert text above a calculation, the easiest way is to position the cursor within the input region to be documented, then click on the **Insert** menu, next on **Text Paragraph Above**. Then type your text. You may also want to insert a calculation above another one. To do this, click on the **Insert** menu, then on **Calculation Above**. This opens a blank input region.

11.2 Symbolic Expressions and Algebra

ASSIGNMENT OPERATOR

Use the assignment operator **:=** to assign a symbolic expression to an identifier named, say, **r**. The identifier **r** can then be used as an abbreviation for the expression. For example,

[> **r:= y^2 + 3*y + 5*x + 7*x^2 + 2*x^3**
[$2x^3 + 7x^2 + 5x + y^2 + 3y$

Notice that the expression has been rearranged in the display.

It is good practice to delete the identifiers of variables, expressions, and functions to avoid conflicts if you later use those identifiers to mean something else. Use the **delete (p)** function to delete **p**. It is also necessary to remove any assumptions placed on a parameter if that parameter will be used in later computations. Examples of this are the **assume** and **unassume** functions, to be discussed later.

Expressions Versus Functions

It is important to realize that an *expression* in MuPAD is not the same thing as a *function*. For example, consider the expression

[> **w := x^6:**

We cannot evaluate the expression at a particular value of *x*, say $x = 3$ by typing **w (3)** or by typing **float (w(3))**.

User-defined functions can be created by defining a *procedure*. Consider the function $f(x) = x^2 + 3x + 7$. We can define it by a procedure as follows:

[> f : = x -> x^2 + 3*x + 7:

The arrow symbol is obtained by typing the hyphen - followed by the greater than sign (>). We can then evaluate the function at given values of x:

[> f(x)
[$x^2 + 3x + 7$

[> f(y)
[$y^2 + 3y + 7$

[> f(a+b)
[$3a + 3b + (a + b)^2 + 7$

[> f(2.5)
[20.75

You can plot expressions or functions in the same way. For example,

[> y := 5*exp(-0.5*x)*sin(3*x):
[> plot (y,x=0..5)

The result is shown in Figure 11.2–1. You could also type **plot(y)** without limits, but MuPAD would select the plot limits so that the plot was centered around $x = 0$. Typing **plot(%)** produces the same result.

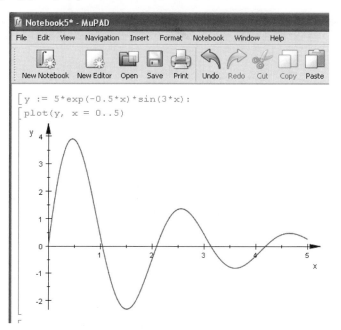

Figure 11.2–1 Creating a plot.

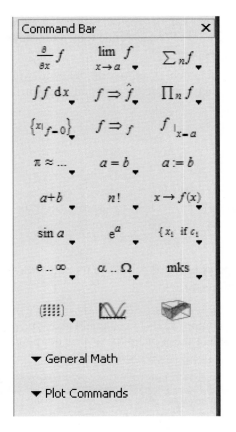

Figure 11.2–2 The Command bar.

Manipulating Expressions

Figure 11.2–2 shows the Command bar. The Command bar provides shortcuts to avoid typing commands into the input region. When you click on an item, that command is inserted at the current cursor location in the notebook. Table 11.2–1 lists the operations that are available from the upper portion of the Command bar. We will discuss these commands throughout the chapter, and you can refer to this table when particular commands are discussed.

The following functions can be used to manipulate expressions by collecting coefficients of like powers, expanding powers, and factoring expressions, for example. You can type functions directly in the input region, or you can select functions from the Command bar, which contains a subset of all the MuPAD functions. The following choices appear when you choose **General Math** from the Command bar:

The General Math Menu	
Expand	Simplify
Factor	Combine
Normalize	Rewrite
Evaluate	Solve

Table 11.2–1 Items on the Command bar

Derivatives	Limits	Sums
Integrals	Rewrite Expressions	Products
Solve Equations	Simplify	Evaluate with $x = a$
Numerical Evaluation and Rounding	Equality Tests	Assignment
Math Operators	Factorials	Function Definition
Trig Functions	Exponentials and Logs	Piecewise Definitions
Reserved Symbols	Greek Letters	Physical Units
Matrices and Vectors	2D Plot	3D Plot

The last five items on the General Math menu contain submenus. We will discuss the **Solve** function in Section 11.3. When you click on any of these items, MuPAD inserts the corresponding function into the input region at the location of the cursor. The function argument will be the place marker #, which you should delete, and then you enter the appropriate expression. In every case you may type in the function if you do not want to use the menu.

Expand   When you select **Expand** from the menu, MuPAD inserts **expand(#)** in the input region. Delete the place marker # and type in your text. The function **expand(p)** expands the expression **p** by using a set of rules appropriate to the given expression. The following example shows how the expansion is carried out by powers.

`[ > p1 := (x-5)^2+(y-3)^2`
$[\ (x - 5)^2 + (y - 3)^2 $

`[ > expand(p1)`
$[\ x^2 - 10x + y^2 - 6y - 34 $

You can reuse expression symbols such as **p1** by typing **delete(p1)** in the input box.

The next example shows how the expansion is carried out using trigonometric identities.

`[ > expand(sin(x + y))`
$[\ \cos(x) \sin(y) + \cos(y) \sin(x) $

Factor   When you select **Factor** from the menu, MuPAD inserts **factor (#)** in the input region. Use the **factor(p)** function to factor the expression **p**.

`[ > factor(x^2-1)`
$[\ (x + 1)(x - 1) $

The **factorout(p,q)** function factors out a given expression **q** from the expression **p**.

[> **factorout((a*b + a*c)/(d*c+c), a/c)**

$$\frac{a}{c}\frac{b+c}{d+1}$$

Normalize Clicking on **Normalize** places the function **normal (#)** in the input area that contains the cursor. The function **normal(p)** finds a common denominator for rational expressions.

[> **p := x/(1 + x) - 3/(1 - x)**

$$\frac{x}{x+1} + \frac{3}{x-1}$$

[> **normal(p)**

$$\frac{x^2 + 2x + 3}{x^2 - 1}$$

Also, **normal** cancels common factors in the numerator and the denominator:

[> **x^2/(x + y) - y^2/(x + y)**

$$\frac{x^2}{x+y} - \frac{y^2}{x+y}$$

[> **normal(%)**

[$x - y$

Evaluate The menu item **Evaluate** has two subchoices: **Numerically** and **Boolean**. Choosing **Numerically** inserts **float (#)** into the input region. We have already seen the **float** function. Choosing **Boolean** inserts **bool (#),** which evaluates a Boolean expression.

[> **bool(float(sqrt(18)) <= float(sqrt(3)*sqrt(6)))**

[TRUE

Simplify The **Simplify** menu item has eight subchoices:

The Simplify Menu	
General	Exponential
Logical	Logarithm
Radical	Sine
Relational	Cosine

Choosing **General** inserts the function **Simplify(#)**. Note the capitalized first letter. You can also type the related function **simplify(p). The simplify** and **Simplify** functions are similar; **Simplify** is more powerful, but may be slower.

The general form of these functions is **simplify(p, <target>)**, where **<target>** restricts the simplification rules to the target function(s). For example, choosing the submenu item **Logical** inserts the function **simplify(#, logic)**. Other choices for **<target>** are **sqrt, relation, exp, ln, sin,** and **cos,** which are obtained by choosing the menu items **Radical, Relational, Exponential, Logarithm, Sine,** and **Cosine,** respectively. For example,

[> **simplify(sqrt(4 + 2*sqrt(3)), sqrt)**
[$\sqrt{3} + 1$

Sometimes the target function is not needed. For example,

[> **simplify(cos(x)*sin(y) + cos(y)*sin(x))**
[$\sin(x + y)$

Combine The **Combine** menu item has eight subchoices:

The Combine Menu	
General	Power
Arc Tangent	Sine/Cosine
Exponential	Sine/Cosine Hyp
Logarithm	Square Root

Choosing **General** inserts the function **combine(#)** into the notebook. The **combine(p)** function combines terms in the expression **p** of the same algebraic structure. The general form of the **combine** function is **combine(p <target>)**. The **<target>** option combines several calls to the target functions(s) in the expression **p** to a single call. Other target functions are available, but are not on the menu; these are **log** and **gamma.**

For example, choosing the submenu item **Sine/Cosine** inserts the function **combine(#, sincos)**. Other choices for **<target>** are **arctan, exp, ln, power, sincos, sinhcosh,** and **sqrt,** which are obtained by choosing the corresponding menu items.

When the second argument is not specified, **combine** groups powers of the same base:

[> **combine(x*y*x^y)**
[$x^{y+1}y$

The function **combine** combines powers with the same exponent in certain cases, but not always:

[> **combine(sqrt(2)*sqrt(3)*y^5*x^5)**
[$\sqrt{6}x^5y^5$

With the target function **sincos,** products of sines and cosines are expressed as sums of sines and cosines.

[> **combine(cos(a)*sin(b) + sin(a)^2, sincos)**
$$\left[\frac{\sin(a + b)}{2} - \frac{\cos(2a)}{2} - \frac{\sin(a - b)}{2} + \frac{1}{2} \right.$$

Some combinations may require certain assumptions to be made. For example,

[> assume(a > 0): assume(b > 0):

[> combine(ln(a) + ln(b), ln)

[$\ln(ab)$

[> unassume(a): unassume(b):

Rewrite   The **Rewrite** menu under **General Math** has eight subchoices:

The Rewrite Menu	
Differential	Heaviside
Exponential	Logarithm
Factorial	Sign
Gamma	Sine/Cosine

Choosing any one of these items inserts the function **rewrite(#, <target>)** into the notebook. The **rewrite(p, <target>)** function transforms an expression **p** to a mathematically equivalent form, trying to express **p** in terms of the specified target function. For example, choosing the submenu item **Sine/Cosine** inserts the function **rewrite(#, sincos).**

[> rewrite(cot(x), sincos)

$$\left[\frac{\cos(x)}{\sin(x)} \right.$$

[> rewrite(arcsinh(x), ln)

[$\ln(x + \sqrt{x^2 + 1})$

Other choices for **<target>** are **diff,** which is discussed in Section 11.5; and **exp, fact, heaviside, ln, sign,** and **sincos,** which are obtained by choosing the corresponding menu items. Many other target functions are available, but are not on the menu; consult the Help for **rewrite** to see the list.

Functions Not on the Toolbar   There are many useful functions that are not on the toolbar. One example is the function **collect(p,x).** It collects coefficients of like powers of **x** in the expression **p.**

[> p2 := (x+3)^2+3*x+5

[$3x + (x + 3)^2 + 5$

[> collect(p2, x)

[$x^2 + 9x + 15$

If there is more than one variable, you can specify which one to collect.

[> p3:= (x+3)^2+5*x+(y-4)^3

[$5x + (x + 3)^2 + (y - 4)^3$

[> collect(p3,x)

[$x^2 + 11x + (y - 4)^3 + 9$

You can perform operations involving more than one expression. For example,

[> collect(p2*p3,x)

$$x^4 + 20x^3 + ((y - 4)^3 + 122)\, x^2 + (9(y - 4)^3 + 225)\, x + 14(y - 4)^3 + 126$$

Another function not on the Command bar is the **subs** function. You can replace a variable **x** in an expression **p** with the variable **y** by typing **subs(p, x = y)**. For example,

[> p4:=3*x^2+5*x+9

$3x^2 + 5x + 9$

[> subs(p4,x=x+1)

$5x + 3(x + 1)^2 + 14$

Note that **p4** has not changed:

[> 2*p4

$6x^2 + 10x + 18$

To obtain a new variable, you must give it a name:

[>p5:=subs(p4,x=x+1)

$5x + 3(x + 1)^2 + 14$

[>2*p5

$10x + 6(x + 1)^2 + 28$

The remaining items on the **General Math** menu are discussed in later sections.

Test Your Understanding

T11.2–1 Given the expressions $E_1 = x^3 - 15x^2 + 75x - 125$ and $E_2 = (x + 5)^2 - 20x$, use MuPAD to

(a) find the product $E_1 E_2$ and express it in its simplest form.
(b) find the quotient E_1/E_2 and express it in its simplest form.
(c) evaluate the sum $E_1 + E_2$ at $x = 7.1$ in symbolic form and in numeric form.

(Answers: (a) $(x - 5)^5$, (b) $x - 5$, (c) $13{,}671/1000$ in symbolic form, 13.6710 in numeric form)

11.3 Algebraic and Transcendental Equations

MuPAD can solve algebraic and transcendental equations as well as systems of such equations. A *transcendental* equation is one that contains one or more transcendental functions, such as $\sin x$, e^x, or $\log x$. The appropriate function to solve such equations is the **solve** function.

The `solve` Function

The function **solve(p)** solves a symbolic expression or equation (an expression containing the equals sign =) represented by the expression **p**. If the expression **p** does not contain an equals sign, MuPAD assumes that the expression is set to 0. For example, to solve the equation $5x + 25 = 0$ using **solve(p)**, type

[> **solve(5*x+25=0)**

[$\{[x = -5]\}$

or

[> **y := solve(5*x+25=0)**

[$\{[x = -5]\}$

[>**3*y**

[$\{3[y = -5]\}$

The expression need not equal 0.

[> **solve(5*x=-25)**

[$\{[x = -5]\}$

Using the version **solve(p,x),** we obtain

[> **solve(5*x+25=0,x)**

[$\{-5\}$

or

[> **y := solve(5*x+25=0,x)**

[$\{-5\}$

MuPAD Libraries

MuPAD has hundreds of functions, which are organized into *libraries.* The Standard library is the most important MuPAD library. It has the most frequently used functions such as **solve,** and **simplify.** Except for the functions of the Standard library, library functions have a prefix separated with ::. For example, the **solve** function in the Standard library solves for analytic solutions, whereas the **solve** function in the Numeric library is called as **numeric::solve(p).**

Library functions can also be called by their short notation if you first export them with the **use** function. One library we will see soon is the **numeric** library; it contains the function **polyroots,** which obtains the roots of polynomials numerically rather than symbolically. We can call this function by typing **numeric:: polyroots**. However, you can "export" this library's functions from the library into MuPAD by typing **use(numeric)**. After exporting, you can type **polyroots** instead of the long form **numeric::polyroots**, for example. To avoid confusion as to whether a library function has been exported, in this chapter we will always use the long form.

There are many libraries, and you can obtain more information about them and their functions by searching for libraries in the Help system. You can obtain a list of the functions in a particular library, say, the **numeric** library, by typing **info(numeric).**

Selecting Functions from the Command Bar

You can type functions directly in the input region of the notebook, or you can select functions from the Command bar, which contains a subset of all the MuPAD functions. The following choices appear when you choose **Solve** under **General Math** from the Command bar:

The Solve Menu	
Exact	Polynomial Diophantine Equation
Numeric	Recurrences
Linear System	ODE

Selecting **Exact** puts **solve(#)** in the input box. You then delete the # and enter the equation to be solved. Instead of using the menu, you can type **solve(p)** or **solve(p,x),** where **p** represents the equation to be solved and **x** is the solution variable. To solve $x^2 - 4x + 6 = 0$, type

[>solve(x^2-4*x+6)
[$\{[x = 2 - \sqrt{2}i], [x = 2 + \sqrt{2}i]\}$

Some nonlinear equations have an infinite number of solutions. MuPAD indicates these in the following way, if it can find them exactly.

[> solve(sin(2*x)-cos(x)=0,x)

$$\left[\left\{ \frac{\pi}{2} + 2\pi k \ \middle| \ k \in \mathcal{Z} \right\} \cup \left\{ \frac{\pi}{6} + \frac{2\pi k}{3} \ \middle| \ k \in \mathcal{Z} \right\} \right.$$

The $\cup$ symbol means that the complete solution can be obtained from the expressions on both sides of the $\cup$ symbol, where the $k \in \mathcal{Z}$ symbols indicate the two solutions given are true for any integer value of k.

To find all the roots of $x^4 = 1$, you type **solve(x^4=1)** to get $x = \pm 1$ and $x = \pm i$. To find only the real roots of $x^4 = 1$, you can restrict the results to be positive by using the option **Type::Real** as follows:

[> assume (x, Type::Real): solve(x^4 = 1, x)
[$\{-1, 1\}$

Be sure to type **unassume(x):** afterward if you intend to use **x** later as another variable. There are other options, such as **Type::Positive**, available to use with **assume** and **solve.** We will see some of these later; see the Help for details.

You need not use the **assume** function. For example, there are an infinite number of solutions to the equation $e^{2x} + 3e^x = 54$, and MuPAD can find them using **solve.** Suppose, however, that you are interested only in the real solutions.

Then you can use the **Real** option as follows to obtain the only real solution, $x = \ln 6$.

```
[ > solve(exp(2*x) + 3*exp(x)=54, x, Real)
```
$[\ \{\ln(6)\}$

You must always be careful in interpreting the results. For example, the equation $x^4 - 5x^2 + 6x - 2 = 0$ is fourth order and therefore has four roots, but MuPAD gives only three roots and does not indicate that the root $x = 1$ is repeated.

```
[ > solve(x^4 - 5*x^2 + 6*x = 2, x)
```
$[\ \{1, -\sqrt{3} - 1, \sqrt{3} - 1\}$

You can avoid this problem by using the option **Multiple**, as follows:

```
[ > solve(x^4-5*x^2 + 6*x = 2, x, Multiple)
```
$[\ \{[1, 2], [1, -\sqrt{3} - 1, \sqrt{3}-1]\}$

The result [1, 2] indicates that the root $x = 1$ is repeated twice.

You can specify the interval over which you seek a solution. For example, the equation $\cos(\pi x/5) = 0$ has an infinite number of solutions, which can be found using the **solve** function. However, if you want only the solutions on the interval $-10 \le x \le 10$, type

```
[ >solve(cos(x*PI/5) = 0, x = -10..10)
```
$$\left[\left\{ -\frac{15}{2}, -\frac{5}{2}, \frac{5}{2}, \frac{15}{2} \right\} \right.$$

MuPAD cannot always find an exact solution. In such cases, MuPAD simply displays the command. You can use the **float** function to force MuPAD to produce a numerical solution. For example,

```
[ > solve(cos(x) = x^2, x)
```
$[\ \text{solve}(\cos(x) - x^2 = 0, x)$

```
[ > float(%)
```
$[\ -0.8241323123$

You can also solve inequalities. For example, the inequality $x^2 + 3x > 4$ has solutions on the intervals $-\infty < x < -4$ and $1 < x < \infty$.

```
[ > solve (x^2 + 3*x > 4, x)
```
$[\ (1,\infty) \bigcup (-\infty,-4)$

You can instruct MuPAD to make assumptions about the problem. For example, the quadratic equation $ax^2 + bx + c = 0$ has the following types of solutions, depending on the values of the coefficients:

1. $x = \dfrac{-b \pm \sqrt{b^2 - 4ac}}{2a}$ if $a \ne 0$

2. $x = -\dfrac{c}{b}$ if $a = 0$ and $b \ne 0$

3. There is no solution if $a = b = 0$ but $c \neq 0$.

4. There are an infinite number of solutions if $a = b = c = 0$.

You can eliminate the last three cases by telling MuPAD to assume that $a \neq 0$, as follows.

[> solve(a*x^2 + b*x + c, x, assume (a<>0))

$$\left[\left\{ -\frac{b + \sqrt{-4ac + b^2}}{2a}, -\frac{b - \sqrt{-4ac + b^2}}{2a} \right\} \right.$$

which is equivalent to the solution given previously. You can relax the assumption on a for later solutions by typing **unassume(a):** in the input box. Do not forget to delete variables if you intend to reuse their names. The names **x** and **y** are commonly used, so you should delete them before continuing.

Numeric

Selecting **Numeric** from the **solve** menu puts **numeric::solve(#)** in the input box. The results are returned as floating-point numbers.

[> numeric::solve(x^2-4*x+6,x)

[{2 + 1.414213562i, 2 − 1.414213562i}

Note that MuPAD might not be able to solve any given nonlinear equation, and if so, you must use the numerical solver. However, in contrast to polynomial equations, the numerical solver computes at most one solution of a nonpolynomial equation.

[> numeric::solve(x+2*exp(-x)-3)

[{[x = −0.583073876]}

Sometimes we can direct the search for a solution. Suppose we are interested only in a solution over the range $0 \leq x \leq 10$. The session is

[> numeric::solve(x+2*exp(-x)-3, x = 0..10)

[{[x = 2.888703356]}

Sets of Equations

The **solve** function can also solve sets of equations. The solutions can be expressed as a set of substitutions or as vectors. For example, to solve the set $x + 2y = 1, 3x + 7y = 5$, you type

[> solve ([x + 2*y = 1, 3*x + 7*y = 5], [x, y])

[{[x = −3, y = 2]}

or

[> solve([x + 2*y = 1, 3*x + 7*y = 5], [x, y], VectorFormat)

$$\left[\left\{ \begin{pmatrix} -3 \\ 2 \end{pmatrix} \right\} \right.$$

Consider another example:

[> **eqs := {x + y = a, x - a*y = b}:**
[> **solve(eqs, [x, y], IgnoreSpecialCases)**

$$\left[\ \left\{ \left[x = \frac{a^2 + b}{a + 1}, y = \frac{a - b}{a + 1} \right] \right\} \right.$$

Here there are two symbolic parameters a and b, and thus we need to specify the symbols representing the variables to be found (here, x and y). The solution is incorrect if $a = -1$, and the option **IgnoreSpecialCases** tells MuPAD to ignore this possibility.

The **solve** function can also handle sets of nonlinear equations. For example,

[> **solve([x^2 + y = 1, x + 2*y = 1], [x, y])**

$$\left[\ \left\{ [x = 1, y = 0], \left[x = -\frac{1}{2}, y = \frac{3}{4} \right] \right\} \right.$$

Test Your Understanding

T11.3–1 Use MuPAD to solve the equation $\sqrt{1 - x^2} = x$. (Answer: $x = \sqrt{2}/2$)

T11.3–2 Use MuPAD to solve the equation set $x + 6y = a, 2x - 3y = 9$ in terms of the parameter a. (Answer: $x = (a + 18)/5, y = (2a - 9)/15$)

Here is a somewhat more involved application.

EXAMPLE 11.3–1   Intersection of Two Circles

We want to find the intersection points of two circles. The first circle has a radius of 2 and is centered at $x = 3, y = 5$. The second circle has a radius b and is centered at $x = 5$, $y = 3$. See Figure 11.3–1.

(a) Find the (x, y) coordinates of the intersection points in terms of the parameter b.
(b) Evaluate the solution for the case where $b = \sqrt{3}$.

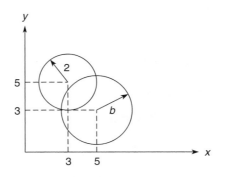

Figure 11.3–1 Intersection points of two circles.

■ **Solution**

(a) The intersection points are found from the solutions of the two equations for the circles. These equations are

$$(x - 3)^2 + (y - 5)^2 = 4$$

for the first circle and

$$(x - 5)^2 + (y - 3)^2 = b^2$$

The session to solve these equations is the following.

[> p1:=(x-3)^2 + (y-5)^2-4
[$(x - 3)^2 + (y - 5)^2 - 4$

[> p2:=(x-5)^2 + (y-3)^2-b^2
[$(x - 5)^2 + (y - 3)^2 - b^2$

[> ans1:=solve([p1=0,p2=0],[x,y])

$$\left[\left\{ \left[x = \frac{9}{2} - \frac{b^2}{8} + \sigma_1, \quad y = \frac{b^2}{8} + \sigma_1 + \frac{7}{2} \right] \right\} \left\{ \left[x = -\sigma_1 - \frac{b^2}{8} + \frac{9}{2}, \quad y = -\sigma_1 + \frac{b^2}{8} + \frac{7}{2} \right] \right\} \right\}$$

[*where*

$$\left[\sigma_1 = -\frac{\sqrt{-\dfrac{b^4}{16} + \dfrac{3b^2}{2} - 1}}{2} \right.$$

These are the x and y coordinates of the two intersection points.

(b) Continue the above session by substituting $b = \sqrt{3}$ into the expression for x.

[> ans2:=subs(ans1,b=sqrt(3))

$$\left[\left\{ \left[x = \frac{33}{8} - \frac{\sqrt{47}}{8}, \quad y = \frac{31}{8} - \frac{\sqrt{47}}{8} \right] \right\} \left\{ \left[x = -\frac{\sqrt{47}}{8} + \frac{33}{8}, \quad y = \frac{\sqrt{47}}{8} + \frac{31}{8} \right] \right\} \right\}$$

[> float(ans2)
[$\{[x = 4.982, \quad y = 4.732]\} \{[x = 3.268, \quad y = 3.018]\}$

These are the two pairs of coordinates for the two intersection points if $b = \sqrt{3}$.

Test Your Understanding

T11.3–3 Find the x and y coordinates of the two intersection points in Example 11.3–1 using $b = \sqrt{5}$. (Answer: $x = 4.986$, $y = 5.236$; $x = 2.764$, $y = 3.014$)

T11.3–4 Obtain the exact solution for x from the following equations.

$$x + \sqrt{3}y = e^2$$
$$3x - \pi y = 4$$

(Answer: $x = (\pi e^2 + 4\sqrt{3})/(\pi + 3\sqrt{3})$)

EXAMPLE 11.3–2   Positioning a Robot Arm

Figure 11.3–2 shows a robot arm having two joints and two links. The angles of rotation of the motors at the joints are θ_1 and θ_2. From trigonometry we can derive the following expressions for the (x, y) coordinates of the hand.

$$x = L_1 \cos\theta_1 + L_2 \cos(\theta_1 + \theta_2)$$
$$y = L_1 \sin\theta_1 + L_2 \sin(\theta_1 + \theta_2)$$

Suppose that the link lengths are $L_1 = 4$ ft and $L_2 = 3$ ft.
    Compute the motor angles required to position the hand at $x = 6$ ft, $y = 2$ ft.

■ Solution
Substituting the given values of L_1, L_2, x, and y into the above equations gives

$$6 = 4 \cos\theta_1 + 3 \cos(\theta_1 + \theta_2)$$
$$2 = 4 \sin\theta_1 + 3 \sin(\theta_1 + \theta_2)$$

The following session solves these equations. The variables **th1** and **th2** represent θ_1 and θ_2.

[>**numeric::solve([4*cos(th1)+3*cos(th1+th2)=6,4*sin(th1)+3*sin(th1+th2)=2],[th1,th2])**

[{[*th1* = −0.05756292117, *th2* = 0.8956647939]}

Thus the solution in degress is $\theta_1 = -3.2981°$, $\theta_2 = 51.3178°$.

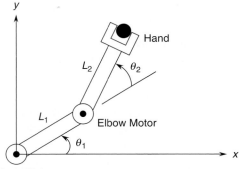

Figure 11.3–2 A robot arm having two joints and two links.

Selecting **Linear System** from the **Solve** menu inserts **linsolve**(#) into the input box. This function solves sets of linear equations numerically rather than symbolically.

Polynomial Diophantine Equations

The next item on the **Solve** menu is **Polynomial Diophantine Equation.** It inserts the function **solvelib::pdioe**(#, #) into the input box. Such an equation is one whose solutions are restricted to be polynomials. An important type of this class has the form

$$au + bv = c$$

where a, b, and c are given polynomial functions of x, and u and v are the unknowns. This type of equation can be solved with the **solve::pdioe(a,b,c,x)** function, for example,

$$(x^2 + 1)u + (x^3 + 1)v = 2x$$

The solution is found as follows.

[> **solvelib::pdioe((x^2 + 1), (x^3 + 1), 2*x, x)**
[$-x^2 + x + 1, x - 1$

Thus one solution is $u = -x^2 + x + 1$ and $v = x - 1$. Polynomial diophantine equations do not have unique solutions. Any multiple wab can be used to transform u and v into another solution u_1, v_1, where $u_1 = u + wb$ and $v_1 = v - wa$. For example, the choice $w = -1$ leads to the solution $u_1 = -x^3 - x^2 + x$, $v_1 = x^2 + x$.

Recurrence Relations

An example of a recurrence relation, sometimes called a *difference equation*, is

$$y(n + 2) - 2y(n + 1) + y(n) = 5$$

Such equations can be solved using the menu item **Recurrences** on the **Solve** menu. For example, with the given initial conditions $y(0) = 0$, $y(1) = 1$,

[> **eqn := y(n + 2) - 2*y(n + 1) + y(n) = 5:**
[> **solve (rec(eqn, y(n), {y(0) = 0, y(1) = 1}))**
$\left[\left\{ \dfrac{5n^2}{2} - \dfrac{3n}{2} \right\} \right.$

Thus the solution is

$$y(n) = \frac{5n^2}{2} - \frac{3n}{2} \qquad n = 2, 3, \ldots$$

The last menu item under the **Solve** menu is **ODE,** which stands for the ordinary differential equation solver. This is discussed in Section 11.6.

Optimization Problems

There are other solvers that deal with sets of linear *inequalities*, rather than linear equations. One class of such problems is called *linear programming*. A linear programming problem is one in which you need to find the values of a set of n variables, denoted $x_1, x_2, \ldots, x_n$ required to either maximize or minimize a function $J(x_1, x_2, \ldots, x_n)$ subject to a set of equalities or inequalities that are linear functions of the variables $x_1, x_2, \ldots, x_n$. Such problems have found widespread applications in industry, including optimizing airline schedules and many types of production problems. They can be solved using the functions in the **linopt** library.

An example of such a problem is the following. Suppose we want to find the values of x, y, and z to maximize the function

$$J = -x + y + 2z$$

subject to the constraints

$$3x + 4y - 3z \leq 20$$
$$6x - 4y - 3z = 10$$
$$7x + 4y + 11z \leq 30$$

In MuPAD we would type

```
[ > eq1 := 3*x + 4*y - 3*z <= 20
[ > eq2 := 6*x - 4*y - 3*z = 10
[ > eq3 := 7*x + 4*y + 11*z <= 30
[ > J := [{eq1,eq2,eq3},- x + y + 2*z]
[ > linopt::maximize(J)
```

$$\left[\left[\text{OPTIMAL}, \left\{ x = \frac{200}{87}, y = 0, z = \frac{110}{87} \right\}, \frac{20}{87} \right] \right]$$

The answer gives the values of x, y, and z that maximize J subject to the three constraints. The resulting maximum possible value of J is given as $20/87$.

Sometimes we must specify constraints that are not immediately obvious. For example, Example 8.3–4, Production Planning, has the constraint equations

$$5x + 3y + 3z = 40$$
$$3x + 3y + 4z = 30$$

where x, y, and z are the number of tons of each product to be produced each week. The profit to be maximized is

$$P = 400x + 600y + 100z$$

Without additional constraints, the solver may give physically unrealistic values. Here x, y, and z cannot be negative, so we add these three additional constraints. The session is

```
[ > eqn1 := 5*x+3*y+3*z=40
[ > eqn2 := 3*x+3*y+4*z=30
```

[> eqn3 := x >= 0: eqn4:= y >= 0: eqn5 := z >= 0:
[> J := [eqn1,eqn2,eqn3,eqn4,eqn5, 400*x+600*y+100*z]:
[> linopt::maximize(J)
[$[OPTIMAL, \{x = 5, y = 5, z = 0\}, 5000]$

The maximum profit is $5000.

Functions can be minimized with the **linopt::minimize(J)** function. The **linopt** library has many more functions that are useful for linear programming problems.

11.4 Linear Algebra

The **linalg** library contains over 40 functions dealing with linear algebra. Type **info(linalg)** to see a list of the functions. In this section we will use only a few of these. A reminder: library functions can also be called by their short notation if you first export them with the **use** function, for example, **use(linalg)**. After exporting, you can type **charpoly** instead of the long form **linalg::charpoly,** for example. To avoid confusion as to whether a function has been exported, we will always use the long form.

Matrix Operations

Linear algebra deals with matrices. The easiest way to create a matrix is the function **matrix**, which is in the Standard library. For example,

[> A := matrix([[3, 5], [-4, 2]]),B := matrix([[2,1], [3, -2]])

$$\left[\begin{pmatrix} 3 & 5 \\ -4 & 2 \end{pmatrix}, \begin{pmatrix} 2 & 1 \\ 3 & -2 \end{pmatrix} \right]$$

Note that MuPAD surrounds matrices with a pair of parentheses, whereas square brackets are used in MATLAB and much of the technical literature. You can add, subtract, multiply, or divide matrices using the standard arithmetical operators. Adding or subtracting a scalar has the effect of adding or subtracting that scalar from only the *diagonal* elements in the array. Note that this is different from MATLAB, in which the scalar is subtracted from *every* element.

[> A * A, 1/A, 2*A,A-3

$$\left[\begin{pmatrix} -11 & 25 \\ -20 & -16 \end{pmatrix}, \begin{pmatrix} \frac{1}{13} & -\frac{5}{26} \\ \frac{2}{13} & \frac{3}{26} \end{pmatrix}, \begin{pmatrix} 6 & 10 \\ -8 & 4 \end{pmatrix}, \begin{pmatrix} 0 & 5 \\ -4 & -1 \end{pmatrix} \right]$$

The determinant and rank are calculated as follows.

[> linalg::det(A),linalg::rank(A)
[26, 2

Matrix entries can be accessed with the index operator [ ].

[> **A[2,1]**

[−4

You can multiply matrices.

[> **A*B**

$$\left[\begin{pmatrix} 21 & -7 \\ -2 & -8 \end{pmatrix} \right.$$

You can create symbolic matrices.

[> **C := matrix([[a, b], [c, d]])**

$$\left[\begin{pmatrix} a & b \\ c & d \end{pmatrix} \right.$$

Create a diagonal matrix as follows:

[> **D := matrix(3, 3, [2, -2, 5], Diagonal)**

$$\left[\begin{pmatrix} 2 & 0 & 0 \\ 0 & -2 & 0 \\ 0 & 0 & 5 \end{pmatrix} \right.$$

Consider the rotation matrix $\mathbf{R}(a)$ for a coordinate system rotated through an angle a:

[> **R := matrix([[cos(a),sin(a)],[-sin(a),cos(a)]])**

$$\left[\begin{pmatrix} \cos(a) & \sin(a) \\ -\sin(a) & \cos(a) \end{pmatrix} \right.$$

One property of coordinate rotations is that if we rotate the coordinate system twice by the same angle to produce a third coordinate system, the result is the same as a single rotation with twice the angle. Thus, $\mathbf{R}(a)\ \mathbf{R}(a)$ should equal $\mathbf{R}(2a)$. Let us check this with MuPAD as follows.

[> **simplify(R*R)**

$$\left[\begin{pmatrix} \cos(2a) & \sin(2a) \\ -\sin(2a) & \cos(2a) \end{pmatrix} \right.$$

which is the same as $\mathbf{R}(2a)$:

[> **subs(R, a=2*a)**

$$\left[\begin{pmatrix} \cos(2a) & \sin(2a) \\ -\sin(2a) & \cos(2a) \end{pmatrix} \right.$$

To evaluate a symbolic matrix numerically, use the **float** and **subs** functions.

[> **float(subs(R, a=2))**

$$\left[\begin{pmatrix} -0.4161468365 & 0.9092974268 \\ -0.9092974268 & -0.4161468365 \end{pmatrix} \right.$$

Note that this operation did not change the symbolic matrix **R**.

A rotation in the negative direction is represented by **R**(−*a*), which equals the inverse of **R**(*a*). The inverse of a matrix can be found from its reciprocal **1/A** or with the **inverse** function, which is in the Standard library. You get the message *FAIL* if the matrix is not invertible.

[> **simplify(inverse(R))**

$$\left[\begin{pmatrix} \cos(a) & -\sin(a) \\ \sin(a) & \cos(a) \end{pmatrix} \right]$$

This equals **R**(−*a*), as can be shown by typing **simplify(subs(R,a=−a))**.

The Characteristic Polynomial, Eigenvalues, and Eigenvectors

The function **linalg::charpoly(A, x)** computes the characteristic polynomial of the matrix **A.** Using the matrix **A** created previously, we have

[> **p := linalg::charpoly(A,x)**

[$x^2 - 5x + 26$

The eigenvalues are the roots of the characteristic polynomial and can be found as follows.

[> **solve(p,x)**

$$\left[\left\{ \frac{5}{2} - \frac{\sqrt{79}i}{2}, \frac{5}{2} + \frac{\sqrt{79}i}{2} \right\} \right]$$

If you do not need the characteristic polynomial, you can calculate the eigenvalues of **A** directly with the function **linalg::eigenvalues(A)**. This gives the exact solution for the eigenvalues. This might not be possible for large matrices; if so, you can use the **numeric::eigenvalues(A)** function.

The function **linalg::eigenvectors(A)** computes both the eigenvalues and the eigenvectors of the matrix. The function returns a list of sublists; each sublist consists of an eigenvalue of **A,** its algebraic multiplicity, and a basis vector, which is an eigenvector normalized so that the last element is 1. If a basis vector cannot be computed, the message *FAIL* is returned. For example,

[> **A := matrix([[3, 5], [-4, 2]]):**
[> **linalg::eigenvectors(A)**

$$\left[\left[\left[\frac{5}{2} - \frac{\sqrt{79}i}{2}, 1, \left[\begin{pmatrix} -\frac{1}{8} + \frac{\sqrt{79}i}{8} \\ 1 \end{pmatrix} \right] \right], \left[\frac{5}{2} + \frac{\sqrt{79}i}{2}, 1, \left[\begin{pmatrix} -\frac{1}{8} - \frac{\sqrt{79}i}{8} \\ 1 \end{pmatrix} \right] \right] \right] \right]$$

Matrix Solution of Linear Equations

The function **linalg::matlinsolve(A, b)** computes the general solution of the equation **Ax** = **b**, where **A** is an *m* × *n* matrix, **b** is an *m* × 1 vector, and the solution **x** is an *n* × 1 vector. Consider the set

$$2x + y = 11$$
$$3x - 5y = -16$$

It can be solved as follows.

```
[ > A := matrix([[2,1], [3, -5 ]]):b:= matrix ([11, 16]):
[ > x := linalg::matlinsolve(A, b)
```

$$\left[\begin{pmatrix} 3 \\ 5 \end{pmatrix} \right]$$

This set has a unique solution $x = 3$, $y = 5$, whereas the set

$$2x + 3y = 14$$
$$4x + 6y = 10$$

has no solution. This is indicated by an empty set of brackets as the result.

```
[ > A := matrix ([[2,3], [4,6]]): b := matrix([14, 10]):

[ > x := linalg:: matlinsolve(A, b)

[ [ ]
```

For this set of equations the rank of **A** equals 1, which is less than the rank of the matrix [**A, b**], which is created by typing **A.b**. Thus there is no solution. These ranks can be computed as follows.

```
[ > linalg::rank(A)

[  1

[ > C := A.b:

[ > linalg::rank(C)

[  2
```

See Chapter 8 for discussion of the existence and uniqueness of solutions of linear algebraic equations.

Now consider the following set, which has the same matrix **A** but an infinite number of solutions, $x = 7 - 3y/2$.

$$2x + 3y = 14$$
$$4x + 6y = 28$$

The MuPAD session is

```
[ > A := matrix([[2,3], [4,6]]); b := matrix([14, 28]):

[ > C:= A.b:

[ > linalg::rank(A)

[  1

[ > linalg::rank(C)

[  1

[ > x := linalg::matlinsolve(A, b)
```

$$\left[\left[\begin{pmatrix} 7 \\ 0 \end{pmatrix}, \left[\begin{pmatrix} -\dfrac{3}{2} \\ 1 \end{pmatrix} \right] \right] \right]$$

Because the rank of **A** equals the rank of [**A**, **b**], but is less than the number of equations, there are an infinite number of solutions. MuPAD expresses this fact indirectly in vector form. The first vector on the solution, $(7, 0)$, says that one solution of the set is $y = 0$ and $x = 7$. The square brackets enclosing the second vector $(-3/2, 1)$ may be thought of as an eigenvector that expresses the slope of x versus y in the solution $x = 7 - 3y/2$.

Equation sets can be solved symbolically. Consider the set

$$2x + ay = 1$$
$$3x - 6y = 3$$

It can be solved as follows.

```
[ > A:=matrix([[2,a],[3,-6]])
[ > b := matrix([[1],[3]])
[ > linalg::matlinsolve(A,b)
```

$$\left[\begin{pmatrix} \dfrac{6a+12}{6a+24} \\ -\dfrac{6}{6a+24} \end{pmatrix} \right]$$

Use the function **numeric::matlinsolve** to solve a linear system numerically.

Test Your Understanding

T11.4–1 Consider three successive coordinate rotations using the same angle a. Show that the product **RRR** of the rotation matrix **R**(a) equals **R**($3a$).

T11.4–2 Find the characteristic polynomial and roots of the following matrix.

$$\mathbf{A} = \begin{bmatrix} -2 & 1 \\ -3k & -5 \end{bmatrix}$$

(Answers: $s^2 + 7s + 10 + 3k$ and $s = (-7 \pm \sqrt{9 - 12k})/2$)

T11.4–3 Use the matrix inverse and the **matlinsolve** function to solve the following set.

$$-4x + 6y = -2$$
$$7x - 4y = 23$$

(Answer: $x = 5, y = 3$)

11.5 Calculus

In Chapter 9 we discussed techniques for performing numerical differentiation and numerical integration; this section covers differentiation and integration of symbolic expressions to obtain closed-form results for the derivatives and integrals.

Differentiation

The **diff(f,x)** function is used to obtain the symbolic ordinary derivative of the *expression* **f** with respect to **x**. Although this function has the same name as the function used to compute numerical differences (see Chapter 9), MuPAD uses the symbolic form. The differential operator **D(f),** abbreviated as **f′,** computes the derivative of the univariate *function* **f**. The **D** operator is used for *functions,* not expressions. We will not use it here.

For example, the derivatives

$$\frac{d(x^n)}{dx} = nx^{n-1}$$

$$\frac{d(\ln x)}{dx} = \frac{1}{x}$$

$$\frac{d(\sin^2 x)}{dx} = 2\sin x \cos x$$

$$\frac{d(\sin y)}{dy} = \cos y$$

$$\frac{\partial[\sin(xy)]}{\partial x} = y\cos(xy)$$

are obtained with the following session.

[> **diff(x^n, x)**
[nx^{n-1}

[> **diff(ln(x),x)**
$$\frac{1}{x}$$

[> **diff(sin(x)^2,x)**
[$2\cos(x)\sin(x)$

[> **diff(sin(y),y)**
[$\cos(y)$

When there is more than one variable, the **diff** function computes the *partial* derivative. For example, if

$$f(x, y) = \sin(xy)$$

then

$$\frac{\partial f}{\partial x} = y\cos(xy)$$

The corresponding session is

[> **diff(sin(x*y),x)**
[$y\cos(xy)$

Higher derivatives can be obtained by successive application of the **diff** command. For example,

[> **diff(diff(ln(x),x),x)**

$$-\frac{1}{x^2}$$

A shorter form is

[> **diff(ln(x),x,x)**

The mixed second partial derivative may be calculated as follows:

[> **diff((x²*y³ + 6*x²*y), y, x)**
[$6xy^2 + 12x$

Max-Min Problems

The derivative can be used to find the maximum or minimum of a continuous function, say $f(x)$, over an interval $a \le x \le b$. A *local* maximum or local minimum (one that does not occur at one of the boundaries $x = a$ or $x = b$) can occur only at a *critical point*, which is a point where either $df/dx = 0$ or df/dx does not exist. If $d^2f/dx^2 > 0$, the point is a relative minimum; if $d^2f/dx^2 < 0$, the point is a relative maximum. If $d^2f/dx^2 = 0$, the point is neither a minimum nor a maximum, but is an *inflection point*. If multiple candidates exist, you must evaluate the function at each point to determine the *global* maximum and global minimum.

Topping the Green Monster

EXAMPLE 11.5–1

The "Green Monster" is a wall 37 ft high in left field at Fenway Park in Boston. The wall is 310 ft from home plate down the left field line. Assuming that the batter hits the ball 4 ft above the ground, and neglecting air resistance, determine the *minimum* speed the batter must give to the ball in order to hit it over the Green Monster. In addition, find the angle at which the ball must be hit. (see Figure 11.5–1).

■ Solution

The equations of motion for a projectile launched with a speed v_0 at an angle θ relative to the horizontal are

$$x(t) = (v_0 \cos \theta)t \qquad y(t) = -\frac{gt^2}{2} + (v_0 \sin \theta)t$$

where $x = 0$, $y = 0$ is the location of the ball when it is hit. Because we are not concerned with the time of flight in this problem, we can eliminate t and obtain an equation for y in terms of x. To do this, solve the x equation for t and substitute this into the y equation to obtain

$$y(t) = -\frac{g}{2}\frac{x^2(t)}{v_0^2 \cos^2 \theta} + x(t) \tan \theta$$

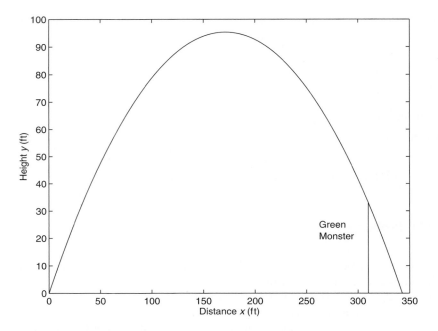

Figure 11.5–1 A baseball trajectory to clear the Green Monster.

(You could use MuPAD to do this algebra if you wish. We will use MuPAD to do the more difficult task to follow.)

Because the ball is hit 4 ft above the ground, the ball must rise $37 - 4 = 33$ ft to clear the wall. Let h represent the relative height of the wall (33 ft). Let d represent the distance to the wall (310 ft). Use $g = 32.2$ ft/sec^2. When $x = d$, $y = h$. Thus the previous equation gives

$$h = -\frac{g}{2}\frac{d^2}{v_0^2\cos^2\theta} + d\tan\theta$$

which can easily be solved for v_0^2 as follows.

$$v_0^2 = \frac{g}{2}\frac{d^2}{\cos^2\theta(d\tan\theta - h)}$$

Because $v_0 > 0$, minimizing v_0^2 is equivalent to minimizing v_0. Note also that $gd^2/2$ is a multiplicative factor in the expression for v_0^2. Thus the minimizing value of θ is independent of g and can be found by minimizing the function

$$f = \frac{1}{\cos^2\theta(d\tan\theta - h)}$$

The session to do this is as follows. The variable **th** represents the angle θ of the ball's velocity vector relative to the horizontal. The first step is to calculate the derivative $df/d\theta$ and solve the equation $df/d\theta = 0$ for θ. We know θ must be less than $\pi/2 \approx 1.57$ rad.

[> **d := 310:h:=33:g:=32.2:**

[> **f := 1/(((cos(th))^2)*(d*tan(th)-h)):**

[> **dfdth := diff(f, th):**

[> **thmin := numeric::realroot(dfdth,th = 0..1.57)**
[0.838424752

So the solution candidate is $\theta = 0.8384$ rad, or about 48°. To verify that this is a minimum solution, and not a maximum or an inflection point, we can check the second derivative $d^2f/d\theta^2$. If this derivative is positive, the solution represents a minimum. To check this and to find the speed required, continue the session as follows.

[> **second := diff(diff(f,th),th):**

[> **subs(second, th = thmin)**

[> **float(%)**
[0.03209697518

So the second derivative is positive, and the solution represents the minimum. To find the speed required, continue the session as follows.

[> **v2:=(g*d^2/2)*f**

[> **v2min := subs(v2, th = thmin):**

[> **vmin :=sqrt(v2min)**
[105.3612757

Thus the minimum speed required is about 105 ft/sec, or about 72 mph. A ball hit with this speed will clear the wall only if it is hit at an angle of approximately 48°.

Test Your Understanding

T11.5–1 Given that $y = \sinh(3x)\cosh(5x)$, use MuPAD to find dy/dx at $x = 0.2$. (Answer: 9.2288)

T11.5–2 Given that $z = 5\cos(2x)\ln(4y)$, use MuPAD to find $\partial z/\partial y$. (Answer: $5\cos(2x)/y$)

Integration

The **int(f,x)** function is used to integrate a symbolic *expression* **f** with respect to **x**. It is possible that the integral does not exist in closed form, or that MuPAD cannot find the integral even if it exists. For example, you can obtain the following integrals with the session shown below.

$$\int \frac{1}{x}\, dx = \ln x$$

$$\int \cos x\, dx = \sin x$$

$$\int \sin y \, dy = -\cos y$$

$$\int x^n \, dx = \frac{x^{n+1}}{n+1}$$

[> int(1/x,x)
[$\ln(x)$

[> int(cos(x),x)
[$\sin(x)$

[> int(sin(y),y)
[$-\cos(y)$

[> assume(n <>-1):int(x^n,x)
[$\dfrac{x^{n+1}}{n+1}$

If we had not forced MuPAD into assuming that $n \neq -1$, MuPAD would have given us a second solution, ln *x*, as well.

Definite integrals can be computed by giving the limits after the integrand. For example, the integral

$$\int_0^{10} x^3 \, dx$$

is computed as follows:

[> int(x^3, x=0..10)
[2500

The following session gives an example for which no integral can be found. The indefinite integral exists, but the definite integral does not exist if the limits of integration include the singularity at $x = 1$. The integral is

$$\int \frac{1}{x-1} \, dx = \ln|x - 1|$$

The session is

[> int(1/(x-1),x)
[$\ln(x - 1)$

[> int(1/(x-1),x=0..2
[*undefined*

Test Your Understanding

T11.5–3 Given that $y = x \sin(3x)$, use MuPAD to find $\int y \, dx$. (Answer: $(\sin(3x) - 3x \cos(3x))/9$)

T11.5–4 Given that $z = 6y^2 \tan(8x)$, use MuPAD to find $\int z \, dy$. (Answer: $2y^3 \tan(8x)$)

T11.5–5 Use MuPAD to evaluate

$$\int_{-2}^{5} x \sin(3x)\, dx$$

(Answer: 0.6672)

Taylor Series

The **taylor(f,x=a,n)** function gives the first $n - 1$ terms in the Taylor series for the function defined in the expression f, evaluated at the point $x = a$. The parameter n is optional. If the parameter **a** is omitted, the function returns the series evaluated at $x = 0$. Some common examples of the Taylor series are

$$e^x = 1 + x + \frac{x^2}{2!} + \frac{x^3}{3!} + \frac{x^4}{4!} + \cdots \qquad -\infty < x < \infty$$

$$\sin x = x - \frac{x^3}{3!} + \frac{x^5}{5!} - \frac{x^7}{7!} + \cdots \qquad -\infty < x < \infty$$

where $a = 0$ in both examples.

Here are some examples.

[> **taylor(exp(x), x)**

$$\left[\; 1 + x + \frac{x^2}{2} + \frac{x^3}{6} + \frac{x^4}{24} + \frac{x^5}{120} + O(x^6) \right.$$

[> **taylor(exp(x), x, 8)**

$$\left[\; 1 + x + \frac{x^2}{2} + \frac{x^3}{6} + \frac{x^4}{24} + \frac{x^5}{120} + \frac{x^6}{720} + \frac{x^7}{5040} + O(x^8) \right.$$

[> **taylor(sin(x),x,6)**

$$\left[\; x - \frac{x^3}{6} + \frac{x^5}{120} + O(x^7) \right.$$

A linear approximation to e^x about the point $x = 2$ is found as follows.

[> **taylor(exp(x),x=2,2)**
[$e^2 + e^2(x - 2) + O((x - 2)^2)$

The latter expression corresponds to

$$e^x \approx e^2 [1 + (x - 2)] \qquad \text{for } x \approx 2$$

Sums

The **sum(f,k = a..b)** function returns the sum of the expression **f** as the symbolic variable varies from **a** to **b**. Here are some examples. The summations

$$\sum_{k=0}^{10} k = 0 + 1 + 2 + 3 + \cdots + 9 + 10 = 55$$

$$\sum_{k=0}^{n-1} k = 0 + 1 + 2 + 3 + \cdots + n - 1 = \frac{n(n-1)}{2}$$

$$\sum_{k=1}^{4} k^2 = 1 + 4 + 9 + 16 = 30$$

are given by

[> **sum(k,k = 0..10)**
[55
[> **sum(k, k = 0..n-1)**
$$\left[\frac{n(n-1)}{2} \right.$$
[> **sum(k², k = 1..4)**
[30

Limits

The function **limit(f,x = a)** returns the limit

$$\lim_{x \to a} f(x)$$

For example, the limits

$$\lim_{x \to 3} \frac{x-3}{x^2-9} = \frac{1}{6}$$

$$\lim_{x \to 0} \frac{\sin(x+h) - \sin(x)}{h} = \cos x$$

are found from

[> **limit((x-3)/(x^2-9), x=3)**
$$\left[\frac{1}{6} \right.$$

[> **limit((sin(x + h)-sin(x))/h,h=0)**
[cos(x)

The forms **limit(f, x = a, Left)** and **limit(f, x = a, Right)** specify the direction of the limit. For example,

$$\lim_{x \to 0-} \frac{1}{x} = -\infty$$

$$\lim_{x \to 0+} \frac{1}{x} = \infty$$

are given by

[> **limit(1/x, x=0, Left)**
[$-\infty$
[> **limit(1/x, x=0, Right)**
[∞

Test Your Understanding

T11.5–6 Use MuPAD to find the first three nonzero terms in the Taylor series for $\cos x$.

(Answer: $1 - x^2/2 + x^4/24$)

T11.5–7 Use MuPAD to find a formula for the sum

$$\sum_{m=0}^{m-1} m^3$$

(Answer: $m^4/4 - m^3/2 + m^2/4$)

T11.5–8 Use MuPAD to evaluate

$$\sum_{n=0}^{7} \cos(\pi n)$$

(Answer: 0)

T11.5–9 Use MuPAD to evaluate

$$\lim_{x \to 5} \frac{2x - 10}{x^3 - 125}$$

(Answer: $2/75$)

11.6 Ordinary Differential Equations

Methods for obtaining a numerical solution to differentiation, integration, and differential equation problems were covered in Chapter 9. However, we prefer to obtain an analytical solution whenever possible, because it is more general, and thus more useful for designing engineering devices or processes.

MuPAD provides the **ode** and **solve** functions for solving ordinary differential equations in closed form. These functions are different from the numerical ODE solvers in MATLAB (such as `ode45` and the symbolic ODE solver `dsolve` in the Symbolic Math toolbox).

The syntax of the MuPAD **ode** and **solve** functions differs somewhat according to whether they are used to solve single equations or sets of equations, and whether boundary conditions are specified. MuPAD also has the **numeric::solve** function for solving differential equations numerically, but we will not cover this topic because the MATLAB `ode` solvers are sufficient for our purposes.

Solving a Single Differential Equation

The MuPAD **ode** function is used to describe the equation and its boundary conditions, if any, but it does not solve the equation. The **solve** function is then used to solve the problem. The arbitrary constants in the solution are denoted by $C1$, $C2$, and so on. The number of such constants is the same as the order of the equation. Derivatives are denoted by primes regardless of the independent variable. For example,

$$y' = \frac{dy}{dx} = \frac{dy}{dt} \qquad y'' = \frac{d^2y}{dx^2} = \frac{d^2y}{dt^2}$$

and so on.

For example, the equation

$$\frac{dy}{dt} + 2y = 12 \tag{11.6--1}$$

has the solution

$$y(t) = 6 + C_1 e^{-2t}$$

The solution can be found with the following session. We give the equation the arbitrary name **eqn** and tell MuPAD that the independent variable is t and the dependent variable is $y(t)$. If the colon is not used after the command, MuPAD will display the equation.

```
[ > eqn := ode(y'(t)+2*y(t)=12, y(t)):
[ > solve (eqn)
```

$$\left[\; \left\{ \frac{C1}{e^{2t}} + 6 \right\} \right.$$

which is the correct solution, although it is expressed in a less than conventional form.

Suppose the initial condition is $y(0) = 5$. We find the solution as follows.

```
[ > eqn := ode ({y'(t)+2*y(t)=12, y(0)=5}, y(t)):
[ > solve(eqn)
```

$$\left[\; \left\{ 6 - \frac{1}{e^{2t}} \right\} \right.$$

Note how the braces are used to group the equation and the initial condition.

You need not use t as the independent variable or y as the dependent variable. We could just as easily express Eq. (11.6–1) as

$$\frac{dw}{dx} + 2w = 12$$

and the MuPAD session would be

[> **eqn := ode(w'(x)+2*w(x)=12, w(x)):**
[> **solve(eqn)**

$$\left[\left\{ \frac{C1}{e^{2x}} + 6 \right\} \right.$$

There can be symbolic constants in the equation and boundary conditions, although the more complicated the equation, the more complicated the solution will be, perhaps to the point of not being useful. Here is a second-order example.

$$\frac{d^2y}{dt^2} = c^2y$$

The session is

[> **eqn := ode(y''(t)=c^2*y(t), y(t)):**
[> **solve(eqn)**

$$\left[\left\{ \frac{C1}{e^{ct}} + C2e^{ct} \right\} \right.$$

The solution is $y(t) = C_1 e^{-ct} + C_2 e^{ct}$.

With the initial conditions $y(0) = 1$, $y'(0) = 0$, the session is

[> **eqn := ode({y''(t)=c^2*y(t), y(0)=1, y'(0)=0}, y(t)):**
[> **solve(eqn)**

The solution given by MuPAD is $y(t) = 1$ if $c = 0$ and

$$y(t) = \frac{1}{2}e^{ct} + \frac{1}{2}e^{-ct}$$

if $c \neq 0$.

Here is an example having a forcing function.

$$\frac{dy}{dt} = \sin bt \qquad y(0) = 0$$

The session is

[> **eqn := ode({y'(t)=sin(b*t), y(0)=0}, y(t)):**
[> **solve(eqn)**

$$\left[\left\{ \frac{1}{b} - \frac{\cos(bt)}{b} \right\} \right.$$

The solution is

$$y(t) = \frac{1}{b}(1 - \cos bt)$$

General boundary conditions may also be specified. For example, consider the problem

$$\frac{d^2y}{dt^2} + 9y = 0 \qquad y(0) = 1, \qquad \dot{y}(\pi) = 2$$

The session to find the solution and plot it is

[> eqn := ode({y''(t) + 9*y(t)=0, y(0)=1, y'(PI)=2}, y(t)):
[> solve(eqn)

$$\left[\quad \left\{ \cos(3t) - \frac{2\sin(3t)}{3} \right\} \right.$$

[> plot(float(%))

So the solution is

$$y(t) = \cos 3t - \frac{2}{3}\sin 3t$$

If a plot of the solution is required, then it is better to make use of the powerful MATLAB `plot` function.

Higher-order equations can also be solved. Consider, for example,

$$\frac{d^4y}{dt^4} - y = 0$$

with the initial conditions $y(0) = 1$ and

$$\left.\frac{dy}{dt}\right|_{t=0} = \left.\frac{d^2y}{dt^2}\right|_{t=0} = \left.\frac{d^3y}{dt^3}\right|_{t=0} = 0$$

The session is

[> eqn := ode({y''''(t)-y(t)=0, y(0)=1, y'(0)=0, y''(0)=0, y'''(0)=0}, y(t)):
[> solve(eqn)

$$\left[\quad \left\{ \frac{1}{4e^t} + \frac{\cos(t)}{2} + \frac{e^t}{4} \right\} \right.$$

So the solution is

$$y(t) = \frac{1}{4}e^{-t} + \frac{1}{2}\cos t + \frac{1}{4}e^t$$

Solving Sets of Equations

Sets of equations can be solved with **solve**. Use a set of square brackets to denote the set of equations and another set of brackets to denote the dependent variables. For example, consider the set

$$\frac{dx}{dt} = 3x + 4y$$

$$\frac{dy}{dt} = -4x + 3y$$

The MuPAD session is

[> eqn := ode([x′(t)=3*x(t)+4*y(t), y′(t)= −4*x(t)+3*y(t)], [x(t),y(t)]):
[> solve(eqn)
[{[y(t) = −C1 cos(4t)e^{3t} − C2 sin(4t)e^{3t}, x(t) = C2 cos(4t)e^{3t} − C1 sin(4t)e^{3t}]}

Thus the solution is

$$y(t) = -C_1 e^{3t}\cos 4t - C_2 e^{3t}\sin 4t$$
$$x(t) = C_2 e^{3t}\cos 4t - C_1 e^{3t}\sin 4t$$

Note that MuPAD first displays the solution for the second variable $y(t)$.

Sets of equations with specified boundary conditions can be solved as follows. For example, consider the equation set just given, with the initial conditions $x(0) = 0$, $y(0) = 1$. The session is

[> eqn:=ode({x′(t)=3*x(t)+4*y(t),y′(t)=−4*x(t)+3*y(t),x(0)=0, y(0)=1} [x(t),y(t)]):
[{y(t) = cos(4t)e^{3t}, x(t) = sin(4t)e^{3t}}

Note how when boundary conditions are specified, braces are used instead of brackets to group the equations and the boundary conditions.

Solving Nonlinear Equations

MuPAD can solve some nonlinear differential equations. For example, the problem

$$\frac{dy}{dt} = y^2 \qquad y(0) = 1 \qquad\qquad (11.6\text{--}2)$$

can be solved with the following session.

[> eqn:= ode({y′(t)=(y(t))², y(0)=1}, y(t))):
[> solve(eqn)
$$\left[\left\{ -\frac{1}{t-1} \right\} \right.$$

So the solution is

$$y(t) = \frac{1}{1 - t}$$

Not all nonlinear equations can be solved in closed form. An example is the following equation, which is the equation of motion of a pendulum: $L\theta'' + g \sin\theta = 0$. If you try to solve this equation in MuPAD, the result is expressed in terms of integrals that would need to be evaluated numerically.

Test Your Understanding

T11.6–1 Use MuPAD to solve the problem

$$\frac{d^2y}{dt^2} + ay = 0 \qquad y(0) = 1 \qquad \dot{y}(0) = 0$$

for $a > 0$. Check the answer by hand or with MuPAD. (Answer: $y(t) = \cos(\sqrt{at})$)

11.7 Laplace Transforms

The Laplace transform $\mathcal{L}[y(t)]$ of a function $y(t)$ is defined to be

$$\mathcal{L}[y(t)] = \int_0^\infty y(t)e^{-st}\,dt = Y(s) \tag{11.7–1}$$

and can be obtained in MuPAD by typing **transform::laplace(y,t,s),** where **y** is the function of **t**. The result is a function of **s**. The **laplace** command is in the Transform library, which also contains the Fourier and z transforms.

Here is a MuPAD session with some examples. The functions are t^3, e^{-bt}, and $\sin bt$.

[> transform::laplace(t^3, t, s)

$$\frac{6}{s^4}$$

[> transform::laplace(exp(−b*t), t, s)

$$\frac{1}{b + s}$$

[> transform::laplace(sin(b*t), t, s)

$$\frac{b}{b^2 + s^2}$$

The *inverse Laplace transform* $\mathcal{L}^{-1}[Y(s)]$ is that time function $y(t)$ whose transform is $Y(s)$, that is, $y(t) = \mathcal{L}^{-1}[Y(s)]$. Inverse transforms can be found using the function **transform::invlaplace(Y, s, t)**. Note the reversed order of s and t. For example,

[> **transform::invlaplace(6/s^4, s, t)**
[t^3

[> **transform::invlaplace(1/(s+b), s, t)**

$$\left[\frac{1}{e^{bt}} \right.$$

Depending on what computations have preceded the command in the notebook, you may get different, but still correct, answers. For example, sometimes, but not always, you will get an unreduced answer like the following.

[> **transform::invlaplace(b/(s^2+b^2), s, t)**
[$\sin(\sqrt{b^2}\, t)$

This occurs because MuPAD considers the possibility that b may be negative. Note that MuPAD does not reduce $\sqrt{b^2}$ to b. If in fact $b > 0$, you can avoid this by typing

[> **assume(b>0):**
[> **transform::invlaplace(b/(s^2+b^2), s, t)**
[$\sin(bt)$

Because the transform is an integral, its has the properties of integrals. In particular, it has the *linearity property*, which states that if a and b are not functions of t, then

$$\mathcal{L}[af_1(t) + bf_2(t)] = a\mathcal{L}[f_1(t)] + b\mathcal{L}[f_2(t)] \tag{11.7-2}$$

The transforms of derivatives are useful for solving differential equations. Applying integration by parts to the definition of the transform, we obtain

$$\mathcal{L}\left(\frac{dy}{dt}\right) = \int_0^\infty \frac{dy}{dt} e^{-st}\, dt = y(t)e^{-st}\Big|_0^\infty + s\int_0^\infty y(t)e^{-st}\, dt$$

$$= s\mathcal{L}[y(t)] - y(0) = sY(s) - y(0) \tag{11.7-3}$$

This procedure can be extended to higher derivatives. For example, the result for the second derivative is

$$\mathcal{L}\left(\frac{d^2y}{dt^2}\right) = s^2Y(s) - sy(0) - \dot{y}(0) \tag{11.7-4}$$

The general result for any order derivative is

$$\mathcal{L}\left(\frac{d^ny}{dt^n}\right) = s^nY(s) - \sum_{k=1}^{n} s^{n-k}g_{k-1} \tag{11.7-5}$$

where

$$g_{k-1} = \frac{d^{k-1}y}{dt^{k-1}}\Big|_{t=0} \tag{11.7-6}$$

Application to Differential Equations

The derivative and linearity properties can be used to solve equations such as

$$a\dot{y} + y = bv(t) \tag{11.7-7}$$

Application of the transform gives

$$Y(s) = \frac{ay(0)}{as + 1} + \frac{b}{as + 1}V(s) \tag{11.7-8}$$

The free response is given by

$$\mathcal{L}^{-1}\left[\frac{ay(0)}{as + 1}\right] = \mathcal{L}^{-1}\left[\frac{y(0)}{s + 1/a}\right] = y(0)e^{-t/a}$$

The forced response is given by

$$\mathcal{L}^{-1}\left[\frac{b}{as + 1}V(s)\right] \tag{11.7-9}$$

This cannot be evaluated until $V(s)$ is specified.

Suppose $v(t)$ is a unit-step function, which is also called the *Heaviside function*. In MuPAD this is called by the command **heaviside(t)**. For example,

[> transform::laplace(heaviside(t), t, s)

$$\left[\begin{array}{l} \frac{1}{s} \end{array}\right.$$

Thus the transform of the unit-step function is $1/s$, and so $V(s) = 1/s$. Then Equation (11.7–9) becomes

$$\mathcal{L}^{-1}\left[\frac{b}{s(as + 1)}\right]$$

To find the inverse transform, which is $y(t)$, enter

[> transform::invlaplace(b/(s*(a*s+1)), s, t)

$$\left[\begin{array}{l} b - \dfrac{b}{e^{\frac{t}{a}}} \end{array}\right.$$

which can be expressed as $b(1 - e^{-t/a})$. This is the forced response to a unit-step input.

Consider the second-order model

$$\ddot{x} + 2\dot{x} + x = f(t) \tag{11.7-10}$$

Transforming this equation gives

$$[s^2X(s) - sx(0) - \dot{x}(0)] + 2[sX(s) - x(0)] + X(s) = F(s)$$

Solve for $X(s)$.

$$X(s) = \frac{x(0)s + \dot{x}(0) + 2x(0)}{s^2 + 2s + 1} + \frac{F(s)}{s^2 + 2s + 1}$$

The free response is obtained from

$$x(t) = \mathcal{L}^{-1}\left[\frac{x(0)s + \dot{x}(0) + 2x(0)}{s^2 + 2s + 1}\right]$$

Suppose the initial conditions are $x(0) = 2$ and $\dot{x}(0) = 3$. Then the free response is obtained from

$$x(t) = \mathcal{L}^{-1}\left[\frac{2s + 7}{s^2 + 2s + 1}\right] \qquad (11.7\text{–}11)$$

It can be found in MuPAD by typing

[> **transform::invlaplace((2*s+7)/(s^2+2*s+1), s, t)**

$$\left[\frac{2}{e^t} + \frac{5t}{e^t}\right.$$

So the free response is

$$x(t) = 2e^{-t} + 5te^{-t}$$

The forced response is obtained from

$$x(t) = \mathcal{L}^{-1}\left[\frac{F(s)}{s^2 + 2s + 1}\right]$$

If $f(t)$ is a unit-step function, $F(s) = 1/s$, and the forced response is

$$x(t) = \mathcal{L}^{-1}\left[\frac{1}{s(s^2 + 2s + 1)}\right]$$

To find the forced response, enter

[> **transform::invlaplace(1/(s*(s^2+2*s+1)),s,t)**

$$\left[1 - \frac{t}{e^t} - \frac{1}{e^t}\right.$$

So the forced response is

$$x(t) = 1 - te^{-t} - e^{-t} \qquad (11.7\text{–}12)$$

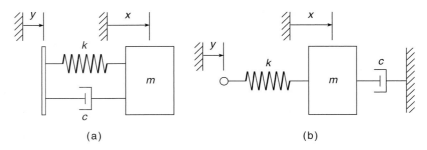

Figure 11.7–1 Two mechanical systems. The model for (a) contains the derivative of the input $y(t)$; the model for (b) does not.

Input Derivatives

Two similar mechanical systems are shown in Figure 11.7–1. In both cases the input is a displacement $y(t)$. Their equations of motion are

$$m\ddot{x} + c\dot{x} + kx = ky + c\dot{y} \qquad (11.7\text{–}13)$$

$$m\ddot{x} + c\dot{x} + kx = ky \qquad (11.7\text{–}14)$$

The only difference between these systems is that the system in Figure 11.7–1a has an equation of motion containing the derivative of the input function $y(t)$. Both systems are examples of the more general differential equation

$$m\ddot{x} + c\dot{x} + kx = dy + g\dot{y} \qquad (11.7\text{–}15)$$

We now demonstrate how to use the Laplace transform to find the step response of equations containing derivatives of the input. Suppose the initial conditions are zero. Then transforming Equation (11.7–15) gives

$$X(s) = \frac{d + gs}{ms^2 + cs + k} Y(s) \qquad (11.7\text{–}16)$$

Let us compare the unit-step response of Equation (11.7–16) for two cases using the values $m = 1$, $c = 2$, and $k = d = 1$, with zero initial conditions. The two cases are $g = 0$ and $g = 5$.

With $Y(s) = 1/s$, Equation (11.7–16) gives

$$X(s) = \frac{1 + gs}{s(s^2 + 2s + 1)} \qquad (11.7\text{–}17)$$

The response for the case $g = 0$ was found earlier in Equation (11.7–14). The response for $g = 5$ is found by typing

[> **transform::invlaplace((1+5s*)/(s*(s²+2*s+1)), s, t)**

$$\left[\frac{4t}{e^t} - \frac{1}{e^t} + 1 \right.$$

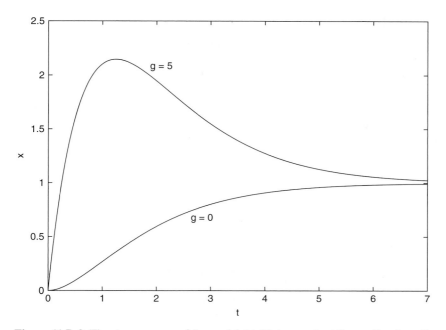

Figure 11.7–2 The step response of the model $\ddot{x} + 2\dot{x} + x = u + g\dot{u}$ for $g = 0$ and $g = 5$.

or

$$x(t) = 4te^{-t} - e^{-t} + 1$$

The two response are plotted in Figure 11.7–2. The effect of differentiating the input is an increase in the response's peak value.

Partial fraction expansions are useful when dealing with inverse Laplace transforms and for other applications. The **partfrac(f,x)** function creates the partial fraction expansion of the expression **f(s)** in terms of its variable **s**.

[> **partfrac((6*s+5)/(s^3+14*s^2+59*s+70),s)**

$$\left[\frac{25}{6(s+5)} - \frac{7}{15(s+2)} - \frac{37}{10(s+7)} \right.$$

Test Your Understanding

T11.7–1 Find the Laplace transform of the functions $1 - e^{-at}$ and $\cos bt$. Use the **ilaplace** function to check your answers.

T11.7–2 Use the Laplace transform to solve the problem $5\ddot{y} + 20\dot{y} + 15y = 30u - 4\dot{u}$, where $u(t)$ is a unit-step function and $y(0) = 5$, $\dot{y}(0) = 1$. (Answer: $y(t) = -1.6e^{-3t} + 4.6e^{-t} + 2$)

11.8 Special Functions

Differential equations that do not have closed-form solutions can often be solved in terms of *special functions*. Many such ordinary differential equations arise in the solution of partial differential equations. Examples include the Chebyshev polynomials of the first kind $T(n, x)$, the Hermite polynomials $H_n(x)$, and several types of Bessel functions. Other examples are the Airy functions $Ai(x)$ and $Bi(x)$, which are the two independent solutions to Airy's equation $y'' - xy = 0$. Table 11.8–1 lists some of the special functions available in MuPAD.

Consider a specific case of Legendre's equation $(1 - x^2)y'' - 2xy' + 6y = 0$ with the initial conditions $y(0) = 1$, $y'(0) = 0$. It can be solved in MuPAD with the following session:

[> **eqn1:= ode({(1-x^2)*y″ (x)-2*x*y′ (x)+6*y(x)=0, y(0)=1, y′(0)=0}, y(x)):**
[> **solve(eqn1)**

[$\{1 - 3x^2\}$

The result is a finite polynomial of degree 2.

Airy's equation $y'' - xy = 0$ with the same initial conditions can be solved with the following session:

[> **eqn2:= ode({y″ (x)-x*y(x)=0, y(0)=1, y′(0) = 0}, y(x)):**

[> **solve(eqn2)**

$$\left[\left\{ \frac{3^{2/3}\Gamma\left(\frac{2}{3}\right)\text{airyAi}(x,0)}{2} + \frac{3^{1/6}\Gamma\left(\frac{2}{3}\right)\text{airyBi}(x,0)}{2} \right\} \right.$$

where $\Gamma(x)$ is another special function called the *gamma function*. The gamma function is evaluated by typing **gamma(x)**. Table 11.8–2 shows that the **float** function is used to evaluate special functions numerically. For example,

[> **float(gamma(2/3))**
[1.354117939
[> **float(airy Ai(2,0))**
[0.03492413042

Table 11.8–1 Special function calls in MuPAD

Name and Symbol	Function Call
Airy, $Ai(x)$	**airy Ai(x)**
Airy, $Bi(x)$	**airy Bi(x)**
Chebyshev of first kind, $T(n, x)$	**chebyshev1 (n, x)**
Gamma, $\Gamma(x)$	**gamma(x)**
Hermite, $H_n(x)$	**hermite (n,x)**
Bessel I, $I_n(x)$	**besselI(n,x)**
Bessel J, $J_n(x)$	**besselJ(n,x)**
Bessel K, $K_n(x)$	**besselK(n,x)**
Bessel Y, $Y_n(x)$	**besselY (n,x)**
Laguerre, $L(n, a, x)$	**laguerreL(n,a,x)**
Legendre, $P_n(x)$	**legendre(n,x)**

Table 11.8–2 Evaluation of special functions in MuPAD

Result	Code
Symbolic finite series	**orthpoly::**
Symbolic infinite series	**series**
Numeric result	**float**

If an explicit series solution is preferred, you can use the **series** option as follows. Using the particular Legendre's equation $(1 - x^2)y'' - 2xy' + 12y = 0$ as an example, solving it with the **solve** function, even with simple initial conditions such as $y(0) = 1$ and $y'(0) = 0$, results in a complicated expression that is too detailed to display here (try it). However, even with arbitrary initial conditions, the series solution is not very complicated. The session is

[> eqn3 := (1-x^2)*y"(x)-2*y'(x)+12*y(x)=0:

[> ode::series({eqn3, y(0)=a, y'0)=b},y(x), x=0)

$$\left[\left\{ a + bx - 6ax^2 - \frac{5b}{3}x^3 + 3ax^4 + O(x^6) \right\} \right.$$

To evaluate a special function symbolically as a *finite series*, use either the **orthpoly::** option or the **series** function. The **orthpoly** package provides some standard orthogonal polynomials. Call the package functions by using the package name **orthpoly** and the name of the function. For example, the fifth-order Legendre polynomial is obtained as follows:

[> orthpoly::legendre(5, x)

$$\left[\left\{ \mathrm{poly}\left(\frac{63x^5}{8} - \frac{35x^3}{4} + \frac{15x}{8}, [x] \right) \right\} \right.$$

[> orthpoly::legendre(5,2)

$$\left[\frac{743}{4} \right.$$

[> float (orthpoly::legendre(5,2))

[185.75

To evaluate symbolically a special function that consists of an *infinite series*, use the **series** function. For example,

[> series(besselJ(0,x), x)

$$\left[1 - \frac{x^2}{4} + \frac{x^4}{64} + O(x^6) \right.$$

You can plot special functions just as any other functions. For example,

[> plot(airyAi(x),x=0..5)

You can plot the solution of a differential equation by following this example based on Airy's equation.

```
[ > eqn4 := ode({y''(x)=x*y(x), y(0)=1, y'(0)=-0.1}, y(x)):
[ > solve(eqn4):
[ > y4 := op(%):
[ > plotfunc2d(y4, x=0..1)
```

The **op** function removes the braces surrounding the solution obtained by the **solve** function.

Test Your Understanding

T11.8–1 Solve Legendre's equation for the given initial conditions.
$$(1 - x^2)y'' - 2xy' + 6y = 0 \qquad y(0) = 12 \qquad y'(0) = 0$$
(Answer: $y(x) = 12 - 36x^2$)

T11.8–2 Evaluate the solution to Airy's equation at $x = 3$ for the given initial conditions.
$$y'' - xy = 0 \qquad y(0) = 1 \qquad y'(0) = 5$$
(Answer: $y(3) = 89.6423632$)

11.9 Summary

This chapter covers a subset of the capabilities of MuPAD. Now that you have finished this chapter, you should be able to use MuPAD to do the following:

- Create symbolic expressions and manipulate them algebraically.
- Obtain symbolic solutions to algebraic and transcendental equations.
- Perform symbolic differentiation and integration.
- Evaluate limits and series symbolically.
- Obtain symbolic solutions to ordinary differential equations.
- Obtain and apply Laplace transforms.
- Perform symbolic linear algebra operations, including obtaining expressions for determinants, matrix inverses, eigenvectors, and eigenvalues.
- Evaluate the special functions of mathematics.

Key Terms with Page References

Assignment operator, 472
Input region, 468
MuPAD Library, 480

Output region, 468
Reserved symbols, 470
Text region, 468

Problems

You can find the answers to problems marked with an asterisk at the end of the text.

Section 11.2

1. Use MuPAD to prove the following identities:
    a. $\sin^2 x + \cos^2 x = 1$
    b. $\sin(x + y) = \sin x \cos y + \cos x \sin y$
    c. $\sin 2x = 2 \sin x \cos x$
    d. $\cosh^2 x - \sinh^2 x = 1$

2.* Two polynomials in the variable x are represented by the coefficient vectors `p1 = [6,2,7,-3]` and `p2 = [10,-5,8]`.
    a. Use MuPAD to find the product of these two polynomials; express the product in its simplest form.
    b. Use MuPAD to find the numeric value of the product if $x = 2$.

3.* The equation of a circle of radius r centered at $x = 0$, $y = 0$ is

    $$x^2 + y^2 = r^2$$

    Use MuPAD functions to find the equation of a circle of radius r centered at the point $x = a$, $y = b$. Rearrange the equation into the form $Ax^2 + Bx + Cxy + Dy + Ey^2 = F$ and find the expressions for the coefficients in terms of a, b, and r.

Section 11.3

4.* The law of cosines for a triangle states that $a^2 = b^2 + c^2 - 2bc \cos A$, where a is the length of the side opposite the angle A, and b and c are the lengths of the other sides.
    a. Use MuPAD to solve for b.
    b. Suppose that $A = 60°$, $a = 5$ m, and $c = 2$ m. Determine b.

5. Use MuPAD to solve the polynomial equation $x^3 + 8x^2 + ax + 10 = 0$ for x in terms of the parameter a, and evaluate your solution for the case $a = 17$. Use MuPAD to check the answer.

6.* The equation for an ellipse centered at the origin of the Cartesian coordinates (x, y) is

    $$\frac{x^2}{a^2} + \frac{y^2}{b^2} = 1$$

    where a and b are constants that determine the shape of the ellipse.
    a. In terms of the parameter b, use MuPAD to find the points of intersection of the two ellipses described by

    $$x^2 + \frac{y^2}{b^2} = 1$$

and

$$\frac{x^2}{100} + 4y^2 = 1$$

b. Evaluate the solution obtained in part *a* for the case $b = 2$.

7. The equation

$$r = \frac{p}{1 - \epsilon \cos \theta}$$

describes the polar coordinates of an orbit with the coordinate origin at the sun. If $\epsilon = 0$, the orbit is circular; if $0 < \epsilon < 1$, the orbit is elliptical. The planets have orbits that are nearly circular; comets have orbits that are highly elongated with ϵ nearer to 1. It is of obvious interest to determine whether a comet's or an asteroid's orbit will intersect that of a planet. For each of the following two cases, use MuPAD to determine whether orbits A and B intersect. If they do, determine the polar coordinates of the intersection point. The units of distance are AU, where 1 AU is the mean distance of the Earth from the sun.
a. Orbit A: $p = 1$, $\epsilon = 0.01$. Orbit B: $p = 0.1$, $\epsilon = 0.9$.
b. Orbit A: $p = 1$, $\epsilon = 0.01$. Orbit B: $p = 1.1$, $\epsilon = 0.5$.

Section 11.4

8. Show that $\mathbf{R}^{-1}(a)\mathbf{R}(a) = \mathbf{I}$, where $\mathbf{I}$ is the identity matrix and $\mathbf{R}(a)$ is the rotation matrix. This equation shows that the inverse coordinate transformation returns you to the original coordinate system.

9. Show that $\mathbf{R}^{-1}(a) = \mathbf{R}(-a)$. This equation shows that a rotation through a negative angle is equivalent to an inverse transformation.

10.* Find the characteristic polynomial and roots of the following matrix:

$$\mathbf{A} = \begin{bmatrix} -6 & 2 \\ 3k & -7 \end{bmatrix}$$

11.* Use the matrix inverse and the matrix division method to solve the following set for x and y in terms of c:

$$4cx + 5y = 43$$
$$3x - 4y = -22$$

12. The currents i_1, i_2, and i_3 in the circuit shown in Figure P12 are described by the following equation set if all the resistances are equal to R.

$$\begin{bmatrix} 2R & -R & 0 \\ -R & 3R & -R \\ 0 & R & -2R \end{bmatrix} \begin{bmatrix} i_1 \\ i_2 \\ i_3 \end{bmatrix} = \begin{bmatrix} v_1 \\ 0 \\ v_2 \end{bmatrix}$$

Here v_1 and v_2 are applied voltages; the other two currents can be found from $i_4 = i_1 - i_2$ and $i_5 = i_2 - i_3$.

a. Use both the matrix inverse method and the matrix division method to solve for the currents in terms of the resistance R and the voltages v_1 and v_2.

b. Find the numerical values for the currents if $R = 1000\ \Omega$, $v_1 = 100$ V, and $v_2 = 25$ V.

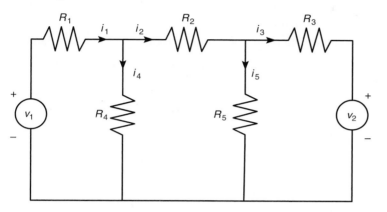

Figure P12

13. The equations for the armature-controlled dc motor shown in Figure P13 follow. The motor's current is i, and its rotational velocity is ω.

$$L\frac{di}{dt} = -Ri - K_e\omega + v(t)$$

$$I\frac{d\omega}{dt} = K_Ti - c\omega$$

where L, R, and I are the motor's inductance, resistance, and inertia; K_T and K_e are the torque constant and back-emf constant; c is a viscous damping constant; and $v(t)$ is the applied voltage.

a. Find the characteristic polynomial.

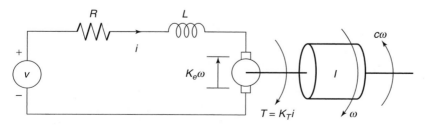

Figure P13

b. Use the values $R = 0.8\ \Omega$, $L = 0.003$ H, $K_T = 0.05$ N $\cdot$ m/A, $K_e = 0.05$ V $\cdot$ s/rad, and $I = 8 \times 10^{-5}$ kg $\cdot$ m^2. The damping constant c is often difficult to determine with accuracy. For these values find the expressions for the two characteristic roots in terms of c.

c. Using the parameter values in part b, determine the roots for the following values of c (in N $\cdot$ m $\cdot$ s): $c = 0$, $c = 0.01$, $c = 0.1$, and $c = 0.2$. For each case, use the roots to estimate how long the motor's speed will take to become constant; also discuss whether the speed will oscillate before it becomes constant.

14. Solve the following recurrence relation for the given initial conditions.
$$y(n + 2) - 0.3y(n + 1) + 0.02y(n) = 10 \qquad y(0) = 2 \qquad y(1) = 0$$

15. Solve the following optimization problem. Minimize
$$J = x + 3y + 2z$$

subject to the constraints
$$3x + 3y - 3z \le 20$$
$$6x - 4y - 3z = 20$$
$$7x + 4y + 11z \le 50$$

Section 11.5

16. Use MuPAD to find all the values of x where the graph of $y = 3^x - 2x$ has a horizontal tangent line.

17.* Use MuPAD to determine all the local minima and local maxima and all the inflection points where $dy/dx = 0$ of the following function:
$$y = x^4 - \tfrac{16}{3}x^3 + 8x^2 - 4$$

18. The surface area of a sphere of radius r is $S = 4\pi r^2$. Its volume is $V = 4\pi r^3/3$.
a. Use MuPAD to find the expression for dS/dV.
b. A spherical balloon expands as air is pumped into it. What is the rate of increase in the balloon's surface area with volume when its volume is 30 in.^{3}?

19. Use MuPAD to find the point on the line $y = 2 - x/3$ that is closest to the point $x = -3$, $y = 1$.

20. A particular circle is centered at the origin and has a radius of 5. Use MuPAD to find the equation of the line that is tangent to the circle at the point $x = 3$, $y = 4$.

21. Ship A is traveling north at 6 mi/hr, and ship B is traveling west at 12 mi/hr. When ship A was dead ahead of ship B, it was 6 mi away. Use MuPAD to determine how close the ships come to each other.

22. Suppose you have a wire of length L. You cut a length x to make a square and use the remaining length $L - x$ to make a circle. Use MuPAD to find the length x that maximizes the sum of the areas enclosed by the square and the circle.

23.* A certain spherical street lamp emits light in all directions. It is mounted on a pole of height h (see Figure P23). The brightness B at point P on the sidewalk is directly proportional to $\sin\theta$ and inversely proportional to the square of the distance d from the light to the point. Thus

$$B = \frac{c}{d^2}\sin\theta$$

where c is a constant. Use MuPAD to determine how high h should be to maximize the brightness at point P, which is 30 ft from the base of the pole.

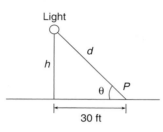

Figure P23

24.* A certain object has a mass $m = 100$ kg and is acted on by a force $f(t) = 500[2 - e^{-t}\sin(5\pi t)]$ N. The mass is at rest at $t = 0$. Use MuPAD to compute the object's velocity v at $t = 5$ s. The equation of motion is $m\dot{v} = f(t)$.

25. A rocket's mass decreases as it burns fuel. The equation of motion for a rocket in vertical flight can be obtained from Newton's law and is

$$m(t)\frac{dv}{dt} = T - m(t)g$$

where T is the rocket's thrust and its mass as a function of time is given by $m(t) = m_0(1 - rt/b)$. The rocket's initial mass is m_0, the burn time is b, and r is the fraction of the total mass accounted for by the fuel. Use the values $T = 48\ 000$ N, $m_0 = 2200$ kg, $r = 0.8$, $g = 9.81$ m/s², and $b = 40$ s.

   a. Use MuPAD to compute the rocket's velocity as a function of time for $t \le b$.

   b. Use MuPAD to compute the rocket's velocity at burnout.

26. The equation for the voltage $v(t)$ across a capacitor as a function of time is

$$v(t) = \frac{1}{C}\left[\int_0^t i(t)\,dt + Q_0\right]$$

where $i(t)$ is the applied current and Q_0 is the initial charge. Suppose that $C = 10^{-6}$ F and that $Q_0 = 0$. If the applied current is $i(t) = [0.01 + 0.3e^{-5t}\sin(25\pi t)]10^{-3}$ A, use MuPAD to obtain the voltage $v(t)$.

27. The power P dissipated as heat in a resistor R as a function of the current $i(t)$ passing through it is $P = i^2R$. The energy $E(t)$ lost as a function of time is the time integral of the power. Thus

$$E(t) = \int_0^t P(t)\,dt = R\int_0^t i^2(t)\,dt$$

If the current is measured in amperes, the power is in watts and the energy is in joules (1 W = 1 J/s). Suppose that a current $i(t) = 0.2[1 + \sin(0.2t)]$ A is applied to the resistor.
a. Determine the energy $E(t)$ dissipated as a function of time.
b. Determine the energy dissipated in 1 min if $R = 1000\ \Omega$.

28. The *RLC* circuit shown in Figure P28 can be used as a *narrowband filter*. If the input voltage $v_i(t)$ consists of a sum of sinusoidally varying voltages with different frequencies, the narrowband filter will allow to pass only those voltages whose frequencies lie within a narrow range. The magnification ratio M of a circuit is the ratio of the amplitude of the output voltage $v_o(t)$ to the amplitude of the input voltage $v_i(t)$. It is a function of the radian frequency ω of the input voltage. Formulas for M are derived in elementary electrical circuits courses. For this particular circuit, M is given by

$$M = \frac{RC\omega}{\sqrt{(1 - LC\omega^2)^2 + (RC\omega)^2}}$$

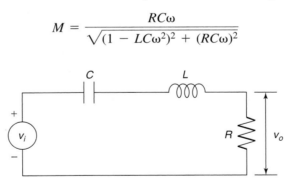

Figure P28

The frequency at which M is a maximum is the frequency of the desired carrier signal. Determine this frequency as a function of R, C, and L.

29. The shape of a cable hanging with no load other than its own weight is a *catenary* curve. A particular bridge cable is described by the catenary $y(x) = 10\cosh[(x - 20)/10]$ for $0 \le x \le 50$, where x and y are the horizontal and vertical coordinates measured in feet. (See Figure P29.) It is desired to hang plastic sheeting from the cable to protect passersby while

the bridge is being repainted. Use MuPAD to determine how many square feet of sheeting are required. Assume that the bottom edge of the sheeting is located along the x axis at $y = 0$.

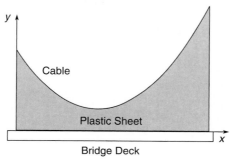

Figure P29

30. The shape of a cable hanging with no load other than its own weight is a catenary curve. A particular bridge cable is described by the catenary $y(x) = 10 \cosh[(x - 20)/10]$ for $0 \le x \le 50$, where x and y are the horizontal and vertical coordinates measured in feet.

   The length L of a curve described by $y(x)$ for $a \le x \le b$ can be found from the following integral:

$$L = \int_b^a \sqrt{1 + \left(\frac{dy}{dx}\right)^2}\, dx$$

   Determine the length of the cable.

31. Use the first five nonzero terms in the Taylor series for e^{ix}, $\sin x$, and $\cos x$ about $x = 0$ to demonstrate the validity of Euler's formula $e^{ix} = \cos x + i\sin x$.

32. Find the Taylor series for $e^x \sin x$ about $x = 0$ in two ways:
   a. By multiplying the Taylor series for e^x and that for $\sin x$.
   b. By using the **taylor** function directly on $e^x \sin x$.

33. Integrals that cannot be evaluated in closed form sometimes can be evaluated approximately by using a series representation for the integrand. For example, the following integral is used for some probability calculations (see Section 7.2):

$$I = \int_0^1 e^{-x^2}\, dx$$

   a. Obtain the Taylor series for e^{-x^2} about $x = 0$ and integrate the first six nonzero terms in the series to find I. Use the seventh term to estimate the error.

b. Compare your answer with that obtained with the MuPAD **erf(t)** function, defined as

$$\text{erf}(t) = \frac{2}{\sqrt{\pi}} \int_0^t e^{-t^2} dt$$

34.* Use MuPAD to compute the following limits.

a. $\displaystyle \lim_{x \to 1} \frac{x^2 - 1}{x^2 - x}$

b. $\displaystyle \lim_{x \to -2} \frac{x^2 - 4}{x^2 + 4}$

c. $\displaystyle \lim_{x \to 0} \frac{x^4 + 2x^2}{x^3 + x}$

35. Use MuPAD to compute the following limits.

a. $\displaystyle \lim_{x \to 0+} x^x$

b. $\displaystyle \lim_{x \to 0+} (\cos x)^{1/\tan x}$

c. $\displaystyle \lim_{x \to 0+} \left(\frac{1}{1 - x} \right)^{-1/x^2}$

d. $\displaystyle \lim_{x \to 0-} \frac{\sin x^2}{x^3}$

e. $\displaystyle \lim_{x \to 5-} \frac{x^2 - 25}{x^2 - 10x + 25}$

f. $\displaystyle \lim_{x \to 1+} \frac{x^2 - 1}{\sin[(x - 1)^2]}$

36. Use MuPAD to compute the following limits.

a. $\displaystyle \lim_{x \to \infty} \frac{x + 1}{x}$

b. $\displaystyle \lim_{x \to -\infty} \frac{3x^3 - 2x}{2x^3 + 3}$

37. Find the expression for the sum of the geometric series

$$\sum_{k=0}^{n-1} r^k$$

for $r \neq 1$.

38. A particular rubber ball rebounds to one-half its original height when dropped on a floor.

a. If the ball is initially dropped from a height h and is allowed to continue to bounce, find the expression for the total distance traveled by the ball after the ball hits the floor for the nth time.

b. If it is initially dropped from a height of 10 ft, how far will the ball have traveled after it hits the floor for the eighth time?

Section 11.6

39. The equation for the voltage y across the capacitor of an RC circuit is

$$RC\frac{dy}{dt} + y = v(t)$$

where $v(t)$ is the applied voltage. Suppose that $RC = 0.2$ s and that the capacitor voltage is initially 2 V. If the applied voltage goes from 0 to 10 V at $t = 0$, use MuPAD to determine the voltage $y(t)$.

40. The following equation describes the temperature $T(t)$ of a certain object immersed in a liquid bath of temperature $T_b(t)$:

$$10\frac{dT}{dt} + T = T_b$$

Suppose the object's temperature is initially $T(0) = 70°F$ and the bath temperature is 170°F. Use MuPAD to answer the following questions:
a. Determine $T(t)$.
b. How long will it take for the object's temperature T to reach 168°F?

41.* This equation describes the motion of a mass connected to a spring with viscous friction on the surface

$$m\ddot{y} + c\dot{y} + ky = f(t)$$

where $f(t)$ is an applied force. The position and velocity of the mass at $t = 0$ are denoted by x_0 and v_0. Use MuPAD to answer the following questions.
a. What is the free response in terms of x_0 and v_0 if $m = 3$, $c = 18$, and $k = 102$?
b. What is the free response in terms of x_0 and v_0 if $m = 3$, $c = 39$, and $k = 120$?

42. The equation for the voltage y across the capacitor of an RC circuit is

$$RC\frac{dy}{dt} + y = v(t)$$

where $v(t)$ is the applied voltage. Suppose that $RC = 0.2$ s and that the capacitor voltage is initially 2 V. If the applied voltage is $v(t) = 10[2 - e^{-t}\sin(5\pi t)]$, use MuPAD to obtain the voltage $y(t)$.

43. The following equation describes a certain dilution process, where $y(t)$ is the concentration of salt in a tank of freshwater to which salt brine is being added:

$$\frac{dy}{dt} + \frac{2}{10 + 2t}y = 4$$

Suppose that $y(0) = 0$. Use MuPAD to obtain $y(t)$.

44. This equation describes the motion of a certain mass connected to a spring with viscous friction on the surface

$$3\ddot{y} + 18\dot{y} + 102y = f(t)$$

where $f(t)$ is an applied force. Suppose that $f(t) = 0$ for $t < 0$ and $f(t) = 10$ for $t \geq 0$.
a. Use MuPAD to obtain $y(t)$ if $y(0) = \dot{y}(0) = 0$.
b. Use MuPAD to obtain $y(t)$ if $y(0) = 0$ and $\dot{y}(0) = 10$.

45. This equation describes the motion of a certain mass connected to a spring with viscous friction on the surface

$$3\ddot{y} + 39\dot{y} + 120y = f(t)$$

where $f(t)$ is an applied force. Suppose that $f(t) = 0$ for $t < 0$ and $f(t) = 10$ for $t \geq 0$.
a. Use MuPAD to obtain $y(t)$ if $y(0) = \dot{y}(0) = 0$.
b. Use MuPAD to obtain $y(t)$ if $y(0) = 0$ and $\dot{y}(0) = 10$.

46. The equations for an armature-controlled dc motor follow. The motor's current is i and its rotational velocity is ω.

$$L\frac{di}{dt} = -Ri - K_e\omega + v(t)$$

$$I\frac{d\omega}{dt} = K_T i - c\omega$$

where L, R, and I are the motor's inductance, resistance, and inertia; K_T and K_e are the torque constant and back-emf constant; c is a viscous damping constant; and $v(t)$ is the applied voltage.
Use the values $R = 0.8\ \Omega$, $L = 0.003$ H, $K_T = 0.05$ N·m/A, $K_e = 0.05$ V·s/rad, $c = 0$, and $I = 8 \times 10^{-5}$ kg · m².
Suppose the applied voltage is 20 V. Use MuPAD to obtain the motor's speed and current versus time for zero initial conditions. Choose a final time large enough to show the motor's speed becoming constant.

Section 11.7

47. The *RLC* circuit described in Problem 28 and shown in Figure P28 has the following differential equation model:

$$LC\ddot{v}_o + RC\dot{v}_o + v_o = RC\dot{v}_i(t)$$

Use the Laplace transform method to solve for the unit-step response of $v_0(t)$ for zero initial conditions, where $C = 10^{-5}$ F and $L = 5 \times 10^{-3}$ H. For the first case (a broadband filter), $R = 1000\ \Omega$. For the second case (a narrowband filter), $R = 10\ \Omega$. Compare the step responses of the two cases.

48. The differential equation model for a certain speed control system for a vehicle is

$$\ddot{v} + (1 + K_p)\dot{v} + K_I v = K_p \dot{v}_d + K_I v_d$$

where the actual speed is v, the desired speed is $v_d(t)$, and K_p and K_I are constants called the *control gains*. Use the Laplace transform method to find the unit-step response [that is, $v_d(t)$ is a unit-step function]. Use zero initial conditions. Compare the response for three cases.

a. $K_p = 9, K_I = 50$
b. $K_p = 9, K_I = 25$
c. $K_p = 54, K_I = 250$

49. The differential equation model for a certain position control system for a metal cutting tool is

$$\frac{d^3x}{dt^3} + (6 + K_D)\frac{d^2x}{dt^2} + (11 + K_p)\frac{dx}{dt} + (6 + K_I)x$$

$$= K_D\frac{d^2x_d}{dt^2} + K_p\frac{dx_d}{dt} + K_I x_d$$

where the actual tool position is x; the desired position is $x_d(t)$; and K_p, K_I, and K_D are constants called the control gains. Use the Laplace transform method to find the unit-step response [that is, $x_d(t)$ is a unit-step function]. Use zero initial conditions. Compare the response for three cases.

a. $K_p = 30, K_I = K_D = 0$
b. $K_p = 27, K_I = 17.18, K_D = 0$
c. $K_p = 36, K_I = 38.1, K_D = 8.52$

50.* The differential equation model for the motor torque $m(t)$ required for a certain speed control system is

$$4\ddot{m} + 4K\dot{m} + K^2 m = K^2 v_d$$

where the desired speed is $v_d(t)$ and K is a constant called the control gain.
a. Use the Laplace transform method to find the unit-step response [that is, $v_d(t)$ is a unit-step function]. Use zero initial conditions.

b. Use symbolic manipulation in MuPAD to find the value of the peak torque in terms of the gain K.

Section 11.8

51. Solve Legendre's equation for the given initial conditions.

$$(1 - x^2)y'' - 2xy' + 6y = 0 \qquad y(0) = 5 \qquad y'(0) = 0$$

52. Evaluate the solution to Airy's equation at $x = 2$ for the given initial conditions.

$$y'' - xy = 0 \qquad y(0) = 0 \qquad y'(0) = 1$$

53. Obtain a series solution to the following Legendre equation for the given initial conditions.

$$(1 - x^2)\, y'' - 2xy' + 6y = 0 \qquad y(0) = a \qquad y'(0) = b$$

Guide to Commands and Functions in This Text

Operators and special characters

Item	Description	Pages
+	Plus; addition operator.	8
−	Minus; subtraction operator.	8
*	Scalar and matrix multiplication operator.	8
.*	Array multiplication operator.	66
^	Scalar and matrix exponentiation operator.	8
.^	Array exponentiation operator.	66
\	Left division operator.	8, 66
/	Right division operator.	8, 66
.\	Array left division operator.	66
./	Array right division operator.	66
:	Colon; generates regularly spaced elements and represents an entire row or column.	12, 57
( )	Parentheses; enclose function arguments and array indices; override precedence.	9, 117
[ ]	Brackets; enclose array elements.	19, 55
{ }	Braces; enclose cell elements.	91
. . .	Ellipsis; line-continuation operator.	12
,	Comma; separates statements, and elements in a row of an array.	12
;	Semicolon; separates columns in an array, and suppresses display.	12, 55
%	Percent sign; designates a comment, and specifies formatting.	27, 549
'	Quote sign and transpose operator.	55, 57
.'	Nonconjugated transpose operator.	57
=	Assignment (replacement) operator.	10
@	Creates a function handle.	124

Logical and relational operators

Item	Description	Pages
==	Relational operator: equal to.	155
~=	Relational operator: not equal to.	155
<	Relational operator: less than.	155
<=	Relational operator: less than or equal to.	155
>	Relational operator: greater than.	155
>=	Relational operator: greater than or equal to.	155
&	Logical operator: AND.	158
&&	Short-circuit AND.	158
|	Logical operator: OR.	158
||	Short-circuit OR.	158
~	Logical operator: NOT.	158

Special variables and constants

Item	Description	Pages
ans	Most recent answer.	14
eps	Accuracy of floating-point precision.	14
i,j	The imaginary unit $\sqrt{-1}$.	14
Inf	Infinity.	14
NaN	Undefined numerical result (not a number).	14
pi	The number π.	14

Commands for managing a session

Item	Description	Pages
clc	Clears Command window.	12
clear	Removes variables from memory.	12
doc	Displays documentation.	38
exist	Checks for existence of file or variable.	12
global	Declares variables to be global.	124
help	Displays Help text in the Command window.	38
helpwin	Displays Help text in the Help browser.	38
lookfor	Searches Help entries for a keyword.	38
quit	Stops MATLAB.	12
who	Lists current variables.	12
whos	Lists current variables (long display).	12

System and file commands

Item	Description	Pages
cd	Changes current directory.	23
date	Displays current date.	122
dir	Lists all files in current directory.	23
load	Loads workspace variables from a file.	21
path	Displays search path.	23
pwd	Displays current directory.	23
save	Saves workspace variables in a file.	21
type	Displays contents of a file.	38
what	Lists all MATLAB files.	23
wk1read	Reads .wk1 spreadsheet file.	138
xlsread	Reads .xls spreadsheet file.	135

Input/output commands

Item	Description	Pages
disp	Displays contents of an array or string.	31
format	Controls screen display format.	14, 31
fprintf	Performs formatted writes to screen or file.	549
input	Displays prompts and waits for input.	31, 171
menu	Displays a menu of choices.	31
;	Suppresses screen printing.	12

Numeric display formats

Item	Description	Pages
format short	Four decimal digits (default).	14, 31
format long	16 decimal digits.	14, 31
format short e	Five digits plus exponent.	14, 31
format long e	16 digits plus exponent.	14, 31
format bank	Two decimal digits.	14, 31
format +	Positive, negative, or zero.	14, 31
format rat	Rational approximation.	14, 31
format compact	Suppresses some line feeds.	14, 31
format loose	Resets to less compact display mode.	14, 31

Array functions

Item	Description	Pages
cat	Concatenates arrays.	64
find	Finds indices of nonzero elements.	60, 161
length	Computes number of elements.	19
linspace	Creates regularly spaced vector.	60
logspace	Creates logarithmically spaced vector.	60
max	Returns largest element.	60
min	Returns smallest element.	60
size	Computes array size.	60
sort	Sorts each column.	60, 296
sum	Sums each column.	60

Special matrices

Item	Description	Pages
eye	Creates an identity matrix.	83
ones	Creates an array of 1s.	83
zeros	Creates an array of 0s.	83

Matrix functions for solving linear equations

Item	Description	Pages
det	Computes determinant of an array.	333
inv	Computes inverse of a matrix.	333
pinv	Computes pseudoinverse of a matrix.	342
rank	Computes rank of a matrix.	335
rref	Computes reduced row echelon form.	345

Exponential and logarithmic functions

Item	Description	Pages
exp(x)	Exponential; e^x.	21, 114
log(x)	Natural logarithm; ln x.	114
log10(x)	Common (base 10) logarithm; $\log x = \log_{10} x$.	114
sqrt(x)	Square root; $\sqrt{x}$.	114

Complex functions

Item	Description	Pages
abs(x)	Absolute value; $|x|$	114
angle(x)	Angle of a complex number x.	114
conj(x)	Complex conjugate of x.	114
imag(x)	Imaginary part of a complex number x.	114
real(x)	Real part of a complex number x.	114

Numeric functions

Item	Description	Pages
ceil	Rounds to the nearest integer toward ∞	114
fix	Rounds to the nearest integer toward zero.	114
floor	Rounds to the nearest integer toward $-\infty$	114
round	Rounds toward the nearest integer.	114
sign	Signum function.	114

Trigonometric functions

Item	Description	Pages
acos(x)	Inverse cosine; $\arccos x = \cos^{-1}x$.	21, 118
acot(x)	Inverse cotangent; $\text{arccot } x = \cot^{-1}x$.	118
acsc(x)	Inverse cosecant; $\text{arccsc } x = \csc^{-1}x$.	118
asec(x)	Inverse secant; $\text{arcsec } x = \sec^{-1}x$.	118
asin(x)	Inverse sine; $\arcsin x = \sin^{-1}x$.	21, 118
atan(x)	Inverse tangent; $\arctan x = \tan^{-1}x$.	21, 118
atan2(y,x)	Four-quadrant inverse tangent.	118
cos(x)	Cosine; $\cos x$.	21, 118
cot(x)	Cotangent; $\cot x$.	118
csc(x)	Cosecant; $\csc x$.	118
sec(x)	Secant; $\sec x$.	118
sin(x)	Sine; $\sin x$.	118
tan(x)	Tangent; $\tan x$.	118

Hyperbolic functions

Item	Description	Pages
acosh(x)	Inverse hyperbolic cosine; $\cosh^{-1}x$.	119
acoth(x)	Inverse hyperbolic cotangent; $\coth^{-1}x$.	119
acsch(x)	Inverse hyperbolic cosecant; $\text{csch}^{-1}x$.	119
asech(x)	Inverse hyperbolic secant; $\text{sech}^{-1}x$.	119
asinh(x)	Inverse hyperbolic sine; $\sinh^{-1}x$.	119
atanh(x)	Inverse hyperbolic tangent; $\tanh^{-1}x$.	119
cosh(x)	Hyperbolic cosine; $\cosh x$.	119
coth(x)	Hyperbolic cotangent; $\cosh x / \sinh x$.	119
csch(x)	Hyperbolic cosecant; $1/\sinh x$.	119
sech(x)	Hyperbolic secant; $1/\cosh x$.	119
sinh(x)	Hyperbolic sine; $\sinh x$.	119
tanh(x)	Hyperbolic tangent; $\sinh x / \cosh x$.	119

Polynomial functions

Item	Description	Pages
conv	Computes product of two polynomials.	86
deconv	Computes ratio of polynomials.	86
eig	Computes the eigenvalues of a matrix.	397
poly	Computes polynomial from roots.	86
polyfit	Fits a polynomial to data.	273
polyval	Evaluates polynomial.	225
roots	Computes polynomial roots.	86

Logical functions

Item	Description	Pages
any	True if any elements are nonzero.	161
all	True if all elements are nonzero.	161
find	Finds indices of nonzero elements.	161
finite	True if elements are finite.	161
ischar	True if elements are a character array.	161
isinf	True if elements are infinite.	161
isempty	True if matrix is empty.	161
isnan	True if elements are undefined.	161
isreal	True if all elements are real.	161
isnumeric	True if elements have numeric values.	161
logical	Converts a numeric array to a logical array.	161
xor	Exclusive OR.	161

Miscellaneous mathematical functions

Item	Description	Pages
cross	Computes cross products.	85
dot	Computes dot products.	85
function	Creates a user-defined function.	119
nargin	Number of function input arguments.	170
nargout	Number of function output arguments.	170

Cell and structure functions

Item	Description	Pages
cell	Creates cell array.	90
fieldnames	Returns field names in a structure array.	94
isfield	Identifies a structure array field.	94
isstruct	Identifies a structure array.	94
rmfield	Removes a field from a structure array.	94
struct	Creates a structure array.	93

Basic xy plotting commands

Item	Description	Pages
axis	Sets axis limits and other axis properties.	222, 225
cla	Clears the axes.	539
fplot	Intelligent plotting of functions.	223, 225
ginput	Reads coordinates of the cursor position.	25
grid	Displays grid lines.	225
plot	Generates xy plot.	23, 232
print	Prints plot or saves plot to a file.	225
title	Puts text at top of plot.	225
xlabel	Adds text label to x axis.	225
ylabel	Adds text label to y axis.	225

Plot enhancement commands

Item	Description	Pages
colormap	Sets the color map of the current figure.	539
gtext	Enables label placement by mouse.	232
hold	Freezes current plot.	232
legend	Places legend by mouse.	230
subplot	Creates plots in subwindows.	232
text	Places string in figure.	232

Specialized plot functions

Item	Description	Pages
bar	Creates bar chart.	235, 295
loglog	Creates log-log plot.	235
plotyy	Enables plotting on left and right axes.	235
polar	Creates polar plot.	235
semilogx	Creates semilog plot (logarithmic abscissa).	235
semilogy	Creates semilog plot (logarithmic ordinate).	235
stairs	Creates stairs plot.	235
stem	Creates stem plot.	235

Three-dimensional plotting functions

Item	Description	Pages
contour	Creates contour plot.	248
mesh	Creates three-dimensional mesh surface plot.	248
meshc	Same as mesh with contour plot underneath.	248
meshgrid	Creates rectangular grid.	247
meshz	Same as mesh with vertical lines underneath.	249
plot3	Creates three-dimensional plots from lines and points.	246
shading	Specifies type of shading.	539
surf	Creates shaded three-dimensional mesh surface plot.	248
surfc	Same as surf with contour plot underneath.	248
surfl	Same as surf with lighting.	539
view	Sets the angle of the view.	539
waterfall	Same as mesh with mesh lines in one direction.	249
zlabel	Adds text label to z axis.	248

Program flow control

Item	Description	Pages
break	Terminates execution of a loop.	176
case	Provides alternate execution paths within `switch` structure.	188
continue	Passes control to the next iteration of a `for` or `while` loop.	276
else	Delineates alternate block of statements.	166
elseif	Conditionally executes statements.	168
end	Terminates `for`, `while`, and `if` statements.	166
for	Repeats statements a specific number of times.	172
if	Executes statements conditionally.	165
otherwise	Provides optional control within a `switch` structure.	188
switch	Directs program execution by comparing input with `case` expressions.	188
while	Repeats statements an indefinite number of times.	183

Optimization and root-finding functions

Item	Description	Pages
fminbnd	Finds the minimum of a function of one variable.	128
fminsearch	Finds the minimum of a multivariable function.	128
fzero	Finds the zero of a function.	128

Histogram functions

Item	Description	Pages
bar	Creates a bar chart.	235, 295
hist	Aggregates the data into bins.	300

Statistical functions

Item	Description	Pages
cumsum	Computes the cumulative sum across a row.	303
erf	Computes the error function $erf(x)$.	305
mean	Calculates the mean.	296
median	Calculates the median.	296
std	Calculates the standard deviation.	304

Random number functions

Item	Description	Pages
rand	Generates uniformly distributed random numbers between 0 and 1; sets and retrieves the state.	309
randn	Generates normally distributed random numbers; sets and retrieves the state.	309
randperm	Generates random permutation of integers.	309

Polynomial functions

Item	Description	Pages
poly	Computes the coefficients of a polynomial from its roots.	86
polyfit	Fits a polynomial to data.	273
polyval	Evaluates a polynomial and generates error estimates.	273
roots	Computes the roots of a polynomial from its coefficients.	86

Interpolation functions

Item	Description	Pages
interp1	Linear and cubic spline interpolation of a function of one variable.	320
interp2	Linear interpolation of a function of two variables.	317
spline	Cubic spline interpolation.	320
unmkpp	Computes the coefficients of cubic spline polynomials.	320

Numerical integration functions

Item	Description	Pages
dblquad	Numerical integration of a double integral.	372
polyint	Integration of a polynomial.	371
quad	Numerical integration with adaptive Simpson's rule.	371
quadl	Numerical integration with Lobatto quadrature.	371
trapz	Numerical integration with the trapezoidal rule.	371
triplequad	Numerical integration of a triple integral.	371

Numerical differentiation functions

Item	Description	Pages
del2	Computes the Laplacian from data.	382
diff(x)	Computes the differences between adjacent elements in the vector x.	329, 382
gradient	Computes the gradient from data.	382
polyder	Differentiates a polynomial, a polynomial product, or a polynomial quotient.	382

ODE solvers

Item	Description	Pages
ode45	Nonstiff, medium-order solver.	385, 395
ode15s	Stiff, variable-order solver.	385, 395
odeset	Creates integrator options structure for ODE solvers.	395

LTI object functions

Item	Description	Pages
ss	Creates an LTI object in state-space form.	400
ssdata	Extracts state-space matrices from an LTI object.	400
tf	Creates an LTI object in transfer-function form.	400
tfdata	Extracts equation coefficients from an LTI object.	400

LTI ODE solvers

Item	Description	Pages
impulse	Computes and plots the impulse response of an LTI object.	401
initial	Computes and plots the free response of an LTI object.	401
lsim	Computes and plots the response of an LTI object to a general input.	401
step	Computes and plots the step response of an LTI object.	401

Predefined input functions

Item	Description	Pages
gensig	Generates a periodic sine, square, or pulse input.	408

Manipulating expressions in **MuPAD**

Item	Description	Pages
collect	Collects terms with the same powers.	478
combine	Combines terms of the same algebraic structure.	477
delete	Deletes the value of an identifier.	472
expand	Expands an expression.	475
factor	Factors a polynomial into irreducible polynomials.	475
normal	Returns the normal form of a rational expression.	476
rewrite	Transforms an expression in terms of a target function.	478
simplify	Simplifies an expression.	476

Solution of nonlinear algebraic and differential equations in **MuPAD**

Item	Description	Pages
assume	Specifies constraints on the solution variable.	472, 478
ode	Specifies a differential equation and its boundary conditions.	502
polyroots	Numerically solves for all the roots of a polynomial.	480
solve	Solves nonlinear algebraic and differential equations.	480
unassume	Removes constraints imposed by assume.	478

Numerical evaluation in **MuPAD**

Item	Description	Pages
DIGITS	Specifies the number of displayed digits.	471
float	Evaluates the result as a floating-point number.	471

Matrix operations in **MuPAD**

Item	Description	Pages
charpoly	Finds the characteristic polynomial of a matrix.	489
det	Finds the determinant of a matrix.	489
eigenvalues	Finds the eigenvalues of a matrix.	491
eigenvectors	Finds the eigenvectors of a matrix.	491
matlinsolve	Solves the matrix equation **Ax = b.**	491
matrix	Creates a matrix.	489
rank	Finds the rank of a matrix.	489

Symbolic calculus in **MuPAD**

Item	Description	Pages
diff	Finds the first derivative of an expression.	494
int	Finds the single integral of an expression.	497
limit	Finds the limit of an expression.	500
sum	Finds the sum of terms in an expression.	500
taylor	Finds the Taylor series expansion of an expression.	499

Laplace transforms in **MuPAD**

Item	Description	Pages
invlaplace	Finds the inverse Laplace transform of an expression.	507
laplace	Finds the Laplace transform of an expression.	506

Animation functions

Item	Description	Pages
drawnow	Initiates immediate plotting.	541
getframe	Captures current figure in a frame.	538
movie	Plays back frames.	538
moviein	Initializes movie frame memory.	540
pause	Pauses the display.	541

Sound functions

Item	Description	Pages
sound	Plays a vector as sound.	546
soundsc	Scales data and plays as sound.	547
wavplay	Plays recorded sound.	547
wavread	Reads Microsoft WAVE file.	547
wavrecord	Records sound from input device.	548
wavwrite	Writes Microsoft WAVE file.	548

B

Animation and Sound in MATLAB

B.1 Animation

Animation can be used to display the behavior of an object over time. Some of the MATLAB demos are M-files that perform animation. After completing this section, which has simple examples, you may study the demo files, which are more advanced. Two methods can be used to create animations in MATLAB. The first method uses the `movie` function. The second method uses the Erase-Mode property.

Creating Movies in MATLAB

The `getframe` command captures, or takes a snapshot of, the current figure to create a single frame for the movie. The `getframe` function is usually used in a `for` loop to assemble an array of movie frames. The `movie` function plays back the frames after they have been captured.

To create a movie, use a script file of the following form.

```
for k = 1:n
  plotting expressions
  M(k) = getframe; % Saves current figure in array M
end
movie(M)
```

For example, the following script file creates 20 frames of the function $te^{-t/b}$ for $0 \le t \le 100$ for each of 20 values of the parameter b from $b = 1$ to $b = 20$.

```
% Program movie1.m
% Animates the function t*exp(-t/b).
```

```
t = 0:0.05:100;
for b = 1:20
  plot(t,t.*exp(-t/b)),axis([0 100 0 10]),xlabel('t');
  M(:,b) = getframe;
end
```

The line `M(:,b) = getframe;` acquires and saves the current figure as a column of the matrix `M`. Once this file is run, the frames can be replayed as a movie by typing `movie(M)`. The animation shows how the location and height of the function peak changes as the parameter *b* is increased.

Rotating a 3D Surface

The following example rotates a three-dimensional surface by changing the viewpoint. The data are created using the built-in function `peaks`.

```
% Program movie2.m
% Rotates a 3D surface.
[X,Y,Z] = peaks(50);    % Create data.
surfl(X,Y,Z)    % Plot the surface.
axis([-3 3 -3 3 -5 5])% Retain same scaling for each frame.
axis vis3d off % Set the axes to 3D and turn off tick marks,
               % and so forth.
shading interp     % Use interpolated shading.
colormap(winter)   % Specify a color map.
for k = 1:30 % Rotate the viewpoint and capture each frame.
        view(-37.5+0.5*(k-1),30)
        M(k) = getframe;
end
cla      % Clear the axes.
movie(M)     % Play the movie.
```

The `colormap(map)` function sets the current figure's color map to `map`. Type `help graph3d` to see a number of color maps to choose for `map`. The choice `winter` provides blue and green shading. The `view` function specifies the 3D graph viewpoint. The syntax `view(az,el)` sets the angle of the view from which an observer sees the current 3D plot, where `az` is the azimuth or horizontal rotation and `el` is the vertical elevation (both in degrees). Azimuth revolves about the *z* axis, with positive values indicating counterclockwise rotation of the viewpoint. Positive values of elevation correspond to moving above the object; negative values move below. The choice $az = -37.5$, $el = 30$ is the default 3D view.

Extended Syntax of the `movie` Function

The function `movie (M)` plays the movie in array `M` once, where `M` must be an array of movie frames (usually acquired with `getframe`). The function movie(M,n) plays the movie n times. If n is negative, each "play" is once forward and once backward. If n is a vector, the first element is the number of times to play the movie, and the remaining elements are a list of frames to play in the movie. For example, if M has four frames, then n = [10 4 4 2 1] plays the movie 10 times, and the movie consists of frame 4 followed by frame 4 again, followed by frame 2 and finally frame 1.

The function movie (M,n,fps) plays the movie at fps frames per second. If fps is omitted, the default is 12 frames per second. Computers that cannot achieve the specified fps will play the movie as fast as they can. The function movie(h,...) plays the movie in object h, where h is a handle to a figure or an axis. Handles are discussed in Section 2.2.

The function movie(h,M,n,fps,loc) specifies the location to play the movie, relative to the lower left corner of object h and in pixels, regardless of the value of the object's Units property, where loc = [x y unused unused] is a four-element position vector, of which only the *x* and *y* coordinates are used, but all four elements are required. The movie plays back using the width and height in which it was recorded.

Note that for code to be compatible with versions of MATLAB prior to Release 11 (5.3), the moviein(n) function must be used to initialize movie frame memory for n frames. To do this, place the line M = moviein(n); before the for loop that generates the plots.

The disadvantage of the movie function is that it might require too much memory if many frames or complex images are stored.

Animation with the EraseMode Property

One form of extended syntax for the plot function is

plot(...,'PropertyName','PropertyValue',...)

This form sets the plot property specified by PropertyName to the values specified by PropertyValue for all line objects created by the plot function. One such property name is EraseMode. This property controls the technique MATLAB uses to draw and erase line objects and is useful for creating animated sequences. The allowable values for the EraseMode property are the following.

■ normal This is the default value for the EraseMode property. By typing

plot(...,'EraseMode','normal')

the entire figure, including axes, labels, and titles, is erased and redrawn using only the new set of points. In redrawing the display, a three-dimensional analysis is performed to ensure that all objects are rendered correctly. Thus, this mode produces the most accurate picture but is the slowest. The other modes are faster but do not perform a complete redraw and are therefore less accurate. This method may cause blinking between each frame because everything is erased and redrawn. This method is therefore undesirable for animation.

- ■ `none` When the EraseMode property value is set to `none`, objects in the existing figure are not erased, and the new plot is superimposed on the existing figure. This mode is therefore fast because it does not remove existing points, and it is useful for creating a "trail" on the screen.

- ■ `xor` When the EraseMode property value is set to `xor`, objects are drawn and erased by performing an exclusive OR with the background color. This produces a smooth animation. This mode does not destroy other graphics objects beneath the ones being erased and does not change the color of the objects beneath. However, the object's color depends on the background color.

- ■ `background` When the EraseMode property value is set to `background`, the result is the same as with `xor` except that objects behind the erased objects are destroyed. Objects are erased by drawing them in the axes' background color or in the figure background color if the axes `Color` property is set to `none`. This damages objects that are behind the erased line, but lines are always properly colored.

The `drawnow` command causes the previous graphics command to be executed immediately. If the `drawnow` command were not used, MATLAB would complete all other operations before performing any graphics operations and would display only the last frame of the animation.

Animation speed depends of the intrinsic speed of the computer and on what and how much is being plotted. Symbols such as `o`, `*`, or `+` will be plotted slower than a line. The number of points being plotted also affects the animation speed. The animation can be slowed by using the `pause(n)` function, which pauses the program execution for n seconds.

Using Object Handles

An expression of the form

`p = plot(...)`

assigns the results of the plot function to the variable p, which is a figure identifier called a *figure handle*. This stores the figure and makes it available for future use. Any valid variable name may be assigned to a handle. A figure handle is a specific type of *object handle*. Handles may be assigned to other types of objects. For example, later we will create a handle with the `text` function.

The `set` function can be used with the handle to change the object properties. This function has the general format

`set(object handle, 'PropertyName', 'PropertyValue', ...)`

If the object is an entire figure, its handle also contains the specifications for line color and type, marker size, and the value of the EraseMode property. Two of the properties of the figure specify the data to be plotted. Their property names are `XData` and `YData`. The following example shows how to use these properties.

Animating a Function

Consider the function $te^{-t/b}$, which was used in the first movie example. This function can be animated as the parameter b changes with the following program.

```
% Program animate1.m
% Animates the function t*exp(-t/b).
t = 0:0.05:100;
b = 1;
p = plot(t,t.*exp(-t/b),'EraseMode','xor');...
   axis([0 100 0 10]),xlabel('t');
for b = 2:20
    set(p,'XData',t,'YData',t.*exp(-t/b)),...
    axis([0 100 0 10]),xlabel('t');
    drawnow
    pause(0.1)
end
```

In this program the function $te^{-t/b}$ is first evaluated and plotted over the range $0 \leq t \leq 100$ for $b = 1$, and the figure handle is assigned to the variable p. This establishes the plot format for all following operations, for example, line type and color, labeling, and axis scaling. The function $te^{-t/b}$ is then evaluated and plotted over the range $0 \leq t \leq 100$ for $b = 2, 3, 4, \ldots$ in the for loop, and the previous plot is erased. Each call to set in the for loop causes the next set of points to be plotted. The EraseMode property value specifies how to plot the existing points on the figure (i.e., how to refresh the screen), as each new set of points is added. You should investigate what happens if the EraseMode property value is set to none instead of xor.

Animating Projectile Motion

This following program illustrates how user-defined functions and subplots can be used in animations. The following are the equations of motion for a projectile launched with a speed s_0 at an angle θ above the horizontal, where x and y are the horizontal and vertical coordinates, g is the acceleration due to gravity, and t is time.

$$x(t) = (s_0 \cos \theta)t \qquad y(t) = -\frac{gt^2}{2} + (s_0 \sin \theta)t$$

By setting $y = 0$ in the second expression, we can solve for t and obtain the following expression for the maximum time the projectile is in flight t_{max}.

$$t_{max} = \frac{2s_0}{g} \sin\theta$$

The expression for $y(t)$ may be differentiated to obtain the expression for the vertical velocity:

$$v_{vert} = \frac{dy}{dt} = -gt + s_0 \sin\theta$$

The maximum distance x_{max} may be computed from $x(t_{max})$, the maximum height y_{max} may be computed from $y(t_{max}/2)$, and the maximum vertical velocity occurs at $t = 0$.

The following functions are based on these expressions, where s0 is the launch speed s_0 and th is the launch angle θ.

```
function x = xcoord(t,s0,th);
% Computes projectile horizontal coordinate.
x = s0*cos(th)*t;

function y = ycoord(t,s0,th,g);
% Computes projectile vertical coordinate.
y = -g*t.^2/2+s0*sin(th)*t;

function v = vertvel(t,s0,th,g);
% Computes projectile vertical velocity.
v = -g*t+s0*sin(th);
```

The following program uses these functions to animate the projectile motion in the first subplot, while simultaneously displaying the vertical velocity in the second subplot, for the values $\theta = 45°$, $s_0 = 105$ ft/sec, and $g = 32.2$ ft/sec^2. Note that the values of xmax, ymax, and vmax are computed and used to set the axes scales. The figure handles are h1 and h2.

```
% Program animate2.m
% Animates projectile motion.
% Uses functions xcoord, ycoord, and vertvel.
th = 45*(pi/180);
g = 32.2; s0 = 105;
%
tmax = 2*s0*sin(th)/g;
xmax = xcoord(tmax,s0,th);
ymax = ycoord(tmax/2,s0,th,g);
vmax = vertvel(0,s0,th,g);
w = linspace(0,tmax,500);
%
subplot(2,1,1)
plot(xcoord(w,s0,th),ycoord(w,s0,th,g)),hold,
h1 = plot(xcoord(w,s0,th),ycoord(w,s0,th,g),'o','EraseMode','xor')
axis([0 xmax 0 1.1*ymax]),xlabel('x'), ylabel('y')
subplot(2,1,2)
```

```
plot(xcoord(w,s0,th),vertvel(w,s0,th,g)),hold,
h2 = plot(xcoord(w,s0,th),vertvel(w,s0,th,g),'s','EraseMode','xor');
axis([0 xmax 0 1.1*vmax]),xlabel('x'),...
   ylabel('Vertical Velocity')
for t = 0:0.01:tmax
   set(h1,'XData',xcoord(t,s0,th),'YData',ycoord(t,s0,th,g))
   set(h2,'XData',xcoord(t,s0,th),'YData',vertvel(t,s0,th,g))
   drawnow
   pause(0.005)
end
hold
```

You should experiment with different values of the `pause` function argument.

Animation with Arrays

Thus far we have seen how the function to be animated may be evaluated in the `set` function with an expression or with a function. A third method is to compute the points to be plotted ahead of time and store them in arrays. The following program shows how this is done, using the projectile application. The plotted points are stored in the arrays x and y.

```
% Program animate3.m
% Animation of a projectile using arrays.
th = 70*(pi/180);
g = 32.2; s0=100;
tmax = 2*s0*sin(th)/g;
xmax = xcoord(tmax,s0,th);
ymax = ycoord(tmax/2,s0,th,g);
%
w = linspace(0,tmax,500);
x = xcoord(w,s0,th);y = ycoord(w,s0,th,g);
plot(x,y),hold,
h1 = plot(x,y,'o','EraseMode','xor');
axis([0 xmax 0 1.1*ymax]),xlabel('x'),ylabel( 'y')
%
kmax = length(w);
for k =1:kmax
    set(h1,'XData',x(k),'YData',y(k))
    drawnow
    pause(0.001)
end
hold
```

Displaying Elapsed Time

It may be helpful to display the elapsed time during an animation. To do this, modify the program `animate3.m` as shown in the following. The new lines are indicated in bold; the line formerly below the line `h1 = plot(...` has been deleted.

```
% Program animate4.m
% Like animate3.m but displays elapsed time.
th = 70*(pi/180);
g = 32.2; s0 = 100;
%
tmax = 2*s0*sin(th)/g;
xmax = xcoord(tmax,s0,th);
ymax = ycoord(tmax/2,s0,th,g);
%
t = linspace(0,tmax,500);
x = xcoord(t,s0,th);y = ycoord(t,s0,th,g);
plot(x,y),hold,
h1 = plot(x,y, 'o','EraseMode','xor');
text(10,10,'Time = ')
time = text(30,10,'0','EraseMode','background')
    axis([0 xmax 0 1.1*ymax]),xlabel('x'),ylabel('y')
%
kmax = length(t);
for k = 1:kmax
  set(h1,'XData',x(k),'YData',y(k))
  t_string = num2str(t(k));
  set(time,'String',t_string)
  drawnow
  pause(0.001)
end
hold
```

The first new line creates a label for the time display using the `text` statement, which writes the label once. The program must not write to that location again. The second new statement creates the handle `time` for the text label and creates the string for the first time value, which is 0. By using the `background` value for `EraseMode`, the statement specifies that the existing display of the time variable will be erased when the next value is displayed. Note that the numerical value of time `t(k)` must be converted to a string, by using the function `num2str`, before it can be displayed. In the last new line, in which the `set` function uses the `time` handle, the property name is `'String'`, which is not a variable but a property associated with text objects. The variable being updated is `t_string`.

B.2 Sound

MATLAB provides a number a functions for creating, recording, and playing sound on the computer. This section gives a brief introduction to these functions.

A Model of Sound

Sound is the fluctuation of air pressure as a function of time t. If the sound is a *pure tone*, the pressure $p(t)$ oscillates sinusoidally at a single frequency, that is,

$$p(t) = A \sin(2\pi f t + \phi)$$

where A is the pressure amplitude (the 'loudness"), f is the sound frequency in cycles per second (Hz), and ϕ is the phase shift in radians. The *period* of the sound wave is $P = 1/f$.

Because sound is an analog variable (one having an infinite number of values), it must be converted to a finite set of numbers before it can be stored and used in a digital computer. This conversion process involves *sampling* the sound signal into discrete values and *quantizing* the numbers so that they can be represented in binary form. Quantization is an issue when you are using a microphone and analog-to-digital converter to capture real sound, but we will not discuss it here because we will produce only simulated sounds in software.

You use a process similar to sampling whenever you plot a function in MATLAB. To plot the function you should evaluate it at enough points to produce a smooth plot. So, to plot a sine wave, we should "sample" or evaluate it many times over the period. The frequency at which we evaluate it is the *sampling frequency*. So, if we use a time step of 0.1 s, our sampling frequency is 10 Hz. If the sine wave has a period of 1 s, then we are "sampling" the function 10 times every period. So we see that the higher the sampling frequency, the better is our representation of the function.

Creating Sound in MATLAB

The MATLAB function `sound(sound_vector,sf)` plays the signal in the vector `sound_vector`, created with the sampling frequency `sf`, on the computer's speaker. Its use is demonstrated with the following user-defined function, which plays a simple tone.

```
function playtone(freq,sf,amplitude,duration)
% Plays a simple tone.
% freq = frequency of the tone (in Hz).
% sf = sampling frequency (in Hz).
% amplitude = sound amplitude (dimensionless).
% duration = sound duration (in seconds).
t = 0:1/sf:duration;
sound_vector = amplitude*sin(2*pi*freq*t);
sound(sound_vector,sf)
```

Try this function with the following values: `freq = 1000`, `sf = 10000`, `amplitude = 1`, and `duration = 10`. The `sound` function truncates or "clips" any values in `sound_vector` that lie outside the range −1 to +1. Try using `amplitude = 0.1` and `amplitude = 5` to see the effect on the loudness of the sound.

Of course, real sound contains more than one tone. You can create a sound having two tones by adding two vectors created from sine functions having different frequencies and amplitudes. Just make sure that they are sampled with the same frequency and, have the same number of samples, and that their sum lies in the range −1 to +1. You can play two different sounds in sequence by concatenating them in a row vector, as `sound([sound_vector_1, sound_vector_2], sf)`. You can play two different sounds simultaneously in stereo by concatenating them in a column vector, as `sound([sound_vector_1', sound_vector_2'], sf)`.

MATLAB includes some sound files. For example, load the MAT-file `chirp.mat` and play the sound as follows:

```
>>load chirp
>>sound(y,Fs)
```

Note that the sound vector has been stored in the MAT-file as the array `y` and the sampling frequency has been stored as the variable `Fs`. You can also try the file `gong.mat`.

A related function is `soundsc(sound_vector,sf)`. This function scales the signal in `sound_vector` to the range −1 to +1 so that the sound is played as loudly as possible without clipping.

Reading and Playing Sound Files

The MATLAB function `wavread('filename')` reads a Microsoft WAVE file having the extension `.wav`. The syntax is

```
[sound_vector, sf, bits] = wavread('filename')
```

where `sf` is the sampling frequency used to create the file and `bits` is the number of bits per sample used to encode the data. To play the file, use the `wavplay` function as follows:

```
>>wavplay(sound_vector, sf)
```

Most computers have WAVE files to play bells, beeps, chimes, etc., to signal you when certain actions occur. For example, to load and play the WAVE file chimes.wav located in C:\windows\media on some PC systems, you type

```
>>[sound_vector, sf] = wavread('c:\windows\media\chimes.wav');
>>wavplay(sound_vector, sf)
```

You can also play this sound using the sound command, as `sound(y,sf)`, but the `wavplay` function has more capabilities than the `sound` function. See the MATLAB Help for information about the extended syntax of the `wavplay` function.

Recording and Writing Sound Files

You can use MATLAB to record sound and write sound data to a WAVE file. The `wavrecord` function records sound from a PC-based audio input device. Its basic syntax is

```
sound_vector = wavrecord(n,sf)
```

where n is the number of samples, sampled at the rate sf. The default value is 11,025 Hz. For example, to record 5 s of audio from channel 1 sampled at 11,025 Hz, speak into the audio device while the following program runs.

```
>>sf = 11025;
>>sound_vector = wavrecord(5*sf, sf);
```

Play back the sound by typing `wavplay(sound_vector,sf)`.

You can use the `wavwrite` function to write sound stored in the vector `sound_vector` to a Microsoft WAVE file. One syntax is `wavwrite(sound_vector, sf, 'filename')`, where the sampling frequency is sf Hz and the data are assumed to be 16-bit data. The function clips any amplitude values outside the range -1 to $+1$.

Formatted Output in MATLAB

The `disp` and `format` commands provide simple ways to control the screen output. However, some users might require greater control over the screen display. In addition, some users might want to write formatted output to a data file. The `fprintf` function provides this capability. Its syntax is `count = fprintf(fid,format,A,...)`, which formats the data in the real part of matrix A (and in any additional matrix arguments) under control of the specified format string `format`, and writes the data to the file associated with file identifier `fid`. A count of the number of bytes written is returned in the variable `count`. The argument `fid` is an integer file identifier obtained from `fopen`. (It may also be 1 for standard output—the screen—or 2 for standard error. See `fopen` for more information.) Omitting `fid` from the argument list causes output to appear on the screen, and is the same as writing to standard output ($fid = 1$). The string `format` specifies notation, alignment, significant digits, field width, and other aspects of output format. It can contain ordinary alphanumeric characters, along with escape characters, conversion specifiers, and other characters, organized as shown in the following examples. Table C.1 summarizes the `basic` syntax of `fprintf`. Consult MATLAB Help for more details.

Suppose the variable `Speed` has the value 63.2. To display its value using three digits with one digit to the right of the decimal point, along with a message, the session is

```
>>fprintf('The speed is: %3.1f\n',Speed)
The speed is:   63.2
```

Here the "field width" is 3, because there are three digits in 63.2. You may want to specify a wide enough field to provide blank spaces or to accommodate an unexpectedly large numerical value. The % sign tells MATLAB to interpret the

Table C.1 Display formats with the `fprintf` function

Syntax	Description
`fprintf('format',A, ...)`	Displays the elements of the array A, and any additional array arguments, according to the format specified in the string `'format'`.
`'format'` structure	`%[-][number1.number2]C`, where `number1` specifies the minimum field width, `number2` specifies the number of digits to the right of the decimal point, and `C` contains control codes and format codes. Items in brackets are optional. `[-]` specifies left-justified.

Control codes		Format codes	
Code	**Description**	**Code**	**Description**
`\n`	Start new line.	`%e`	Scientific format with lowercase e.
`\r`	Beginning of new line.	`%E`	Scientific format with uppercase E.
`\b`	Backspace.	`%f`	Decimal format.
`\t`	Tab.	`%g`	`%e` or `%f`, whichever is shorter.
`' '`	Apostrophe.		
`\\`	Backslash.		

following text as codes. The code `\n` tells MATLAB to start a new line after displaying the number.

The output can have more than one column, and each column can have its own format. For example,

```
>>r = 2.25:20:42.25;
>>circum = 2*pi*r;
>>y = [r;circum];
>>fprintf('%5.2f %11.5g\n',y)
  2.25       14.137
 22.25        139.8
 42.25       265.46
```

Note that the `fprintf` function displays the *transpose* of the matrix y.

Format code can be placed within text. For example, note how the period after the code `%6.3f` appears in the output at the end of the displayed text.

```
>>fprintf('The first circumference is %6.3f.\n',circum(1))
The first circumference is 14.137
```

An apostrophe in displayed text requires two single quotes. For example:

```
>>fprintf('The second circle''s radius %15.3e is large.\n',r(2))
The second circle's radius       2.225e+001 is large.
```

A minus sign in the format code causes the output to be left-justified within its field. Compare the following output with the preceding example:

```
>>fprintf('The second circle''s radius %-15.3e is large.\n',r(2))
The second circle's radius 2.225e+001          is large.
```

Control codes can be placed within the format string. The following example uses the tab code (\t).

```
>>fprintf('The radii are:%4.2f \t %4.2f \t %4.2f\n',r)
The radii are:   2.25     22.25    42.25
```

The disp function sometimes displays more digits than necessary. We can improve the display by using the fprintf function instead of disp. Consider the program:

```
p = 8.85; A = 20/100^2;
d = 4/1000; n = [2:5];
C = ((n - 1).*p*A/d);
table (:,1) = n';
table (:,2) = C';
disp (table)
```

The disp function displays the number of decimal places specified by the format command (4 is the default value).

If we replace the line disp(table) with the following three lines

```
E='';
fprintf('No.Plates Capacitance (F) X e12 %s\n',E)
fprintf('%2.0f \t \t \t %4.2f\n',table')
```

we obtain the following display:

```
2          4.42
3          8.85
4          13.27
5          17.70
```

The empty matrix E is used because the syntax of the fprintf statement requires that a variable be specified. Because the first fprintf is needed to display the table title only, we need to fool MATLAB by supplying it with a variable whose value will not display.

Note that the fprintf command truncates the results, instead of rounding them. Note also that we must use the transpose operation to interchange the rows and columns of the table matrix in order to display it properly.

Only the real part of complex numbers will be displayed with the `fprintf` command. For example,

```
>>z = -4+9i;
>>fprintf('Complex number:     %2.2f \n',z)
Complex number:     -4.00
```

Instead you can display a complex number as a row vector. For example, if $w = -4+9i$,

```
>>w = [-4,9];
>>fprintf('Real part is %2.0f. Imaginary part is %2.0f. \n',w)
Real part is -4. Imaginary part is 9.
```

APPENDIX D

References

[Brown, 1994] Brown, T. L.; H. E. LeMay, Jr.; and B. E. Bursten. *Chemistry: The Central Science*. 6th ed. Upper Saddle River, NJ: Prentice-Hall, 1994.

[Eide, 2008] Eide, A. R.; R. D. Jenison; L. L. Northup; and S. Mickelson. *Introduction to Engineering Problem Solving*. 5th ed. New York: McGraw-Hill, 2008.

[Felder, 1986] Felder, R. M., and R. W. Rousseau. *Elementary Principles of Chemical Processes*. New York: John Wiley & Sons, 1986.

[Garber, 1999] Garber, N. J., and L. A. Hoel. *Traffic and Highway Engineering*. 2nd ed. Pacific Grove, CA: PWS Publishing, 1999.

[Jayaraman, 1991] Jayaraman, S. *Computer-Aided Problem Solving for Scientists and Engineers*. New York: McGraw-Hill, 1991.

[Kreyzig, 2009] Kreyzig, E. *Advanced Engineering Mathematics*. 9th ed. New York: John Wiley & Sons, 1999.

[Kutz, 1999] Kutz, M., editor. *Mechanical Engineers' Handbook*. 2nd ed. New York: John Wiley & Sons, 1999.

[Palm, 2010] Palm, W. *System Dynamics*. 2nd ed. New York: McGraw-Hill, 2010.

[Rizzoni, 2007] Rizzoni, G. *Principles and Applications of Electrical Engineering*. 5th ed. New York: McGraw-Hill, 2007.

[Starfield, 1990] Starfield, A. M.; K. A. Smith; and A. L. Bleloch. *How to Model It: Problem Solving for the Computer Age*. New York: McGraw-Hill, 1990.

Answers to Selected Problems

Chapter 1

2. (a) -13.3333; (b) 0.6; (c) 15; (d) 1.0323.

8. (a) $x + y = -3.0000 - 2.0000i$; (b) $xy = -13.0000 - 41.0000i$; (c) $x/y = -1.7200 + 0.0400i$.

18. $x = -15.685$ and $x = 0.8425 \pm 3.4008i$.

Chapter 2

3.
$$\mathbf{A} = \begin{bmatrix} 0 & 6 & 12 & 18 & 24 & 30 \\ -20 & -10 & 0 & 10 & 20 & 30 \end{bmatrix}$$

7. (a) Length is 3. Absolute values $=$ [2 4 7];
(b) Same as (a); (c) Length is 3. Absolute values $=$ [5.8310 5.0000 7.2801].

11. (b) The largest elements in the first, second, and third layers are 10, 9, and 10, respectively. The largest element in the entire array is 10.

12. (a)
$$\mathbf{A} + \mathbf{B} + \mathbf{C} = \begin{bmatrix} -6 & -3 \\ 23 & 15 \end{bmatrix}$$
(b)
$$\mathbf{A} - \mathbf{B} + \mathbf{C} = \begin{bmatrix} -14 & 7 \\ -1 & 19 \end{bmatrix}$$

13. (a) A.*B = [784, -128; 144,32];
(b) A/B = [76, -168; -12, 32];
(c) B.^3 = [2744, -64;216, -8].

14. (a) F.*D = [1200, 275, 525, 750, 3000] J; (b) sum(F.*D) = 5750 J.

25. (a) A*B = [-47, -78; 39, 64];
(b) B*A = [-5, -3, 48, 22].

28. 60 tons of copper, 67 tons of magnesium, 6 tons of manganese, 76 tons of silicon, and 101 tons of zinc.

31. $M = 675$ N·m if $\mathbf{F}$ is in newtons and $\mathbf{r}$ is in meters.

38. [q,r] = deconv([14,-6,3,9], [5,7,-4]), q = [2.8, -5.12],r = [0, 0, 50.04, -11.48]. The quotient is $2.8x - 5.12$ with a remainder of $50.04x - 11.48$.

39. 2.0458.

Chapter 3

1. (a) 3, 3.1623, 3.6056;
(b) 1.7321i, 0.2848 + 1.7553i, 0.5503 + 1.8174i;
(c) 15 + 21i, 22 + 16i, 29 + 11i;
(d) $-0.4 - 0.2i$, $-0.4667 - 0.0667i$, $-0.5333 + 0.0667i$.

2. (a) $|xy| = 105$, $\angle xy = -2.6$ rad;
(b) $|x/y| = 0.84$, $\angle x/y = -1.67$ rad.

3. (a) 1.01 rad (58°); (b) 2.13 rad (122°);
(c) -1.01 rad ($-58°$); (d) -2.13 rad ($-122°$).

7. $F_1 = 197.5217$ N.

10. 2.7324 sec while ascending; 7.4612 sec while descending.

Chapter 4

4. (*a*) $z = 1$; (*b*) $z = 0$; (*c*) $z = 1$; (*d*) $z = 1$.

5. (*a*) $z = 0$; (*b*) $z = 1$; (*c*) $z = 0$; (*d*) $z = 4$; (*e*) $z = 1$; (*f*) $z = 5$; (*g*) $z = 1$; (*h*) $z = 0$.

6. (*a*) z = [0, 1, 0, 1, 1];
(*b*) z = [0, 0, 0, 1, 1];
(*c*) z = [0, 0, 0, 1, 0];
(*d*) z = [1, 1, 1, 0, 1].

11. (*a*) z = [1, 1, 1, 0, 0, 0];
(*b*) z = [1, 0, 0, 1, 1, 1];
(*c*) z = [1, 1, 0, 1, 1, 1];
(*d*) z = [0, 1, 0, 0, 0, 0].

13. (*a*) \$7300; (*b*) \$5600; (*c*) 1200 shares; (*d*) \$15,800.

28. Best location: $x = 9$, $y = 16$. Minimum cost: \$294.51. There is only one solution.

34. After 33 years, the amount will be \$1,041,800.

36. $W = 300$ and $T = [428.5714, 471.4286, 266.6667, 233.3333, 200, 100]$

48. Weekly inventory for cases (*a*) and (*b*):

Week	1	2	3	4	5
Inventory (*a*)	50	50	45	40	30
Inventory (*b*)	30	25	20	20	10
Week	6	7	8	9	10
Inventory (*a*)	30	30	25	20	10
Inventory (*b*)	10	5	0	0	(<0)

Chapter 5

1. Production is profitable for $Q \geq 10^8$ gal/yr. The profit increases linearly with Q, so there is no upper limit on the profit.

3. $x = -0.4795$, 1.1346, and 3.8318.

5. 37.622 m above the left-hand point, and 100.6766 m above the right-hand point.

10. 0.54 rad (31°).

14. The steady-state value of y is $y = 1$. $y = 0.98$ at $t = 4/b$.

17. (*a*) The ball will rise 1.68 m and will travel 9.58 m horizontally before striking the ground after 1.17 s.

Chapter 6

2. (*a*) $y = 53.5x - 1354.5$;
(*b*) $y = 3582.1x^{-0.9764}$;
(*c*) $y = 2.0622 \times 10^5 (10)^{-0.0067x}$

4. (*a*) $b = 1.2603 \times 10^{-4}$; (*b*) 836 years. (*c*) Between 760 and 928 years ago.

8. If unconstrained to pass through the origin, $f = 0.3998x - 0.0294$. If constrained to pass through the origin, $f = 0.3953x$.

10. $d = 0.0509v^2 + 1.1054v + 2.3571$; $J = 10.1786$; $S = 57,550$; $r^2 = 0.9998$.

11. $y = 40 + 9.6x_1 - 6.75x_2$. Maximum percent error is 7.125 percent.

Chapter 7

7. (*a*) 96%; (*b*) 68%.

11. (*a*) Mean pallet weight is 3000 lb. Standard deviation is 10.95 lb; (*b*) 8.55 percent.

18. Mean yearly profit is \$64,609. Minimum expected profit is \$51,340. Maximum expected profit is \$79,440. Standard deviation of yearly profit is \$5967.

22. The estimated temperatures at 5 P.M. and 9 P.M. are 22.5° and 16.5°.

Chapter 8

2. (*a*) $\mathbf{C} = \mathbf{B}^{-1}(\mathbf{A}^{-1}\mathbf{B} - \mathbf{A})$.
(*b*) C = [-0.8536, -1.6058; 1.5357, 1.3372].

5. (*a*) $x = 3c$, $y = -2c$, $z = c$; (*b*) The plot consists of three straight lines that intersect at (0,0).

8. $T_1 = 19.7596°C$, $T_2 = -7.0214°C$, $T_3 = -9.7462°C$. Heat loss in watts is 66.785.

11. Infinite number of solutions: $x = -1.3846z + 4.9231$, $y = 0.0769z - 1.3846$.

14. Unique solution: $x = 8$ and $y = 2$.

15. Least-squares solution; $x = 6.0928$ and $y = 2.2577$.

Chapter 9

1. 23 690 m.

7. 13.65 ft.

10. 1363 m/s.

25. 150 m/s.

Chapter 11

2. (*a*) $60x^5 - 10x^4 + 108x^3 - 49x^2 + 71x - 24$; (*b*) 2546.

3. $A = 1$, $B = -2a$, $C = 0$, $D = -2b$, $E = 1$, and $F = r^2 - a^2 - b^2$.

4. $(a)\, b = c \cos A \pm \sqrt{a^2 - c^2 \sin^2 A};\, (b)\, b = 5.6904.$

6. $(a)\, x = \pm 10\sqrt{(4b^2 - 1)/(400b^2 - 1)},$

$y = \pm \sqrt{99/(400b^2 - 1)};$

$(b)\, x = \pm 0.9685,\, y = \pm 0.4976.$

10. $s^2 + 13s + 42 - 6k,\, s = (-13 \pm \sqrt{1 + 24k})/2.$

11.

$$x = \frac{62}{16c + 15} \qquad y = \frac{129 + 88c}{16c + 15}$$

23. $\theta = 0.6155$ rad $(35.26°).$

24. 49.6808 m/s.

34. $(a)\, 2;\, (b)\, 0;\, (c)\, 0.$

41. $(a)\, (3x_0/5 + v_0/5)e^{-3t}\sin 5t + x_0 e^{-3t}\cos 5t;$

$(b)\, e^{-5t}(8x_0/3 + v_0/3) + (-5x_0/3 - v_0/3)e^{-8t}$

INDEX

MATLAB Symbols

+ addition, 8
− subtraction, 8
* multiplication, 8
.* array multiplication, 66
^ exponentiation, 8
.^ array exponentiation, 66
\ left division, 8, 66
/ right division, 8, 66
.\ array left division, 66
./ array right division, 66
: colon
   array addressing, 57, 58
   array generation, 12, 54, 55
() parentheses

function arguments, 117
modifying precedence, 9
{ } braces; enclose cell elements, 91
[] brackets, 19, 55
. . . ellipsis, line continuation, 12
, comma
   column separation, 12
   statement separation, 12
; semicolon
   display suppression, 12
   row separation, 55

% percent sign
   comment designation, 27
   format specification, 549
' apostrophe
   complex conjugate transpose, 57
   string designation, 31, 171
   transpose, 55
.' nonconjugated transpose (dot transpose), 57
= assignment or replacement operator, 10
== equal to, 155

~= not equal to, 155
< less than, 155
<= less than or equal to, 155
> greater than, 155
>= greater than or equal to, 155
& AND, 158
&& short-circuit AND, 158
| OR, 158
|| short-circuit OR, 158
~NOT, 158
>> MATLAB prompt, 6
@ creates a function handle, 124

MATLAB Commands

A

abs, 114
acos, 21, 118
acosh, 119
acot, 118
acoth, 119
acsc, 118
acsch, 119
addpath, 23
all, 161
angle, 114
ans, 14
any, 161
asec, 118
asech, 119
asin, 21, 118
asinh, 119
atan, 21, 118
atan2, 118
atanh, 119
axis, 222, 225

B

bar, 235, 295, 300
break, 176
bvp4c, 409

C

case, 188
cat, 64
cd, 23
ceil, 114
cell, 90
celldisp, 91
cellplot, 91
cla, 539
clabel, 249
clc, 12
clear, 12
colormap, 539
conj, 114
continue, 276
contour, 250
conv, 86
cos, 21, 118
cosh, 119
cot, 118
coth, 119
cross, 85
csc, 118
csch, 119
cumsum, 303

D

date, 122
dblquad, 372
dde23, 409
ddesd, 409
deconv, 86
del2, 382
det, 333
deval, 409
diff, 379, 382
dir, 23
disp, 31

doc, 38
dot, 85
drawnow, 541

E

eig, 397
else, 166
elseif, 168
end, 166
eps, 14
erf, 305
exist, 12
exp, 21, 114
eye, 83

F

fieldnames, 94
find, 60, 161
finite, 161
fix, 114
floor, 114
fminbnd, 128
fminsearch, 128
for, 172
format, 15, 31
fplot, 223, 225
fprintf, 549
function, 119
fzero, 128

G

gensig, 408
getframe, 538
ginput, 25
gradient, 382
grid, 25, 222, 225
global, 124
gtext, 25, 232

H

help, 38
helpwin, 38
hist, 296, 300
hold, 231, 232

I

i, 14
if, 165
imag, 114
impulse, 401
Inf, 14
initial, 401
inline, 130
input, 31, 171
interp1, 317, 320
interp2, 317
interpn, 317
inv, 333
ischar, 161
isempty, 161
isfield, 94
isinf, 161
isnan, 161
isnumeric, 161
isreal, 161
isstruct, 94

J

j, 14

L

legend, 232
length, 19, 60
linspace, 56, 60
load, 21
log, 21, 114
log10, 21, 114
logical, 156, 161
loglog, 235
logspace, 56, 60
lookfor, 38
lsim, 401

M

max, 60
mean, 296
median, 296
menu, 31

mesh, 250
meshc, 250
meshgrid, 250
meshz, 250
min, 60
mode, 296
movie, 538
moviein, 538
mupadwelcome, 466

N

NaN, 14
nargin, 170
nargout, 170
norm, 60

O

ode15i, 409
ode15s, 385, 395
ode45, 385, 395
odephase, 409
odeplot, 409
odeprint, 409
odeset, 394, 395
ones, 83
otherwise, 188

P

path, 23
pathtool, 23
pause, 541
pchip, 320, 329
pdeval, 409
pi, 14
pinv, 342
plot, 25, 225, 232
plotyy, 235
plot3, 246
polar, 235
poly, 86
polyder, 382
polyfit, 266, 273
polyint, 371

polyval, 86, 87, 225, 273
print, 221, 225
pwd, 23

Q

quad, 371
quadl, 371
quit, 12
quiver, 382

R

rand, 307, 309
randn, 309, 310
randperm, 309
rank, 335
real, 114
rmfield, 94
rmpath, 23
roots, 20, 86
round, 114
rref, 345

S

save, 21
sec, 118
sech, 119
semilogx, 235
semilogy, 235
shading, 539
sign, 114
simulink, 431
sin, 21, 118
sinh, 119
size, 60
sort, 60, 296
sound, 546
soundsc, 547
spline, 318, 320
sqrt, 21, 114
ss, 399, 400
ssdata, 399, 400
stairs, 235

`std`, 304
`stem`, 235
`step`, 401
`struct`, 94
`subplot`, 232
`sum`, 60
`surf`, 250
`surfl`, 539
`surfc`, 250
`switch`, 188

T

`tan`, 21, 118
`tanh`, 119

`text`, 232
`tf`, 399, 400
`tfdata`, 400
`title`, 25, 225
`trapz`, 371
`triplequad`, 371
`type`, 38

U

`unmkp`, 319, 320

V

`var`, 304
`view`, 539

W

`waterfall`, 250
`wavplay`, 547
`wavread`, 547
`wavrecord`, 548
`wavwrite`, 548
`what`, 23
`which`, 23
`while`, 183
`who`, 12
`whos`, 12
`wk1read`, 138

X

`xlabel`, 25, 225
`xlsread`, 135
`xor`, 159, 161

Y

`ylabel`, 25, 225

Z

`zeros`, 83
`zlabel`, 248

Simulink Blocks

C

Clock, 425
Constant, 432

D

Dead Zone, 438
Derivative, 453

F

Fcn, 454

G

Gain, 420

I

Integrator, 420

L

Look-Up Table, 454

M

MATLAB Fcn, 454
Mux, 425

R

Rate Limiter, 450
Relay, 433

S

Saturation, 450
Scope, 425
Signal Builder, 453
Signal Generator, 443
Sine Wave, 423
State-Space, 427
Step, 429
Subsystem, 443
Summer, 421

T

To Workspace, 425
Transfer Fcn, 437
Transfer Fcn (with
    initial outputs),
    437
Transport Delay, 448
Trigonometric
    Function, 432

MuPAD Symbols and Commands[1]
Symbols

: colon, display
    suppression, 470
:= assignment
    operator, 472
:: library reference,
    480

% reference to
    previous result,
    471
´ derivative notation,
    502
place marker, 475

Commands

A

airyAi(x), 512
airyBi(x), 512
arg, 471
assume, 478

B

besselI, 512
besselJ, 512
besselK, 512
besselY, 512
bool, 476

[1] Most MuPAD symbols and commands are identical to their MATLAB counterparts; for example **cos** and `cos`. Here we list commonly used symbols and commands that have special meaning in MuPAD.

C

charpoly, 489
chebyshev1, 512
collect, 478
combine, 477
conjugate, 471

D

DIGITS, 471
delete, 472
det, 489
diff, 494

E

E, 470
eigenvalues, 491
eigenvectors, 491
expand, 475

F

factor, 475
factorout, 476
float, 471

G

gamma, 512

H

heaviside, 508
hermite, 512

I

I, 470
Im, 471
int, 497
inverse, 491
invlaplace, 507

L

laguerreL, 512
laplace, 506
legendre, 512
limit, 500
log, 470
ln, 470

M

matlinsolve, 491
matrix, 489
maximize, 488
minimize, 489

N

normal, 476

O

ode, 502
op, 514

P

pdioe, 487
PI, 470
plotfunc2d, 514
polyroots, 480

R

rank, 489
Re, 471

realroot, 497
rec, 487
rectform, 472
rewrite, 478

S

series, 513
Simplify, 476
simplify, 476
solve, 480, 502
subs, 479
sum, 500

T

taylor, 499

U

unassume, 478

Topics

A

absolute frequency,
    297
absolute value, 61
algebraic equations,
    see also linear
        algebraic equations
    polynomial diophan-
        tine equations,
        487
    recurrence relations,
        488
    solving numerically,
        483
    solving sets of,
        483
    solving symboli-
        cally, 480

algorithm, 148
animation, 538
anonymous function,
    130, 132
argument, 8
array, 19
    addition and subtrac-
        tion, 65
    addressing, 58
    cell, 90
    creating an, 55
    division, 69
    empty, 58
    exponentiation, 70
    functions, 60
    index, 19
    multidimensional, 63
    multiplication, 65

operations, 65
    pages, 63
    powers, 70
    size, 56
    structure, 93
ASCII files, 21
assignment operator,
    10
augmented matrix, 335
axis label, 220
axis limits, 251

B

backward differences,
    378
bar plots, 296
Basic Fitting Interface,
    282

bins, 296
block diagram, 420
Block Parameters
    window, 424
Boolean operator, 157
boundary value
    problem, 409
breakpoint, 193

C

Cauchy form, 390
cell indexing, 90
cell array, 90
cell mode, 191
central difference, 379
characteristic roots,
    397
clearing variables, 12

coefficient of
  determination, 275
colors, 228
column vector, 54
command, 8
Command bar
  (MuPAD), 475
Command window, 6
comment, 27
common mathematical
  functions, 21, 114
complex numbers, 16
complex conjugate
  transpose, 57
computer solution
  steps, 41, 149
conditional statement,
  164
content indexing, 90
contour plots, 248
Control System
  toolbox, 398
cubic splines, 317
current directory, 16
curve fit, quality of, 275

D

data files, 21
data markers, 25, 228
Data Statistics tool, 300
data symbol, 251
dead time, 448
dead zone, 437
Debug menu, 192
debugging, 29, 190
definite integral, 370
delay differential
  equation, 409
derivative, *see*
  differentiation
Desktop, 5
determinants, 335
differential equations
  Cauchy form, 391

characteristic roots,
  397
delay, 409
higher order, 390
nonlinear, 383
ordinary, 382
partial, 382
piecewise-linear, 430
solvers, 385
state variable form,
  391
symbolic solution of,
  501
differentiation,
  numerical, 329, 382
partial, 409
polynomial, 382
symbolic, 494
directory, 6
dot transpose, 57

E

Edit menu, 17
Editor/Debugger, 28,
  190
eigenvalue, 397, 491
element-by-element
  operations, 66,
  69, 70
ellipsis, 12
empty array, 58
EraseMode property,
  540
error function, 305
Euclidean norm, 343
Euler method, 383
exporting data, 139
exporting figures, 225
extrapolation, 313

F

field, 92
figure handle, 540

File menu, 17
files
  ASCII, 21
  command, 27
  data, 21
  function, 27
  MAT-files, 20
  M-files, 20
  script, 27
  spreadsheet, 135
  user-defined, 119
flowchart, 150
for loop, 172
forced response, 383
formatting, 15
forward differences,
  379
free response, 383
function argument,
  117
function definition line,
  119
function discovery,
  263
function file, 119
function handle, 124
functions
  anonymous, 131
  argument, 117
  complex, 114
  elementary mathe-
    matical, 113
  exponential, 114
  handle, 124
  hyperbolic, 118
  logarithmic, 124
  minimization of,
    126, 127
  nested, 131
  numeric, 116
  overloaded, 131
  primary, 131
  private, 131
  of random variables,
    311

subfunction, 131
trigonometric, 117
user-defined, 119
zeros of, 124

G

Gauss elimination, 335
Gaussian function, 303
global variable, 124
gradient, 380
Graphics window, 23
grid, 25, 222

H

H1 line, 30
handle, 124
help functions, 38
Help system, 33
histogram, 296
histogram functions,
  300
homogeneous
  equations, 335
hyperbolic functions,
  119

I

identity matrix, 82
ill-conditioned
  problem, 333
implied loop, 177
importing data, 172
importing spreadsheet
  files, 173
Import Wizard, 173
improper integral, 370
indefinite integral,
  370
initial value problem,
  382
input region, 468
input/output
  commands, 31

integral,
  definite, 370
  double, 376
  improper, 370
  indefinite, 370
  singularity, 370
  triple, 377
integration
  numerically, 371
  symbolically, 497
  panel, 370
  trapezoidal, 370
interpolation, 313
  cubic spline, 317
  Hermite polynomials, 329
  linear, 315, 317
  2–D, 316
  polynomial, 320
inverse Laplace
  transform, 507

L

Laplace transform, 506
Laplacian, 382
least squares, 312
left division method, 8,
  335
legend, 230
Library Browser, 421
limits, 500
line continuation, 12
line types, 228
linear algebra
  characteristic polynomial, 491
  eigenvalues, 397,
    491
  eigenvectors, 491
  matrix operations,
    489
linear algebraic
  equations, 26, 84,
  331

application of matrix
  rank, 335
and augmented
  matrix, 335
and Euclidean norm,
  343
homogeneous, 335
ill-conditioned
  system of, 333
and linearity, 336
matrix solution, 492
overdetermined
  system of, 350
and reduced row
  echelon form, 345
singular set of, 333
solution by left division method, 28,
  335
solution by matrix
  inverse, 333
solution by pseudoinverse method,
  342
underdetermined
  system of, 341
linear-in-parameters,
  329
linear interpolation
  functions, 317
linear programming,
  488
local variable, 32, 119
logarithmic plots,
  233
logical arrays, 157
logical functions, 161
logical operators, 158
logical variable, 123,
  155
loop variable, 172
LTI ODE solvers,
  401
LTI object, 399, 400
LTI Viewer, 407

M

M-files, 29
magnitude, 61
managing the work
  session, 12
mask, 180
MAT-files, 20
MathWorks website,
  37
matrix, 56
  augmented, 345
  creating a, 55
  division, 83
  exponentiation, 84
  identity, 82
  inverse, 332
  multiplication,
    75, 80
  null, 82
  operations, 73
  rank, 334
  special, 82
  transpose, 57
  unity, 82
mean, 296
median, 296
methodology
  for developing a
    computer solution,
    42
  for engineering problem solving, 38
minimization and rootfinding functions,
  128
modified Euler
  method, 384
multidimensional
  arrays, 81
multiple linear
  regression, 328
MuPAD libraries, 480
MuPAD menus
  Combine, 477

General math, 475
Rewrite, 478
Simplify, 476
Solve, 481

N

naming variables, 11
nested function, 131,
  135
nested loops, 174
normal distribution, 301
normal function, 303
normally distributed
  numbers, 303
null matrix, 82
numeric display
  formats, 15
numerical
  differentiation,
  329, 382
numerical integration
  functions, 371

O

ODE. *See* differential
  equation, ordinary
operations research, 193
optimization problems,
  488
order of precedence, 9,
  158
output region, 468
overdetermined
  system, 350
overlay plots, 24, 228
overloaded function,
  131

P

pages (in
  multidimensional
  arrays), 81
panel, 471

path, 22
PI controller, 450
plot, 25
  axis label, 220
  bar, 296
  colors, 228
  contour, 248
  data markers, 228
  editor, 241
  enhancement commands, 232
  hints for improving, 220
  grid, 25, 222
  interactive interface, 241
  legend, 230
  line types, 228
  logarithmic, 233
  overlay, 24, 228
  polar, 236
  requirements, 221
  second y-axis, 236
  specialized, 235
  stairs, 230
  stem, 236
  subplots, 226
  surface mesh, 247
  text placement, 232
  three-dimensional, 250
  three-dimensional line, 246
  tick mark label, 220
  tick marks 220
  tick mark spacing, 220
  title, 220, 225
plotting
  complex numbers, 223
  in MuPAD, 473
  polynomials, 87
  with smart function plot command, 227

symbolic expressions, 593
tools, 243
xy plots, 225
polar plot, 235
polynomial, 20
  addition, 86
  differentiation, 380
  division, 86
  functions, 86
  interpolation functions, 320
  multiplication, 86
  plotting, 87
  regression, 273
  roots, 20
precedence 9, 158
predefined constants, 15
predefined input functions, 407
predictor-corrector method, 384
primary function, 131
private function, 137
program documentation, 149
programming style, 30
prompt, 6
pseudocode, 151
pseudoinverse method, 342

Q

quadrature, 373

R

random number functions, 309
random number generator, 307
rank, 334
rate limiter, 450
rectangular integration, 370

reduced form, 390
reduced row echelon form, 345
regression, 271
relational operators, 155
relative frequency, 287
relay, 433
replacement operator, 10
reserved symbols in MuPAD, 470
residuals, 272, 277
right division, 8
row vector, 54
r-squared value, 275
Runge-Kutta methods, 385

S

saturation nonlinearity, 440
saving figures, 225
session, 7
scalar, 8
scalar arithmetic operations, 8
scaled frequency histogram, 301
scaling data, 276
script file, 27
search path, 2
session, 7
short-circuit operators, 160
simulation, 193
simulation diagrams, 420
singular matrix, 333
singularity, 370
smart recall, 13
sound, 546
special functions (MuPAD), 512, 513

special matrices, 83
special variables and constants, 14
spreadsheet files, 138
stairs plots, 235
standard deviation, 303
state of random generator, 307
state transition diagram, 195
state-variable form, 390, 441
statically indeterminate problem, 343
stem plots, 235
step function, 402
step size, 384
string, 31
structure arrays, 92
structure chart, 150
structure functions, 94
structured programming, 148
subdeterminant, 335
subfunction, 131, 134
subplots, 232
subsystems, 443
surface mesh plot, 247
sums, 500
switch structure, 188
symbolic expressions, 472
  collecting, 478
  combining, 477
  evaluating, 476
  expanding, 475
  factoring, 475
  manipulating, 474
  normalizing, 476
  rewriting, 478
  simplifying, 476
  substituting, 479

system, directory, and file commands, 23

T

tab completion, 13
Taylor series, 499
text region, 468
three-dimensional plots, 246
  contour plots, 248
  line plots, 246
  surface mesh plots, 247

toolbox, 5
top-down design, 149
transfer-function form, 437
transport delay, 448
transpose, 55
trapezoidal integration, 370
trigonometric functions, 118, 119
truth table, 159

U

underdetermined system, 341
uniformly distributed numbers, 307
user-defined functions, 119

V

Variable editor, 62
variance, 303
variable, 7

vector, 70
  absolute value of, 61
  cross product, 85
  dot product, 85
  length of, 61
  magnitude of, 61
  multiplication, 74

W

while loop, 183
working directory. *See* current directory
workspace, 11